Algorithms for Intelligent Systems

This book series publishes research on the analysis and development of algorithms for intelligent systems with their applications to various real world problems. It covers research related to autonomous agents, multi-agent systems, behavioral modeling, reinforcement learning, game theory, mechanism design, machine learning, meta-heuristic search, optimization, planning and scheduling, artificial neural networks, evolutionary computation, swarm intelligence and other algorithms for intelligent systems.

The book series includes recent advancements, modification and applications of the artificial neural networks, evolutionary computation, swarm intelligence, artificial immune systems, fuzzy system, autonomous and multi agent systems, machine learning and other intelligent systems related areas. The material will be beneficial for the graduate students, post-graduate students as well as the researchers who want a broader view of advances in algorithms for intelligent systems. The contents will also be useful to the researchers from other fields who have no knowledge of the power of intelligent systems, e.g. the researchers in the field of bioinformatics, biochemists, mechanical and chemical engineers, economists, musicians and medical practitioners.

The series publishes monographs, edited volumes, advanced textbooks and selected proceedings.

Indexed by zbMATH.

All books published in the series are submitted for consideration in Web of Science.

Vladan Devedzic · Basant Agarwal ·
Mukesh Kumar Gupta
Editors

Proceedings of the International Conference on Intelligent Computing, Communication and Information Security

ICICCIS 2022

Springer

Editors
Vladan Devedzic
Serbian Academy of Sciences and Arts
Belgrade, Serbia

Mukesh Kumar Gupta
Department of Computer Science
and Engineering
Swami Keshvanand Institute of Technology
Management and Gramothan
Jaipur, Rajasthan, India

Basant Agarwal
Department of Computer Science
and Engineering
Central University of Rajasthan
Ajmer, Rajasthan, India

ISSN 2524-7565 ISSN 2524-7573 (electronic)
Algorithms for Intelligent Systems
ISBN 978-981-99-1372-5 ISBN 978-981-99-1373-2 (eBook)
https://doi.org/10.1007/978-981-99-1373-2

This Springer imprint is published by the registered company Springer Nature Singapore Pte Ltd.
The registered company address is: 152 Beach Road, #21-01/04 Gateway East, Singapore 189721, Singapore

Executive Committee

General Chair

Prof. G. R. Sinha, Adjunct Professor, International Institute of Information Technology (IIIT), Bangalore, and IEEE Executive Council Member, MP Subsection, India.
Prof. Valentina Emilia Balas, Head of the Intelligent Systems Research Centre, Aurel Vlaicu University of Arad, Romania.

Technical Program Chair

Prof. Vladan Devedzic, Professor, Computer Science and Software Engineering, Department of Software Engineering, University of Belgrade, Serbia.
Dr. Basant Agarwal, Assistant Professor, Department of Computer Science and Engineering, Indian Institute of Information Technology Kota (IIIT Kota), India.
Dr. Mukesh Kumar Gupta, Professor and Head, Department of Computer Science and Engineering, Swami Keshvanand Institute of Technology, Management and Gramothan, Jaipur, India.

Conference Chair

Dr. Priyadarshi Nanda, Senior Lecturer, Faculty of Engineering and Information Technology, University of Technology, Sydney, Australia.
Dr. Karl McCreadie, Lecturer in Data Analytics, School of Computing and Intelligent Systems, UIster University, Magee Derry, UK.

Dr. Tarun Kumar Sharma, Professor and Head of Computer Science and Engineering and Dean—School of Engineering and Technology, Shobhit University, Meerut, India.
Dr. Pramod Gaur, Assistant Professor, Computer Science, BITS Pilani, Dubai Campus, Dubai.

Organizing Chair

Dr. Mukesh Kumar Gupta, Professor and Head, Department of Computer Science and Engineering, Swami Keshvanand Institute of Technology, Management and Gramothan, Jaipur, India.
Dr. Pankaj Dadheech, Associate Professor and Deputy Head, Department of Computer Science and Engineering, Swami Keshvanand Institute of Technology, Management and Gramothan, Jaipur, India.

Preface

The International Conference on 'Intelligent Computing, Communication and Information Security' (ICICCIS-2022) was held during November 25–26, 2022, at the Department of Computer Science and Engineering, Swami Keshvanand Institute of Technology, Management and Gramothan, Jaipur, Rajasthan, India, to explore new solutions in ever-changing and unpredictable environments. Creative solutions, novelties and innovation share an essential feature: often, innovative events do not happen by chance; instead, they seem to be triggered by some previous novelty or innovation. The main focus of this conference is to provide state of the art to research scholars, scientists, industry learners and postgraduates from the various domains of engineering and related fields, as this incorporates latest development in the field of machine learning, artificial intelligence, intelligent systems and communications.

There were nine keynote sessions covering the different areas of the conference: Prof. Vladan Devedzic, Professor (CS), University of Belgrade, Serbia, presented critical review of the state of the art and developments in the field of AI. Dr. Somitra Kr. Sanadhya, Professor (SAIDE) and Dean (DT), Indian Institute of Technology (IIT), Jodhpur, discussed Format-Preserving Encryption Techniques. Professor G. R. Sinha, Adjunct Professor, International Institute of Information Technology (IIIT), Bangalore, presented Intelligent Computing and Intervention for Augmented Mental Health and Community Wellbeing. Dr. Jagdish Chand Bansal, Associate Professor, South Asian University, New Delhi, explained about the Applications, Classification and Challenges in Drone Swarm. Professor (Dr.) C. P. Gupta, Professor (CSE), Rajasthan Technical University (RTU), Kota, briefed the Security Issues in Internet of Things (IoT). Dr. Harish Sharma, Associate Professor (CSE), Rajasthan Technical University (RTU), Kota, talked about Swarm Intelligence-Based Algorithms. Mr. Amit Kumar Gupta, Scientist 'E', DRDO, Hyderabad, discussed Applications of AI in Defense. Professor Poonam Goyal, Professor (CS), Birla Institute of Technology and Science (BITS), Pilani, delivered a session on Anytime Mining of Data Streams. Dr. Anurag Jain, Professor, Coordinator (AICTE Coordination Team), University School of ICT, Guru Gobind Singh Indraprastha University, Delhi, discussed the Research Issues and Challenges in Cyber-Financial Frauds.

The conference witnessed 213 submissions, out of which 36 papers were finally accepted after rigorous reviews. The online presentations, fruitful discussions and exchanges contributed to the conference's success. Papers and participants from various countries made the conference genuinely international in scope. The diversified presenters were academicians, young scientists, research scholars, postdocs and students who brought new perspectives to their fields.

Through this platform, the editors would like to express their sincere appreciation and thanks to our publication partner Springer Singapore, the authors for their contributions to this publication and all the reviewers for their constructive comments on the papers. We would also like to extend our thanks to All India Council for Technical Education (AICTE) for Sponsoring this conference under GOC Scheme.

Belgrade, Serbia — Vladan Devedzic
Ajmer, India — Basant Agarwal
Jaipur, India — Mukesh Kumar Gupta

Acknowledgments Serbian Academy of Sciences and Arts has supported Vladan Devedzic's involvement in this conference as a keynote speaker and a proceedings editor through the project Data analysis in selected domains (Grant No: F-150, 2022).

Contents

About the Editors

Vladan Devedzic is a Professor of Computer Science and Software Engineering at the University of Belgrade, Faculty of Organizational Sciences, Belgrade, Serbia. Since 2021, he is a Corresponding Member of the Serbian Academy of Sciences and Arts (SASA). His long-term professional objective is to bring close together ideas from the broad fields of Artificial Intelligence/Intelligent Systems and Software Engineering. His current professional and research interests include programming education, software engineering, intelligent software systems and technology-enhanced learning (TEL). He is the founder and the chair of the GOOD OLD AI research network. Vladan Devedzic has authored/co-authored more than 370 research papers, published in international and national journals or presented at international and national conferences, as well as six books on intelligent systems and software engineering. Some of his papers have been selected by foreign editors and published in books on artificial intelligence systems.

Dr. Basant Agarwal is currently working as an Associate Professor at Computer Science and Engineering, Central University of Rajasthan. Dr. Basant Agarwal holds a M.Tech. and Ph.D. in Computer Science and Engineering from Malaviya National Institute of Technology (MNIT) Jaipur, India. He has more than 9 years of experience in research and teaching. He has worked as a Postdoc Research Fellow at the Norwegian University of Science and Technology (NTNU), Norway, under the prestigious ERCIM (European Research Consortium for Informatics and Mathematics) fellowship in 2016. He has also worked as a Research Scientist at Temasek Laboratories, National University of Singapore (NUS), Singapore. He has worked on exploring the methods for knowledge extraction from text using deep learning algorithms, Convolutional Neural Network, Recurrent Neural Network etc. He has published more than 60 research papers in reputed conferences and Journals. His research interests include Deep Learning, Natural Language Processing, Machine Learning and Sentiment Analysis.

Mukesh Kumar Gupta is a Professor and Head in the Department of Computer Science and Engineering, at Swami Keshvanand Institute of Technology, Management and Gramothan, Jaipur, India. Altogether he has 20 years of teaching and research experience. Dr. Gupta has received his Ph.D. in Computer Science and Engineering from Malaviya National Institute of Technology (MNIT), Jaipur, India. He has completed M.Tech. in Computer Science and Engineering from Indian Institute of Technology (IIT), Bombay. Dr. Gupta is an active member of Board of Studies (BOS), Department Research Committee for Department of Computer Science and Engineering, and many other committees at Rajasthan Technical University, Kota. He has authored one book, edited three conference proceedings, and published more than 50 technical papers in the referred journals and conference proceedings. The scope of his current research interests encompasses web application security, machine learning, deep learning and Internet of Things.

Chapter 1
Deep Vision: A Robust Dominant Colour Extraction Framework for T-Shirts Based on Semantic Segmentation

R. Kishore Kumar, Kaustav Sengupta, Shalini Sood Sehgal, and Poornima Santhanam

1 Introduction

Over the past two decades, fashion has changed drastically and has become one of the most important aspects of everyone's life. It is a sense of self-expression revealed in a particular time and region through clothing, footwear, lifestyle, accessories, makeup, and body posture. External factors like political, economical, social, and technology also affect fashion and influence trends. Identifying a fashion trend/pattern is one of today's most essential tasks. It helps in better planning and timely delivery. Other elements influencing fashion trends include styles, colour, textile, print, and pattern. Among these elements, colour is one such attribute that plays a vital role in design and communication, influencing the user's sentiments and psychological well-being.

Colour is important as it helps identify consumer behaviour and, in turn, links to one's physiology. Another aspect of colour is the product's or brand's purpose and personality and how they are expressed. For example, the colour blue is associated with characteristics of reliability and trust, thus becoming popular in a corporate setting. As often seen around us, green symbolizes wholesomeness and nature, usually representing wellness brands. On the other hand, red characterizes prosperity and power in certain cultures. As a result, we may conclude that colour can be used

R. Kishore Kumar (✉) · K. Sengupta · P. Santhanam
National Institute of Fashion Technology, Chennai, Chennai, India
e-mail: kishorekumar.ravi@nift.ac.in

K. Sengupta
e-mail: kaustav.sengupta@nift.ac.in

P. Santhanam
e-mail: poornima.santhanam@nift.ac.in

S. S. Sehgal
National Institute of Fashion Technology, Delhi, New Delhi, India
e-mail: shalini.sud@nift.ac.in

V. Devedzic et al. (eds.), *Proceedings of the International Conference on Intelligent Computing, Communication and Information Security*, Algorithms for Intelligent Systems, https://doi.org/10.1007/978-981-99-1373-2_1

as a strategy to position a brand in customers' minds. Moreover, trends in colour are constantly evolving; they are an anchor to an environment's cultural and social context at a given time. Thus, colour is a powerful element that plays a vital role in many aspects, such as understanding the mindsets of individuals to the positioning of a brand. Hence, there is an urge to have an automated system to extract appropriate colours from the apparel images. The extracted colours further guide the tasks of predicting trends, user sentiment detection, suggesting colours to brands, etc. It has motivated us to introduce a novel method of extracting colours from the images.

The objective of this paper is to extract the colours present in the T-shirt of Generation Z in India. Then, the dominant colours in different regions of India are determined by grouping the extracted colours regionally. This study considers T-shirts to extract and predict the dominant colours regionally since they are an essential and flexible item in most people's wardrobes. There are standard techniques to extract colours from apparel images in literature. One such technique is extracting the RGB from each pixel in an image and then mapping it to the preset colours to recognize the colours. Over the years, clustering techniques have emerged to group the extracted RGB to recognize colours automatically. Recently, classification tasks based on deep learning techniques have been adopted to perform colour recognition. The classification techniques attempt to understand the images' engineering properties through feature vectors and map them to the desired colour. These approaches use bounding boxes or region proposals techniques for extracting feature vectors. These feature vectors are extracted from a significant amount of natural skin tone, hair, and a small amount of the background. It will be inappropriate and inadequate for fashion clothing colour preferences or analysis. Therefore, this paper introduces a new framework for accurate colour extraction exclusively for T-shirt images by first segmenting the clothing area. The proposed framework uses the U-Net deep learning model to segment the T-shirt contained in apparel images. Then, the RGB of each pixel in the T-shirt segment is extracted from the segmented images. The RGB features of each image is clustered using the k-means clustering algorithm to discover the broader colour groups and their percentages. By aggregating the T-shirts based on broader colours and their percentages state-wise, the evolving colour preferences of generation Z in various geographical locations are predicted. The proposed framework of extracting colour from T-shirt images helps the artists, designers, image retrieval [34, 42], fashion forecast [3, 39], examine the psychology of an individual or a group, etc.

The remainder of the paper is structured as follows: Sect. 2 discusses the research works related to semantic segmentation techniques and colour extraction frameworks. Section 3 discusses the proposed approach of semantic segmentation, evaluation of segmentation models, and the process of colour extraction from the T-shirt images based on different geographical locations in India. Section 4 discusses the conclusion and future directions.

2 Literature Review

The work related to semantic segmentation and colour extraction frameworks are discussed in this section. The process of segmenting images into various segments or objects is known as image segmentation [27, 43]. Image segmentation plays a vital role in many application including colour prediction [23], medical image analysis [33], autonomous driving [18], industrial inspection [10], classification of terrain visible in satellite imagery, etc. The image segmentation process is broadly classified into three categories (i) semantic segmentation [38], (ii) instance segmentation [22], and (iii) panoptic segmentation [26]. Semantic segmentation is categorizing each pixel of an image to a specific class label. On the other hand, instance segmentation performs object detection and assigns a unique label to each object in the image. Panoptic segmentation combines two distinct approaches to image segmentation. It is a hybrid of instance and semantic segmentation techniques. In this work, we attempt to segment the T-shirt from the other elements like background, human skin, hair, and other regions in an apparel image. Moreover, each apparel image contains only one T-shirt. Therefore, the proposed method utilizes the semantic segmentation approach, and hence our discussions are limited to semantic segmentation techniques.

Initially, several neighbouring pixels of the same shades falling within a certain threshold are grouped to perform semantic segmentation [28, 31]. Further, the extracted pixels are fed to the subtractive clustering technique to generate the initial cluster centroids. Then the generated centroids are given as input to the k-means algorithm to segment the image [8]. Later, two-stage processing is carried out on the images to perform the semantic segmentation [32]. First, a simple image classifier is applied to detect the presence of objects in the image, and in the second stage, a segmentation technique is used to locate the objects [32].

Deep learning-based semantic segmentation models have emerged in recent years, significantly outperforming older techniques in terms of performance. Fully convolutional network (FCN) [38] is the first deep learning model for semantic segmentation. The FCN performs the convolutional operation on the input image and produces the segmentation map, which is the same size as the input image. The key difference in the FCN compared to the conventional convolutional neural network (CNN) is that the fully connected layers in CNN are replaced with fully convolutional layers. As a result, the model outputs the spatial segmentation map instead of classification scores. Even though the FCN model achieves state-of-the-art segmentation performance, it has several drawbacks. One such limitation is that the FCN is slow for real-time inference. The model generates segmental mapping with poor boundary precision. The approach ignores the global context of an image and potentially fails to detect scene-level semantic context [25]. ParseNet [21], a semantic segmentation technique was introduced to overcome one of the drawbacks, that is, considering global context information. The ParseNet appends pool6 as the global context feature to improve the performance over the basic FCN. A combination of conventional conditional random field (CRF) and CNN was introduced by Zheng et al. [47]. The method performs well in semantic segmentation by enhancing the segmentation at the boundaries. Unlike fully convolutional networks, encoder–decoder-based networks were introduced for semantic segmentation. The encoder performs the convolution operations, and the decoder performs the deconvolution operations (i.e., transposed convolution) [29]. SegNet is an encoder–decoder architecture that was specifically designed to perform nonlinear upsampling in the decoder [4] to achieve segmentation. It aids the upsampling by using the pooling indices obtained from the encoder's max-pooling step. It eliminates the necessity for upsampling learning. Following that, other segmentation techniques were introduced for a wide range of applications, taking inspiration from FCNs and encoder–decoder models. U-Net [35], an encoder–decoder model was first developed for segmenting biological microscopy images. This network can segment images with higher accuracy, trained with very few annotated images. The network follows an encoder–decoder with reduced feature maps. The feature map from the encoder to the decoder is concatenated, avoiding loss in mapping pattern information. The model was applied to various segmentation tasks [37]. Similar to U-Net, V-Net [24] network was presented for 3D medical image segmentation. Other networks, such as R-CNN [7], Fast R-CNN [12], Mask R-CNN [13], have demonstrated object detection using a region proposal network (RPN) and also extended to address the problem of instance segmentation (i.e., performing object detection as well as the semantic segmentation). Many studies have been done on the instance and panoptic segmentation; however, the proposed approach uses semantic segmentation to extract the T-shirt from the apparel images.

In the context of colour extraction, a region located within an image containing preset colours was introduced. It is primarily done to index the image based on colour [40]. A skin-colour extraction algorithm is proposed to detect human faces in colour images with the presence of complex backgrounds [15, 41]. Later, the salient region detection technique was used to detect the boundaries of objects in the image, and colours were extracted [17]. Most of the research works utilize the fully convolutional networks to segment the image pixels into hair, face, and background classes. The hair tone (black, blond, brown, red, or white grey) is then determined using a Random Forest classifier [6]. Recently, Mohammed et al. [1] used an instance segmentation approach called Mask-R-CNN to segment each clothing item from the images and extract the primary/dominant colours. Based on the literature, colour extraction has potential in various sectors, including fashion trends and colour prediction.

3 Proposed Method

This research aims to determine the prominent colours present in the T-shirts geographically. Understanding the prominent colour in a particular region might help decode the mindset of the individuals in that region. The proposed approach consists of two stages in achieving the above goal. The first stage performs the image segmentation task based on semantic segmentation. The semantic segmentation process extracts only the T-shirt segments from the apparel images. The U-Net deep learning model is applied to the apparel images to perform the semantic segmentation task.

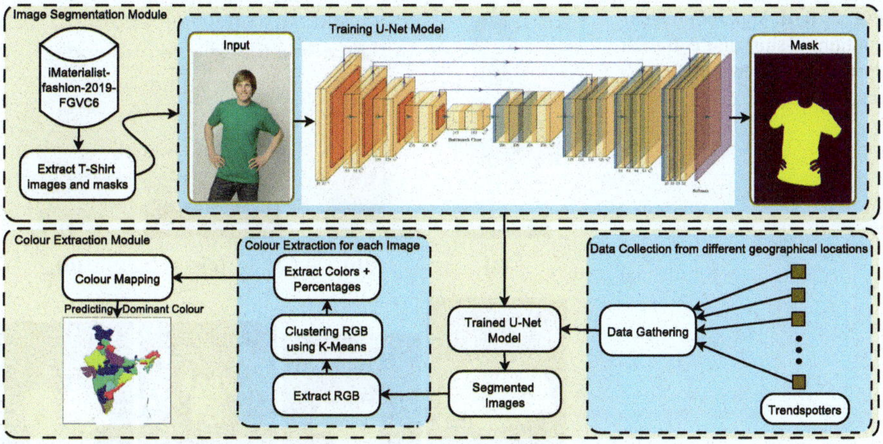

Fig. 1 Block diagram of the proposed image segmentation and colour extraction framework

The U-Net is a traditional convolutional neural network (CNN) architecture trained with apparel images and T-shirt masks. The U-Net architecture attempts to produce a mask that separates the T-shirt from other classes like hair, face, and background. The trained U-Net CNN model is used in the second stage to segment T-shirts from the images collected by the team of Trendspotters from different geographical locations across India. The RGB values of each pixel in the segmented T-shirt area are extracted and grouped using the k-means clustering algorithm to exact colours and their percentages. The extracted colours and percentages are aggregated region-wise. Then, the colour mapping to broader colour family groups is performed, and the primary colours are identified. The dominating colour in each location is determined. Figure 1 shows the overall block diagram of the proposed method.

3.1 Database for Semantic Segmentation

To perform the semantic segmentation task, the iMaterialist (Fashion) 2019 at FGVC6 [16] data set is considered. The data set was primarily created to bridge the gap between fashion and computer vision communities in the field of fine-grained segmentation. The iMaterialist database consists of one million fashion images. The database consists of 27 main apparel objects (coat, dress, jumpsuit, t-shirts, skirts, etc.) and 19 apparel parts (sleeves, collars, etc.). Experts annotated a total of 294 fine-grained attributes for main apparel objects. This study will utilize the database to segment the T-shirts from the apparel images.

To begin with, T-shirts containing apparel images and their associated masks are extracted from the entire data set of 331,213 images. The total number of T-shirt images was 13,725, and an equal number of masks was also acquired. Figure 2a–d shows the apparel images contains T-shirt, and Fig. 2e–h shows their corresponding masks respectively.

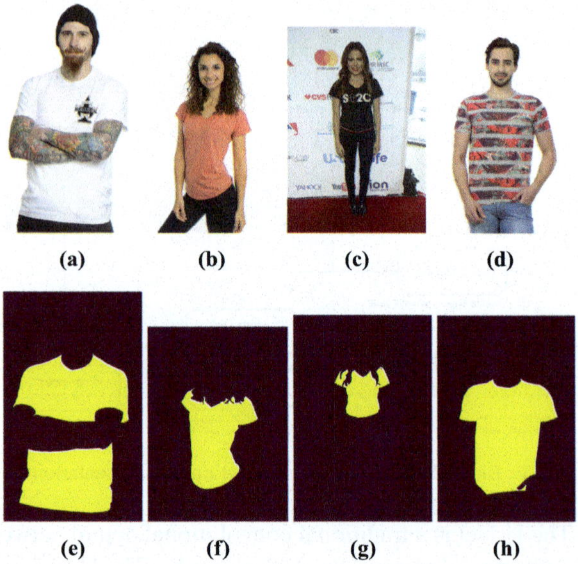

Fig. 2 **a–d** Apparel images contains T-shirt and **e–h** their corresponding masks, respectively

3.2 Semantic Segmentation Approach

Image segmentation is the task of dividing the image into segments of different categories. Semantic segmentation is one such process that aims in grouping the pixels of an image into categories semantically. Object detection/image classification, for example, is a high-level job in computer vision that attempts to classify an image into classes or labels. On the other hand, semantic segmentation attempts to assign a class to each pixel in the image and split it into regions belonging to different categories. This task outputs a high-resolution image of the same size as the original input image, with each pixel classified into a different class. This task is also referred to as dense prediction.

Initially, simple image processing techniques such as thresholding were used to do semantic segmentation [28, 31]. Later, probabilistic models such as conditional random fields (CRFs) [20] is quite time-consuming. CRF, on the other hand, is quite time-consuming [46]. Recently, deep learning-based approaches have emerged to perform semantic segmentation with higher accuracy. The state-of-the-art semantic segmentation model is FCN [38]. Although FCN is widely used, still it has certain drawbacks. Several attempts have been made in the literature to overcome these limitations, and many methods are introduced like ParseNet [21], U-Net [35], V-Net [7, 24], ReSeg [44], etc. One of the best and fastest semantic segmentation models is the U-Net among all the above models. The U-Net deep learning network requires very few training images for model building and yields more precise segmentation [25]. In this study, we utilize the U-Net deep learning network for segmenting the T-shirt from the apparel images.

The U-Net model was created to segment medical images. Later, architecture was applied extensively in other fields apart from medicine. One such study is road surface segmentation [25]. The U-Net deep learning model is a CNN architecture that generates a mask that separates an image into many classes by segmenting the portions of an image into classes. The U-Net architecture consists of two parts: one contracting path (also called encoder) and a symmetric expanding path (also called decoder). The encoder path captures the context of the image, and the decoder path enables precise localization. The encoder path gradually compresses information into lower-dimensional representation using a traditional convolutional and max-pooling layers stack. The decoder part decodes back the information into the original image dimension using transposed convolutions.

The U-Net architecture is depicted in Fig. 3. The network is composed of four different layers: convolutional layers, max-pooling layers, transposed convolution layers, and feature map concatenation (encoder to the decoder). The convolutional layer applies the convolution operation with the filters or kernels to the input image to produce the feature maps. In the first encoding layer, the input image of size 128 $\times$ 128 $\times$ 1 is convolved with 3 $\times$ 3 kernels of 32 feature maps. These feature maps are normalized using batch normalization and followed by an activation function. The rectified linear unit (ReLU) is considered as the activation function. The ReLU activation function adds non-linearity to the network, which aids generalization. The convolution process is repeated in the first encoding layer. Then, 2 $\times$ 2 max-pooling operations are performed with stride 2 for downsampling. The downsampled feature maps are given as input to the second encoding layer. In the second encoding layer, the number of feature channels is doubled. The convolution process continues till it reaches 8 $\times$ 8 $\times$ 512 (see Fig. 3). The decoder or the expanding path now begins the upsampling process. The feature maps are upsampled by 2 $\times$ 2 transposed convolution. As a result, the input array is doubled from 8 to 16. The number of feature channels is reduced half by 256. Again the convolution process is repeated twice with 3 $\times$ 3 kernels. An activation function (ReLU) follows each convolutional process. In addition to these convolution processes, the expanding path has a concatenation of feature map (cropped feature map), derived from the contracting path. The primary goal of this step is to prevent the loss of border pixel information during each convolution operation. The above steps are repeated in the decoder, and in the final decoding layer, an 1-to-1 convolution operation is employed to map each 32 components features to the required number of classes.

3.3 Training

From the iMaterialist (Fashion) 2019 data set, 13,725 apparel images containing T-shirts were considered. Out of which 10,980 images were used for training, and the remaining 2745 were used to test the U-Net model. The Adam optimization algorithm updates network kernels iteratively based on training data. To fine-tune the model, we used the binary cross-entropy loss function. We adopted binary cross-entropy

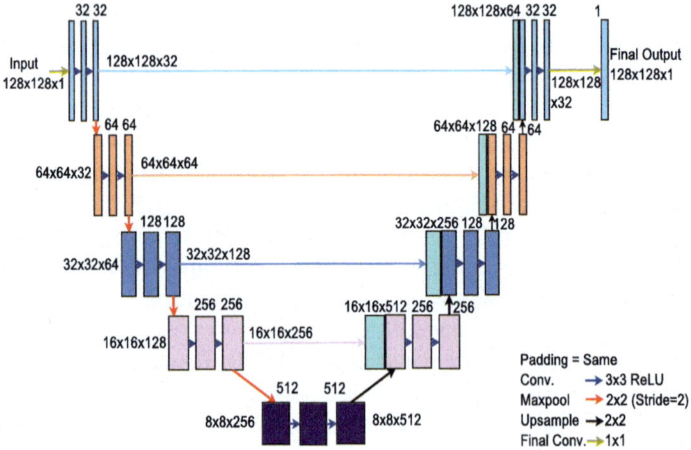

Fig. 3 U-Net architecture

Fig. 4 Training loss

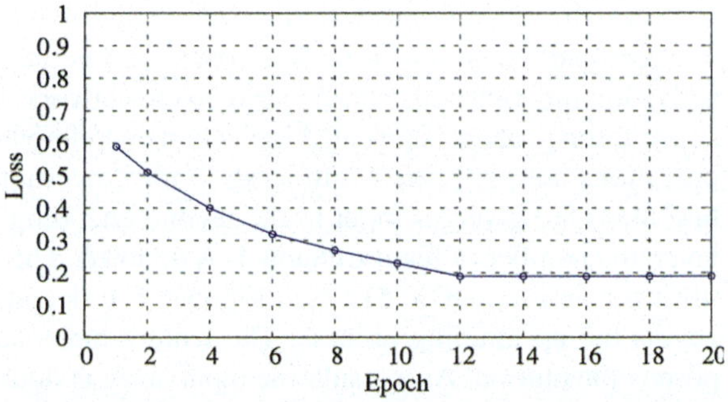

loss since our study has just two classes (T-shirt and non-T-shirt). In each epoch, the model learns the weights, and it is trained for 20 epochs. Figure 4 shows the training loss.

3.4 Testing and Performance Evaluation

The trained U-Net model is considered for segmenting the T-shirt present in the apparel images. The model tries to predict the mask of the T-shirt present in the apparel images. Figure 5a, d shows image taken from the iMaterialist (Fashion) 2019 data set. Figure 5b, e shows the ground truth mask of the images (a) and (d), respectively. Figure 5c, f shows the mask predicted by the trained U-Net model.

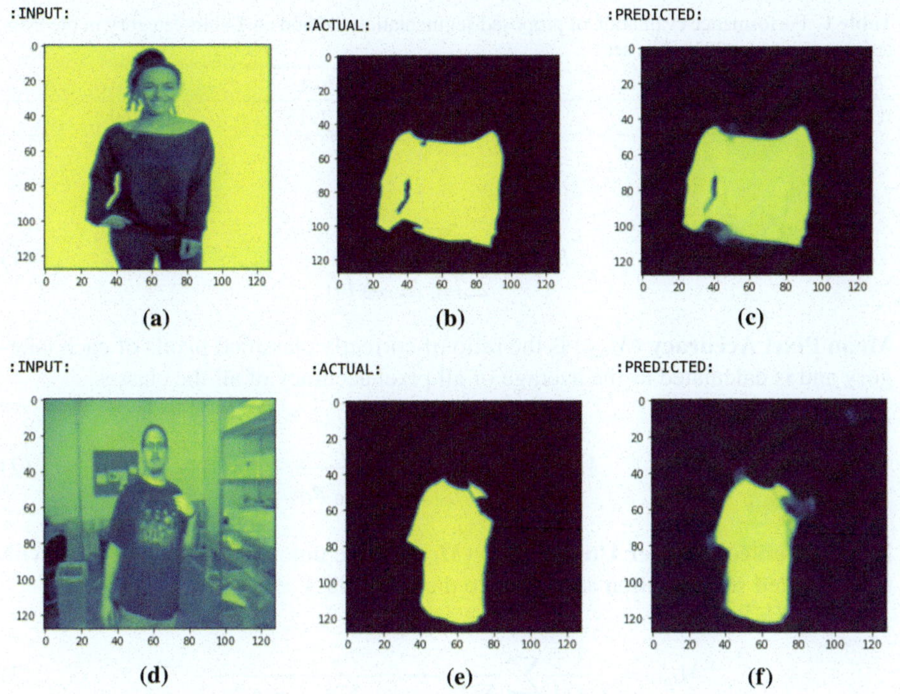

Fig. 5 First column: **a** and **d** show image taken from the iMaterialist (Fashion) 2019 data set. Second column: **b** and **e** are the ground truth mask of the images (**a**) and (**d**), respectively. Third column: **c** and **f** are the mask predicted by the trained U-Net model

3.4.1 Performance Evaluation of the Segmentation Model

To evaluate the segmentation model, we use the standard quantitative metrics discussed in the literature [9, 38]. The quantitative metrics are **Pixel Accuracy** (P_{acc}), **Mean Pixel Accuracy** (M_{acc}), **Mean Intersection over Union (IoU)** (M_{IoU}), and **Frequency Weighted IoU** (FW_{IoU}). Let k represent the number of classes (i.e., T-shirt and non-T-shirt classes). Let p_{ii} represent the number of true positives, that is the number of pixels that belong to the T-shirt class in the ground truth were correctly predicted by the U-Net model. Additionally, the model predicts an amount of pixels as T-shirt class whereas in ground truth they are not. These pixels are false positives and denoted as p_{ij}. The non-T-shirt pixels which are correctly classified as non-T-shirt pixels are false negatives and represented as p_{ji}. With the notations p_{ii}, p_{ij}, p_{ji}, the quantitative metrics such as P_{acc}, M_{acc}, M_{IU}, and FW_{IU} are described as follows:

Pixel Accuracy (P_{acc}) is the ratio of number of pixels that are correctly classified to an individual class, divided by the total number of pixels.

Table 1 Performance evaluation of proposed segmentation method on T-shirt category in the iMaterialist (Fashion) 2019 data set

Approach	P_{acc}	M_{acc}	M_{IoU}	FW_{IoU}
U-Net	0.9664	0.8513	0.7915	0.9469

$$P_{\text{acc}} = \frac{\sum_{i=0}^{k} p_{ii}}{\sum_{i=0}^{k} \sum_{j=0}^{k} p_{ij}} \quad (1)$$

Mean Pixel Accuracy (M_{acc}) is the ratio of correctly classified pixels of each category and is calculated as the average of all pixel accuracy of all the classes.

$$M_{\text{acc}} = \frac{1}{k+1} \sum_{i=0}^{k} \frac{p_{ii}}{\sum_{j=0}^{k} p_{ij}} \quad (2)$$

Mean Intersection over Union (IoU) (M_{IoU}) is the union of the ground truth with the predicted segmentation according to the pixel class.

$$M_{\text{IoU}} = \frac{1}{k+1} \sum_{i=0}^{k} \frac{p_{ii}}{\sum_{j=0}^{k} p_{ij} + \sum_{j=0}^{k} p_{ji} - p_{ii}} \quad (3)$$

Frequency Weighted IoU (FW_{IoU}) is an extension of M_{IoU} in which weights are assigned according to the frequency of each class.

$$\text{FW}_{\text{IoU}} = \frac{1}{\sum_{i=0}^{k} \sum_{j=0}^{k} p_{ij}} \sum_{i=0}^{k} \frac{p_{ii}}{\sum_{j=0}^{k} p_{ij} + \sum_{j=0}^{k} p_{ji} - p_{ii}} \quad (4)$$

The experiment of semantic segmentation is carried out for the test data set, and the masks are predicted. The predicted masks are compared with the ground truth masks, and the quantitative metrics are computed and are given in Table 1.

The trained U-Net model accomplishes the segmentation task with a pixel accuracy of 96.64%, see Table 1. The model also achieved a mean pixel accuracy of 85.13%. It ensures that the proposed method, based on the U-Net model, accurately segments the T-shirt from the apparel images. Additionally, the number of classes in our study is smaller (i.e., segment only the T-shirt class from the apparel images). The U-Net model can learn the weights suitably and obtain an overall mean pixel accuracy of 85.13%. Apart from the pixel accuracy and mean pixel accuracy measures, the mean intersection over union score evaluates the percentage of overlap of ground truth and predicts masks. The scores are 79.15% and 94.69% for the M_{IoU} and FW_{IoU}, respectively. It demonstrates that the proposed method successfully detected the T-shirt in its precise location.

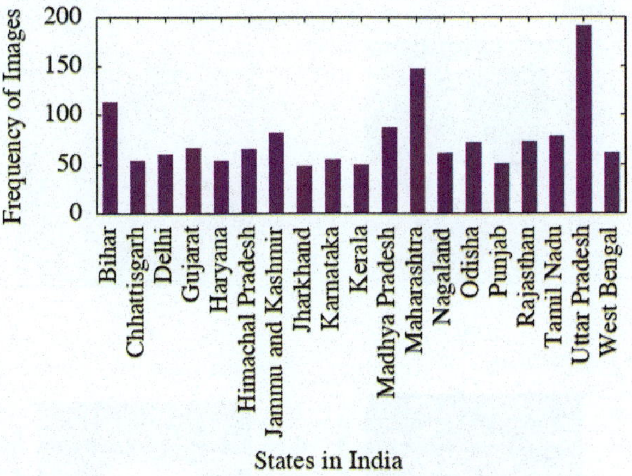

Fig. 6 Distribution of the apparel images (containing T-shirts) collected by the Trendspotters from different parts of India

3.5 Data Collection for Colour Extraction and Analysis

To carry out the subsequent colour extraction and analysis task, a new framework is devised to collect the apparel images (containing T-shirts) from different geographical locations in India. *VisioNxt* [45], a mobile application has been developed in-house to organize the data collection. The application is designed to upload images of different categories for spotting trends. The categories include youth wear, occasion wear, athleisure, Indian clothing, accessories, and footwear. The images are captured by the Trendspotters who are present across India. Trendspotters were trained to capture images and videos in a defined dimension and resolution. The captured images are uploaded to the server and organized based on the category, date, gender, and location. The T-shirt category is enabled in the VisioNxt mobile app for the initial experiment. The Trendspotters are allowed to submit the images. For our initial experiment, the Trendspotters were asked to focus on T-shirts of the Generation Z cohort as most of them also belong to the same. The images were collected in the duration of 27 July to 18 September 2021 and were considered for further analysis. Figure 6 shows the distribution of the T-shirt images collected by the Trendspotters located in different states of India.

From the apparel images collected from the Trendspotters, 50 images from each state are randomly considered and given as input to the trained U-Net model for segmenting the T-shirt. Once T-shirt segmentation is performed, the colour extraction process is carried out. Figure 7a–c shows a subset of T-shirt images collected from the Trendspotters, and Fig. 7d–f shows their corresponding predicted masks, respectively.

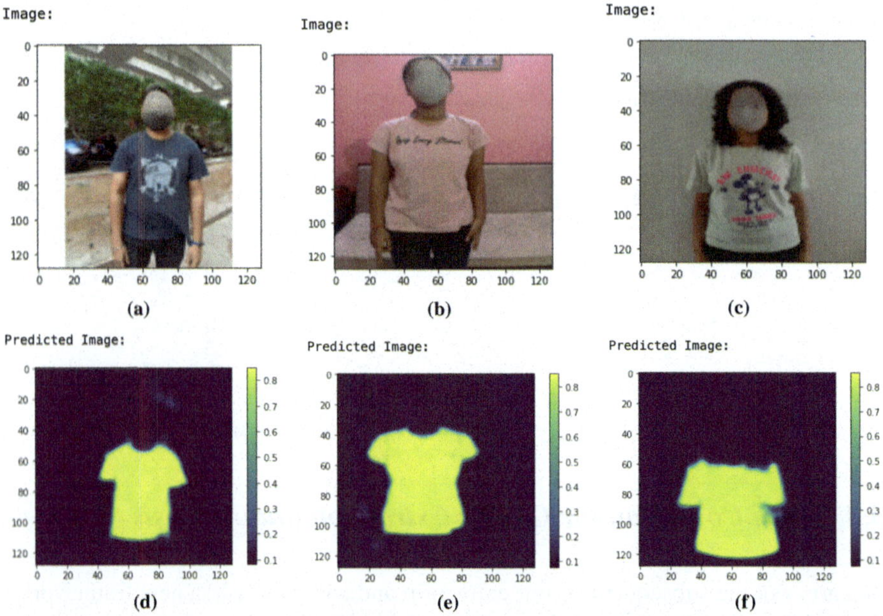

Fig. 7 **a–c** Apparel images collected from the Trendspotters and **d–f** their corresponding predicted T-shirt masks, respectively

3.6 Colour Extraction Method

The colour extraction method begins with the predicted masks for the images collected from the Trendspotters. The masks are used to crop the original T-shirt present in the apparel images. The colours are extracted solely from the pixels present in the T-shirt segment. Figure 7d shows the predicted mask of the apparel image Fig. 7a.

Note that the pixel values of the predicted mask (i.e., Fig. 8a) are in the range of 0–1. It is observed that the pixel values near to one represent the T-shirt, and the pixel values near to zero represent other regions like background, hair, skin, etc. Now, the masks are transformed into the binary matrix for seamless processing. A new technique is devised based on pixel values to binaries the masks. The binarization process is as follows: the pixel values in the mask are taken, and a histogram is computed. A histogram is a graphical representation that depicts the frequency distribution of values occurring in an image. Figure 8b shows the histogram for the predicted mask image (Fig. 8a). The histogram shows that the mask contains the major portion of pixels with values ranging from 0 to 0.2, which contributes to the non-T-shirt area. The histogram also shows that the pixel values in the range 0.7–0.9 are the second highest and contribute to the T-shirt area. Therefore, the gradient of the histogram values is computed to binarise the mask. The gradient recognizes 0.7 as a change in the histogram values, which acts as a threshold for binarizing the mask. Figure 8c shows the binarized mask of the original mask (see Fig. 8a).

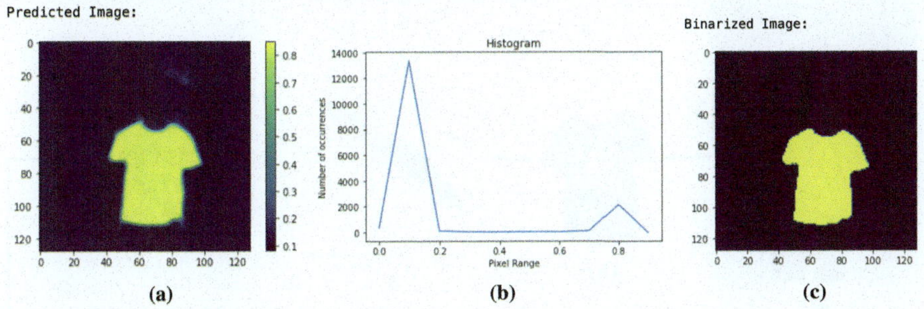

Fig. 8 **a** Predicted mask image, **b** the histogram for the predicted mask image (**a**), and **c** the binarized mask

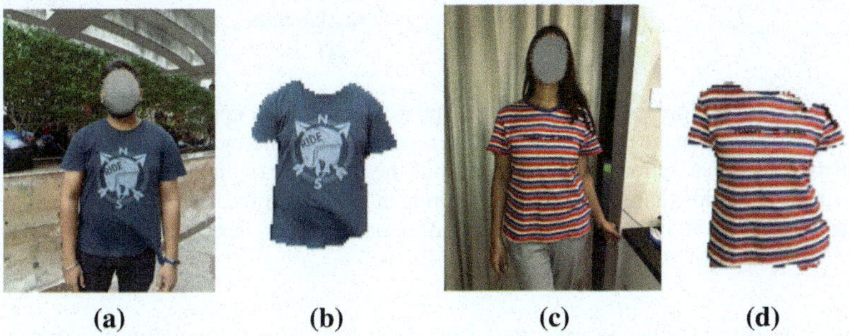

Fig. 9 **a**, **c** Images collected from Trendspotters and **b**, **d** their corresponding segmented images

The procedure is applied to all the predicted masks (i.e., the masks are predicted for all the images collected from Trendspotters) and generates the binarised masks. The binarized masks are resized back to the original size of the image. The apparel images are then cropped to obtain the T-shirt segments. Figure 9a, c shows the apparel images collected from the Trendspotters, and Fig. 9b, d shows their corresponding T-shirt segments.

From the T-shirt segments, the Red, Green, Blue (RGB) of each pixel is extracted. The RGBs of each image are taken as three-dimensional feature vectors. These feature vectors are clustered using the k-means clustering technique to find the most prominent colours. k-means [11] is a straightforward method for classifying a given set of data using a fixed number of clusters (assuming k clusters). The core element is to define k centres. In this study, the k centres are the number of prominent colours. The number of prominent colours (k) for each image should be known before applying k-means clustering. The T-shirt in each apparel image may have many prominent colours, which cannot be determined in advance. Therefore, the elbow method is applied to all the T-shirt segments, obtaining the best k value. From the experiment, it was observed that the best k value is 5. With the k value and the RGB features, the k-means clustering algorithm is processed. The method chooses k random initial cluster centres or centroids and assigns each feature vector to one of them. Each

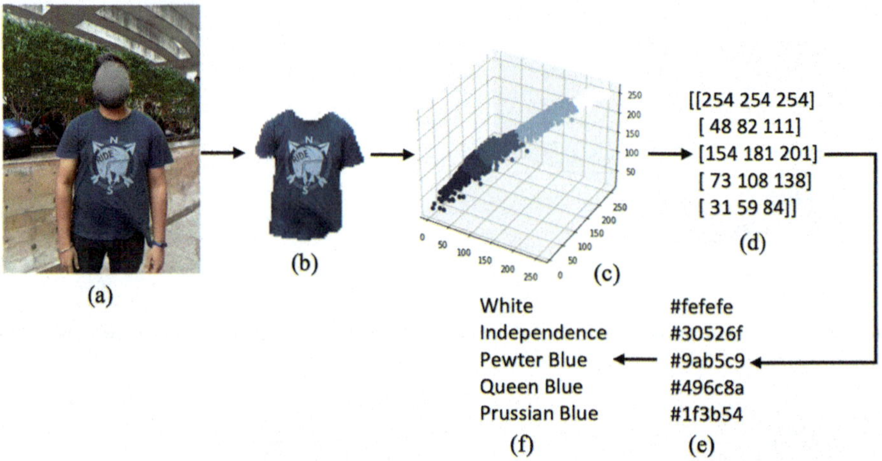

Fig. 10 Process of colour extraction from the apparel images after applying the segmentation task. **a** Depicts the original image collected from the Trendspotter. **b** Obtained after applying the proposed segmentation and binarization approach. **c** Exhibits the clustering of the RGB pixels in the 3D space. **d** The cluster centroids that represents the prominent colours in the T-shirt. **e** Represents the hex codes for the derived RGB's. **f** Their corresponding names of the hex codes

centroid's feature vector is assigned based on the minimum distance scores. Once this process completes, the centroids are updated by taking the mean. This process repeats until the cluster centres do not change their values. Now, the method stops the iterations, and the cluster centres tend to be the prominent colours in each image. Along with the prominent colours, the number of data points associated with each cluster centres is also determined for further study. Figure 10a shows the apparel image, Fig. 10b shows the T-shirt segment, Fig. 10c shows the clustering of RGB features in the 3D space, and Fig. 10d lists the cluster centroids which represent the prominent colours in the T-shirt segment.

Now the prominent RGBs of each image are converted into hex codes. It is done by creating a dictionary consisting of all the RGB and corresponding hex codes. Figure 10e shows the hex codes for the corresponding RGB (see Fig. 10d). Similarly, another dictionary is built to map the hex codes to the colour names. Figure 10f shows the colour names for the corresponding hex codes (see Fig. 10e). While extracting the hex codes or colours names, the data points belonging to each RGB (cluster centres) are acquired to analyze colour patterns geographically. Finally, the colour names or hex codes for each image are mapped to broader colour families by creating the dictionary using encycolorpedia [14]. This encycolorpedia gives the details of the hex code, colour name, and corresponding broader colour family.

The colour extraction method is applied to all the images taken randomly from each state. The broader colours and their percentages state-wise are extracted from the T-shirt segments.

Table 2 Division of states of India into zones

S. No.	Zones	States
1	North	Haryana, Himachal Pradesh, Jammu and Kashmir, Punjab, Rajasthan
2	Center	Chhattisgarh, Delhi, Madhya Pradesh, Uttar Pradesh
3	South	Karnataka, Kerala, Tamil Nadu
4	East	Bihar, Jharkhand, Odisha, West Bengal
5	Northeast	Nagaland
6	West	Gujarat, Maharashtra

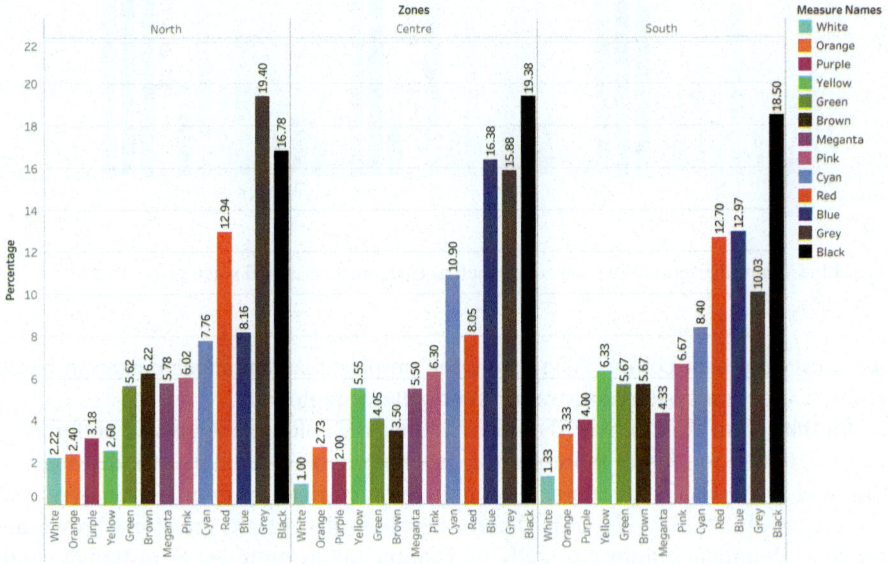

Fig. 11 Colour distribution in zones such as north, center, and south, respectively

3.7 Analysing the Broader Colours Zone-Wise

To analyze the extracted broader colours, we divide the states of India into five zones: north, south, center, east, west, and northeast. Table 2 shows the division of states in India into zones. The colour distribution corresponds to each states are aggregated zone-wise and are depicted as bar chart in Figs. 11 and 12.

Figure 11 shows the colour distribution in India's northern, central, and southern zones, respectively. It is observed from the bar chart; grey is the dominating colour in the northern zone. As we move towards the central and southern Indian states, black is the most dominant colour.

Similarly, Fig. 12 shows the colour distribution of India's east, northeast, and west zones, respectively. It is observed from the bar chart that the dominant colours of the east and the west zones are grey. In the northeastern zone, brown turned out to be

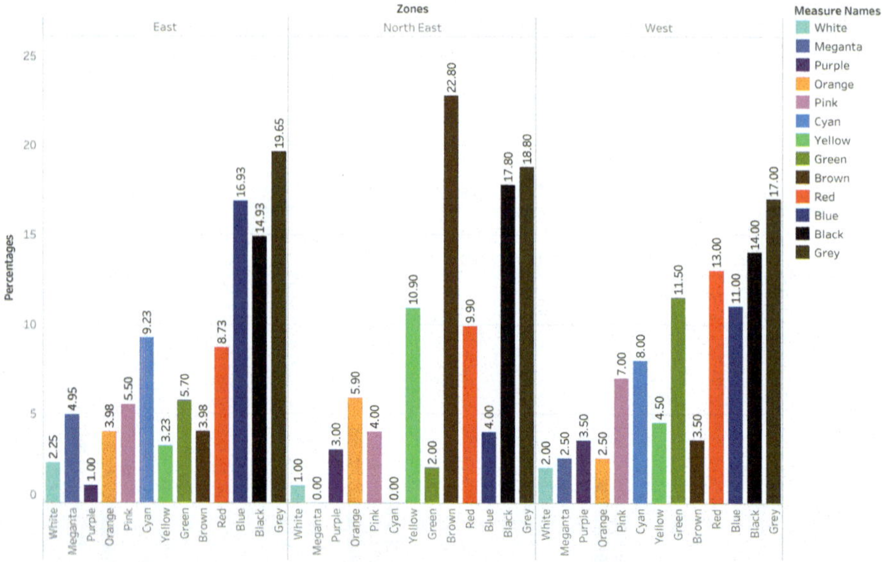

Fig. 12 Colour distribution in each zones such as east, north east, and west, respectively

the most dominant colour. As a result, the proposed semantic segmentation-based colour extraction system discovers colour patterns regionally.

Further, from the discovered colour patterns, the colour analysis and preferences can be studied to extract more precise psycho-graphic outcomes of individuals in different locations. There are extensive research carried out in colour analysis and preferences [2, 5, 19, 30, 36]. Our results show that black, grey, and brown are the most dominant colours. A study on fashion colour preferences in Indian youth category was carried out by Sengupta [36]. The study observed that black colour is preferred more on clothing among the Indian youth in metro cities, with 28% respondents indicating black as their most preferred clothing colour. Hence, the finding of this paper also resonates with the above citations [36].

This proposed framework of extracting colours from apparel images will help understand the emotions, personalities, opinions, attitudes, interests, and lifestyles of different age groups in India. However, a more extensive data set, collected consistently over time, might help us understand the shifts in trends more precisely. The predicted insights can help the retailers make an informed decision at the right time.

4 Conclusion and Future Directions

As the demand for understanding a consumer in businesses is ever-growing, the role of colour in fashion is crucial. Colours dictate moods, attitudes and influence the choice of buying for a consumer. With T-shirts being a staple fashion apparel product in the wardrobes of most Generation Z, it was chosen to understand the changing colour preferences in that cohort. Thus, this paper introduces a framework for colour extraction from the T-shirt images by adopting the semantic segmentation technique. Following that, the colours are extracted from the T-shirt segments to geographically identify the colour patterns. The above task is performed by first training the U-Net deep learning model (semantic segmentation model) with the T-shirt images and masks. The trained U-Net deep learning model is tested with the apparel images and extracts the desired T-shirt segments. The proposed segmentation model achieved the mean pixel accuracy of 85% and mean IoU of 80% on the iMaterialist (Fashion) 2019 data set. The trained U-Net segmentation model is applied to the apparel images collected by the Trendspotters from different parts of India to discover the dominant colour patterns. The U-Net segmentation model extracts the T-shirt segments. The colours are extracted from each pixel and clustered using the K-Means algorithm to extract the prominent colours. The prominent colours are aggregated region-wise (state-wise), and the dominant colour patterns are discovered geographically. In future, the proposed framework can be applied to extensive data collected daily from different geographical locations. Also, this framework can be used to predict the evolving colour preferences and trends in a specified age cohort and location.

Acknowledgements We take immense pleasure to acknowledge the Research and Development (R&D) division of the Ministry of Textile (MoT) for funding the research work via the VisioNxt Project. We are incredibly thankful to Dr. M. A. Rajeev for his valuable and insightful feedback. We also extend our gratitude to all the Trendspotters. They are students of the National Institute of Fashion Technology (NIFT) for support in capturing images from different parts of India.

References

1. Al-Rawi M, Beel J. Probabilistic color modelling of clothing items. In recommender systems in fashion and retail, pp 21–40. http://dx.doi.org/978-3-030-66103-8
2. Akcay O, Dalgin H (2011) Perception of color in product choice among college students: a cross-national analysis of USA, India, China and Turkey. Int J Bus Soc Sci 2(21)
3. Al-Halah Z, Stiefelhagen R, Grauman K (2017) Fashion forward: forecasting visual style in fashion. In: Proceedings of the IEEE international conference on computer vision (ICCV), Oct 2017. IEEE, pp 1–10
4. Badrinarayanan V, Kendall A, Cipolla R (2017) SegNet: a deep convolutional encoder-decoder architecture for image segmentation. IEEE Trans Pattern Anal Mach Intell 39(12):2481–2495
5. Bakker I, van der Voordt T, Vink P, de Boon J, Bazley C (2015) Color preferences for different topics in connection to personal characteristics. Color Res Appl 40(1):62–71. https://onlinelibrary.wiley.com/doi/abs/10.1002/col.21845
6. Borza D, Ileni T, Darabant A (2018) A deep learning approach to hair segmentation and color extraction from facial images. In: Advanced concepts for intelligent vision systems. Springer International Publishing, pp 438–449
7. Chen C, Liu MY, Tuzel O, Xiao J (2017) R-CNN for small object detection. In: Computer vision—ACCV 2016. Springer International Publishing, Cham, pp 214–230
8. Dhanachandra N, Manglem K, Chanu YJ (2015) Image segmentation using k-means clustering algorithm and subtractive clustering algorithm. Procedia Comput Sci 54:764–771

9. Garcia-Garcia A, Orts S, Oprea S, Villena Martinez V, Rodríguez J (2017) A review on deep learning techniques applied to semantic segmentation. arXiv, pp 1–23
10. Gheorghe A, Amza CG, Popescu D (2012) Image segmentation for industrial quality inspection. Fiabil Durabil 1:126–132
11. Ghosh P, Mali K, Das SK (2018) Use of spectral clustering combined with normalized cuts (N-Cuts) in an iterative k-means clustering framework (NKSC) for superpixel segmentation with contour adherence. Pattern Recognit Image Anal 28:400–409
12. Girshick R (2015) Fast R-CNN. In: International conference on computer vision (ICCV), Dec 2015. IEEE, pp 1440–1448
13. He K, Gkioxari G, Dollar P, Girshick R (2017) Mask R-CNN. In: International conference on computer vision (ICCV), Oct 2017. IEEE, pp 2980–2988
14. Hex color codes, paint matching and color picker. https://encycolorpedia.com/
15. Huynh-Thu Q, Meguro M, Kaneko M (2002) Skin-color extraction in images with complex background and varying illumination. In: Workshop on applications of computer vision (WACV). IEEE, pp 280–285
16. iMaterialist (Fashion) 2019 at FGVC6. https://www.kaggle.com/c/imaterialist-fashion-2019-FGVC6
17. Jahanian A, Vishwanathan SVN, Allebach JP (2015) Autonomous color theme extraction from images using saliency. In: Imaging and multimedia analytics in a web and mobile world, SPIE. International Society for Optics and Photonics, pp 57–64. https://doi.org/10.1117/12.2084051
18. Kaymak C, Ucar A (2019) Semantic image segmentation for autonomous driving using fully convolutional networks. In: International artificial intelligence and data processing symposium (IDAP). IEEE, pp 1–8
19. Kodžoman D, Hladnik A, Pavko Čuden A, Čok V (2022) Exploring color attractiveness and its relevance to fashion. Color Res Appl 47(1):182–193. https://onlinelibrary.wiley.com/doi/abs/10.1002/col.22705
20. Lafferty JD, McCallum A, Pereira FCN (2001) Conditional random fields: probabilistic models for segmenting and labeling sequence data. In: Proceedings of the eighteenth international conference on machine learning. Morgan Kaufmann Publishers Inc., San Francisco, CA, pp 282–289
21. Liu W, Rabinovich A, Berg AC (2015) ParseNet: looking wider to see better. arXiv arXiv:1506.04579
22. Li G, Xie Y, Lin L, Yu Y (2017) Instance-level salient object segmentation. In: 2017 IEEE conference on computer vision and pattern recognition (CVPR). IEEE, pp 247–256
23. Mahrishi M, Morwal S, Muzaffar AW, Bhatia S, Dadheech P, Rahmani MKI (2021) Video index point detection and extraction framework using custom YoloV4 darknet object detection model. IEEE Access 9:143378–143391
24. Milletari F, Navab N, Ahmadi SA (2016) V-Net: fully convolutional neural networks for volumetric medical image segmentation. In: Fourth international conference on 3D vision (3DV). IEEE, pp 565–571
25. Minaee S, Boykov YY, Porikli F, Plaza AJ, Kehtarnavaz N, Terzopoulos D (2021) Image segmentation using deep learning: a survey. IEEE Trans Pattern Anal Mach Intell 1–22
26. Mohan R, Valada A (2021) EfficientPS: efficient panoptic segmentation. Int J Comput Vis 129(5):1551–1579
27. Nava R, Fehr D, Petry F, Tamisier T (2021) Tire surface segmentation in infrared imaging with convolutional neural networks and transfer learning. Pattern Recognit Image Anal 31:466–476
28. Nock R, Nielsen F (2004) Statistical region merging. IEEE Trans Pattern Anal Mach Intell 26(11):1452–1458
29. Noh H, Hong S, Han B (2015) Learning deconvolution network for semantic segmentation. arXiv arXiv:1505.04366
30. Okan A, Hasan D (2012) Marketing to teenagers: the influence of colour, ethnicity and gender. Int J Bus Soc Sci 3(22):10–18
31. Otsu N (1979) A threshold selection method from gray-level histograms. IEEE Trans Syst Man Cybern 9(1):62–66

32. Plath N, Toussaint M, Nakajima S (2009) Multi-class image segmentation using conditional random fields and global classification. In: Proceedings of the 26th annual international conference on machine learning. Association for Computing Machinery, pp 817–824
33. Punn NS, Agarwal S (2020) Inception U-Net architecture for semantic segmentation to identify nuclei in microscopy cell images. ACM Trans Multimed Comput Commun Appl 16(1):1–15
34. Radenović F, Tolias G, Chum O (2019) Fine-tuning CNN image retrieval with no human annotation. IEEE Trans Pattern Anal Mach Intell 41(7):1655–1668
35. Ronneberger O, Fischer P, Brox T (2015) U-Net: convolutional networks for biomedical image segmentation. In: International conference on medical image computing and computer-assisted intervention. Springer, pp 234–241
36. Sengupta K (2018) An integrative analysis on values and lifestyle (VALS) of Indian youth in metro cities and it's impact on their clothing colour preference, colour—emotion and colour-image association. PhD dissertation, National Institute of Fashion Technology
37. Sevastopolsky A (2017) Optic disc and cup segmentation methods for glaucoma detection with modification of U-Net convolutional neural network. Pattern Recognit Image Anal 618–624
38. Shelhamer E, Long J, Darrell T (2017) Fully convolutional networks for semantic segmentation. IEEE Trans Pattern Anal Mach Intell 39(4):640–651
39. Silva ES, Hassani H, Madsen DO, Gee L (2019) Googling fashion: forecasting fashion consumer behaviour using Google trends. Soc Sci 8(4):1–23
40. Smith J, Chang SF (1995) Single color extraction and image query. In: International conference on image processing, pp 528–531
41. Srinivasan KS, Karantharaj P, Sainarayanan G (2011) Skin colour segmentation based 2D and 3D human pose modelling using discrete wavelet transform. Pattern Recognit Image Anal 21:740–753
42. Sun K, Zhu J (2022) Searching and learning discriminative regions for fine-grained image retrieval and classification. IEICE Trans Inf Syst E105.D(1):141–149
43. Szeliski R (2011) Computer vision algorithms and applications. Springer Science & Business Media
44. Visin F, Ciccone M, Romero A, Kastner K, Cho K, Bengio Y, Matteucci M, Courville A (2016) ReSeg: a recurrent neural network-based model for semantic segmentation. In: Proceedings of the IEEE conference on computer vision and pattern recognition (CVPR) workshops, June 2016. IEEE
45. VisioNxt—apps on Google Play. https://play.google.com/store/apps/details?id=com.visionxtnift.visionxt
46. Yang Z, Yu H, Sun W, Mao Z, Sun M (2019) Locally shared features: an efficient alternative to conditional random field for semantic segmentation. IEEE Access 7:2263–2272
47. Zheng S, Jayasumana S, Romera-Paredes B, Vineet V, Su Z, Du D, Huang C, Torr PHS (2015) Conditional random fields as recurrent neural networks. In: IEEE international conference on computer vision (ICCV), Dec 2015, pp 1–17. http://dx.doi.org/10.1109/ICCV.2015.179

Chapter 2
Wheel Shaped Defected Ground Structure Microstrip Patch Antenna with High Gain and Bandwidth for Breast Tumor Detection

Sonam Gour, Reena Sharma, Abha Sharma, and Amit Rathi

1 Introduction

Biotelemetry provides link for interconnecting the implantable medical device to the outside of the body structure. An implantable body antenna is used for the transmission toward the external connected devices through the implantable body device. Lots of people are vulnerable to the different disease due to avoidance of the regular body checkup. So many devices designed by the designer for early detection of the disease for the easy and fast recovery [1]. Some reconfigurable devices should also use for designing the antenna for achieving the high gain [2]. To meet the requirement of the biomedical devices, many antennas have used minimum size, low value of SAR, high directivity and high gain [3]. A number of issues have to be considered while designing any implantable device like small size of antenna, wide operating bandwidth and high radiation efficiency and radiation pattern of the antenna [4]. The implantable antenna designed using microstrip patch antenna due its low cost, easily availability and ease fabrication [5]. Different design specification can be done in the designing for achieving biocompatibility and miniaturization like slot cutting, metamaterial uploading [6], ground defected structure [7] and flexible techniques [8, 9]. Some designs used shorting pins for achieving resonance for the two different frequencies. The position of the shorting pin is close to the feed point.

The gain of antenna is improved by using metamaterial with the large value of epsilon. This paper represents a wheel shaped circular design patch antenna for high gain and directivity. The antenna structure is designed for the ISM band because the main aspect of the IMD is to working in the ISM band ranges from 2.4 to

S. Gour · R. Sharma
Department of Computer Engineering, Poornima College of Engineering, Jaipur, India

S. Gour · A. Sharma · A. Rathi (✉)
Department of EC Engineering, Manipal University Jaipur, Jaipur, India
e-mail: amitrathi1978@gmail.com

V. Devedzic et al. (eds.), *Proceedings of the International Conference on Intelligent Computing, Communication and Information Security*, Algorithms for Intelligent Systems, https://doi.org/10.1007/978-981-99-1373-2_2

2.48 GHz [10]. The patch antenna is modified by using defected ground structure and rectangular slots. The size of the antenna is 54 * 54 * 1.64 mm. The main advantage of the antenna is high gain and directivity. FR4 ($\varepsilon_r = 4.3$) has been used for achieving flexibility in the antenna because it is the main requirement while designing any implantable device.

2 Antenna Design

As per the required operation, different design techniques have been used for the design of antenna. The proposed antenna has wheel shaped structure, and circular slots used for enhancing the gain and directivity. The design is simulated by using CST Microwave Studio 2021. The same design is used for detecting the tumor in the breast phantom structure. The parameter of antenna is selected by using the design equation is shown in Table 1.

Further miniaturization has been achieved by doing some design modification in the patch all design specification describes in Fig. 1.

2.1 Design Evaluation

The main purpose for designing of the antenna is to achieve high gain and directivity. The lower tripping and slightly modification in the bandwidth achieved from design a to f. Final reduction in the band achieved by using some defected ground structure. The working frequency of the antenna is lowered by providing current distribution throughout the ground; length path of the antenna is further increased by using electrical slots in the patch [10]. The antenna frequency is lowered by providing the current distribution along the ground, which helps in achieving the dual-band operation. By inserting slots electrical length can be increased. The used defected ground structure provides the electrical freedom in the biocompatibility [7].

The considering final design provides the better performance as per used design parameter and miniaturization techniques. As shown in Fig. 2, the initial implantable antenna is not as per requirement, further design evaluation and consideration in the

Table 1 Design parameters

Parameter	Value (mm)
Ground width (W)	54.4
Ground length (L)	54.4
Height of ground (H_t)	0.035
Height of substrate (H)	1.6
Cylindrical patch radius (a)	17

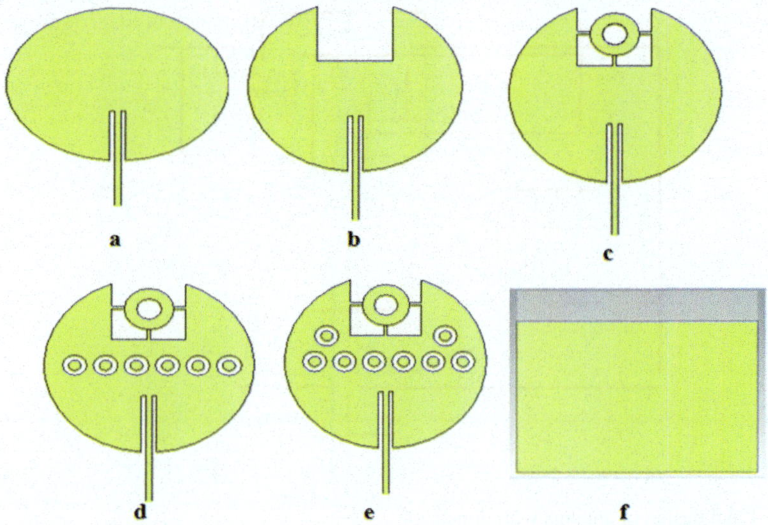

Fig. 1 Design evaluation for the better result

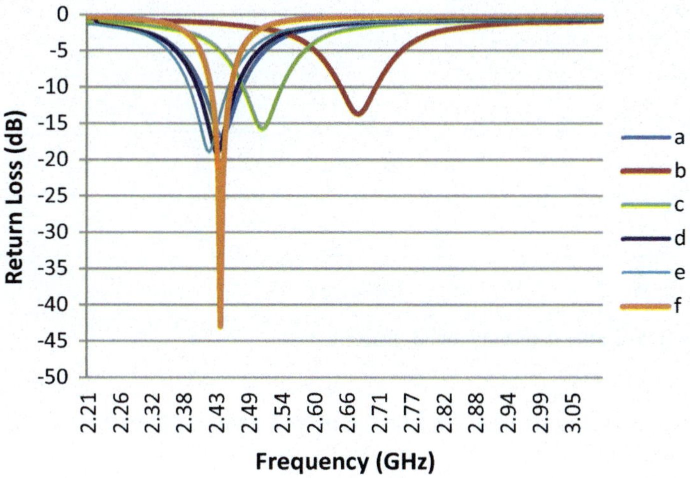

Fig. 2 Changes in the return loss as per design evaluation

patch reduces the operating frequency and as well as shift the return loss to the lowest position as per applicable. The final design with dimension is represents in Fig. 3. The final evaluation in the design with by defected ground structure (discussed in Fig. 1f) provides the better return loss and compatibility in the design.

So, Fig. 4 shows that the antenna operated in the ISM band, therefore it is suitable design for the implantable medical device and also works as the biocompatible design. The simulated achieved S11 bandwidth is 266 MHz ranging from 2.42 to

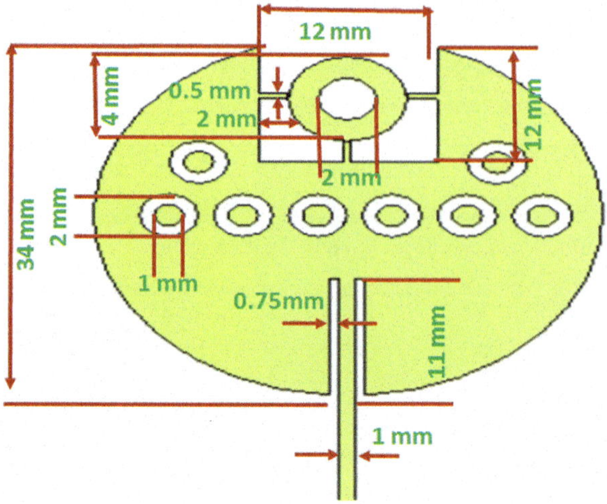

Fig. 3 Final design of antenna with dimension

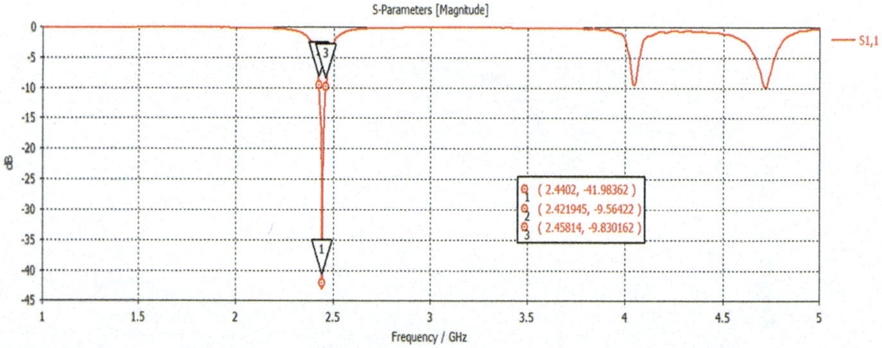

Fig. 4 Return loss after defected ground structure

2.45 GHz. The design is used for enhancing the gain and directivity. The enhanced gain represents in Fig. 5.

The achieved directivity and gain of the design are, respectively, 5.88 dB and 5.36 dB. Furthermore, the design antenna is studied for the biomedical application. The antenna is used for checking the location and presence of the tumor in the breast phantom.

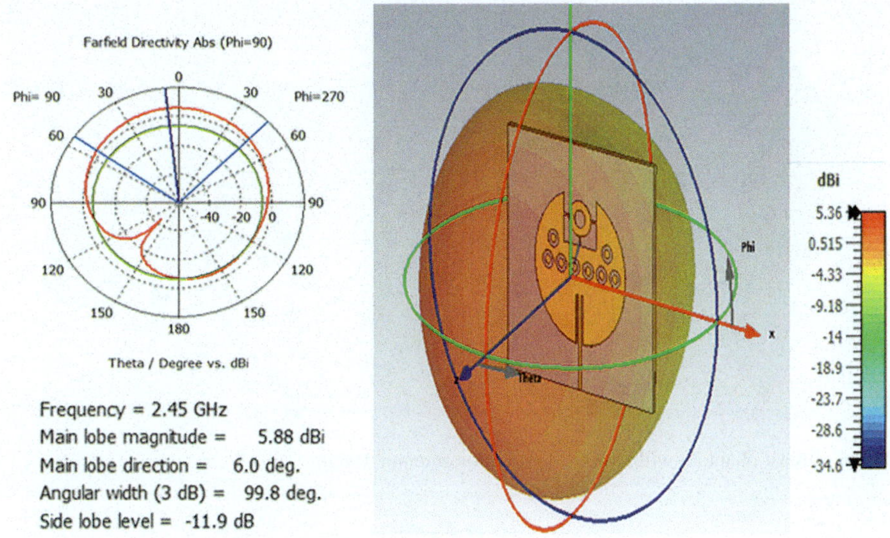

Fig. 5 Enhanced gain and directivity of the antenna

3 Modeling of the Breast Phantom

The antenna is simulated in the CST Design Suite 2021 and checked the performance in the breast phantom. At first, the breast phantom had design for the perfect parametric value for the resonating frequency of 2.45 GHz. Table 2 shows the specification of the body phantom used for designing to breast model.

A breast phantom was created by CST Studio Suite software. The design is simulated and checked the performance of the biocompatibility in terms of SAR. The value of the SAR is to be far less than the maximum allowed limit of 2 W/kg for the 10 g of tissue exposed by the EM radiation [10] and 1.6 W/kg for the 1 g of the tissue [11]. The position of the tumor in the skeletal muscle is $X_{\text{center}} = 10$ mm, $Y_{\text{center}} = 0$ mm and $W_{\text{center}} = 40$ mm. The Dielectric properties of muscle, skin and tumor such as permeability, density (kg/m^3), thermal conductance (W/K/m), heat and diffusivity (m^2/s) were used for creating the body phantom (Figs. 6 and 7).

A breast tissue was created and designed antenna simulated by placing the antenna on the top of the breast phantom. The return loss reduced to − 20 dB and achieved resonating frequency is 2.45 GHz. It shows that design structure keeps the results

Table 2 Dielectric properties of different layers

Layer	Rho (kg/m^3)	Thermal conductivity (W/K/m)	Epsilon	Loss tangent
Muscle	1041	0.53	54.97	0.241
Skin	1100	0.50	41.32	0.272
Tumor	1058	–	54.9	–

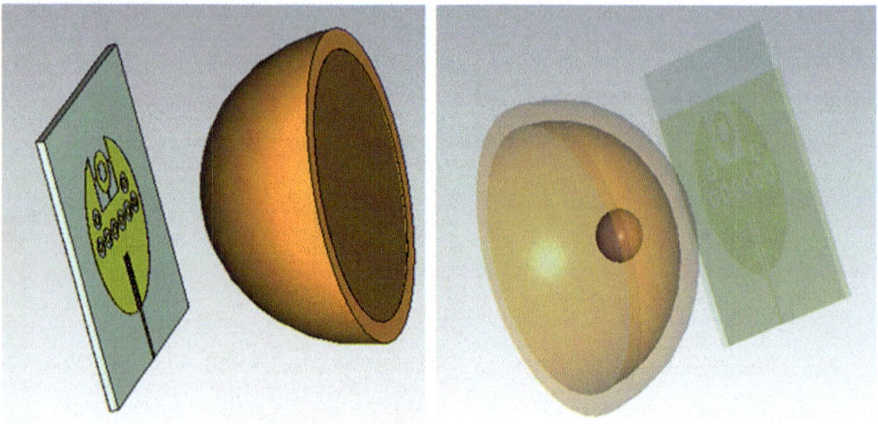

Fig. 6 Breast phantom with and without tumor antenna design

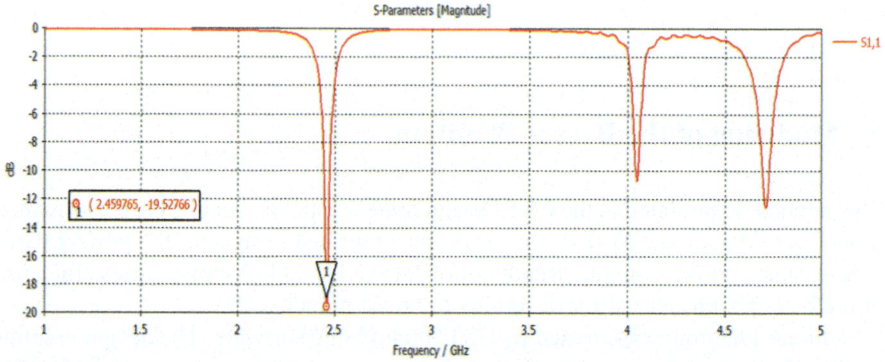

Fig. 7 Return loss of the designed antenna while using near the breast phantom

as it can be implanted in the body phantom. The new directivity of the antenna is 6.33 dB. It represents that the designed structured radiated the more radiation in the particular way. The lower the value of SAR shows the better performance of the antenna in the body applications. The SAR value of the antenna was considered for the 1 and 10 g muscle with and without tumor. The minor variation is achieved in the SAR shows that the tumor available in the breast phantom (Fig. 8).

The value of the SAR is little bit increased due to the presence of extra body fat in the breast. It shows in Fig. 9.

Table 3 shows the comparison of the proposed work with the other previous works. It shows that our design proposed better result as compared to the previous one.

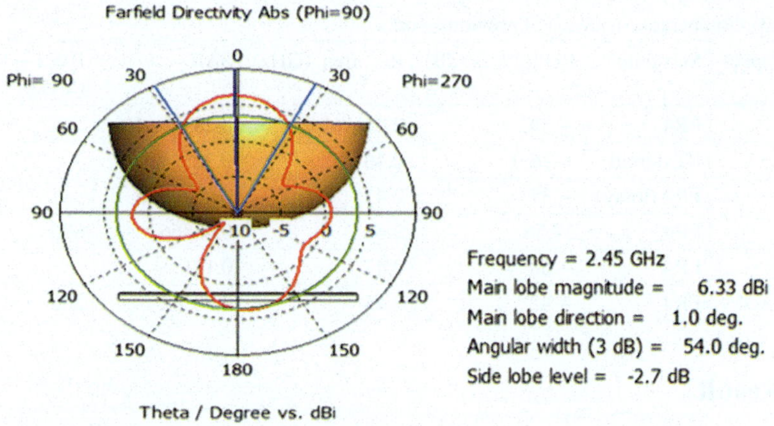

Fig. 8 Directivity of the updated design antenna

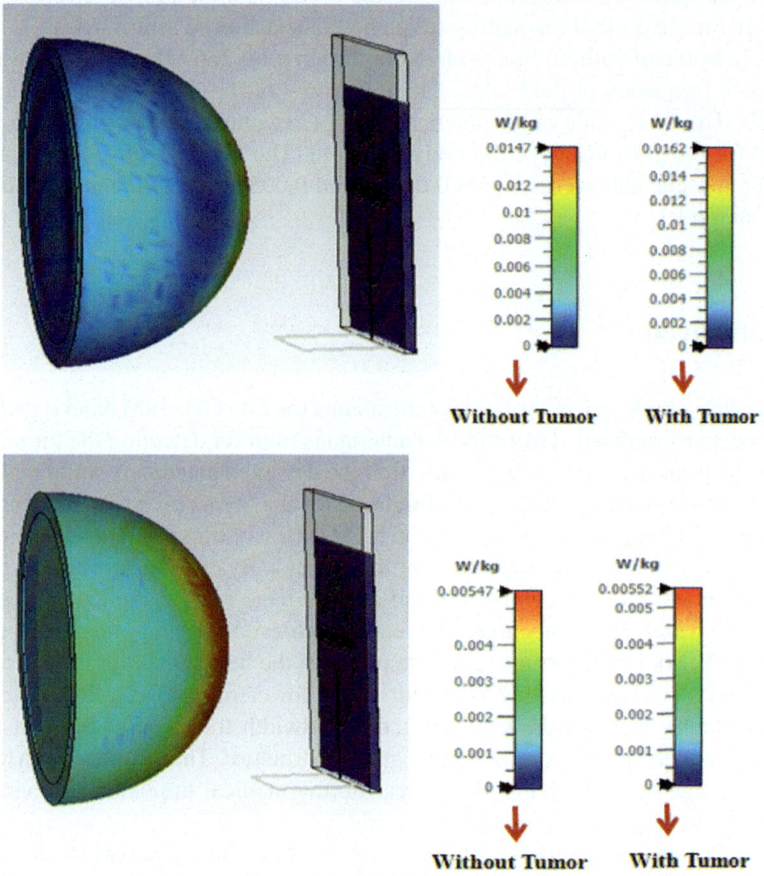

Fig. 9 SAR value of designed antenna with and without tumor for 1 and 10 g muscle

Table 3 Comparative analysis of previous works

References	Substrate	Return loss (dB)	Res. freq. (GHz)	SAR	BW (– 10 dB) (MHz)
[12]	FR4	– 38	2.45	0.283 for 10 g	82
[13]	RT duroid	– 16.9	2.46	0.22	
[14]	FR4 (lossy)	– 37	2.41	0.0000346	
[15]	FR4	37.56	2.48	0.03866	800
[16]	FR4	– 58	2.47	0.111	
Proposed	FR4	– 42	2.45	0.00552	266

4 Result

The designed antenna is simulated with and without breast phantom design. Breast tissue is designed by using skin and muscle tissue with the proper specification of the body for the desired resonating frequency. The achieved return loss is – 41 and – 20 dB with and without breast phantom design with 266 MHz bandwidth for the resonance frequency of 2.45 GHz. The directivity of the final designed antenna is 5.88 dB. The SAR value of the antenna is 0.0147 (without tumor) and 0.0162 (with tumor) for the 1 g muscle of the body phantom. The same design is simulated for the 10 g muscle; achieved result is 0.00547 and 0.00552 for with and without body phantom model.

5 Conclusion

In this letter, a high gain microstrip patch antenna for 2.45 GHz ISM band is proposed, analyzed and simulated. The proposed antenna is used for detecting the tumor in the breast phantom structure. In this work, first the designed antenna is simulated as the normal design and after that it was simulated for the breast phantom structure. The simulated S11 bandwidth is found as 266 MHz. The working mechanism is analyzed by calculating the directivity and gain of the antenna. By creating slots in the design, the resonance frequency can be reduced and the band shift to the lower frequency range. The better impedance matching can be achieved by using the slotted ground structure. The defected ground structure provides the better result of the attenuation output and also shifts the desired result to the lower range of the frequency. The design antenna provides the high gain and bandwidth for the ISM band; it can be further miniaturized by using defected ground structure. The simulated SAR value is in the range defined by the guidelines for any medical implantable device. The tumor effect is represented by the increasing value of the SAR. The insertion of the tumor is little bit increased the value of SAR; it shows that extra part is added in the defined structure.

References

1. Movassaghi S, Abolhasan M, Lipman J, Smith D, Jamalipour A (2014) Wireless body area networks: a survey. IEEE Commun Surv Tutor 16(3):1658–1686
2. Sharma A, Sharma VK, Gour S, Rathi A (2022) Conical shaped frequency reconfigurable antenna using DGS for cognitive radio applications. In: 8th international conference on advanced computing and communication systems, Coimbatore, India, pp 799–804
3. Malik NA, Sant P, Ajmal T, Rehman MU (2021) Implantable antennas for biomedical applications. IEEE J Electromagn RF Microw Med Biol 5(1):84–96
4. Kiourti A, Nikita KS (2012) A review of implantable patch antennas for biomedical telemetry challenges and solutions. IEEE Antennas Propag Mag 54(3):210–228
5. Vijayakumar P et al (2022) Network security using multi-layer neural network. In: 4th RSRI international conference on recent trends in science and engineering, REST Labs, Krishnagiri, Tamil Nadu, India, 27–28 Feb 2021. AIP Conf Proc 2393:020089, 020089-1–020089-5. https://doi.org/10.1063/5.0074089
6. Rawat A, Tiwari A, Gour S, Joshi R (2021) Enhanced performance of metamaterials loaded substrate integrated waveguide antenna for multiband application. In: Proceedings of IEEE international conference on mobile
7. Karthik V, Rama Rao T (2017) Investigations on SAR and thermal effects of a body wearable microstrip antenna. Wireless Pers Commun 3385–3401
8. Sharma A, Saini Y, Singh AK, Rathi A (2020) Recent advancements and technological challenges in flexible electronics: mm wave wearable array for 5G networks. AIP Conf Proc 2294(1):020007. AIP Publishing LLC
9. Mahrishi M et al (2020) Machine learning and deep learning in real time applications. IGI Global. https://doi.org/10.4018/978-1-7998-3095-5. ISBN 13: 9781799830955, ISBN 10: 1799830950, EISBN 13: 9781799830979
10. Ganeshwaran N, Kumar KJ (2019) Design of a dual-band circular implantable antenna for biomedical applications. IEEE Antennas Wireless Propag Lett 99
11. International Commission on Non-Ionizing Radiation Protection (1998) Guidelines for limiting to time varying electric, magnetic, and electromagnetic fields (upto 300 GHz). Health Phys 74(4):494–522
12. Hossain MB, Hossain MF (2021) A dual band microstrip patch antenna with metamaterial superstrate for biomedical applications. In: Proceedings of international conference on electronics, communications and information teleology (ICECIT), Khulna, Bangladesh, p 14
13. Alhuwaidi S, Rashid T (2021) A novel compact wearable microstrip patch antenna for medical application. In: Proceedings of 2020 international conference on communications, signal processing, and their applications
14. Mahbub F, Islam R (2021) Design and implementation of a microstrip patch antenna for the detection of cancers and tumors in skeletal muscle of the human body using ISM band. In: Proceedings of annual information technology, electronics and mobile communication conference
15. Rahayu Y, Saputra R, Reza MH (2021) Microstrip antenna for tumor detection: SAR analysis. In: Proceedings of international conference on smart instrumentation, measurement and applications
16. Soni BK, Singh K, Rathi A, Sancheti S (2022) Performance improvement of aperture coupled MSA through Si micromachining. Int J Circuits Syst Signal Process 16:272–277

Chapter 3
IoT-Based Automated Drip Irrigation and Plant Health Management System

Pandurangan Uma Maheswari, U. S. Praveen Raj, and S. Dinesh Kumar

1 Introduction

The agriculture sector is being the backbone of the Indian Economy. Substantial innovations have been made to enhance the agricultural yield with fewer resources and manual efforts. On the other hand, the growing population rate necessitates the technological revolution. In traditional agriculture, farmers spent 70% of the time for monitoring and understanding the crop states despite of field work. The farmers are facing financial loss because of the wrong weather prediction and incorrect irrigation, delayed disease identification, and blind usage of pesticides. The continuous cultivation of plants with inefficient management of fertilizer inputs has resulted in an effect on consumer malnutrition, environmental concerns, and a decrease in yields—qualitative and quantitative. With the evolution in wireless sensor networking technology, automated field and environmental monitoring and parameters controlling leads to precision agriculture (PA) application [1]. IoT-based smart agriculture provides a remote eye and facilitating crops monitoring constantly with higher accuracy and detect early stages of unwanted state. Major problems being faced by the agriculture sector are variation in weather, lack of sufficient knowledge in farmers, water scarcity and unplanned irrigation model, plant diseases, and security alert. Technological emergence in the agriculture sector still not covering these issues at field level and adoption of technology at farmer level is very less due to lack of awareness and affordability. It is essential to develop a cost-effective solution using current computing and networking technology to address the above-said issues faced day by day and to help the farmer expert assistance to take decisions and activation. The proposed model is the solution which integrates the power of IoT generally

P. Uma Maheswari (✉) · S. Dinesh Kumar
College of Engineering, Guindy, Anna University, Chennai, India
e-mail: dr.umasundar@gmail.com

U. S. Praveen Raj
University of Pennsylvania School of Engineering, Philadelphia, PA, USA

V. Devedzic et al. (eds.), *Proceedings of the International Conference on Intelligent Computing, Communication and Information Security*, Algorithms for Intelligent Systems, https://doi.org/10.1007/978-981-99-1373-2_3

referred to as a wireless sensor and actuator network that are spatially distributed autonomous sensors to monitor field parameters such as temperature, humidity, soil moisture, weather condition, plant health, and the like. These parameters are periodically sensed as per schedule in real time and aggregated at sink nodes and then send to centralized cloud storage. IoT enables technology for low-power wireless and control applications and the elimination of wires provides significant cost savings as well as creating improved reliability for several future monitoring applications. The plant diseases are identified and recognized with the help of advanced image processing techniques and pesticides prescribed by the diagnosing system will be sprayed using a sprayer nozzle integrated with Arduino Uno, ESP8266, and servo base motor. By sensitively applying water, fertilizer, and pesticide to the needy will secure soil health and reduce cost. Irregular irrigation badly affects crop quality, quantity, and soil health. Parameters like crop type, soil type, irrigation method, precipitation, and soil moisture retention are to be accessed to provide an efficient irrigation model. There are many proven water requirement determination approaches. Perhaps, this proposed model used water balance Approach to calculate the required amount of water to be pumped in an irrigation system for a day.

The major advantage of this proposed smart irrigation is the amount of water required for the day that depends upon field parameters and environmental factors. Since the usage of a fixed quantity of water pumped every day in a traditional drip system is eliminated, the model has evidenced a significant quantity of water saving. The proposed model eliminated expert consultation for disease management and also delay in disease treatment and substantially reduces the loss of crop. The model further eliminates blind nitrification of soil which intern spoils soil health. Crop yields can be increased by optimizing agronomic practices to identify disease in the plant for increasing crop yields.

Novelty—This proposed model used water balance approach to calculate the required amount of water to be pumped in an irrigation system for a day in which the day to day field parameters and environmental factors are taken for determining water requirement. Yet another novelty is field health-based nutrification using IoT enabled health monitoring system.

2 Literature Review

There are reasonable efforts that have been reported in portraying the application of the IoT in the agriculture industry. The potential of wireless sensors and IoT in agriculture and the current and future trends of IoT in agriculture and potential research challenges were highlighted [2]. A Remote Monitoring System (RMS) is proposed [3] that gathers continuous information on farming generation conditions that gives simple access to horticultural offices through SMS and advice on climate design, crops, and so on. A WSN-based precision agriculture using Zigbee module is proposed [4] to control drip irrigation through established wireless network in the form that is driven by Zigbee module. IoT-based Agriculture Stick in assisting farmers for efficient environment monitoring has been proposed [5] that enable smart farming and increase their overall yield and quality of products. A time Variant Growth Approximation Model has been proposed [6] for Estimation of Crop Yield and Water Regulation using Environmental Factors. An IoT-based approach for smart monitoring of crops is proposed [7] to monitor crop growth. A methodology for smart farming by linking a smart sensing system and smart irrigation system [8] through wireless communication technology that leads to low cost and efficient wireless sensor network technique to acquire field parameters was proposed. Real-time monitoring drip irrigation system [9] is proposed using NodeMCU that send information and automatic valve controller. An algorithm for image segmentation technique [10] which is used for automatic detection and classification of plant leaf diseases was done by using a genetic algorithm. The use of the Programmable System [1] on Chip Technology (PSoC) has been proposed and analyzed to monitor and control various parameters of the greenhouse as a part of wireless sensor networks (WSN). An IoT based system [11] is proposed that reads soil moisture in real time in an automated way. An IoT-based plant leaf disease diagnosis system is proposed [12] which used segmentation of region of interest by Gray co-occurrence matrix and SVM for classification. The soil moisture-based irrigation was examined [13], and this practice reduced percolation and optimized tomato and green bell pepper yield under varying levels of soil moisture controlled drip irrigation. A model to optimize water use for crops using the WSN network with the GPRS module [14]. An energy-efficient data aggregation and delivery (ESDAD) protocol [15] is proposed to do sensor data aggregation in intermediate nodes to eliminate redundancy. An IoT based monitoring system is designed [16] to analyze the crop environment and the method to improve the efficiency of decision making by analyzing harvest statistics. A hybrid wired/wireless network [17] is proposed in which Zigbee protocols, and control area network is used together to resolve problems in integration. A Nutrients Dispense System for Soil Deficiency using IoT is proposed [18] in which the pH sensor is used to remotely monitor the nutrients level in the soil and based on the pH value the NPK values are obtained. The color and texture feature extraction and classification of plant disease using support vector machine (SVM) and artificial neural network (ANN) classifiers is proposed [19] in which a reduced feature set-based approach for recognition and classification of images of plant diseases is used. Alghazzawi et al. [20] suggested an automatic smart irrigation decision support system to manage irrigation in agriculture that estimates the weekly irrigation needs of a plantation, based on both soil measurements and climatic variables gathered by several autonomous nodes deployed in the field. A study on a cotton farm with wireless soil moisture monitoring equipment is proposed [1] that integrates weather data, soil moisture sensor data, and remotely sensed crop vegetation index to train models to predict future soil moisture. Lysimeter experiment conducted [21] on green gram during Kharif season to determine the seasonal crop evapotranspiration (ETc) and crop coefficient (Kc) of green gram crop for different development stages. An ontology based decision as a result of a machine learning algorithm with an edge server is introduced between the main IoT server and the GSM module proposed

[22]. This method avoids the overburden of the IoT server for data processing but also reduce the latency rate. An effective utilization of excessive water generated from showers to increase the groundwater levels that provides flexibility to farmers for monitoring the farms in real time using the farmer's cockpit is proposed [23, 24] in which actuation and automation are done based on certain imposed constraints to respond as per inputs and outputs generated by various devices. Retrieval of real-time data and use them to determinate the correct amount of water to be used in a garden. Glória et al. [25] is studied that when using data facing a typical irrigation system, the improvements go up to 26% when using just temperature, and 34% when combining more sensor data, such as humidity and soil moisture. The various IoT-based solutions for water management are studied [26] and concluded that all such recent advancements are based on real data sensed by IoT devices hosted on the field and based on decision generated by any machine learning model.

With reference to the detailed survey, it is found that none of them integrated water balance approach in decision making based on traditionally determined proven factors. All modern systems are purely observing real-time data and take decision based on them. Smart Soil nutrification also seems shallow research focus.

3 Materials and Method

Variable rate optimized irrigation based on crop water stress index (CWSI), soil data, weather data, and crop data for water requirement assessment and irrigation index is generated for each site. For this, crop canopy at different periods in different atmospheric conditions is obtained to calculate CWSI. A wireless sensors-based monitoring system to collect the mentioned measurements, then transmit to processing center housed at cloud where the DSS is used to analyze the farm data. This model works according to the soil variability and ultimately improves water usage efficiency. The proposed architecture is depicted in Fig. 1. This architecture is an embedded architecture that uses Arduino Uno 3 which is driven by the Linux operating system and its simple act as a motherboard where all the components are to be connected. It is a single board minicomputer with 64-bit quad-core processor with faster Ethernet support. In the proposed architecture, there are three modules such as (1) sensor module, (2) activator module, and (3) data processing module.

3.1 Detection Module

The detection module comprises wide range of sensors for measuring various soil and plant health parameters of soil moisture, humidity, soil PH, and plant health. SparkFun Soil Moisture Sensor reads the content of water in the soil and presents the moisture value in voltage. BME280 sensor reads the temperature, barometric pressure, and humidity. Soil pH meter measures the activity of the hydrogen ions in

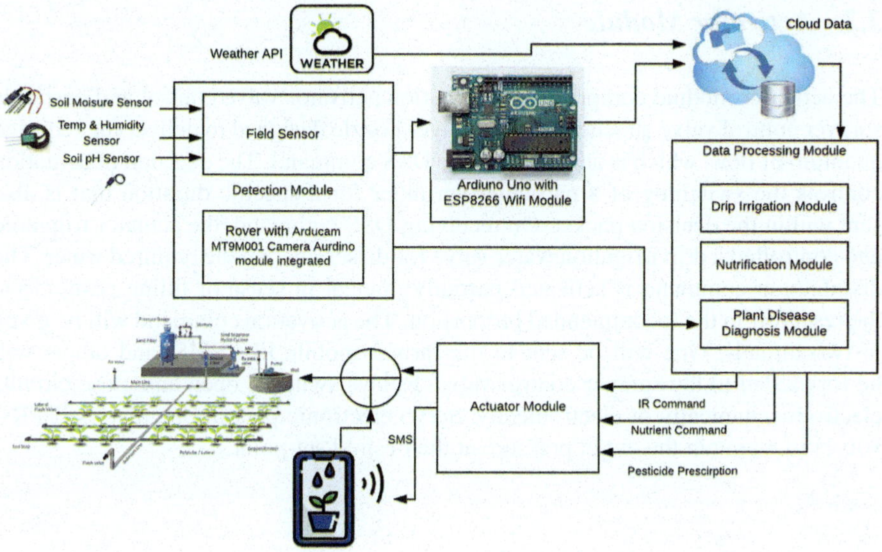

Fig. 1 Proposed system block diagram

the solutions to determine the soil pH. For environmental parameters like rainfall and average temperature, the One Call API provides essential weather data for a specific location which covers Current weather, Minute forecast for 1 h, Hourly forecast for 48 h, Daily forecast for 7 days, and Historical weather data for the previous 5 days. The API response parameters are as given: Requested time, Sunrise time, Sunset time, Temperature, Atmospheric pressure on the sea level, Humidity, Atmospheric temperature, Cloudiness, Midday UV index, Wind speed, Wind direction, Rain Precipitation volume from which the Temperature, Cloudiness, and Rain Precipitation volume alone were taken for analysis.

3.1.1 Locomotory Rover

The plant health parameters are measures visually (computer vision) with the help of the Rocker-Bogie mechanism-based multi-terrain locomotory rover. For capturing the leaf image to classify affected/infected areas for disease measurement, Arducam MT9M001 (Arduino camera with a maximum resolution of 1280 × 1024 at 30 fps having extremely high sensitivity for low-light operation) is used.

Soil, plant health, and environmental parameters that are measured by different sensors and the rover will be sent over an MQTT server for cloud-based data processing, through the integrated ARDUINO UNO—ESP8266 Wi-Fi module.

3.2 Activator Module

The activator module comprised the irrigation activator valve control and pesticide sprayer control valve attached with a sprayer nozzle designed to deliver the required quantum of flow, which is activated by the DSS command. The decision of actuation controls the switching of a particular sprinkler for a specific duration that is also sent within the decision packet. On receiving DSS command, the actuator triggered and controlled a drip irrigation water valve for disseminating the required water. The disinfection command is activated partially manual in sense of filling pesticide in the reservoir in the recommended proportion. The activation command will be given in two threads: One will be sent to the farmer mobile like SMS, and others will be forwarded to the sprayer control valve. Relays switches open and close circuits electro mechanically or electronically. Servo electronic pressure regulators (EPRs) volves to maintain the outlet pressure at the desired set-point.

3.3 Data Processing Module

Cloud-based data storage and processing (Amazon Web services) were used to store sensor data and to analyze them to take a necessary decision in terms of irrigation plan, nutrification, and disease care. This module has a controlled user interface enabled with client authentication and access. This module has a rule-based Decision Sub System (DSS) which regulates the drip irrigation schedule, nutrification mechanism, and plant disease recognition and diagnosis.

3.3.1 Drip Irrigation Automation

On receiving the field and environmental parameters from controller sensor integration, the drip irrigation decision is taken based on predefined rules. These rules are different in a sense of crop and climate. The quantum of water is determined by the water balance approach [27]. In this water balance approach, irrigation is required when crop water demand (ETc) surpasses the supply of water and rainfall, as ETc differs with plant and weather conditions. In irrigation practice, only a percentage of AWC is allowed to be depleted which typically expressed in terms of inches of water per inch of soil depth. General values of AWC are provided concerning soils on the web by authorized departments. In this work, the soil type of Sandy clay is considered for which AWC ranges are Low—0.13, Avg—0.16, and High—0.18. While growing, the crop extracts water from the soil to fulfill its ETc requirement and subsequently the stored soil water is gradually depleted. In general, the net irrigation requirement is the amount of water required to refill the root zone soil water content. This amount is the difference between field capacity and current soil water level that corresponds to the soil water deficit (D) is given in Eq. (1).

$$\text{Dc} = \text{Dp} + \text{ETc} - P - \text{Irr} \tag{1}$$

where Dc—the soil water deficit or Net irrigation requirement for the current day, ETc is the crop evapotranspiration rate, Dp is deficit on previous day, P is the present day gross rainfall, Irr is the net irrigation amount infiltrated into the soil for the present day, Dc is set equal to zero if its value becomes negative. This will occur if water added to the root zone already exceeds field capacity and excess water in the root zone is assumed to be lost. Management allowed depletion (MAD) of the AWC must be specified with reference to the fact sheet, since a percentage of AWC is allowed to be depleted. The MAD can be expressed in terms of depth of water (dMAD, inches of water) using the following Eq. (2).

$$\text{dMAD} = (\text{MAD}/100) * \text{AWC} * \text{Drz} \tag{2}$$

where MAD is management allowed depletion, AWC is available water capacity and Drz is the depth of root zone. Since the proposed system considers vegetative plants, Drz for vegetable is 6–12 at the seedling stage, 18 at vegetative and flowering stage, and 18–24 at the mature stage. Typically, irrigation water should be applied when the soil water deficit (Dc) approaches dMAD, or when Dc $\geq$ dMAD.

Crop evapotranspiration (ETc), in inches per day, is estimated through Eqs. (3)–(5) as given below:

$$\text{ETc} = \text{ETr} * \text{Kc} * \text{Ks} \tag{3}$$

$$\text{Ks} = (\text{TAW} - D)/((1 - \text{MAD}) * \text{TAW})\ (\text{Ks} = 1 \quad \text{if } D < \text{dMAD}) \tag{4}$$

$$\text{TAW} = \text{AWC} * \text{Drz} \tag{5}$$

where ETr is the transpiration rate (inches/day), the Kc coefficient incorporates crop characteristics and averaged effects of evaporation from the soil. For normal irrigation planning and management purposes, for the development of basic irrigation schedules, and for most hydrologic water balance studies, average crop coefficients are relevant and more convenient than the Kc computed on a daily time step using a separate crop and soil coefficient. Hence, in this paper, Kc value is referred from the fact sheet published by authentic source of Fact Sheet published by FAO. Kc is a crop coefficient that ranges from 0 to 1, and Ks is a water stress coefficient that ranges from 0 to 1. Values of Kc typically range from 0.2 for young seedlings to 1.0 for crops at peak vegetative stage. A typical crop coefficient curve (Kc values that change with crop development) is shown in Fig. 2. Ks may be assumed to be 1 as the crop is not experiencing water stress. The irrigation command is passed to the actuator which controls irrigation pump for on/off periods. The irrigation command triggered by DSS consists of time and quantum of water.

Water requirement is calculated as Dc the soil water deficit (in inch) or net irrigation requirement determined from Eq. (6) for current day represented in inches. The quantity of water to be pumped is calculated from Dc in terms of liters by Eq. (7) given below:

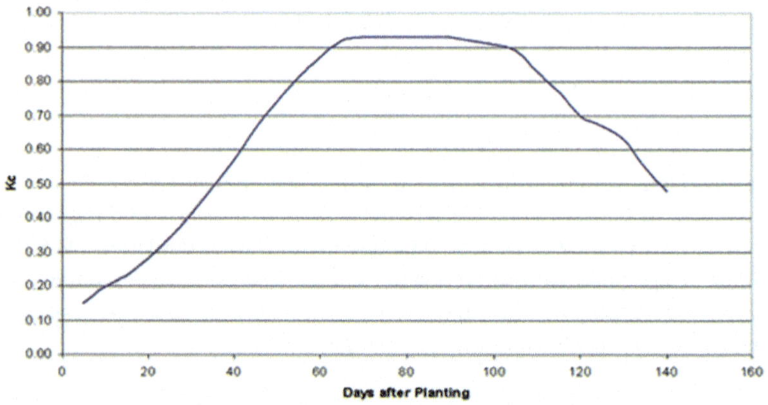

Fig. 2 Crop coefficient curve Kc. *Source* Irrigation scheduling: the water balance approach—4.707 fact sheet

$$\text{Irrigation water Wmm} = \text{Dc} * 25.4\,\text{mm per day} \tag{6}$$

$$\text{Quantity of irrigation water Wlr} = \text{Dc} * 0.0163871\,\text{L per day} \tag{7}$$

The required duration of irrigation is calculated based on hourly irrigation capacity and run-time per day ration of the drip system as given in Eqs. (8) and (9):

$$\text{Dtime} = \text{Wmm}/\text{Hrc}\ \text{h} \tag{8}$$

$$\text{Hrc} = \text{Df}/\text{Dd} * \text{Dl}\ \text{mm/h} \tag{9}$$

Wmm—Total water to irrigate in mm; Hrc—Hourly irrigation capacity; Df—Dripper Flow; Dd—Distance between Drippers; Dl—Distance between drip lines.

On determining water quantity and duration of irrigation, then the irrigation command is passed on to the motor to control the on/off state for the particular time duration.

Irrigation Command: Water Wlr, Duration Dtime.

3.3.2 Soil Nutrification

This module aims at optimal usage of nutrients to uphold the soil health by monitoring and modifying pH level of fertilizers concerning soil parameters. Soil is a major source of nutrients needed by plants for growth. Soil pH is an indication of the acidity or alkalinity of soil and is measured in pH units. The soil pH level is essential because it carries macro-nutrients that of nitrogen (N), potassium (K), and phosphorous (P). Plants cannot utilize these nutrients when pH is too acidic. Plants will take up toxic metals, and some plants eventually die of toxicity (poisoning) in acidic soil. Hence,

Table 1 pH versus nutrient status

pH range	Inference
2–2.9	Nitrogen deficiency
3–5	Nitrogen unavailability
5–5.4	Phosphorous deficiency
5.5–7	Phosphorous unavailability
7.1–8.3	Potassium deficiency
8.4–9.1	Potassium unavailability

it is necessary to monitor the nutrients level in the soil for better crop production. The pH sensor detects the pH value in the soil and sends it to the data processing module via the controller. The PH parameter is taken for soil nutrient level calculation. The major nutrients of soil are nitrogen, phosphorus, and potassium (NPK) which can be determined through PH value concerning Table 1. If the pH range is between 3 and 5, then if the range is between 5.5 and 7, phosphorous content is unavailable in the soil. If the pH range is between 8.4 and 9.1, then potassium content is unavailable in the soil []. The required NPK prescription based on the inference from Table 1 will be sent to farmer's mobile as SMS, which enables the farmer to (re)fill the respective fertilizer (NPK) containers which are further dispensed.

Fertilizer command: Date, Fertilizer name, Quantity.

3.3.3 Plant Disease Recognition and Diagnosis

Plant leaf image is captured by the rover using Arducam MT9M001 camera with a maximum resolution of 1280 × 1024. It has an extremely high sensitivity for low-light operation. Thus, it can be used without additional flash devices. These images undergo various image processing techniques in the pipeline such as noise filtering, image clipping that crop the leaf image to get the interesting image region, image smoothing, and contrast stretching. Then, the histogram equalization is applied that distributes the intensities to enhance the plant disease images. Average or mean filter is a simple, intuitive, and easy to implement the method of smoothing images by reducing the amount of intensity variation, and the mean filter acts as a low-pass frequency filter and, therefore, reduces the spatial intensity derivatives present in the image. The contrast is manipulated using linear mapping. After contrast stretching, the region of interest is identified using K-means clustering [12] to segment the diseased part of the leaf images. Image segmentation was performed to segment affected regions of leaves for which k-mean clustering is applied [10]. Firstly, the source RGB image is converted into L^* a^* b^* color space [10] where L is luminosity, a^* b^* are chromaticity layers which refers to the color fall along red–green axes and blue–yellow axes. The algorithm classifies the colors in a^* b^* and label each pixel. As a result, affected and unaffected parts of the leaf were segmented from each other.

Then, pixel masking is done to mask green colored pixels by computing a threshold Γ' value and comparing each pixel intensity with this threshold. If pixel intensity $Pi < \Gamma'$, then $RGB(Pi) = 0$. Then, the masked cell is removed from the infected inside cluster's boundaries. Obtain useful segments to classify leaf diseases. Since the use of color image features provides more image characteristics, color co-occurrence-based feature extraction method is used in which both the texture and color features of an image are measured. In the color co-occurrence method, RGB images of leaves are converted into HIS color space representation first and then three-color co-occurrence matrices, one for each of H, S, and I are computed. Features called texture features such as contrast, local homogeneity, energy, cluster shade, and cluster prominence are computed for the H image as given.

$$\text{Contrast} = \sum_{i,j=0}^{N-1} (i, j)^2 C(i, j) \tag{10}$$

$$\text{Energy} = \sum_{i,j=0}^{N-1} (i, j)^2 \tag{11}$$

$$\text{Local Homogenity} = \sum_{i,j=0}^{N-1} (i, j) / \left(1 + (i - j)^2\right) \tag{12}$$

$$\text{Entropy} = \sum_{i,j=0}^{N-1} C(i, j) \log C(i, j) \tag{13}$$

Classification Using Multi-Class Support Vector: Support vector machine (SVM) which is a kernel-based administered learning algorithm [10, 12] which it can identify the optimal boundaries between the possible outputs using kernel tricks. The main purpose of SVM is to indicate the class of plant disease by drawing hyperplanes between data sets. The classifiers are trained and tested using benchmark dataset images of plant diseases. Classification using Artificial Neural Network: The ANN model used here is a fully connected backpropagation network with two hidden layers. The input layer consists of five neurons for taking input texture parameters. The activation function used is sigmoid, and the output layer has five neurons to classify four different diseases *Alternaria alternata*, Anthracnose, Bacterial Blight, Cercospora Leaf Spot taken, and one for the healthy leaf. The learning rate is 0.001. For training, 592 images were taken, and 144 images were taken for testing. The training accuracy was measured for the different number of epochs. Once after plant disease is recognized, then the corresponding pesticide is prescribed to the farmer via SMS message comprising Disease name, pesticide recommended, dosage.

4 Results and Discussion

Once the individual components are tested and fully operating, the whole system is integrated. This system is complete and evaluate all types of sensor data and make a decision to control the drip irrigation system. The system enables the farmers to monitor the field conditions from anywhere.

General values of AWC are provided concerning soils on the web by authorized departments. In this work, the soil type of sandy clay is considered for which AWC ranges are Low—0.13, Avg—0.16, and High—0.18 [27]. While growing, the crop extracts water from the soil to satisfy its ETc requirement, and correspondingly, the stored soil water is gradually depleted. Management allowed depletion (MAD) is taken as 60% here since the soil type is sand and clay loam and concerning the standard table, Drz for vegetable is 6–12 at the seedling stage, 18 at vegetative and flowering stage, and 18–24 at mature stage [27]. A proposed model is tested against irrigation by schedule at the identified field to measure the usage of water. It is observed that plants were irrigated based on the current parameters observed by the detection module. The test shows that there is an average savings of 500 L per day when compared to the conventional drip irrigation system which is depicted in Fig. 3. Water requirement determination with respect to conventional and proposed method is depicted in the chart given in Fig. 4.

After estimating water requirement *W*, it is essential to design a drip outlet to supply water *W* in correct proportion to the planted area. For simplicity, it is assumed that two 1/2 gph emitters (lower flow for denser soil) to each plant (a single emitter is fine for a small plant initially). The dripper flow Df is 1 L per hour; the distance between drippers Dd is 0.5 m, and the distance between the dripper lines Dl is 1.0 m [28]. The required duration of irrigation is calculated based on the hourly irrigation capacity of the system as given in Eqs. (6) and (7): The hourly capacity irrigation Hrc = 1/(0.5 * 1) = 2 mm per hour. For example, Duration per day for delivering 13.97 mm of water requirement is: Dtime = Wmm/Hrc = 13.97/2 = 7 h.

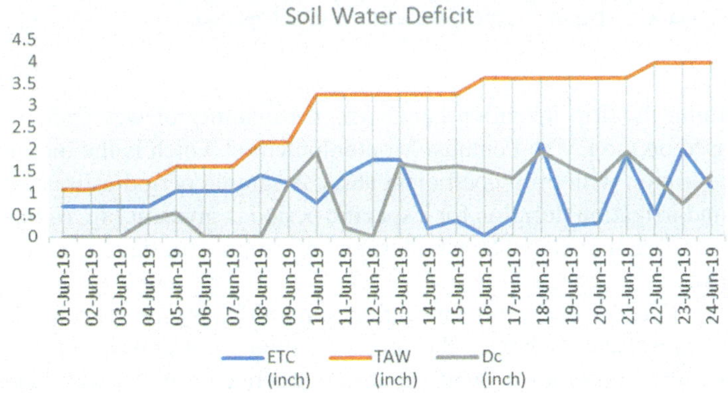

Fig. 3 Soil water deficit (net water requirement) curve with respect to evotranspiration rate and available water capacity

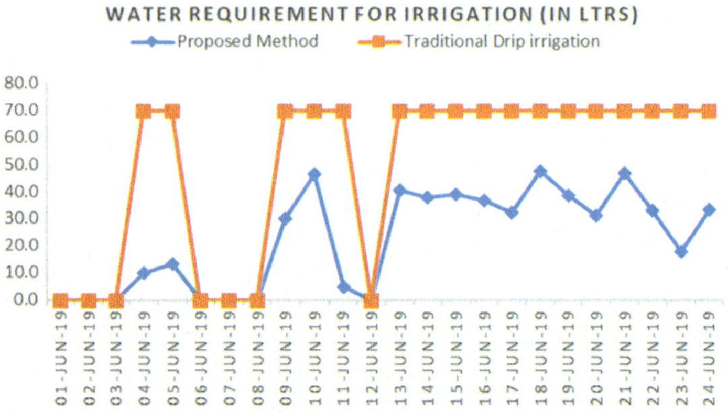

Fig. 4 Water requirement (in L) conventional versus proposed model

Irrigation Water Requirement Estimation - Ist June 2019 to 24th June 2019																
Day	Temp (O^0 C)	Rainfall (inches)	Soil Moist (inches)	D_{rz}	ET_r	K_c	K_s	ET_c (inch/d)	AWC	TAW (inch)	MAD	d_{MAD}	D_p	D_c	D_c (in)	Rtigation water in Litres
01-Jun-19	34	0.9	0.4	6	7	0.1	1	0.7	0.18	1.08	0.6	0.648	0.1	-0.5	0	0.0
02-Jun-19	34	2.3	0.7	6	7	0.1	1	0.7	0.18	1.08	0.6	0.648	0	-2.3	0	0.0
03-Jun-19	35	0.6	0.3	7	7	0.1	1	0.7	0.18	1.26	0.6	0.756	0	-0.2	0	0.0
04-Jun-19	35	0.2	0.1	7	7	0.1	1	0.7	0.18	1.26	0.6	0.756	0	0.4	0.4	9.8
05-Jun-19	35	0.6	0.3	9	7	0.15	1	1.05	0.18	1.62	0.6	0.972	0.4	0.55	0.55	13.4
06-Jun-19	33	2	0.7	9	7	0.15	1	1.05	0.18	1.62	0.6	0.972	0.55	-1.1	0	0.0
07-Jun-19	32	2.6	0.7	9	7	0.15	1	1.05	0.18	1.62	0.6	0.972	0	-2.25	0	0.0
08-Jun-19	35	0.8	0.3	12	7	0.2	1	1.4	0.18	2.16	0.6	1.296	0	0.3	0	0.0
09-Jun-19	34	0	0	12	7	0.2	0.879	1.2306	0.18	2.16	0.6	1.296	0	1.2306	1.2306	30.1
10-Jun-19	31	0.1	0	18	7	0.2	0.547	0.7658	0.18	3.24	0.6	1.944	1.2306	1.8964	1.8964	46.3
11-Jun-19	30	2.4	0.7	18	7	0.2	1	1.4	0.18	3.24	0.6	1.944	1.8964	0.1964	0.1964	4.8
12-Jun-19	31	2.5	0.7	18	7	0.25	1	1.75	0.18	3.24	0.6	1.944	0.1964	-1.2536	0	0.0
13-Jun-19	33.5	0.1	0	18	7	0.25	1	1.75	0.18	3.24	0.6	1.944	0	1.65	1.65	40.3
14-Jun-19	34	0.2	0.1	18	7	0.25	0.108	0.189	0.18	3.24	0.6	1.944	1.65	1.539	1.539	37.6
15-Jun-19	35	0.2	0.1	18	7	0.25	0.2	0.35	0.18	3.24	0.6	1.944	1.539	1.589	1.589	38.8
16-Jun-19	35	0.1	0	20	7	0.3	0.007	0.0147	0.18	3.6	0.6	2.16	1.589	1.5037	1.504	36.7
17-Jun-19	34	0.4	0.2	20	7	0.3	1	0.41	0.18	3.6	0.6	2.16	1.504	1.314	1.314	32.1
18-Jun-19	33.5	1.8	0.5	20	7	0.3	1	2.1	0.18	3.6	0.6	2.16	1.314	1.114	1.935	47.3
19-Jun-19	33.5	0.4	0.2	20	7	0.3	0.114	0.2394	0.18	3.6	0.6	2.16	1.935	1.5744	1.5744	38.5
20-Jun-19	32	0.4	0.2	20	7	0.35	0.122	0.2989	0.18	3.6	0.6	2.16	1.5744	1.2733	1.2733	31.1
21-Jun-19	32.5	0.8	0.4	20	7	0.35	0.7477	1.831865	0.18	3.6	0.6	2.16	1.2733	1.905165	1.905	46.5
22-Jun-19	32	0.7	0.35	22	7	0.35	0.2	0.49	0.18	3.96	0.6	2.376	1.905	1.345	1.345	32.9
23-Jun-19	32.5	1	0.6	22	7	0.4	0.707	1.9796	0.18	3.96	0.6	2.376	1.345	1.7246	0.7246	17.7
24-Jun-19	31.5	0	0.2	22	7	0.4	0.4	1.12	0.18	3.96	0.6	2.376	0.7246	1.6446	1.364	33.3

Fig. 5 Irrigation water requirement estimation for a sample period

Concerning the drip irrigation handbook, the quantity of water required is calculated based on the EVTo Penman–Monteith method which is the measure of pan evapotranspiration with crop coefficient (Kc). A sample period estimation is given in Fig. 5, and irrigation duration for a specific period is given in Fig. 6. Based on the direction, it is determined that 69.6 L of water are required per day in the conventional method [28].

For a period of 24 days, the conventional method consumes 1183.2 L of water (excluding rainy days), whereas the proposed model consumes 537.2 L of water only. Hence, the water requirement is reduced at a rate of 45.4. The proposed work diagnoses plant disease with the help of a leaf image sent to the processing unit. The captured images were processed, and the infected regions were segmented,

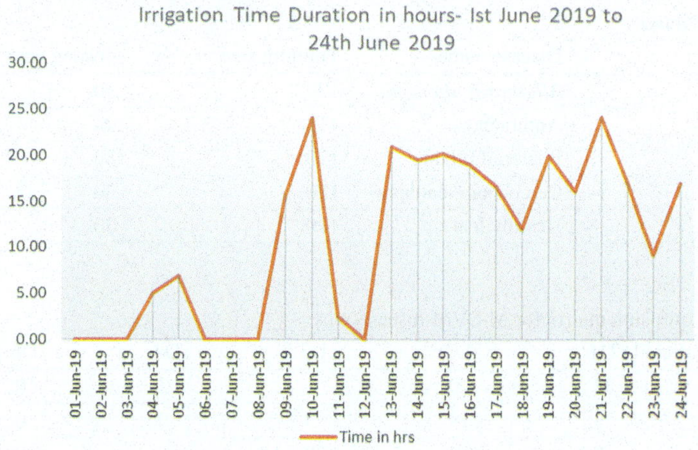

Fig. 6 Irrigation duration in h for a specific period

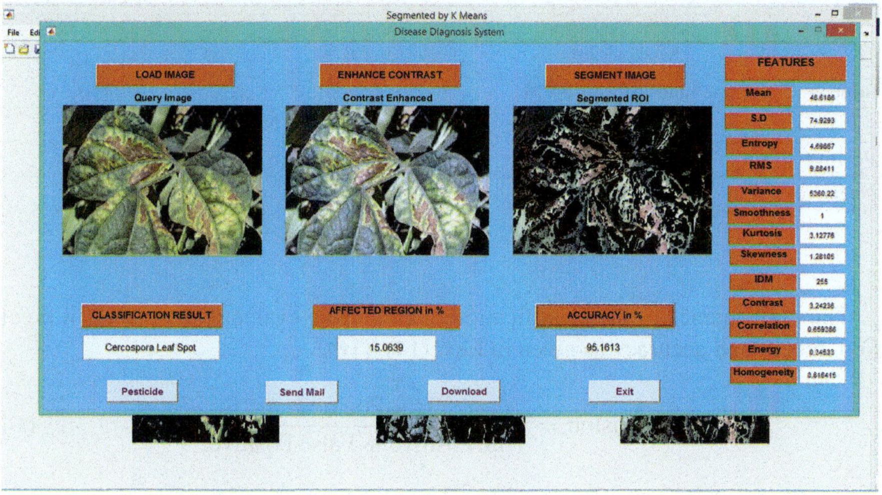

Fig. 7 Segmentation of infected regions

and the infected region features were extracted as shown in Fig. 7. The extracted features were given to the M-SVM classifier. The texture-based and morphology-based feature extraction was done to extract properties like contrast, correlation, energy, homogeneity, mean, standard deviation, and entropy are calculated, and each disease is denoted by a corresponding label (Table 2).

Table 2 Dataset

D-ID	Disease name	Training size	Testing size
D1	*Alternaria alternata*	120	30
D2	Anthracnose	112	24
D3	Bacterial blight	120	30
D4	Cercospora leaf spot	120	30
D5	Healthy leaf	120	30

Table 3 Confusion matrix for M-SVM-disease-wise

Actual/predicted	D1	D2	D3	D4	D5
D1	27	0	1	2	0
D2	3	18	3	0	0
D3	0	0	28	0	2
D4	2	0	0	29	0
D5	0	0	3	3	24

Table 4 Confusion matrix for M-SVM

	Positive	Negative
Positive	126	6
Negative	10	2

The performance of the classification models were evaluated by means of precision, recall, and accuracy as given in Eqs. (14)–(16):

$$\text{Precision} = \frac{\text{True Positive}}{\text{True Positive} + \text{False Positive}} \tag{14}$$

$$\text{Recall} = \frac{\text{True Positive}}{\text{True Positive} + \text{False Negative}} \tag{15}$$

$$\text{Accuracy} = \frac{\text{True Positive} + \text{True Negative}}{\text{True Positive} + \text{True Negative} + \text{False Positive} + \text{False Negative}} \tag{16}$$

The M-SVM classification algorithm is tested with specified number of test images and the confusion matrix was prepared for disease-wise as given in Table 3 and for the entire model as given in Table 4.

The efficiency of the M-SVM classification model is evaluated with precision, recall, and accuracy concerning the confusion matrix given in Table 4 as given below:

Table 5 Confusion matrix for ANN-disease-wise

Actual/predicted	D1	D2	D3	D4	D5
D1	24	1	2	2	1
D2	2	16	3	2	1
D3	0	0	23	4	3
D4	2	2	0	26	0
D5	0	0	3	0	27

Table 6 Confusion matrix for ANN

	Positive	Negative
Positive	116	3
Negative	20	5

$$\text{Recall} = \frac{\text{TP}}{\text{TP} + \text{FN}} = \frac{126}{136} = 0.926 \tag{17}$$

$$\text{Precision} = \frac{\text{TP}}{\text{TP} + \text{FP}} = \frac{126}{132} = 0.954 \tag{18}$$

$$\text{Accuracy} = \frac{\text{TP} + \text{TN}}{\text{TP} + \text{TN} + \text{FP} + \text{FN}} = \frac{128}{144} = 0.88 \tag{19}$$

The ANN classification algorithm is tested with specified number of test images and the confusion matrix was prepared for disease-wise as given in Table 5 and for the entire model as given in Table 6.

The efficiency of the ANN classification model is evaluated with precision, recall, and accuracy concerning the confusion matrix given in Table 6 as given below:

$$\text{Recall} = \frac{\text{TP}}{\text{TP} + \text{FN}} = \frac{116}{136} = 0.853 \tag{20}$$

$$\text{Precision} = \frac{\text{TP}}{\text{TP} + \text{FP}} = \frac{116}{119} = 0.975 \tag{21}$$

$$\text{Accuracy} = \frac{\text{TP} + \text{TN}}{\text{TP} + \text{TN} + \text{FP} + \text{FN}} = \frac{121}{144} = 0.84 \tag{22}$$

Figure 8 shows a comparison of evaluation done for ANN and M-SVM classifier. The accuracy of M-SVM is 88% and ANN is 84%. M-SVM shows better accuracy than ANN for this case.

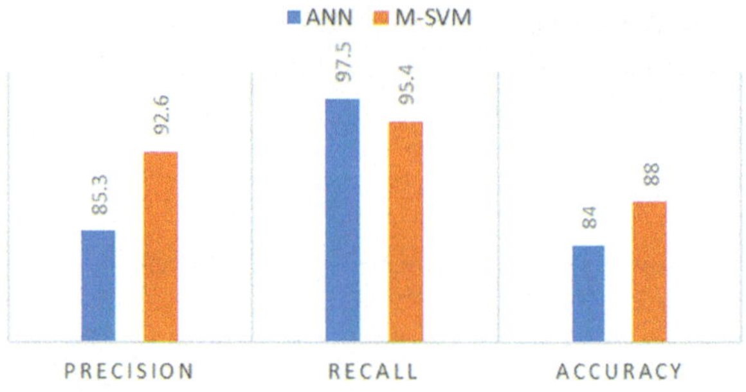

Fig. 8 Classifier performance comparison

5 Conclusion

This smart drip irrigation system automates and regulates the watering without any manual intervention. Field environmental and plant nutrient parameters were monitored, and irrigation was scheduled accordingly. The proposed architecture is cheap and reliable. The selected sensors are compatible with Arduino Uno, and server was able to house all data and display them in a manner that enables its users to visualize the status of the environment and soil around their crops. With the guidance of this system, the farmers can appropriate actions that will result in greater crop yield. The limitation of this design is that the failure of any particular device has to be tested manually. This system is designed as a prototype for small garden farms only. In the future, it can be scalable for large farming fields.

References

1. Mahrishi M, Sharma G, Morwal S, Jain V, Kalla M (2021) Data model recommendations for real-time machine learning applications: a suggestive approach, chap 7. In: Kant Hiran K, Khazanchi D, Kumar Vyas A, Padmanaban S (eds) Machine learning for sustainable development. De Gruyter, Berlin, Boston, pp 115–128. https://doi.org/10.1515/9783110702514-007
2. Ayaz M et al (2019) Internet-of-Things (IoT)-based smart agriculture: toward making the fields talk. IEEE Access PP(99):1
3. Anusha et al (2019) A model for smart agriculture using IoT. Int J Innov Technol Explor Eng 8(6):1656–1659
4. Rajagopal V, Maheswari PU, Deepalakshmi N (2016) Precision irrigation using wireless sensor networks for Indian agriculture—a survey. Asian J Res Soc Sci Humanit 6(7):324–333
5. Nayyar A, Puri V (2016) Smart farming: IoT based smart sensors agriculture stick for live temperature and moisture monitoring using Arduino, cloud computing & solar technology. In: The international conference on communication and computing systems (ICCCS-2016). https://doi.org/10.1201/9781315364094-121
6. Srinivasan R, Uma Maheswari P (2018) Time-variant growth approximation model for estimation of crop yield and water regulation using environmental factors (FGC). Indian J Ecol 45(3):550–554
7. Prema P, Sivasankari B, Kalpana M, Vasanthi R (2019) Smart agriculture monitoring system using IoT. Indian J Pure Appl Biosci 7(4):160–165

8. Lakshmisudha K, Hegde S, Kale N, Iyer S (2012) Smart precision based agriculture using sensors. Int J Comput Appl (0975-8887) 146(11)
9. Kholifah AR, Sarosa KIA, Fitriana R, Rochmawati I, Sarosa M (2019) Drip irrigation system based on Internet of Things (IoT) using solar panel energy. In: 2019 fourth international conference on informatics and computing (ICIC), pp 1–6
10. Singh V, Misra AK (2017) Detection of plant leaf diseases using image segmentation and soft computing techniques. Inf Process Agric 4(1):41–49
11. Chidambaram RMR, Upadhyaya V (2017) Automation in drip irrigation using IoT devices. In: 2017 fourth international conference on image information processing (ICIIP), Shimla, pp 1–5. https://doi.org/10.1109/ICIIP.2017.8313733
12. Pavel MI, Kamruzzaman SM, Hasan SS, Sabuj SR (2019) An IoT based plant health monitoring system implementing image processing. In: 2019 IEEE 4th international conference on computer and communication systems (ICCCS), Singapore, pp 299–303
13. Dukes MD, Muñoz-Carpena R, Zotarelli L, Icerman J, Scholberg JM (2007) Soil moisture-based irrigation control to conserve water and nutrients under drip irrigated vegetable production. In: Giráldez Cervera JV, Jiménez Hornero FJ (eds) Estudios de la Zona No Saturada del Suelo, vol VIII
14. Gutierrez J, Villa-Medina JF, Nieto-Garibay A, Porta-Gándara MÁ (2013) Automated irrigation system using a wireless sensor network and GPRS module. IEEE Trans Instrum Meas 63(1):166–176
15. Mohanty P, Kabat MR (2016) Energy-efficient structure-free data aggregation and delivery in WSN. Egypt Inform J 17(3):273–284
16. Lee M, Hwang J, Yoe H (2013) Automatic irrigation system using IoT. In: IEEE 16th international conference on computational science and engineering
17. Mirabella O, Brischetto M (2011) A hybrid wired/wireless networking infrastructure for greenhouse management. IEEE Trans Instrum Meas 60(2):398–407
18. Brindha S et al (2017) Involuntary nutrients dispense system for soil deficiency using IoT. Int J ChemTech Res 10(14):331–336
19. Pujari JD et al (2017) SVM and ANN based classification of plant diseases using feature reduction technique. Int J Interact Multimed Artif Intell 3(7):6–14
20. Alghazzawi D, Bamasaq O et al (2021) Congestion control in cognitive IoT-based WSN network for smart agriculture. IEEE Access 9:151401–151420. https://doi.org/10.1109/ACCESS.2021.3124791
21. Srinivas B, Tiwari KN (2018) Determination of crop water requirement and crop coefficient at different growth stages of green gram crop by using non-weighing lysimeter. Int J Curr Microbiol Appl Sci 7(09):2580–2589
22. Munir MS, Bajwa IS, Ashraf A, Anwar W, Rashid R (2021) Intelligent and smart irrigation system using edge computing and IoT. Complexity 2021, 16 pages, Article ID 6691571
23. Saqib M, Almohamad TA, Mehmood RM (2020) A low-cost information monitoring system for smart farming applications. Sensors 20(8):2367
24. Koduru S, Padala VPR, Padala P (2019) Smart irrigation system using cloud and internet of things. In: Proceedings of 2nd international conference on communication, computing and networking. Springer, pp 195–203
25. Glória A, Dionisio C, Simões G, Cardoso J, Sebastião P (2020) Water management for sustainable irrigation systems using internet-of-things. Sensors 20(5):1402
26. Obaideen K, Yousef BAA, Al Mallahi MN, Tan YC, Mahmoud M, Jaber H, Ramadan M (2022) An overview of smart irrigation systems using IoT. Energy Nexus 7
27. Chavez JL, Bauder TA (2011) Irrigation scheduling: the water balance approach. In: Andales AA (ed), Fact sheet No. 4.707 (1/15)
28. NETAFIM (2015) Drip irrigation handbook—understanding the basics, V 0.001.02. www.netafim.com
29. Navarro-Hellin H, Torres-Sanchez R, Soto-Valles F, Albaladejo-Perez C, Lopez Riquelme JA, Domingo-Miguel R (2015) A wireless sensors architecture for efficient irrigation water management. Agric Water Manag 151(6):64–74

Chapter 4
An Integrated Approach for Pregnancy Detection Using Canny Edge Detection and Convolutional Neural Network

Nishu Bansal, Swimpy Pahuja, and Inderjeet Kaur

1 Introduction

Deep learning (DL) [1] is an artificial intelligence (AI) technique which models the decision-making skills of the human brain and is now gaining traction in a number of healthcare areas, including pregnancy detection [2, 3]. Rapid development of various DL approaches can reveal valuable information hidden in vast volumes of healthcare data, which can then be utilized for data processing and pattern development for decision-making, growth analysis, regular monitoring, and illness prevention [4, 5]. Deep learning [6] is a branch of machine learning in which neural networks are used to learn unsupervised from unstructured or unlabeled data. Computer vision makes it simple to predict visual outputs. Processing graphical content necessitates the construction of an algorithm framework. The advent of computer vision makes it possible to generate artificial data via simulators in order to augment the training set. Edge detection in computer vision is quite useful in cases where a picture needs to be processed and treated. In a basic manner, this method may forecast when an image point's brightness level changes. Consumers can choose from several different variants of this method right now. Picture distortion detection, sample identification, image partitioning, and image removal are just few of the applications available. In most cases, edges in a picture are represented as step edge, ramp edge, roof edge, or line edge [7].

The body of a pregnant woman changes substantially during the course of the pregnancy's 42 weeks. Sometimes, these modifications can be lethal. High BP, maternal

N. Bansal · I. Kaur
Department of CSE, Ajay Kumar Garg Engineering College, Ghaziabad, India
e-mail: bansalnishu@akgec.ac.in

S. Pahuja (✉)
School of Computing and Information Technology, Reva University, Bangalore, India
e-mail: swimpy.pahuja@reva.edu.in

V. Devedzic et al. (eds.), *Proceedings of the International Conference on Intelligent Computing, Communication and Information Security*, Algorithms for Intelligent Systems, https://doi.org/10.1007/978-981-99-1373-2_4

infection, abortion complications, hypertension, weight gain, and gestation diabetes are all issues that pregnant women may experience. As a result, the linked gynecologist assesses fatal problems on a frequent basis in order to provide timely treatment. Human chorionic gonadotropin (hCG), a hormone present solely in pregnant women, is detected in pregnancy tests [8]. It is generated by uterus cells and is important for directing the ovaries to release estrogen and progesterone to aid in the development of your fetus. The hCG levels will continue to rise as the pregnancy proceeds, and these levels can be detected using both urine and blood pregnancy tests. In medical science, various intelligent devices have been built in the past to monitor pregnancy by reorganizing the patterns obtained from ultrasound reports [9–11]. The fertility services market is expected to rise to $21 billion by 2020, owing to the importance of predicting and facilitating conception. The proposed research is making use of deep learning and image processing for pregnancy detection [8]. The Canny edge detection mechanism was utilized to minimize the size of the image dataset and the amount of time it took to detect it. To find sharp edges in an image, Canny edge detection uses a Gaussian filter. The remaining part of this paper is organized as follows. Section 2 introduces related works discussion about pregnancy determination using ultrasound images. The dataset employed is described in detail in Sect. 3. Section 4 explains the suggested architecture as well as the various parameter strategies/methods we used for pregnancy detection, followed by a review of the experimental outcomes in Sect. 5. Finally, the paper concludes with several suggestions for further research.

2 Related Work

In the literature, several deep learning algorithms have been proposed to monitor pregnancy by reorganizing patterns collected from ultrasound pictures to address image classification, object recognition, and tissue segmentation [3, 12, 13]. In order to reduce the size of the image and limit the comparison time, the proposed work uses deep learning techniques in conjunction with edge detection mechanisms. Many deep learning algorithms [8, 14] based research has previously been employed in pregnancy diagnosis. Machine learning has been the focus of certain previous studies [15–18]. In some past studies, a machine learning approach was utilized to predict pregnancy outcomes after IVF therapy [19–21].

Several studies on edge detection techniques, such as image edge detection method based on multi-sensor data fusion [7], were taken into account. For edge detection of depth images, some authors refined the depth local binary pattern [9]. Edge detection has been made by superposed-spiral phase filter [10] and hybrid dynamic binary threshold [11] in some areas as shown in Table 1. A unique adaptive threshold-based edge detection approach has been proposed in [22]. Also, there have also been studies on the design and implementation of an embedded cloth edge detection system [23].

Two-dimensional (2D) ultrasonography is likely the most widely used tool for prenatal assessment in pregnancy due to its wide availability and high quality [24]. Due to the low image quality and reliance on the operator, most devices now

Table 1 Application of edge detection in various fields

Authors	Approach	Limitations
Hui et al. [7]	Multi-sensor data fusion	Research considered limited application
Navdeep et al. [9]	Local binary pattern procession	Research has not focused on expert system
Li et al. [10]	Superposed-spiral phase filter	This work needs to improve accuracy
Gijandren [11]	Hybrid dynamic binary threshold	The process is found time consuming
Mo et al. [22]	Adaptive threshold	Need to improve the quality of work
Cao et al. [23]	Embedded cloth edge detection	Need to do more work on accuracy and performance

feature three-dimensional (3D) probes and algorithms for detecting fetus structural anomalies [25–28].

There are a number of limitations, including limited soft-tissue acoustic contrast, poor image quality in certain situations such as low amniotic fluid and beam attenuation induced by adipose tissue, because the ultrasound testing reports are highly reliant on the sonographer's training, skills, and experience [29]. The performance sensitivity of existing techniques for fetus detection utilizing ultrasound images ranges from 27 to 96% for various medical organizations.

Because this procedure takes so long, it cannot be used as a population-based screening technique [30]. To overcome this, image processing automation is essential for better and more accurate outcomes.

In this paper, the authors examine current advances in deep learning as they are applicable to ultrasound images for pregnancy identification and offer a new strategy that incorporates edge detection into a typical CNN approach. Pregnancy detection systems have been discovered to have performance difficulties, and graphical examples take up a lot of space. Decision-making is a time-consuming process as the useless pixels are processed in existing researches. The delay in decision-making may lead to an increase in the mortality rate. There is a need to analyze different mechanisms to develop the learning environment for pregnancy detection. The goal of the proposed study is to address the issue of space and time complexity, as well as to propose a technique to enhance the learning environment for pregnancy detection.

In study, deep learning strategies for ultrasound in pregnancy are being used as mentioned in Table 2. In clinical practice, ultrasound is among the most widely used imaging technologies. It is the most often used imaging method in pregnancy since it is inexpensive, involves no radiation exposure, and can be performed at the patient's bedside. Having these benefits, it has a number of disadvantages, including poor image quality, limited contrast, and a top rate of unpredictability. Automating the analysis of ultrasound pictures is challenging given these constraints. On the other hand, a successful automated structure detection method inside three-dimensional ultrasound images has the ability to revolutionize clinical practice.

Table 2 Deep learning approaches for pregnancy detection

Authors	Approach	Limitations
Diniz et al. [19]	Deep learning	The learning mechanism is time consuming
Diniz [31]	Deep learning	Need to improve the performance of system
Liu et al. [32]	Large scale data processing	Need to integrate edge detection to improve performance
Gayathri et al. [33]	Health care monitoring system	System is dependent on IoT resources
Mu et al. [2]	Deep learning	There is need to improve the accuracy, f-score, precision during prediction
Hassan et al. [8]	Machine learning	Lack of technical feasibility
Pruthi [34]	Machine learning	Difficult to adopt on regular basis

3 Materials Used

The image set of 50 pregnancy images and 50 non-pregnancy images has been taken from the real-time system and considered for training. The learning mechanism has trained a network considering this dataset and predicts the cases of pregnancy and non-pregnancy. The edge detection approach was applied after training and testing with these normal images in order to reduce the size of the image and prediction time.

4 Methods Used

The proposed study will be based on the integration of CNN technique with edge detection method in order to establish a learning environment for pregnancy detection. In order to resolve issues such as time and space consumption, the proposed mechanism has been introduced to provide a smart and quick solution to pregnancy detection by integrating an edge detection mechanism in the deep learning environment. The ultrasound image samples are processed with an edge detector at the initial stage, and then, a network model is trained to detect the pregnancy.

4.1 Detailed Description of Proposed Approach

As depicted in Fig. 1, following steps are involved in the work:

Step 1 This step involves the input of ultrasound image samples from the dataset for the purpose of confirming the pregnancy.

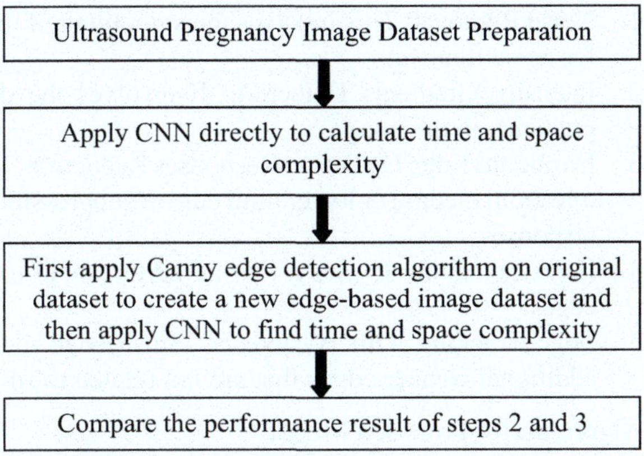

Fig. 1 Stepwise description of proposed approach

Step 2 Apply the standard neural network classifier on images to assess the time and space consumed by the system.

Step 3 On the image set, the Canny edge detection mechanism is used to reduce the image size first. This aids in the removal of superfluous data from photos. Then, on the acquired edge-based images, a convolutional neural network (CNN) is used, followed by assessment of space and time complexity.

Step 4 The time consumption and space consumption from Steps 2 and 3 are compared in this step followed by evaluation of system accuracy.

4.2 *Integration of Edge Detection in Proposed Work*

To reduce the size of the image, the suggested method employs edge detection. Edge detection is an image processing approach that detects brightness discontinuities to find the borders of objects within images. Edge detection is used for picture segmentation and data extraction in disciplines such as image processing, computer vision, and machine vision. In this study, the Canny edge detection technique has been utilized for the detection of edges. Edge detection must catch as much of the picture's edges as feasible with a low error rate. The operator's edge location must be exact while identifying the edge's centroids. Picture noise should not produce erroneous edges, and each image edge should be recorded only once.

Retrieving architectural information from a variety of graphic elements using Canny edge detection substantially minimizes the amount of data that needs to be evaluated. It has been implemented in several computer vision and image processing applications. Canny demonstrated that the requirements for employing edge detection on different vision-based systems were identical. As a consequence, an edge detection technology that satisfies these requirements can be applied to a variety of scenarios. The Canny edge detection approach operates as follows:

1. Noise Removal: This involves the smoothing of the image with a Gaussian filter for noise removal.
2. Intensity Gradients Detection: It involves the determination of the intensity gradients in the image.
3. Erroneous Edge Detection Responses Reduction: It leverages gradient magnitude threshold method or lower limit cut-off suppression to reduce false edge detection responses.
4. Edge Detection: It uses a two-tiered threshold to determine the most probable edges.
5. Edge Tracking with Hysteresis: Finish edge identification by suppressing all additional weaker edges that are not related to strong edges.

Algorithm for Proposed Model

As depicted in Fig. 2, the algorithmic steps for the proposed work are described as under:

Step 1 Collection of ultrasound samples.
Step 2 Selection of characteristic features for the purpose of training the dataset.
Step 3 Setting of training and testing ratio to 70% and 30%, respectively.
Step 4 Application of convolutional neural network.
Step 5 Performing classification.

Fig. 2 Flowchart of proposed work

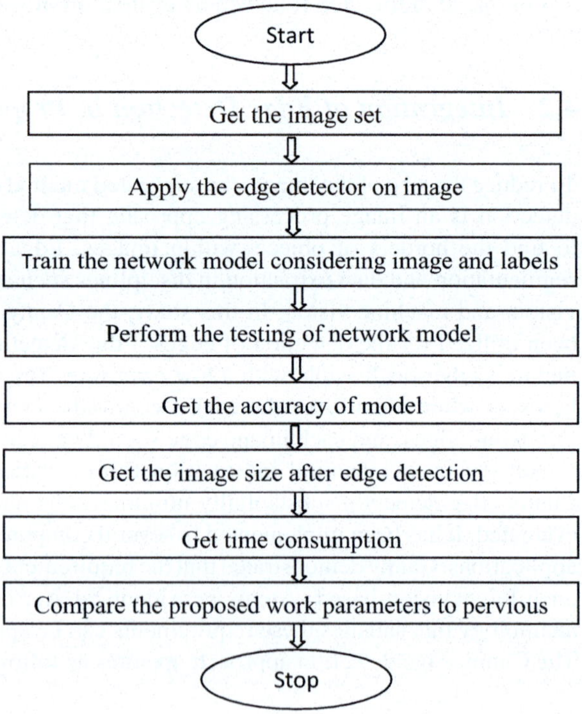

5 Results and Discussion

The ultrasound image dataset is placed in the relevant folder. The patient's sample is then preserved for pregnancy detection. These samples are matched to preserved records using convolution neural network technique in order to forecast pregnancy using attributes retrieved from the current dataset. The size of the dataset as well as the comparison time are taken into account.

The fraction of matching is also considered while determining the accuracy of the model. To avoid using information from the graphical dataset, the edge detection approach is used to the same dataset. As a result, the dataset's snapshot size has been reduced. The dataset is then fed to the convolution neural network method, and the dataset size, processing time, and matching percentage are calculated. The same dataset of ultrasound images has been considered in research as shown in Fig. 3.

The image sample has been categorized in "pregnancy confirmed" and "pregnancy is not confirmed" with edge detection and without edge detection as depicted in Fig. 4. Non-edge-based categories will take longer to simulate than edge-based categories.

A CNN classifier module has been developed to detect pregnancy based on the training and image dataset. The entire implementation work involves the following steps:

Step 1 Dataset Acquisition: This step involves obtaining a dataset containing ultrasound images from various classes. This dataset has been divided into training and testing images such that the classes have no overlap, and they are mutually exclusive.

Step 2 Pixel Normalization: The pixel values have been normalized to a range of 0–1 by using the statement
train_images, test_images = train_images/255.0, test_images/255.0

Step 3 Data Verification: To check the dataset is legitimate, plot the images from the training dataset and exhibit the class labels as indicated:

*L*1 Pregnancy has been verified.
*L*2 Pregnancy has not been verified.

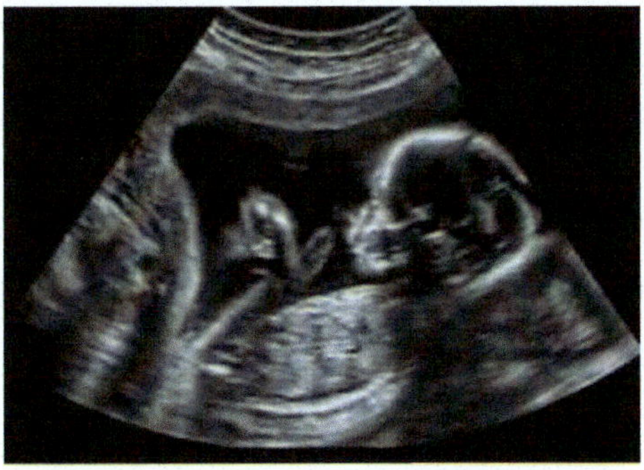

Fig. 3 Ultrasound image [35]

	Before edge detection	**After edge detection**
Ultra sound in case of pregnancy	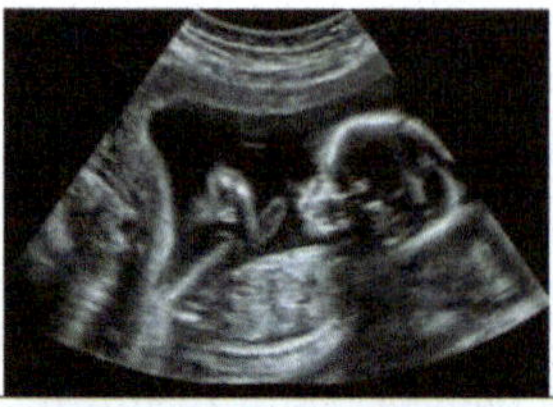	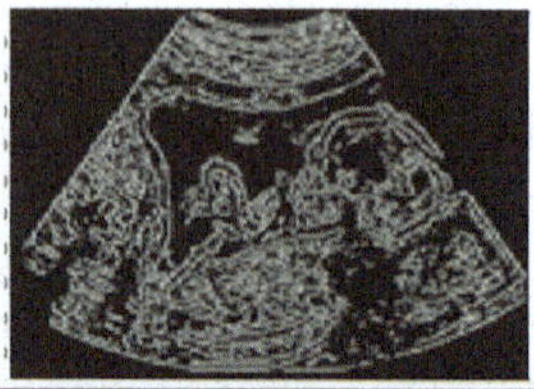
Ultra sound in case of non-pregnancy	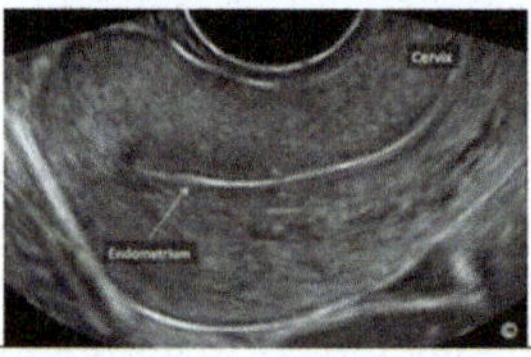	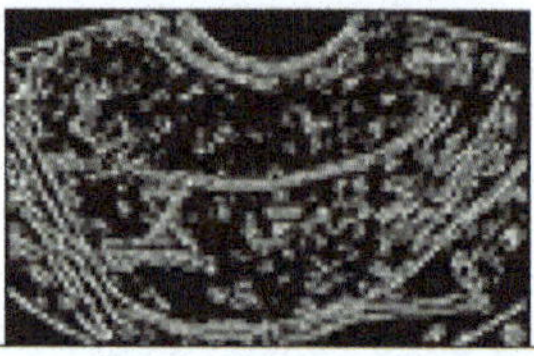

Fig. 4 Image classification

Step 4 Convolutional Base Formation: This step involves the creation of base for applying CNN.

Step 5 Dense Layers Addition: This phase entails adding dense layers to the top. Input the final output tensor of the convolutional base (shape (4, 4, 64)) into one or more dense layers for classification to complete the model. These dense layers generate a three-dimensional tensor from one-dimensional vector input. After the 3D output has been flattened (or unrolled) to 1D, add layers to one or more thick layers. Finally, because CIFAR has ten output classes, utilize a dense layer with ten outputs.

Step 6 Complete architecture of model.

Step 7 Train the model.

Step 8 Analyze the model.

During simulation, the ultrasound image set is taken into account. Figure 5 depicts the state of samples prior to the application of edge detection.

An edge-based image dataset illustrated in Fig. 6 is obtained after applying the Canny edge detector technique to the above dataset.

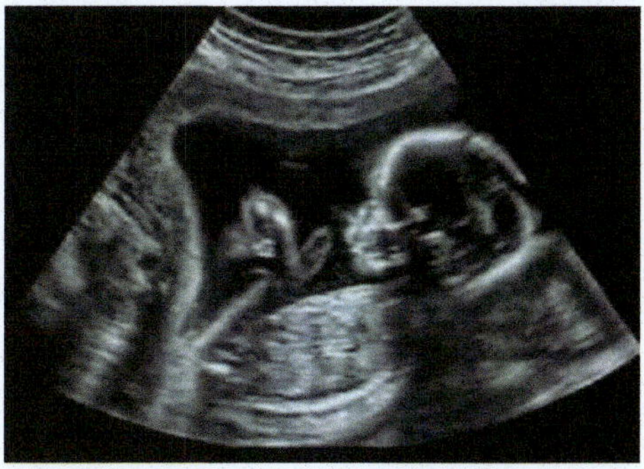

Fig. 5 Before edge detection

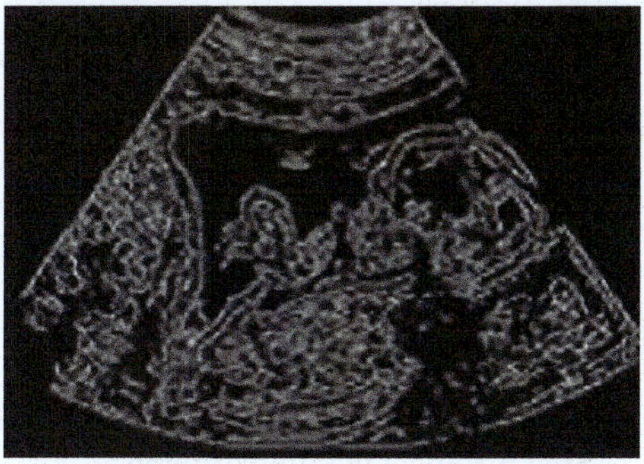

Fig. 6 After edge detection

5.1 Time Consumption Analysis

Table 3 shows the comparison of time consumption for traditional and proposed approach using different sample size. The step by step procedure for time simulation is as follows:

Step 1 Take a sample image set and apply CNN technique on that image set to identify the status of pregnancy. During this process, make a note of the total time $T1$ consumed.

Table 3 Time utilization in the traditional and proposed techniques

Number of ultrasound images taken	Time utilization in traditional CNN method ($T1$)	Time utilization in proposed integrated method ($T2$)
20	21.60	3.47
40	41.36	7.25
60	63.08	8.89
80	83.78	14.21
100	103.34	17.67

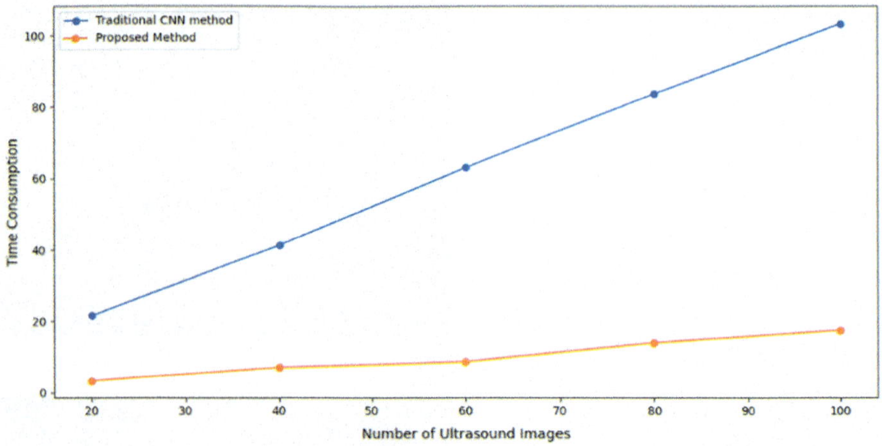

Fig. 7 Time usage in the traditional and proposed scenarios

Step 2 Use the edge detection method on the same pregnancy detection sample taken in Step 1.
Step 3 Apply CNN technique on the result of Step 2 and record the time $T2$ consumed during this detection process.
Step 4 Calculate the proportion of $T1$ and $T2$ reported times.
Step 5 Reflect the time difference.

The graph in Fig. 7 depicts a comparison of time consumption in the traditional and recommended approaches. From Table 3, it can be concluded that the time consumed in the proposed work is less in comparison with the traditional work with respect to different number of samples.

5.2 *Space Analysis*

In the proposed work, space utilization throughout CNN processing is reduced since edge detection has been performed. Step by step procedure for finding the total space consumption is as follows:

Step 1 Take a sample image set and apply CNN technique on that image set to identify the status of pregnancy. During this process, make a note of the total size $S1$ of the image sample taken.
Step 2 Use the edge detection method on the same pregnancy detection sample taken in Step 1.
Step 3 Apply CNN technique on the result of Step 2 and make a note of the sample size $S2$.
Step 4 Calculate the proportion of recorded sizes $S1$ and $S2$.

Table 4 Space utilization in traditional and proposed approach

Number of images taken	Traditional CNN method ($S1$)	Proposed integrated method ($S2$)
20	2200	1200
40	3900	1900
60	5800	2800
80	7500	3400
100	9500	4500

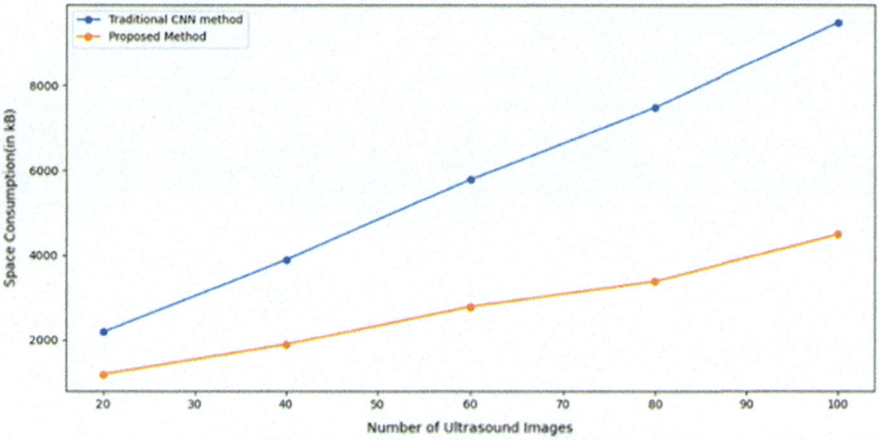

Fig. 8 Space utilization in conventional CNN approach and proposed approach

Step 5 Reflect the size difference.

From Table 4, it can be concluded that space required in the proposed work is less as compared to the traditional work with respect to different number of samples. Figure 8 shows comparison of space usage in the context of conventional and proposed procedures.

5.3 Accuracy Analysis

Accuracy in detection of pregnancy using CNN method and using the proposed integrated CNN method is shown in Table 5.

Table 5 concludes that accuracy in the proposed work is more in comparison with the traditional work with respect to different number of samples. Chart in Fig. 9 depicts the difference in accuracy values of traditional and proposed approaches.

Table 5 Comparison of accuracy in traditional and proposed approach

Number of images taken	Traditional CNN method	Proposed integrated method
20	20.56	24.56
40	41.75	46.23
60	52.45	58.43
80	84.57	90.67
100	90.76	98.89

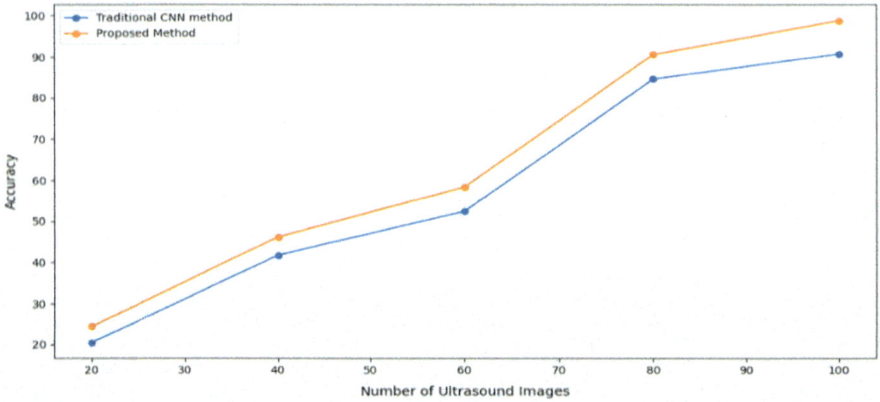

Fig. 9 Accuracy comparison for traditional and proposed approach

6 Conclusion

During the research, a dataset of ultrasound has been taken from patients for pregnancy prediction. Research work compared stored datasets to detect pregnancy using convolution neural network mechanisms. Because the purpose of the study is to save time, space, and enhance the efficiency of the suggested work as compared to past work, the dataset size, as well as the comparison time and percent of matching, are taken into account. After trials with an edge detection-based conventional neural network, the dataset size, operation time, and matching percentage are calculated. The usage of Canny edged detection lowered the amount of time spent on the analysis when compared to the current CNN approach. Furthermore, the size of the graphical application in medical science would boost the convolution neural network's decision-making capabilities. The proposed work is also more accurate than the usual mode, according to the simulation results. This accuracy, however, may vary depending on image size and image dataset modifications.

7 Future Scope

In the future, optimization mechanisms may be used. Furthermore, the edge identification algorithm could be tweaked to improve MSE. Alternative compression strategies could be employed in the future experiments to reduce the amount of data.

References

1. Muthiah A, Ajitha S, Thangam KSM, Vikram K, Kavitha K, Marimuthu R (2019) Maternal ehealth monitoring system using LoRa technology. In: 2019 IEEE 10th international conference on awareness science and technology (iCAST), Morioka, Japan, pp 1–4
2. Mu Y, Feng K, Yang Y, Wang J (2018) Applying deep learning for adverse pregnancy outcome detection with prepregnancy health data. MATEC Web Conf 189:10014. EDP Sciences
3. Huang C, Xiang Z, Zhang Y, Tan DS, Yip CK, Liu Z, Tu W (2021) Using deep learning in a monocentric study to characterize maternal immune environment for predicting pregnancy outcomes in the recurrent reproductive failure patients. Front Immunol 12
4. Fanelli A (2010) Prototype of a wearable system for remote fetal monitoring during pregnancy. In: 2010 annual international conference of the IEEE engineering in medicine and biology, Buenos Aires, pp 5815–5818. https://doi.org/10.1109/IEMBS.2010.5627470
5. Xiaofeng L, Hongshuang J, Yanwei W (2020) Edge detection algorithm of cancer image based on deep learning. Bioengineered 11(1):693–707. https://doi.org/10.1080/21655979.2020.1778913
6. Bobrova YO (2018) The development of a remote fetal activity monitoring system. In: Third international conference on human factors in complex technical systems and environments (ERGO)s and environments (ERGO), St. Petersburg, pp 170–172
7. Hui C, Xingcan B, Mingqi L (2020) Research on image edge detection method based on multisensor data fusion. In: IEEE international conference on artificial intelligence and computer applications (ICAICA), Dalian, China, pp 789–792. https://doi.org/10.1109/ICAICA50127.2020.9182548
8. Hassan MD, Al-Insaif S, Hossain M, Kamruzzaman J (2020) A machine learning approach for prediction of pregnancy outcome following IVF treatment. Neural Comput Appl 32. https://doi.org/10.1007/s00521-018-3693-9
9. Navdeep, Singh V, Rani A, Goyal S (2020) Improved depth local binary pattern for edge detection of depth image. In: 7th international conference on signal processing and integrated networks (SPIN), Noida, India, pp 447–452. https://doi.org/10.1109/SPIN48934.2020.9070820
10. Li Z, Zhao S, Wang L, Zheng B (2020) Edge detection by superposed-spiral phase filter. In: International conference on wireless communications and signal processing (WCSP), Nanjing, China, pp 337–341. https://doi.org/10.1109/WCSP49889.2020.9299865
11. Gijandren A (2020) Edge detection using hybrid dynamic binary threshold. In: International conference on smart electronics and communication (ICOSEC), Trichy, India, pp 126–131. https://doi.org/10.1109/ICOSEC49089.2020.9215349
12. van den Heuvel TLA, Petros H, Santini S, de Korte CL, van Ginneken B (2019) Automated fetal head detection and circumference estimation from free-hand ultrasound sweeps using deep learning in resource-limited countries. Ultrasound Med Biol 45(3):773–785
13. Sobhaninia Z, Rafiei S, Emami A, Karimi N, Najarian K, Samavi S, Soroushmehr SR (2019) Fetal ultrasound image segmentation for measuring biometric parameters using multi-task deep learning. In: 41st annual international conference of the IEEE engineering in medicine and biology society (EMBC). IEEE, pp 6545–6548

14. Garcia-Canadilla P, Sanchez-Martinez S, Crispi F, Bijnens B (2020) Machine learning in fetal cardiology: what to expect. Fetal Diagn Ther 47(5):363–372
15. Chen J, Huang H, Hao W, Xu J (2020) A machine learning method correlating pulse pressure wave data with pregnancy. Int J Numer Methods Biomed Eng 36(1):e3272
16. Lu X, Wu Y, Yan R, Cao S, Wang K, Mou S, Cheng Z (2018) Pulse waveform analysis for pregnancy diagnosis based on machine learning. In: IEEE 3rd advanced information technology, electronic and automation control conference (IAEAC). IEEE, pp 1075–1079
17. Caly H, Rabiei H, Coste-Mazeau P, Hantz S, Alain S, Eyraud JL, Ben-Ari Y (2021) Machine learning analysis of pregnancy data enables early identification of a subpopulation of newborns with ASD. Sci Rep 11(1):1–14
18. Chavez-Badiola A, Farias AFS, Mendizabal-Ruiz G, Garcia-Sanchez R, Drakeley AJ, Garcia-Sandoval JP (2020) Predicting pregnancy test results after embryo transfer by image feature extraction and analysis using machine learning. Sci Rep 10(1):1–6
19. Diniz PHB, Yin Y, Collins S (2020) Deep learning strategies for ultrasound in pregnancy. EMJ Reprod Health 6(1):73–80
20. Jhala D, Ghosh S, Pathak A, Barhate D (2020) Predicting embryo viability to improve the success rate of implantation in IVF procedure: an AI-based prospective cohort study. In: Computational vision and bio-inspired computing. Springer, Singapore, pp 383–400
21. David DS et al (2022) Enhanced detection of glaucoma on ensemble convolutional neural network for clinical informatics. CMC-Comput Mater Contin 70(2):2563–2579. https://doi.org/10.32604/cmc.2022.020059
22. Mo S, Gan H, Zhang R, Yan Y, Liu X (2020) A novel edge detection method based on adaptive threshold. In: IEEE 5th information technology and mechatronics engineering conference (ITOEC), Chongqing, China, pp 1223–1226. https://doi.org/10.1109/ITOEC49072.2020.9141577
23. Cao R, Jiang B, Tang D (2020) Design and implementation of embedded cloth edge detection system. In: IEEE international conference on advances in electrical engineering and computer applications (AEECA), Dalian, China, pp 361–365. https://doi.org/10.1109/AEECA49918.2020.9213542
24. van den Heuvel TL, de Bruijn D, de Korte CL, Ginneken BV (2018) Automated measurement of fetal head circumference using 2D ultrasound images. PLoS ONE 13(8):e0200412
25. Lopez BDB, Aguirre JAA, Coronado DAR, Gonzalez PA (2018) Wearable technology model to control and monitor hypertension during pregnancy. In: 13th Iberian conference on information systems and technologies (CISTI), Caceres, pp 1–6
26. Hata T, Tanaka H, Noguchi J, Hata K (2011) Three-dimensional ultrasound evaluation of the placenta. Placenta 32(2):105–115
27. Mahrishi M, Morwal S, Dahiya N et al (2021) A framework for index point detection using effective title extraction from video thumbnails. Int J Syst Assur Eng Manag
28. Bega G, Lev-Toaff A, Kuhlman K, Berghella V, Parker L, Goldberg B, Wapner R (2000) Three-dimensional multiplanar transvaginal ultrasound of the cervix in pregnancy. Ultrasound Obstet Gynecol 16(4):351–358
29. Shegokarl PS, Paswan RS (2017) Women health monitoring: a survey. IJARCCE 6(5). https://doi.org/10.17148/IJARCCE.2017.65149
30. Raphael R (2018) Can silicon valley get you pregnant? Fast Company [Google Scholar]
31. Diniz PH (2020) Deep learning strategies for ultrasound in pregnancy. Reprod health
32. Liu B, Shi S, Wu Y, Thomas D, Symul L, Pierson E, Leskovec J (2019) Predicting pregnancy using largescale data from a women's health tracking mobile application. In: The world wide web conference, pp 2999–3005
33. Gayathri S, Bharathi T, Devleena Jerusha AR, Ajay Kumar A (2018) Pregnant women health care monitoring system based on IoT. Int J Eng Technol Manag Sci 1:11–14
34. Pruthi J (2019) A walkthrough of prediction for pregnancy complications using machine learning: a retrospective
35. Konnaiyan KR, Cheemalapati S, Pyayt A, Gubanov M (2016) mHealth dipstick analyzer for monitoring of pregnancy complications. IEEE Sens J 1–3

Chapter 5
Ontology-Based Profiling by Hierarchical Cluster Analysis for Forecasting on Patterns of Significant Events

Saurabh Ranjan Srivastava, Yogesh Kumar Meena, and Girdhari Singh

1 Introduction

Catastrophic events [1] lay a disastrous impact on the lives and properties. This impact has always motivated researchers of computational analytics to develop systems for preemptive forecasting of globally occurring catastrophic events. Their research includes conceptualization, development and integration of forecasting approaches with event processing mechanisms. However, various limitations have been encountered in this line of research. Here, the rigid representation of events data and stern recording mechanisms presents the first limitation. Apparently, the data with a limited set of dimensions and exclusive formatting will require specialized measures for access and analysis. Further, the processing will get even more complex due to cross analysis if the event data to be analyzed has been retrieved from different sources. This difference of sources will obviously restrict the possibilities of linking and analyzing. In light of these points, it can be deduced that a framework for assembling the data of various catastrophic events under a single comprehensive format is required. Such a framework will be decisive in enriching the pattern mapping efficiency and forecasting the performance of the target application. Hence, this paper proposes a generic event ontology for re-cataloging catastrophic event features. In simpler terms, the

S. R. Srivastava (✉) · Y. K. Meena · G. Singh
Department of Computer Science and Engineering, Malaviya National Institute of Technology, Jaipur, Jaipur, India
e-mail: saurabhranjansrivastava@gmail.com

Y. K. Meena
e-mail: ymeena.cse@mnit.ac.in

G. Singh
e-mail: gsingh.cse@mnit.ac.in

V. Devedzic et al. (eds.), *Proceedings of the International Conference on Intelligent Computing, Communication and Information Security*, Algorithms for Intelligent Systems, https://doi.org/10.1007/978-981-99-1373-2_5

ontology facilitates the redistribution of event features under 7 concepts: geography, timestamp, actor, activity, component, impact, and details. This redistribution enables the assembling of feature values of such significant events belonging to divergent types and sources under a single comprehensive structure.

2 Literature Survey

For the proposed work, we have referenced a sizable volume of technical literature from July 2017 to July 2022. The literature is exclusively available in the public domain, and therefore, no conflicts of interest have been discovered.

Before moving to forecasts, we first discuss spatiotemporal events and the challenges inherent in their forecasting process. Spatiotemporal events can be defined as occurrences of interest to specific stakeholders that happen at a particular time and location. Examples of significant spatiotemporal events include natural disasters, epidemics, and criminal events [2]. The methods for the discovery of knowledge and patterns from spatiotemporal data are collectively referred to as spatiotemporal data mining [3]. Here, the uncertainty, heterogeneity, and interdisciplinary complexity of spatiotemporal data [4] are the major challenges in its processing and analysis. Though, domain-specific applications for catastrophes like crimes [5] and civil unrest [6] have been proposed. But, the aforementioned challenges still restrict the thorough and collective analysis of catastrophic spatiotemporal events of significance. To resolve these challenges, we propose a comprehensive framework for assembling and managing the data of such significant catastrophic events under a single inclusive format. But before moving ahead, we first discuss the concepts and terminologies relevant to the proposed work.

Ontology. An ontology can be defined as an explicit and formal depiction [7] of a shared semantics and visual conceptualization of a domain to enable its sharing and reuse [8] for utilization in knowledge-based systems. Numerous ontologies have been devised for domains such as network security, text analytics, e-commerce, and so on [9]. Systems for analyzing terrorist activities [10], cyber-attacks [11], and consumer demand hotspots [12] also employ ontologies. However, contrary to these domain-specific applications of ontologies, the need for a comprehensive ontology to capture data of any spatiotemporal disaster of significance is still persistent. In the proposed framework, we will be harnessing the merits of ontologies for the categorization of such significant event features and re-assembling of their values under an inclusive format.

Profile analysis. Profile analysis is another approach for the exploration of a domain. According to Meriam Webster dictionary [13], analysis of a subject's characteristics, behavior or tendencies for inducing information about it can be termed profiling or profile analysis. The most frequent usage of profiling is witnessed in recommender systems [14] for modeling of customer behavior [15] profiles. Extended applications of customer behavior profiling include profile generation of mall users [16] and television viewers [17]. Profile analysis is also employed in domains like mining web users' data [18] and behavior profiling of terrorist groups [19]. In the same league, association patterns for products can also be efficiently discovered by similarity-based profiling [20] of data items.

However, the approaches conferred above are obviously constrained to limited types of domains addressing the demands of explicit data specifications. Therefore, our proposed framework facilitates a generic platform for the generation of Event Profiles for different significant catastrophic events of divergent types and details.

3 Proposed Work

In this section, we discuss the concepts and the methodology of the proposed work. The section illustrates the workflow of the proposed system (Fig. 1) by discussing its components at each level. Here, we will elaborate the input of event features and assembling of event data under the proposed event ontology.

3.1 Event Ontology: Categorization of Event Features

Our approach aims at mapping of patterns in data of significant events. Hence, we propose an ontology that facilitates the assembling of globally occurring disasters of various types from different sources under a single exhaustive format. This assembling enables the mapping of useful patterns from events of analogous types.

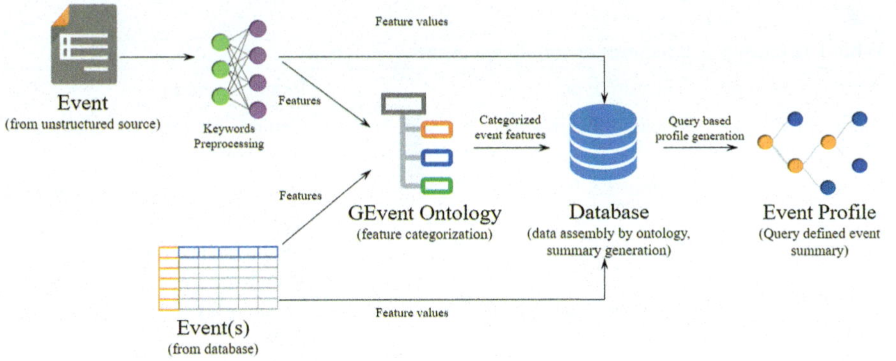

Fig. 1 Workflow of the proposed system

3.1.1 Conceptualization

Generally, catastrophic events share various feature values in common that can be assembled under a standard benchmark for analysis. Here, the event can belong to an existing database of events like terrorist attacks [21], natural calamities [22], or accidents [23]. It may also have been retrieved from an unstructured source like news media or a report. Hence, inspired by the design of multiple event databases, we have summarized their commonly shared features under the following 7 major concepts (Table 1).

Here, the 7 concepts given as geography, timestamp, actor, activity, component, impact, and details serve as benchmark components for cataloging of event features. This implies that features of any catastrophe can be categorized under these concepts. For instance, spatial features like country, province, city, street, suburb, and even scene of an event can be captured under the concept of 'geography'. Similarly, component features like 'type of vehicle' in an accident and 'type of weapon' used in a terrorist attack can be expressed as a 'component' of a catastrophic event. Likewise, participants in an event such as volunteers, patients, victims, or attackers will be counted as 'actors', while the type of event will be covered under the concept of 'activity'. The concept 'impact' captures the numerical damage caused by a catastrophe like the count of dead, infected, and injured people or the cost of destruction.

Any additional details like summary, tags, or labels deduced from existing feature values can be covered under the 'details' concept.

Here, the group of months with maximum rainfall computed from a temporal feature like 'months' already listed under the concept 'timestamp' can be an example of such details. In other cases, purposefully added labels or tags can also be part of these details. For instance, the name 'Katrina' of a hurricane originating in the Gulf

Table 1 Concepts and features of significant events

Name of concept	Feature type	Example features
Geography	Spatial features	Country, city, latitude-longitude, street, suburb
Timestamp	Temporal features	Date, day, month, year, semester, quarter, time
Actor	Actor features	Victim, attacker, patients, volunteers
Activity	Activity features	Flood, pandemic, terrorist attack, theft
Component	Component features	Weapon type, vehicle type, crash intensity
Impact	Impact features	Number of dead, injured, infected, cost of damage
Details	Pre-computed or other details over above features	Any/all of the above features, additional specifications

Fig. 2 Outline of ontology classes for significant events

of Mexico [24] can be expressed as a 'label' or 'tag' under the concept 'details' of a calamity.

Hence, the proposed ontology facilitates an exhaustive set of concepts for cataloging the component features of any global catastrophe as a significant event. These concepts will be implemented as component classes and the cataloged features as their corresponding subclasses by the event ontology. The outline of the proposed event ontology consisting of the 7 concepts described in Table 1 has been presented in Fig. 2.

3.1.2 Ontology Design

Now, we present a formal illustration of the proposed event ontology in semi-computable format. The classes form the focal point of ontologies and can be expanded into subclasses for the representation of more detailed concepts. Here, the proposed event ontology denotes an event by a global superclass G. This superclass (G) expresses every event as a combination of 7 base classes (B) discussed above. The features of the input event will be categorized under these base classes as input subclasses (I) (Table 2).

A 'has' relation connects these classes and subclasses at different levels. The global superclass (G) is linked to each base class (B) by a 'has' relation. Similarly, every input class (event feature) also connects to its parent base class through a 'has' relation. In this way, 'has' enables a detailed expression of any significant event under the proposed event ontology. For further elaboration, we categorize event features for a terrorist attack [25] with the concepts of the proposed event ontology (Fig. 3).

Here, every 'Event' has a 'Timestamp'. Also, each 'Timestamp' has a 'Date'. These relations are presented as follows:

Table 2 Elements of the event ontology

Construct	Type	Syntax	Explanation
Global	Superclass	G	Superclass that represents an event as a whole
Base class	Class	B	Base class from the 7 concepts
Input class	Subclass	I	Input subclass categorized under the base class
Has	Relation	Has	Association for linking classes of 2 levels

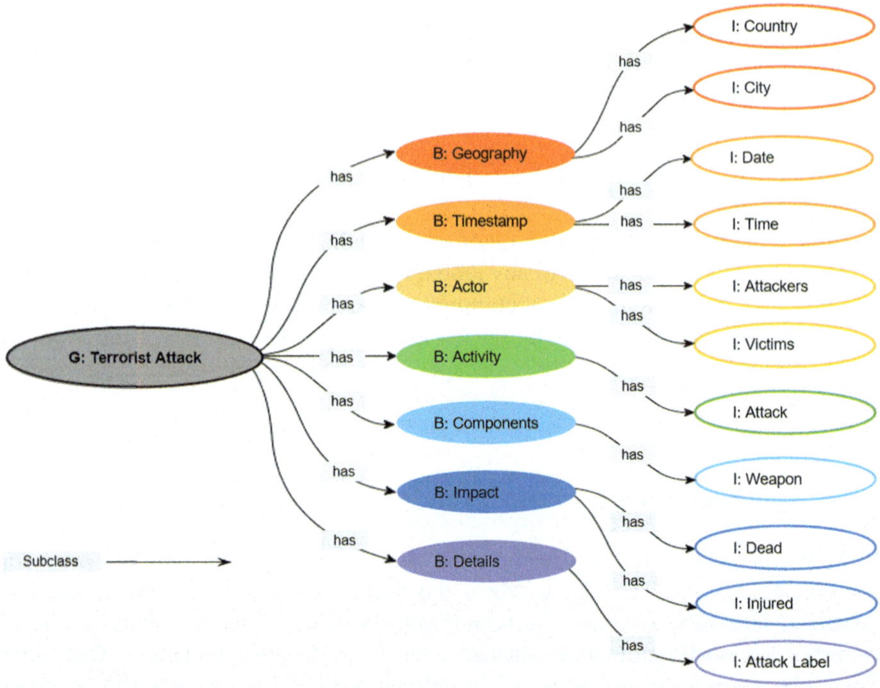

Fig. 3 Schematic architecture of 'GEvent' ontology elaborated with example of terrorist attack event

```
for (G: Terrorist attack, B: Timestamp, I: Date)
        G: Terrorist attack ------ has ------ B: Timestamp
        B: Timestamp ------------- has ------ I: Date
```

Other features can be also captured in a similar hierarchy by the proposed ontology. Such a hierarchy of features for each concept helps to achieve the appropriate detailing in the targeted significant event profile. The availability of features under each of these concepts is subjective to their presence in their source database or document. For instance, every natural calamity may not have a name as an additional 'detail', unlike terrorist attacks having added labels as '9/11 attacks' [26] and so on. This implies that the presence of categorized features under each of the concepts is not mandatory. But more features can be added as per the availability of data and requirements of analysis.

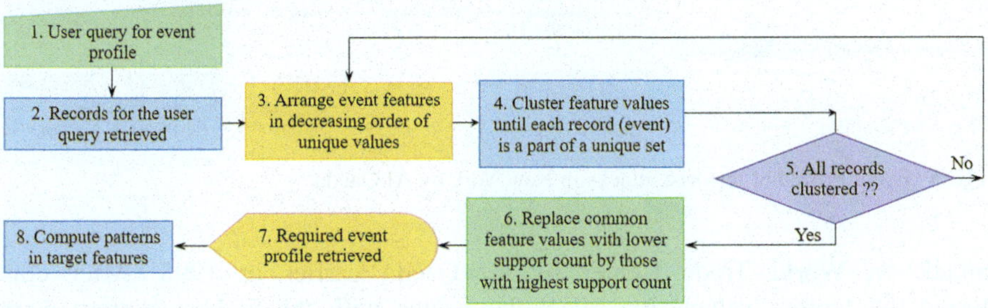

Fig. 4 Flowchart of event profile generation

3.2 *Event Profile: Pattern Discovery Among Events*

For profile generation, generally similar feature values of data items are clustered to generate their groups or sets. These sets of commonly shared feature values of data items are popularly defined as itemsets in data mining [27, 28]. Itemsets are applied in several applications, from crime pattern mining [29] to terrorism research [30]. Similarly, itemsets retrieved by hierarchical clustering [31] of common feature values of spatiotemporal events can also be utilized to form a template of details termed an event profile. This template will act as a summarized assembly of similar event values generally belonging to a common geographical specification for the analysis and forecasting of events. The generation of event profiles in our approach is query-dependent. This implies that any profile of catastrophic events will be generated from itemsets retrieved against a user query. We have depicted the process of profile generation in Fig. 4 as a flowchart.

Once the records required for profile generation are retrieved against the user query, their features will be arranged in decreasing order of unique values to maximize the readability of the tree structure of the target event profile. After organizing the features, we perform agglomerative hierarchical clustering over feature values until every record becomes a part of a unique cluster or itemset. Post clustering, we will replace the common feature values in each itemset having lower support counts with the one having the highest count. Finally, we can retrieve the event profile and compute patterns in target feature values. Now, we discuss the implementation of the above procedure on our example of terrorist attacks. For a query to retrieve details of terrorist attacks performed by the Al Qaeda terrorist group in the New York City of USA, we can have the sample set of records as:

```
{USA (14), New York (01)}{Al Qaeda (01)}{BOMBING (01)}{WTC Bombing}{Bomb (01)}{26-02-1993}{06, 1000}
{USA (15), New York (02)}{Al Qaeda (02)}{SUICIDE (01)}{9/11 Attacks}{Airplane (01)}{11-09-2001}{2996,25000}
```

Here, the second record presents the details of the terrorist attack on the World Trade Towers in New York on September 11, 2001, that caused the death of 2996 people [26]. However, the first record represents another terrorist

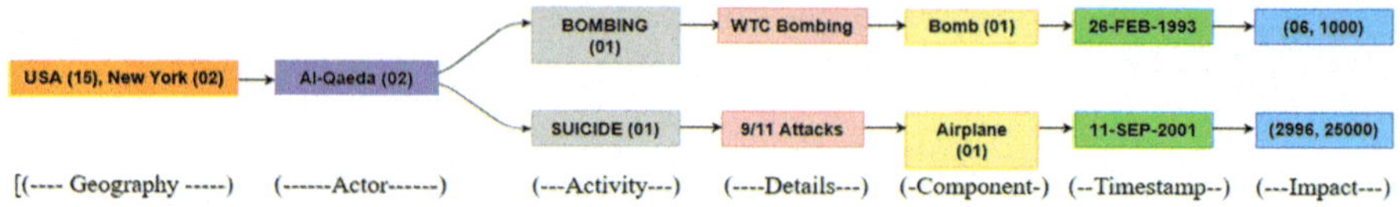

Fig. 5 Event profile of terrorist attacks in New York by Al Qaeda

attack on World Trade Center by Al Qaeda earlier in 1993. After clustering the feature values, we select the values with the highest support count and replace the remaining ones with them. As in the above example, we will replace {`USA (14), New York (01)`}{`Al Qaeda (01)`} with {`USA (15), New York (02)`}{`Al Qaeda (02)`} to obtain this set:

```
{USA (15), New York (02)}{Al Qaeda (02)}{BOMBING (01)}{WTC
Bombing}{Bomb (01)}{26-02-1993}{06, 1000}
{USA (15), New York (02)}{Al Qaeda (02)}{SUICIDE (01)}{9/11
Attacks}{Airplane (01)}{11-09-2001}{2996,25000}
```

This will be the required profile of terrorist attacks performed by Al Qaeda in New York, USA. We have presented a diagrammatic representation of this profile in Fig. 5. The profile can be interpreted as a cluster of 02 itemsets of attack events which are a subset of total 15 attacks executed by Al Qaeda in 'USA'. Each value can now be analyzed against its relative event concept presented at the bottom instead of its original parent feature. The profile can be extended further to cover more details if required.

4 Experimental Results

In this section, we demonstrate the pattern mapping performance of our approach in integration to other computational models. To prove the compatibility of our approach with any type of spatiotemporal events, we have used 3 event databases belonging to disasters of divergent natures. Among them, the RAND database of worldwide terrorism incidents (RDWTI) [32] is a database of terrorist attack events from 1968 to 2009 that represents intentional human disasters. Further, the Airplane crashes and fatalities database [23] is a repository that belongs to the category of unintended human errors.

Similarly, event records of natural disasters in Afghanistan from 2012 to 2018 are organized in the Afghanistan natural disaster incidents database [22] provided by the United Nations Office for the Coordination of Humanitarian Affairs (UNOCHA) in Afghanistan. We have generated event profiles for 4 geographical locations from each of these event databases. The records of each database have been partitioned into a ratio of 70–30 as training and testing data. Further, we have computed the profiles of these locations by using the aforementioned approach. From the computed event profiles, we have plotted the frequency count of the occurrence months of catastrophes and formed sets of common occurrence counts. Sets with the highest occurrence counts are presented ahead in Tables 3, 4 and 5.

Here, the constituent (**Const. %**) percentage is the share of these months among the total number of records in training data. For example, the set of 'Apr, Jul, Aug (03)' implies that there have been 3 occurrences of disasters in the months of April, July, and August each, summing up to 9 records out of a total 16 airplane crashes recorded in Bogota. The forecasted occurrence (**For. %**) of these months in test data have been presented under as the forecast accuracy. We have also plotted a range of fatalities as the impact of these catastrophes for each location. As visible in results, a percentage of 66.67–75% in forecasts is achievable by computing only the frequency count of the occurrence months. In Fig. 6a–c, we have presented sample event profile of a location from each of the event databases.

Table 3 Component and forecast percentages for highest occurrence month sets of 4 locations from the Airplane crash database

Country	Location	Aircrash month details			Fatality
		Largest itemset	Const. %	For. %	
Colombia	Bogota	Apr, Jul, Aug (03)	56.25	56.25	04–26
Russia	Moscow	Jan, Mar, Dec (02)	54.54	54.54	05–34
Sudan	Khartoum	Jun (04)	66.67	66.67	07–50
USA	Chicago	Jul, Sep, Dec (02)	60.00	60.00	00–37

Table 4 Component and forecast percentages for highest occurrence month sets of 4 locations from the RDTWI terrorist attack database

Country	Location	Attack month details			Fatality
		Largest itemset	Const. %	For. %	
Israel	Kfar Azza	Dec (03)	21.43	75.00	00–00
Philippines	Davao	Feb, Mar, Apr, Nov (01)	100.00	66.67	00–24
Spain	Getxo	Jan, Jun, Nov (02)	85.71	66.67	00–00
Turkey	Hakkari	Aug (05)	31.25	28.57	00–06

Table 5 Component and forecast percentages for highest occurrence month sets of 4 locations from the Afghanistan natural disaster database

Province	Location	Disaster month details			Fatality
		Largest itemset	Const. %	For. %	
Badghis	Qala-e-Naw	Apr (07)	31.81	22.22	00–15
Badakhshan	Yaftal Sufla	Apr (04)	36.36	40.00	00–08
Daykundi	Khadir	Feb (04)	44.44	50.00	00–03
Kunar	Dara-e-Pech	Feb (06)	28.57	50.00	00–03

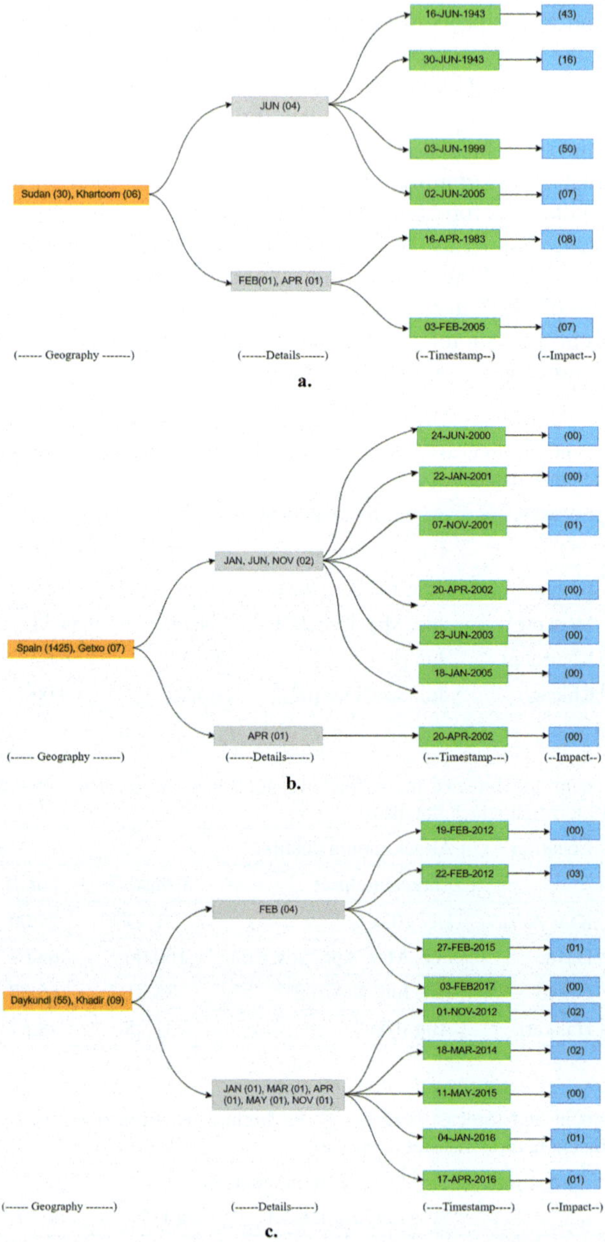

Fig. 6 **a** Event profile of occurrence month sets of 'Sudan' from airplane crash database. **b** Event profile of occurrence month sets of 'Getxo' from RDWTI terrorist attack database. **c** Event profile of occurrence month sets of 'Khadir' from Afghanistan natural disaster database

5 Discussions

We now discuss issues about the proposed approach and their plausible justifications. In all the profiles generated, a clear bias for geographical features has been purposefully introduced to ensure the stability of the event profile. Generally, every event database has a limited number of geographical feature values that remain steady with upcoming event details. For example, the number of dead and injured people may differ in 2 events of 'flood' but the values of 'Daykundi' for district and 'Khadir' for location remain stable.

Section 3.1 conceptualized the proposed event ontology with an example of a terrorist attack to capture the maximum number of possible features under each concept. However, in the examples discussed later, only fewer selected features are presented under each profile. This change in the number of features proves the flexibility of the proposed approach to accommodate features according to the nature and details of their description provided in a user query.

The prominent reason of using minimal computational models for analysis is to showcase the potential of the proposed approach. With more data and added details, computational methods of higher complexity as machine learning and deep learning networks can be utilized for enhanced results.

6 Conclusion and Future Work

In this paper, we presented a novel approach for the discovery of patterns from catastrophic spatiotemporal events. The proposed approach is based on an event ontology and a template of event details termed event profile. The ontology facilitates an inclusive benchmark for redistribution of the event features under 7 concepts of geography, timestamp, actor, activity, component, impact, and details. This redistribution enables the assembling of catastrophic events belonging to divergent types and sources under a single comprehensive structure. The event profiles expedite the exploration of valuable patterns of significant catastrophic events that integrated computational models can further improve. We expect to add text mining-based methods for automated creation and update of features from a provided event or event database. Furthermore, the addition of computational methods of higher complexity as deep learning models can also improve its forecast performance.

Acknowledgements The authors thank the editors and the anonymous reviewers for their helpful comments and suggestions. This work has been supported by the Science and Engineering Research Board (SERB) under Department of Science and Technology, Government of India, under the project 'Forecasting Significant Social Events by Predictive Analytics Over Streaming Open Source Data' (Project File Number: EEQ/2019/000697).

References

1. Subramanian D, Stoll RJ (2006) Events, patterns, and analysis forecasting international conflict in the twenty-first century. In: Programming for peace. Springer, Dordrecht, pp 145–160
2. Yu M, Bambacus M, Cervone G, Clarke K, Duffy D, Huang Q, Li J, Li W, Li Z, Liu Q, Resch B (2020) Spatiotemporal event detection: a review. Int J Digital Earth 13(12):1339–1365
3. Rao KV, Govardhan A, Rao KC (2012) Spatiotemporal data mining: issues, tasks and applications. Int J Comput Sci Eng Surv 3(1):39
4. Hamdi A, Shaban K, Erradi A, Mohamed A, Rumi SK, Salim FD (2022) Spatiotemporal data mining: a survey on challenges and open problems. Artif Intell Rev 55(2):1441–1488
5. Wang X, Brown DE (2012) The spatio-temporal modeling for criminal incidents. Secur Inform 1(1):1–7
6. Cadena J, Korkmaz G, Kuhlman CJ, Marathe A, Ramakrishnan N, Vullikanti A (2015) Forecasting social unrest using activity cascades. PLoS ONE 10(6):e0128879
7. Gruber TR (1993) A translation approach to portable ontology specifications. Knowl Acquis 5(2):199–220
8. Frantz A, Franco M (2005) A semantic web application for the air tasking order. Air Force Research Lab Rome NY Information Directorate
9. Gómez-Pérez A, Corcho O (2002) Ontology languages for the semantic web. IEEE Intell Syst 17(1):54–60
10. Mannes A, Golbeck J (2007) Ontology building: a terrorism specialist's perspective. In: 2007 IEEE aerospace conference. IEEE, pp 1–5
11. Busch M, Wirsing M (2015) An ontology for secure web applications. Int J Softw Inform 9(2):233–258
12. Chang HW, Tai YC, Chen HW, Hsu JY, Kuo CP (2008) iTaxi: context-aware taxi demand hotspots prediction using ontology and data mining approaches. In: Proceedings of TAAI
13. Profiling (2022) Merriam-Webster.com Dictionary, Merriam-Webster. https://www.merriam-webster.com/dictionary/profiling. Retrieved 22 Aug 2022
14. Nadee W (2016) Modelling user profiles for recommender systems. Doctoral dissertation. Queensland University of Technology
15. Adomavicius G, Tuzhilin A (2001) Using data mining methods to build customer profiles. Computer 34(2):74–82
16. Delgado MR, Mata NC, Yepes-Baldó M, Montesinos JVP, Olmos JG (2013) Data mining and mall users profile. Univ Psychol 12(1):195–207
17. Jiyani A et al (2021) NAM: a nearest acquaintance modeling approach for VM allocation using R-Tree. Int J Comput Appl 43(3):218–225
18. Stermsek G, Strembeck M, Neumann G (2007) A user profile derivation approach based on log-file analysis. In: IKE 2007, pp 258–264
19. Raghavan V, Galstyan A, Tartakovsky AG (2013) Hidden Markov models for the activity profile of terrorist groups. In: The annals of applied statistics, pp 2402–2430
20. Yoo JS (2012) Temporal data mining: similarity-profiled association pattern. In: Data mining: foundations and intelligent paradigms. Springer, Berlin, Heidelberg, pp 29–47
21. Bhatia S et al (2022) An efficient modular framework for automatic LIONC classification of MedIMG using unified medical language. In: Frontiers in public health, section digital public health, Manuscript ID: 926229, pp 1–21
22. Afghanistan—Natural Disaster Incidents. https://data.humdata.org/dataset/afghanistan-natural-disaster-incidents-in-2020. Retrieved 22 Aug 2022
23. Airplane Crashes and Fatalities Since 1908. https://data.world/data-society/airplane-crashes. Retrieved 22 Aug 2022
24. Robertson IN, Riggs HR, Yim SC, Young YL (2007) Lessons from Hurricane Katrina storm surge on bridges and buildings. J Waterw Port Coast Ocean Eng 133(6):463–483
25. Global Terrorism Database (GTD) (2018) Codebook: inclusion criteria and variables. University of Maryland, pp 1–53. https://www.start.umd.edu/gtd/downloads/Codebook.pdf. Retrieved 22 Aug 2022

26. Ilardi GJ (2009) The 9/11 attacks—a study of Al Qaeda's use of intelligence and counterintelligence. Stud Conflict Terrorism 32(3):171–187
27. Luna JM, Fournier-Viger P, Ventura S (2019) Frequent itemset mining: a 25 years review. Wiley Interdiscip Rev Data Min Knowl Discov 9(6):e1329
28. Agrawal R, Srikant R (1994) Fast algorithms for mining association rules. In: Proceedings of 20th international conference on very large data bases, VLDB, vol 1215, pp 487–499
29. Ng V, Chan S, Lau D, Ying CM (2007) Incremental mining for temporal association rules for crime pattern discoveries. In: Proceedings of the 18th conference on Australasian database, vol 63, pp 123–132
30. Srivastava SR, Meena YK, Singh G (2020) Itemset mining based episode profiling of terrorist attacks using weighted ontology. In: International conference on advanced machine learning technologies and applications. Springer, Singapore, pp 337–348
31. Murtagh F, Contreras P (2012) Algorithms for hierarchical clustering: an overview. Wiley Interdiscip Rev Data Min Knowl Discov 2(1):86–97
32. RAND Corporation (2014) RAND database of worldwide terrorism incidents

Chapter 6
Meta-algorithm Development to Identify Specific Domain Datasets in Social Science Education and Business Development

Gurpreet Singh, Korakod Tongkachok, K. Kiran Kumar, and Amrita Chaurasia

1 Introduction

Various machine learning algorithms in recent years. They must use the everyday vast volumes of digital data. Developing high-quality predictive and descriptive models is challenging, iterative, and time-consuming due to these algorithms' complexity and nature. Automate model creation technique choosing. We're defining a framework for social science intelligent systems development.

Machine learning technologies are used to extract meaning from large and small data sets by social scientists. We demonstrate how machine learning in the social sciences demands us to rethink both applications and recommended practices. Machine learning is used to identify new ideas, evaluate their prevalence, assess causal effects, and make predictions when applied to social scientific data. The plethora of data and tools promotes a shift from deductive to sequential, interactive, and inductive social science reasoning. We illustrate how an approach to social science machine learning problems may advance to the fallowing topics.

G. Singh (✉)
University Institute of Computing, Chandigarh University, Mohali, India
e-mail: gps.srm@gmail.com

K. Tongkachok
Department of Faculty of Law, Thaksin University, Mueang Songkhla District, Thailand
e-mail: tokorakod@tsu.ac.th

K. Kiran Kumar
Chalapathi Institute of Engineering and Technology, Lam, Guntur, Andhra Pradesh, India

A. Chaurasia
School of Commerce Finance and Accounting, Christ University NCR Campus, Ghaziabad, India
e-mail: amrita.chaurasia@christuniversity.in

V. Devedzic et al. (eds.), *Proceedings of the International Conference on Intelligent Computing, Communication and Information Security*, Algorithms for Intelligent Systems, https://doi.org/10.1007/978-981-99-1373-2_6

You may export and examine these data sets in analytic tools using the information in this section of the tutorial. Microdata may be selected based on narrative descriptions and data documentation in general. Look at the aggregate data tab for data in tables that may be exported and aggregated for analysis. The following headings organize the information provided on this page: Data on governance, international election data portals, US social and opinion data collections, and the VT Library's data services are only some of the resources that may find in this overview of data-finding challenges. Various data sources, as well as information on how to cite the origins of the data you utilize.

Obtaining and working with data may be a challenge. The VT libraries have a group of informatics experts available to assist you with your research technique, interpretation, visualization, and data management/curation. Members of the library's data service division are responsible for the upkeep of a few of the tabs included in this guide.

When selecting a learning algorithm, a meta-learning technique is presented. It uses a simple process to recommend the best algorithm for a specific learning task. Algorithms' performance on past datasets forecasts how well they will perform on a new dataset. Must address several challenges in creating meta-learning systems for algorithm recommendation, including those at the meta and base levels (Fig. 1). A meta-target feature (or meta-target, for short) must first be selected, which is a suggestion to be pro-meta-target sent to users. I decided to use rankings of basic algorithms to make recommendations in the system described in the previous chapter. When it comes to meta-algorithms, the kind of meta-target dictates which meta-algorithms may be employed, which defines the type of meta-knowledge that may be acquired.

It is vital to create a sufficient meta-database before engaging in meta-learning. A collection of meta-examples, such as datasets or a more comprehensive meta-database, is required. Meta-examples are difficulties in learning—repositories, such as the UCI repository—may be a source of (classification) learning issues. Dataset attributes and algorithm performance in a meta-knowledge are the procedure's purpose. Because of this, it is crucial to identify and build meta-features for the most critical attributes for characterizing these data collections. For example, the number of classes in a classification dataset is one meta-feature.

Additionally, the meta-database must hold information about the performance of the base algorithms on the datasets chosen for analysis. An algorithmic recommendation system's initial step is to select a foundation of algorithms on which it will build its recommendations. Must establish metrics used to assess the algorithms' performance. We may use measures like classification accuracy and the area under the ROC curve for diverse purposes.

Meta-learning relies on high-quality data just like any other machine learning activity. Meta-data issues, such as missing values or noise might impair the quality of the suggestions generated by the meta-learning system, as they do in any dataset.

Social science empirical work has historically been scarce. Finding data, conducting surveys, and storing records proved difficult. Limited and costly computer

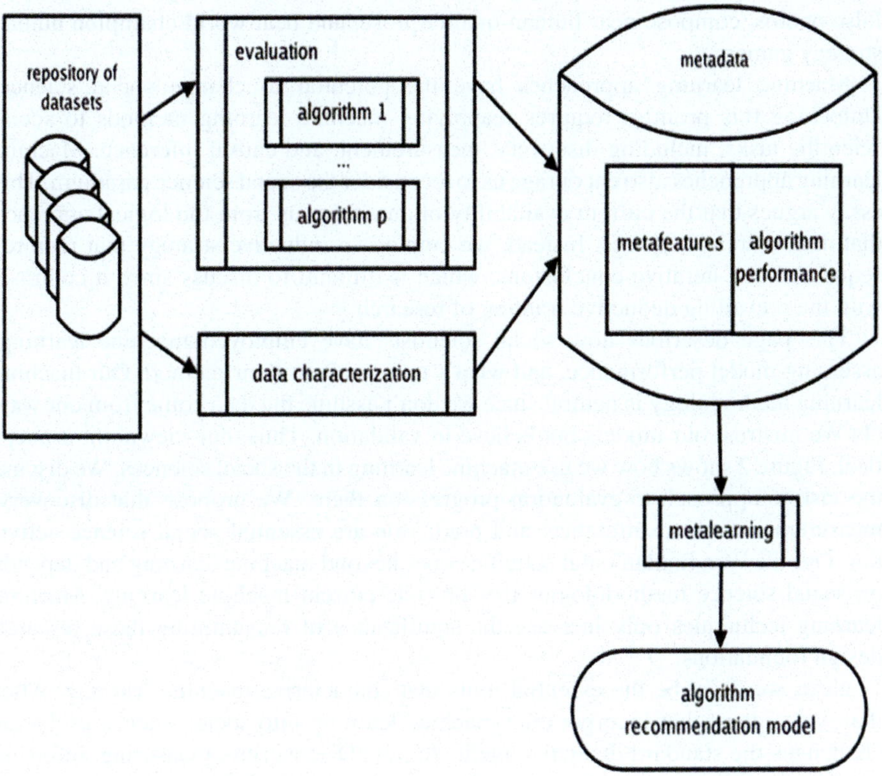

Fig. 1 Algorithm selection process

time exacerbated the bottleneck. Due to data and computer capacity shortages, social scientists created and depended on statistical approaches.

Abundance characterizes social sciences. Rapid data growth has altered evidence. Election academics used to depend on national election surveys; today, they utilize millions of voter data. Declassified state department cables may supplement archival research for international relations researchers. Scale isn't the only distinction. Tracking the removal of social media postings in real time reveals how authoritarian countries manage public information. Personal computers can analyze millions of rows of data, and cloud computing services are robust.

Social scientists use machine learning to maximize this new abundance. Machine learning uses algorithms and statistics to predict and reduce dimensions. Machine learning experts favor quantitative benchmarks, including overtly predictive tasks like categorizing emails as spam or forecasting who would click on an ad and activities susceptible to quantitative feedback, such as compressing picture data or operating a robot. The community has made rapid progress by optimizing such tasks. The outcomes include more accurate spam filters and algorithms that can create realistic false photos, compose near-human-quality prose, and beat world champion human strategy gamers.

Machine learning approaches have the potential to change social science. Unlocking this promise requires reapplying machine learning methods to social scientific tasks, including discovery, measurement, and causal inference. Machine learning approaches also encourage us to reconsider the social science paradigm. This essay argues that the current availability of data frees us from the logical approach that data scarcity requires. Instead, we employ an inductive strategy that requires sequential and iterative conclusions, which is difficult to discuss since it contrasts with the prevailing deductive framing of research.

This page describes how social scientists have employed machine learning, assessing model performance, and what's unique about their method. Our machine learning methodology is neutral since we don't assume the data come from our way. (1) We mistrust our models but believe in validation. Thus, our viewpoint is skeptical. Figure 2 shows how we use machine learning in the social sciences. We discuss the task-to-approach-to-evaluation progression there. We propose that discovery, measurement, causal inference, and prediction are essential social science activities. Our view on fundamental activities goes beyond machine learning and depends on social science methodologies that precede current machine learning. Machine learning techniques only increase the significance of reexamining these research design foundations.

Next, we describe the essential tools that characterize machine learning. After that, we suggest that incorporating machine learning into social science problems challenges the standard deductive methodology. Discovering, measuring, inferring causality, and forecasting are our primary jobs. We finish with a look at social science's machine learning horizons.

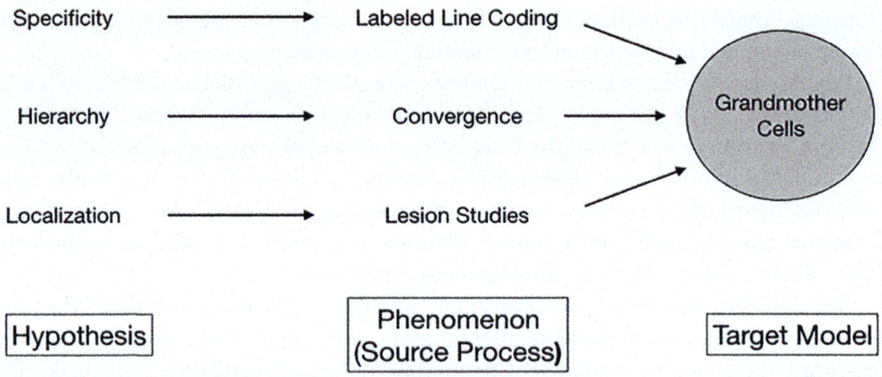

Fig. 2 Approach—machine learning with social science

2 Data Collection Process with Methodology

Why study Social science? Digital systems' interactions with humans provide socially meaningful data. Social science data. In social science, algorithms and tools for data collection and analysis make it feasible to capture previously missed or non-existent occurrences. Social data are dynamic, computerized, and include social interaction. Data gathering, processing, and distribution via these platforms are social actions, whether a company plan or a public database on entrepreneurial activity. Must be considered dataset features before doing an analysis. Two social science domains—Education and business—are merged in this framework to make a whole. Throughout elementary, middle, and high school, students' and teachers behavior data is collected and used to predict academic achievement. Data transformation into structured datasets that machine learning algorithms may use is a critical part of educational predictive modeling. There are a lot of issues in the data analysis process, and one of them is the absence of model assessment and a holistic approach to picking the best predictive models. Predictive models in education should take these correlated, spatial, and fragmented qualities into account while building them since doing so improves the model's readability and precision.

As fuel for our investigation, education is critical to the success of the corporate world. Students join the workforce after graduation. There are a variety of significant data sources in the field of business. The GEM database is a valuable source of entrepreneurial knowledge. An extensive database like GEM may hinder data analysis. Entrepreneurial analytics introduces methodological problems. Bergmann et al. assert there is no systematic study of entrepreneurial dataset aspects and methodological skills. Entrepreneurial activity researchers should evaluate "rare occurrences" (one dependent variable value is more often than the other—class imbalance). These questions inspired the proposal's aims.

3 Intelligent-Frame Work

Intelligent data analysis is a multi-phase process that aims to glean information and insights from "raw data" that might be of value. So far, with this method, various standards have been established. For example, the CRISP-DM standard specifies the following stages: domain understanding, data comprehension, data preparation, modeling, model assessment, and utilization. The modeling phase is critical to the standard since it identifies the patterns used in the evaluation. The protocol described here may be used to solve challenges in various fields. Depending on the domain, the data might have multiple traits and attributes unique to that domain. This framework aims to discover specific properties of datasets from the education and commercial sectors to construct an intelligent system for automatic algorithm selection based on these qualities—the structure in Fig. 3.

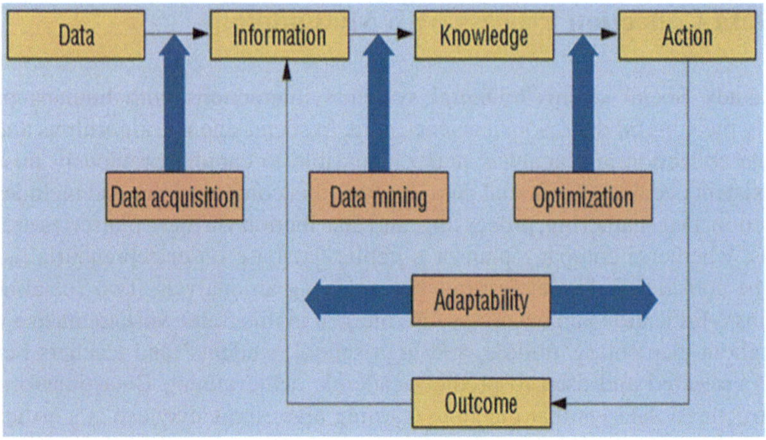

Fig. 3 Intelligent framework

During the first step, known as understanding the domain, the objectives and needs of the field to data mining challenges, a strategy for attaining these goals.

First-phase domain experts will explain each domain's demands. The second phase begins with gathering the initial data and includes examining its quality and dispersion. At this stage, education and business information will be acquired. Educational data sources include e-learning systems, student polls, and public repositories. Business will use Global Entrepreneurship Monitor, World Bank Group Entrepreneurship Survey, and life quality data sources. Second-phase data description focuses on examining datasets meta-characteristics. A literature review identifies pertinent meta-features. Each meta-feature will be categorized. Each domain requires 1024 datasets (a total of 2048) to capture all feature combinations by picking ten sections with at least two categories. If a given combination of features isn't in the original, a new dataset is formed. Each dataset contains machine learning models. Supervised machine learning algorithms include: probability, error, data, and similarities-based algorithms. Minimum of four machine learning algorithms, or 8192 analyses. During the evaluation phase, the model's correctness and reliability (confusion matrix, precision, responsiveness, RSquare) are tested. The meta-dataset includes one meta-example per feature, a meta-feature, and a meta-target characteristic. Meta-models will be built using correlations, rankings, multi-criteria modeling, and supervised learning. Once assessed, the intelligent system's final version will incorporate the best meta-model. Considering the domain and dataset attributes, a smart system for automatic machine learning algorithm proposal will be built.

4 Result

All models developed will be interpreted following the knowledge of the area and the criteria for success. This framework contributes in two ways:

(i) methodologically, via the construction of meta-models and intelligent systems, and
(ii) domain-based, by identifying new information, knowledge, and patterns in each area covered.

The implementation of this framework will have the following scientific contributions:

(1) systematic knowledge in the area of dataset meta-features important to social sciences,
(2) repository of datasets in education and business sectors with identified particular meta-features,

developing descriptive models for education and business, as well as predictive models reliant on the particular meta-features of education and business data.

We constructed and assessed meta-models based on the dataset's meta-features for selecting machine learning algorithms.

Created an intelligent system for automatically selecting a machine learning algorithm based on the job and meta-features, and generated recommendations for educational and corporate decision-makers. (6) Developed.

On the other hand, this meta-learning framework clarifies the process for creating social science models that can be relied upon. We're focusing on designing models that can solve real-world educational and corporate challenges. Many unanswered concerns remain about how to use machine learning in the social sciences, as shown by the literature study, which shows that intelligent data analysis is currently underapplied. We hope that this framework will help us make an impact in this area. In addition to its scientific benefits, framework application also yields a slew of societal benefits. These issues will be addressed in intelligent system development, and we think society will greatly benefit from them.

5 Conclusion

The societal ramifications of the issues being addressed in education are enormous. Both students and schools are interested in predicting academic achievement, and models of student self-assessment and suggestion creation might help enhance online education, which is exploding around the globe. Because of people's interest in quality-of-life concerns, business problems solved by the intelligent system have a significant societal impact. Aside from making a significant contribution to the field of data science, our findings help us address important social issues and connect us with various stakeholders to tackle socio-economic objectives and transfer the knowledge we have gained through developing software applications and recommendations. Several people might benefit from such a system, including:

1. For starters, we have data scientists. Data scientists utilize an intelligent recommender system to choose which approaches to apply in data analysis based on the kind and features of their data collection.
2. Experienced data scientists may benefit from the system, as can domain experts with limited background in data analytics.
3. Predictive student success models are crucial at all three levels of higher education administration because they explain academic progress and prevent high student drop-outs.
4. The predictive models of entrepreneurial activity highlight the current trends in entrepreneurship and provide fresh insight into the entrepreneurial objectives of the managers involved. By following commercial models, governments and policymakers may make better decisions.
5. This framework will be used in future research to construct intelligent systems.

References

1. Sarker IH (2021) Machine learning: algorithms, real-world applications and research directions. SN Comput Sci 2:160. https://doi.org/10.1007/s42979-021-00592-x
2. National Research Council, Division of Behavioral and Social Sciences and Education, Commission on Behavioral and Social Sciences and Education, Committee on Basic Research in the Behavioral and Social Sciences, Gerstein DR, Luce RD, Smelser NJ et al (eds) (1988) Methods of data collection, representation, and analysis. In: The behavioral and social sciences: achievements and opportunities. National Academies Press, Washington, DC, USA (Chapter 5). Available from: https://www.ncbi.nlm.nih.gov/books/NBK546485/
3. Grimmer J, Roberts M, Stewart B (2021) Machine learning for social science: an agnostic approach. Annu Rev Polit Sci 24. https://doi.org/10.1146/annurev-polisci-053119-015921
4. Singh H, Rai V et al (2022) An enhanced whale optimization algorithm for clustering. Multimedia Tools Appl 1–20. https://doi.org/10.1007/s11042-022-13453-3
5. Malpani et al (2016) A novel framework for extracting GeoSpatial information using SPARQL query and multiple header extraction sources. In: Afzalpulkar N, Srivastava V, Singh G, Bhatnagar D (eds) Proceedings of the international conference on recent cognizance in wireless communication & image processing. Springer
6. Chen Y, Wu X, Hu A et al (2021) Social prediction: a new research paradigm based on machine learning. J Chin Sociol 8:15. https://doi.org/10.1186/s40711-021-00152-z
7. Bavel JJV, Baicker K, Boggio PS et al (2020) Using social and behavioural science to support COVID-19 pandemic response. Nat Hum Behav 4:460–471. https://doi.org/10.1038/s41562-020-0884-z
8. Paullada A, Raji ID, Bender EM, Denton E, Hanna A (2021) Data and its (dis)contents: a survey of dataset development and use in machine learning research. Patterns 2(11):100336. ISSN 2666-3899. https://doi.org/10.1016/j.patter.2021.100336. https://www.sciencedirect.com/science/article/pii/S2666389921001847
9. Best K, Gilligan J, Baroud H et al (2022) Applying machine learning to social datasets: a study of migration in southwestern Bangladesh using random forests. Reg Environ Change 22:52. https://doi.org/10.1007/s10113-022-01915-1

10. Gorard S (2012) The increasing availability of official datasets: methods, limitations and opportunities for studies of education. Br J Educ Stud 60(1):77–92. https://doi.org/10.1080/00071005.2011.650946
11. Chen N-C, Drouhard M, Kocielnik R, Suh J, Aragon C (2018) Using machine learning to support qualitative coding in social science: shifting the focus to ambiguity. ACM Trans Interact Intell Syst 8:1–20. https://doi.org/10.1145/3185515
12. Heiberger RH (2022) Applying machine learning in sociology: how to predict gender and reveal research preferences. Köln Z Soziol 74:383–406. https://doi.org/10.1007/s11577-022-00839-2
13. Hymavathi J, Kumar TR, Kavitha S, Deepa D, Lalar S, Karunakaran P (2022) Machine learning: supervised algorithms to determine the defect in high-precision foundry operation. J Nanomater
14. Bhujade S, Kamaleshwar T, Jaiswal S, Babu DV (2022) Deep learning application of image recognition based on self-driving vehicle. In: International conference on emerging technologies in computer engineering. Springer, Cham, pp 336–344
15. Singh C, Rao MS, Mahaboobjohn YM, Kotaiah B, Kumar TR (2022) Applied machine tool data condition to predictive smart maintenance by using artificial intelligence. In: International conference on emerging technologies in computer engineering. Springer, Cham, pp 584–596
16. Rahman RA, Masrom S, Zakaria NB, Nurdin E, Abd Rahman AS (2021) Prediction of earnings manipulation on Malaysian listed firms: a comparison between linear and tree-based machine learning. Int J Emerg Technol Adv Eng 11(8):111–120. https://doi.org/10.46338/IJETAE0821_13

Chapter 7
Methods for Medical Image Registration: A Review

Payal Maken and Abhishek Gupta

1 Introduction

Medical imaging has some vital applications in clinical practices. There are two categories of the medical imaging modalities such as, anatomical which can depict the morphology of the anatomical structures ("for example, X-rays, cone beam computed tomography (CBCT), computed tomography (CT) [1], ultrasound, etc."), and functional modalities which represents the information regarding the process of the underlying anatomy (example, "Single-photon emission computed tomography (SPECT), positron emission tomography (PET"), electroencephalography (EEG), etc.). Volumetric image segmentation is a well-known application of image registration [2] among many [3].

Various representations of the same scene containing useful data acquired during the clinical track of events require proper integration. The information obtained from the two images might be in a complementary form. The first step applied before image analysis is image integration usually referred to as image registration. It is the process of combining two or more images acquired using various sensors, at various times, or with different modalities. The two images aligned or registered are called sensed and present images. Image registration is a significant step for image analysis. The outcome of image analysis is the combination of image registration, image fusion, image restoration, and detection. The terms registration, fusion, integration, or correlation are polysemy in the literature. Image registration could be monomodal

P. Maken (✉)
School of Computer Science and Engineering, Shri Mata Vaishno Devi University, Kakryal, Katra, Jammu & Kashmir 182320, India
e-mail: payalmaken33@gmail.com

A. Gupta
Biomedical Application Division, CSIR-Central Scientific Instruments Organisation, Chandigarh 160030, India
e-mail: abhishekgupta10@yahoo.co.in

V. Devedzic et al. (eds.), *Proceedings of the International Conference on Intelligent Computing, Communication and Information Security*, Algorithms for Intelligent Systems, https://doi.org/10.1007/978-981-99-1373-2_7

[4] or multi-modal. An example of multi-modality medical image registration can be an MRI brain image with PET brain images (axial and sagittal view) [5] where both PET and MRI are used. An example of monomodal medical image registration could be treatment verification, such as a comparison of pre- and post-treatment images, or monitoring of the growth of MR scans time series on tumors and X-rays of specific bones [6].

Image-guided surgery (IGS) primarily depends upon the registration of the preoperative, post-operation, and 2D to 3D registration accuracy. Image registration finds the mapping relationship transformation [7] in the spatial domain between two or more images of the same or different scenes.

1.1 Applications of Medical Image Registration

Registration of medical images has numerous applications ranging from clinical diagnosis to image. Some of the applications are mentioned below [8]:

- Quantification of the changes between the measurements at different time points. The reason for differences can be bulk patient motion, organ motion, growth, disease evolution, etc.
- Segmentation: Exceeding the quantity of data produced for expert visual analysis by imaging modalities. It increases the need for automatic image analysis. One of the common methods of segmentation is the segmentation based on an atlas or reference image, where an "anatomical atlas is registered to an individual's anatomical image".
- Statistical Shape model: In building a statistical shape model, a set of landmarks that are defined unambiguously in each shape are required. The manual identification of landmarks is difficult and time consuming. Thus, this problem of finding correspondences for landmark identification can be solved via registration. One of the methods that register an atlas to another image shape and reproduces the atlas' landmarks to all other shapes has been proposed by Frangi et al. [9]. Surface-based or image-based registration can be utilized for this purpose.

1.2 Uses of Medical Image Registration

Registration of medical images has a very significant function in many clinical applications such as Computer-assisted diagnosis (CAD), Computer-assisted surgery (CAS), and Computer aided therapy (CAT). Some of the image registration uses are [8]:

- There are several medical image modalities, and each one offers a different set of details about the area of interest. The matching anatomical or functional regions between two images or more of the same or a different subject gathered using the same or different imaging modalities at the same or different times are found through image registration [8].

- With the advancements in digitization, medical images could transfer into digital information to process with digital computers. This integration of imaging information media requires some aid from medical image registration techniques. Registration procedures are required to maintain consistency between the points provided in all images. It involves the process of correlating images from various temporal and spatial domains. It especially matches anatomical points to allow a physician to have improved and detailed information [10].
- Medical image registration is the basis for methods like image-guided radiation therapy, surgery, and minimally invasive treatments [10].
- Along with the aforementioned purpose, image registration can also be applied to retinal imaging to provide a thorough description of the retinal morphology [10].

This paper aims to focus on a survey of some of the methods for image registration in medical. This review paper is organized as follows. Section 2 consists of a review of the few techniques used by different authors for medical image registration with their limitations. Section 3 describes the framework of registration. Section 4 explains the analysis of the techniques. Sections 5 and 6 contain a discussion and conclusion. This review paper will provide a short survey of the various registration methods recently in trend and being used by the authors in their work.

2 Literature Review

Various authors have worked and introduced various methods for medical image registration, some of which are shown in Table 1 with their results. This section highlights a few methods used by different authors for medical image registration, along with their objectives, modality used, results, and remarks.

Some of the methods used by authors for medical image registration are mentioned below.

2.1 Normalized Cross-Correlation Based on Sobel Operator

Liu et al. [11] used the normalized cross-correlation technique by combining gradient vector angle and normalized cross-correlation [12]. The angle between the reference image and the DRR image was compared to accomplish the registration. The angle is shown in Eq. (1) [11]. The relationship between the 2D reference image and the DRR image increases with a reducing gradient vector angle. The precise registration between a 2D reference image and a 3D image also depends upon the angle between the gradient vectors.

Table 1 Studies using different registration methods in the literature review

Reference and year	Objective / Technique used	Modality used	Results	Limitations
Liu et al., 2022 [11]	2D/3D medical image registration method is developed and employed based on the normalized correlation method	Projected DRR and CT brain images	**Single-resolution experimental results**: Mean absolute error (MAE) *(should be less than 5)* = **Rotation**: NCCS = 0.49550, NCCL = 0.31890 and NCC = 1.20830 **Translation**: NCCS = 0.9883 mm, NCCL = 0.7750 mm and NCC = 1.4079 mm	• Time consuming • Some tissues of the human body, as well as the spaces between bones, will deform to some amount in therapeutic applications. As a result, rigid body transformation-based 2D/3D registration cannot fully meet clinical needs
	• "Normalized cross-correlation based on the Sobel operator (NCCS)" • "Normalized cross-correlation based on the log operator (NCCL)"		Mean target registration error (MTRE) = NCCS = 1.94117 mm, NCCL = 1.4759 mm and NCC = 3.19295 mm	
Qian et al., 2021 [13]	Developed a medical image registration approach based on the PI-SURF algorithm	CT/MR medical images	**PI-SURF**: **Mutual information (MI) =** 2.2577, normalized correlation coefficient **(NCC) =** 0.9219, mean square difference **(MSD) =** 0.0082 and normalized mutual information **(NMI) =** 1.2161	• Loss in detecting some key feature points • Feature points of the image are mismatched after mapping and extracting features
	SURF algorithm with coarse and fine registration			

(continued)

Table 1 (continued)

Reference and year	Objective Technique used	Modality used	Results	Limitations
Islam et al., 2021 [14]	Developed a registration framework that accurately and quickly registers multi-modal images	CT/MR head medical images	**DSC (Dice similarity coefficient)** = 0.9885 **Jaccard similarity coefficient (JSC) =** 0.9765 **Registration precision (Rp) =** 0.9830 **Registration sensitivity (Rs) =** 0.9870 **Contour matching score (CMS)** = 0.9875 **Structural similarity index measure (SSIM) =** 0.9685 **Execution time, E_t =** 02.80	Need larger dataset
	Convolutional neural network			
Blendowski et al., 2020 [15]	Employed an independent multi-modal medical image registration technique mapping to a shape using an encoder-decoder network	CT/MR medical images	**Dice** = 0.653 with $n = 15$	Need larger dataset
	Convolutional autoencoder (CAE) architecture			
Bashiri et al., 2019 [16]	Proposed a multi-modal to monomodal transformation approach in both full (full) and partial (partial) overlap scenarios	CT/MR human brain images	**Mean absolute error (MAE)**: *CT-T1 rectified*: 0.7424 ± 0.22 *CT-T2 rectified*: 0.9294 ± 0.23	• The current approach is not recommended for 3D volume registration as it is computationally inefficient • For learning the structure of the input images, the suggested pipeline necessitates additional processing time
	Intensity-based and Fourier–Mellin-based image registration			

(continued)

Table 1 (continued)

Reference and year	Objective / Technique used	Modality used	Results	Limitations
Ferrante et al., 2013 [17]	Devised a new mapping method between a 3D volume and 2D images that explore a linear plane transformation and an in-plane dense deformation at the same time "Non-rigid registration using MRFs (Markov Random Fields)"	CT/MR medical images	**Average DICE coefficient** = 0.93 **Average DICE increment after registration** = 0.05 **Average contour mean distance (CMD) decrement** is 0.4 mm	• One is linked to the label space dimensionality (which can be accommodated due to the small 2D grid size) • The other to the approximate coplanarity constraint imposed by the suggested over-parameterization

$$\tau(x, y) = \arccos \frac{\nabla I_R(x, y)\,\nabla I_D(x, y)}{(|\nabla I_R(x, y)|\,|\nabla I_D(x, y)|)} \tag{1}$$

where $\tau(x, y)$ is the angle between reference image I_R and DRR image I_D at pixel (*x*, *y*) and $\nabla I_R(x, y)$, $\nabla I_D(x, y)$ are the gradient vectors of I_R and I_D at pixel coordinates (*x*, *y*).

2.2 *Scale Invariant Feature Transform (SIFT)*

SIFT was first established in 2004 [18]. It combines the advantages of several approaches to discover significant point qualities that are invariant to modifications. SIFT uses the image's greyscale values for feature identification. Another image format such as RGB should be converted to grayscale images before further processing. The core five processes for identifying SIFT features in the images to be registered and establishing related pixels are as follows:

1. Detection of scale-space extrema, 2. Identifying key points, 3. The orientation task, 4. Formation of descriptors, and 5. Matching descriptors.

Since its beginning, SIFT algorithm has been extensively explored and used for a variety of applications, with or without changes to its methodology.

2.3 *Speeded Up Robust Features (SURF)*

Qian et al. [13] used the SURF algorithm (PI-SURF) on progressive images to create a coarse-to-fine medical image registration approach. The SURF algorithm [19] extracts features from numerous progressive images. The SIFT technique [18] was used to find features that are scaled and noise insensitive. It was built on similar features, but with a lower level of complexity. It was used to perform a repeatable orientation utilizing information from a circular zone surrounding the relevant spot. A square region that was aligned to a specified orientation was used to extract the SURF descriptor. It is three times faster than SIFT technique.

2.4 *Non-rigid Registration Using Markov Random Fields (MRF)*

Ferrante et al. [17] proposed a non-rigid registration technique using Markov Random Fields (MRF). "A 2D-2D in-plane local deformation field $\hat{T}_D$ and the plane J (i.e., a bi-dimensional slice from the volume J)" were searched. It was used to minimize the objective function shown in Eq. (2) [17]. Given a 2D input image I and a 3D target volume J in the most general case:

$$\hat{T}_D, \hat{\pi} = \underset{T_{D,\pi}}{\arg\min}\, M(IoT_D(x), \pi[J](x)) + R(T_{D,\pi}) \tag{2}$$

where M is the data term and R is the regularization term

3 Framework of Registration

Anatomical or functional corresponding points are required in the two images for image registration [4]. It can be performed on multiple images received from different imaging modalities of the same subject called multi-modal image registration or on serial image registration that is on the same subject at varying times. These are examples of intra-subject registration where the images are taken from the same subject. When the images acquired are from diverse subjects for registration it is known as inter-subject registration. Due to the variety of images from different modalities and numerous types of degradation, it is thus possible to design a common method of registration that can apply to all. The framework for registration is shown in Fig. 1 [20]. The general steps of image registration are follows.

3.1 Feature Detection

Detecting notable and distinct features is known as feature detection (edges, contours, corners, intersections, etc.). It can be done manually or automatically. These detected features or objects are represented by control points (CPs) (such as, line endpoints, and center of gravity) [20].

3.2 Feature Matching

It matches the features detected to establish a correspondence between the original image and the referenced image (template). Various similarity metrics and feature

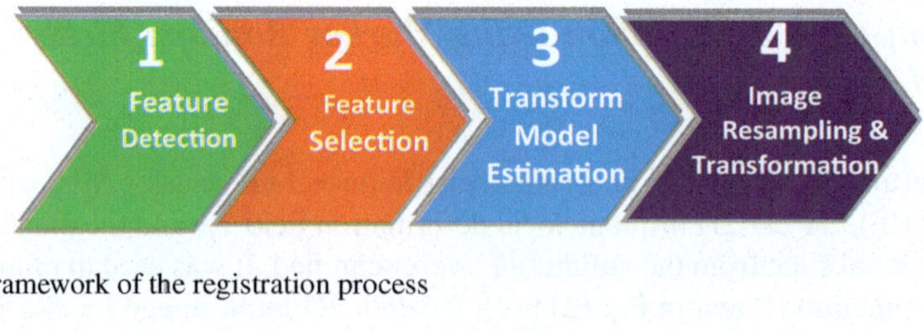

Fig. 1 Framework of the registration process

descriptors along with the spatial relationship between the features are used for finding the correspondence [21].

3.3 Transform Model Estimation

The parameters and type of mapping functions are used to align the estimated and reference images. Feature correspondences are used for the estimation of mapping function parameters [20].

3.4 Image Resampling and Transformation

The mapping functions computed in feature matching are used to transform the referenced image. Transformation applied can be rigid or non-rigid [21]. The transformation can be applied in a forward or backward method. In the forward method, each pixel of a sensed image is transformed directly using estimated mapping functions. Its implementation is complicated because it can lead to holes or can overlap the image (which is due to "discretization and rounding"). In a backward method, the estimated mapping function inverse and coordinates of the image are used to determine image data.

Each registration process has to deal with some issues. First, to recognize which features are suitable for the task. The features should be readily recognizable elements and should be able to interpret physically. These are often scattered throughout the images. The detectable features should have enough common elements to be used in mapping even when the images are not of the same scene or have occlusions or other unanticipated changes. The detected features should be accurate at their location. It should not be affected by image degradation. The detected feature detection algorithm should be able to perceive the same particular features in all the projections.

Image degradations or improper feature recognition can cause problems in the feature matching process. Different imaging conditions and spectral sensitivity of the sensors can cause physically similar characteristics to differ. When choosing a feature description and a similarity measure, these factors must be taken into account. In the presence of the expected degradations, the feature descriptors should be robust. They must be discriminable enough to discriminate between various features while also being stable enough to be unaffected by tiny feature changes and noise.

Prior knowledge should be used to determine the kind of mapping function. The prior knowledge will be an acquisition procedure and image degradations. The model developed should be flexible to handle and manage any degradations that may occur, if no prerequisite data is available.

Finally, the optimal type of resampling technique is determined by the trade-off between the required interpolation accuracy and computing complexity. In most circumstances, nearest-neighbor or bilinear interpolation will suffice; nevertheless, some applications will necessitate more exact approaches.

4 Analysis

This section provides the analysis of the papers added in the literature review section. The analysis is based on the techniques used in its evaluation methods. Figure 2 shows the analysis for the mentioned. The metrics used by the authors in research papers for result analysis are Dice similarity coefficient (DSC), Mean absolute error (MAE), Mean square difference (MSD), Normal correlation coefficient (NCC), Mutual information (MI), Registration sensitivity (Rs) and Registration precision (Rp), etc. These all can be used as performance evaluation metrics of registration frameworks. DSC is a spatial similarity overlap and reproducibility validation matrix. Its value ranges from 0 to 1, 0 being the lowest (no overlap) and 1 refers the highest overlap (fully overlapped). MAE is the absolute difference between the distances of the pixel in one image to the pixel in another image. NCC is the evaluation metric used in image registration, its formula is given in Eq. (3) [13].

$$\text{NCC}\,(R, F) = \frac{\sum_{x=1}^{m}\sum_{y=1}^{n}\left(R(x, y) - \overline{R}\right)\left(F(x, y) - \overline{F}\right)}{\sqrt{\sum_{x=1}^{m}\sum_{y=1}^{n}\left(R(x, y) - \overline{R}\right)^2}\sqrt{\sum_{x=1}^{m}\sum_{y=1}^{n}\left(F(x, y) - \overline{F}\right)^2}} \tag{3}$$

$R(x, y)$ and $F(x, y)$ are pixel (x, y) intensity in the referenced image R and f is the floating image of size $m * n$.

Figure 2a shows that the convolutional neural network (CNN) in Islam et al. [14] shows the highest Dice value which is 98.85% as compared to convolutional autoencoder architecture (CAE) and Markov random fields.

Among the research which used MAE as the evaluation metric in Fig. 2b, the intensity-based method used by Bashiri et al. [16] overcomes the other models used that are "Normalized cross-correlation based on the Log operator (NCCL) and Normalized cross-correlation based on the Sobel operator (NCCS)". Figure 2c shows that with normalized cross-correlation as an evaluation metric, the template and image are highly correlated using the Log operator in Liu et al. [11].

5 Discussion

Image registration is a significant task of integrating and interpreting data acquired from various sources. It plays a key role in the stages involving, image fusion, change detection, super-resolution imaging, etc. This survey examines registration methods

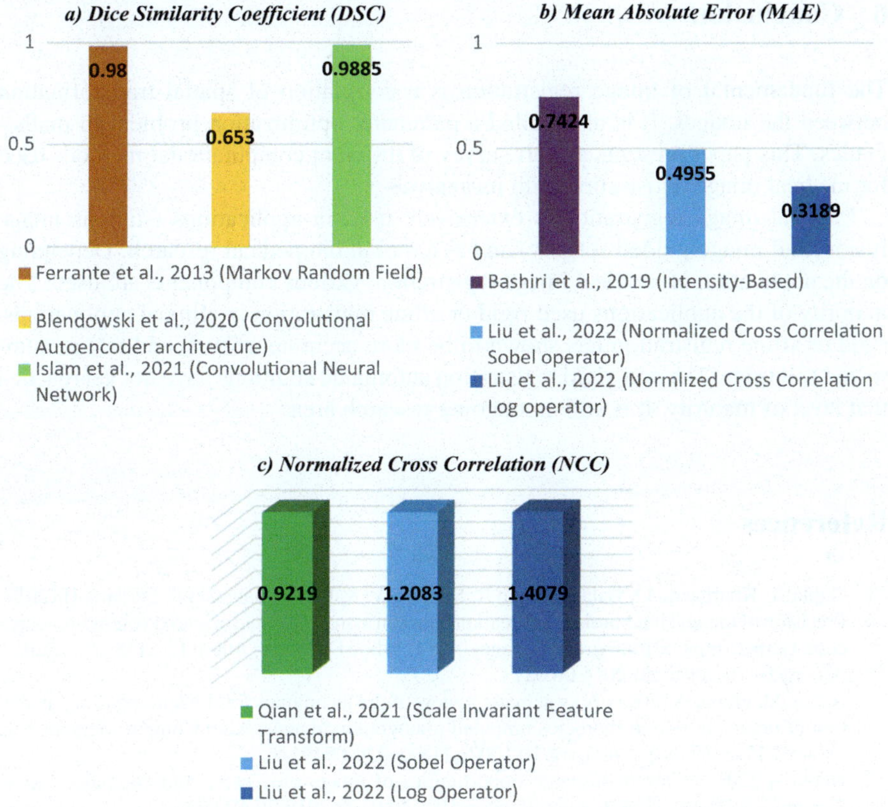

Fig. 2 Results analysis of different registration methods as available in the literature

and provides a state-of-art critical analysis of the methods mentioned in the literature. The paper also includes a general framework of image registration namely, feature extraction, selection, model estimation, and transformation. Among these, feature selection and extraction are the main steps. Since the feature chosen for registration needs to be a good representative of an area of interest to be registered on both images. Although much progress has been made, still automatic image registration remains a challenge. Once, registration has been obtained, a question arises, *how accurate is the computed registration?* The answer is non-trivial because a gold standard is lacking a clinical practice. So, a registration's validation entails more than just checking its accuracy. "Precision, accuracy, Robustness, reliability, clinical use, and many more" are some of the performance evaluation methods in medical image registration techniques. There could be other criteria for the evaluation of registration methods depending on the challenge. Future work in this field could be to pay more attention to different areas of registration methods and develop methods that are less complex, cost-effective, and time-efficient.

6 Conclusion

The fundamental of image registration is a correlation of spatial transformation between the images. It is also called a parameter optimization problem in mathematics. This paper presents a short survey of different computational methods used for medical image registration with its analysis.

Medical image registration is extensively used in applications (such as image fusion, and image-guided surgery) and as a tool in biomedical research. Depending on the application of medical image registration, various components are used. The majority of the applications used rigid or affine registration in clinical applications, rigid or affine registration has shown to be more accurate, robust, and fast in automated solutions. The non-rigid registration automated solutions have not yet reached that level of maturity. It is still an ongoing research area.

References

1. Gupta A, Kharbanda O, Balachandran R, Sardana V, Kalra S, Chaurasia S, Sardana H (2017) Precision of manual landmark identification between as-received and oriented volume-rendered cone-beam computed tomography images. Am J Orthod Dentofac Orthop 151:118–131. https://doi.org/10.1016/j.ajodo.2016.06.027
2. Ashok M, Gupta A (2021) A systematic review of the techniques for the automatic segmentation of organs-at-risk in thoracic computed tomography images. Arch Comput Methods Eng 28(4):3245–3267. https://doi.org/10.1007/s11831-020-09497-z
3. Gupta A (2019) Current research opportunities of image processing and computer vision. Comput Sci 20(4):387–410. https://doi.org/10.7494/csci.2019.20.4.3163
4. Gupta A (2022) RegCal: registration-based calibration method to perform linear measurements on PA (posteroanterior) cephalogram—a pilot study. Multimed Tools Appl. https://doi.org/10.1007/s11042-021-11609-1
5. Nag SJA (2017) Image registration techniques: a survey. abs/1712.07540
6. Maintz JBA, Viergever MA (1998) A survey of medical image registration. Med Image Anal 2(1):1–36. https://doi.org/10.1016/S1361-8415(01)80026-8
7. Maken P, Gupta A (2022) 2D-to-3D: a review for computational 3d image reconstruction from X-ray images. Arch Comput Methods Eng. https://doi.org/10.1007/s11831-022-09790-z
8. Rueckert D, Schnabel JA (2011) Medical image registration. In Deserno TM (ed) Biomedical image processing. Springer, Berlin, Heidelberg, pp 131–154
9. Frangi AF, Rueckert D, Schnabel JA, Niessen WJ (2002) Automatic construction of multiple-object three-dimensional statistical shape models: application to cardiac modeling. IEEE Trans Med Imaging 21(9):1151–1166. https://doi.org/10.1109/TMI.2002.804426
10. Gothai E et al (2022) Design features of grocery product recognition using deep learning. Intell Autom Soft Comput 34(2):1231–1246. https://doi.org/10.32604/iasc.2022.026264
11. Liu S, Yang B, Wang Y, Tian J, Yin L, Zheng W (2022) 2D/3D multimode medical image registration based on normalized cross-correlation. Adv Artif Intell Percept Augment Reason 12(6):16
12. Yoo J-C, Han TH (2009) Fast normalized cross-correlation. Circuits Syst Signal Process 28(6):819. https://doi.org/10.1007/s00034-009-9130-7
13. Zheng Q, Wang Q, Ba X, Liu S, Nan J, Zhang S (2021) A medical image registration method based on progressive images. Comput Math Methods Med 2021:4504306. https://doi.org/10.1155/2021/4504306

14. Islam KT, Wijewickrema S, O'Leary S (2021) A deep learning based framework for the registration of three dimensional multi-modal medical images of the head. Sci Rep 11(1):1860. https://doi.org/10.1038/s41598-021-81044-7
15. Blendowski M, Bouteldja N, Heinrich MP (2020) Multimodal 3D medical image registration guided by shape encoder–decoder networks. Int J Comput Assist Radiol Surg 15(2):269–276. https://doi.org/10.1007/s11548-019-02089-8
16. Bashiri FS, Baghaie A, Rostami R, Yu Z, D'Souza RM (2019) Multi-modal medical image registration with full or partial data: a manifold learning approach. J Imaging 5(1):5
17. Nankani H et al (2021) A formal study of shot boundary detection approaches—comparative analysis. In: Sharma TK, Ahn CW, Verma OP, Panigrahi BK (eds) Soft computing: theories and applications. Advances in intelligent systems and computing. Springer. https://doi.org/10.1007/978-981-16-1740-9
18. Lakshmi KD, Vaithiyanathan V (2017) Image registration techniques based on the scale invariant feature transform. IETE Tech Rev 34(1):22–29. https://doi.org/10.1080/02564602.2016.1141076
19. Bay H, Tuytelaars T, Van Gool L (2006) SURF: speeded up robust features. Springer, Berlin, Heidelberg
20. Zitová B, Flusser J (2003) Image registration methods: a survey. Image Vis Comput 21:977–1000. https://doi.org/10.1016/S0262-8856(03)00137-9
21. Subramanian P, Faizal Leerar K, Hafiz Ahammed KP, Sarun K, Mohammed ZI (2016) Image registration methods. Int J Chem Sci 14:825–828

Chapter 8
Early Diabetes Prediction Using Deep Ensemble Model and Diet Planning

Anjali Jain and Alka Singhal

1 Introduction

A healthy personality gains respect from the society. People are busy in present world and have no time to look after health and lack nutritious diet. People tend to skip meals at particular time, and need of proper diet plan is increasing. There are lot of factors like sleep, nutrition, heredity, etc., which impact health of an individual. Development in technologies contributes to quick disease diagnosis [1]. In present scenario diabetes is one of the life-threatening diseases, early prediction, intervention and control is required. As Internet is flooded with lots of information, people used to take diabetes prediction; risk assessment and diet recommendation to control diabetes through the various sources form the internet. In the survey done in year 2020, mass number of death has happened due to diabetes with an estimated 1.5 million deaths which happened directly due to diabetes. Among every 11 Indians, one is diagnosed with diabetes. The symptoms of diabetes can be delayed using healthy diet healthy diet, physical activity, calorie intake racking and maintaining a normal body weight. Diet recommendation for diabetic patient is the need of the hour as people are looking for it, and lot of research is going on for recommendation of diet.

In most of the cases, the major cause of blindness, kidney failure, heart attacks, stroke and lower limb amputation is diabetes. There are several factors on which diet of a person depends like nutrition, availability, disease, and technology. A diabetic person requires special diet and fixed calorie intake when wants to maintain healthy

A. Jain (✉) · A. Singhal
Jaypee Institute of Information Technology, Noida, India
e-mail: jainanjali4u@gmail.com

A. Singhal
e-mail: alka.sinhal@mail.jjit.ac.in

A. Jain
Jaypee Institute of Information Technology, Delhi-NCR, Noida, India

V. Devedzic et al. (eds.), *Proceedings of the International Conference on Intelligent Computing, Communication and Information Security*, Algorithms for Intelligent Systems, https://doi.org/10.1007/978-981-99-1373-2_8

lifestyle. When diet of a person changes, it impacts morphological changes in the body. People used to skip meals and do fasts and intermittent fasting to lose weight and that shoots up their diabetes. So, suggesting proper and healthy balanced diet plan with recommended calories is the need so that one can follow and stick to it. Diabetes occurs mainly due to (i) when the body is notable to use the insulin it produces effectively. (ii) When the pancreas fail to produce enough insulin.

There are various factors which are taken into consideration while prediction whether a person is diabetic or not like pregnancy, glucose, BMI, insulin, age skin thickness, diabetes pedigree function, etc. As the requirement and intake of proper diet including calories requirement for every diabetic person is different, the recommendation of diet should be done keeping those factors in mind. In this paper, author is going to use machine learning algorithm for diabetic prediction. In earlier times, detection and prediction of the disease were used to done manually by a medical assistant or by taking input in electronic automatic device. All the diagnosis has some benefits and limitations. It's noticed that manually none of the medical expert can find or diagnose the diabetes at early stage because of some hidden side effects produced on the body of human. In this paper, machine learning models have been used for early prediction of diabetes and how it can be controlled with intelligent recommendation. People who are diabetic have to follow strict diet and routine with a calorie count. In this paper, if a person is diabetic, its calorie need per day is calculated based on BMI/BMR according to the categories of weight underweight, healthy, and overweight.

2 Literature Survey

A lot of research has been done for diagnosing diabetes disease. All these authors used various machine learning algorithms, classifiers, and artificial intelligence. Healthcare data can be easily collected using artificial intelligence. The data is collected from the healthcare center, and human disease can be easily predicted using multidisciplinary diabetes. For early diabetes detection, Aminah et al. used k-nearest classifier model and compared the results with support vector machines with accuracy of 85.6% [2]. The paper focused on different machine learning classifiers diabetes prediction, and results were compared some other support vector machine for authentication. In the study done by medical practitioner, it is proved that type 2 diabetes is more dangerous, and its early prediction is required. A. Mohebbi et al. designed a convolutional neural network-based model for the diabetes prediction and compared with multilayer perceptron and linear regression. The accuracy of this model achieved is 77.5% in area under curve [3]. N. Sneha and T. Gangil introduced a model based on SVM for diabetes detection with feature selection and compared the results with random forest, K-Nearest Neighbor, and decision tree (DT). The proposed model has scored the 77.73% accuracy and the feature selection technique helped to reduce systems computational capability and improved accuracy [4]. Number of machine learning models were used to check the authenticity and result comparison. Along with computational tools, bloodless techniques were used by P. Sah et al. for the prediction of diabetes and achieved the 91.67% accuracy using this technique. There are lot of other diseases S. Karthikeyan et al. used AI model for the detection of multiclass retinal disease and used CNN classifier, and the accuracy was 92% [5].

For the prediction of diabetes and cardiovascular disease, A. Dinh et al. [6] have used data-driven approach with machine learning, and an extreme gradient boost is implemented; then, results were compared with the other models like Logistic Regression, Support Vector Machine, Random Forest, and weighted ensemble model. The model achieved an accuracy of 95.7%. For type-2 diabetes prediction, B. P. Nguyen et al. have adopted an ensemble classification model and achieved accuracy 82.2% in the area under curve [7]. A novel methodology was invented by R. Rajalakshmi et al., i.e., diabetes detection based on smartphone [8]. In this paper, the dataset used for diagnosis contains images. Another method was introduced by M. Chen et al. for the measurement of the blood glucose level in patients based on microcontroller-based agent [9]. One of the authors, P. Choudhary et al., used sensor integrated therapy to monitor glucose levels in a diabetic patient [10]. There are lots of recommendation techniques which have been used for diet recommendation like similar user behavior. Osadchiy et al. mentioned user which has similar behavior and item preference chosen by user to recommend the results using association rules and inverse filtering methods [11]. Yuan et al. used K-means clustering along with user behaviors-based similar interest for diet recommendation [12]. The paper considered highly recommended recipes present in cluster to meets the nutritional requirement of user. This model achieved to 70% that is more accurate and effective than others. C. Yuen et al. proposed an ingredient-based recipe recommendation based on user preference for similar items [13]. Sookrah et al. proposed dietary approaches to stop hypertension based on item features for similar item recommendation and machine learning for hypertensive patients [14]. In this paper, author has reviewed various healthcare datasets and analyzed their result using various machine learning methods and predictions and techniques. D. Saravana et al. implemented a technique for analysis of diabetic data using system using Hadoop and Map Reduce. The system was used to predict the type of diabetes and risks associated with life-threatening disease. The Hadoop-based system is economical for any healthcare organization [15]. A cloud-based system for food recommendation was proposed by Rehman et al. that used ant colony optimization for generating optimized food list, and recommendation was done after viewing the user's pathological reports, and a better accuracy was attained by increasing number of ants [16].

Singh et al. introduced diet recommendation using analytic hierarchy method along with fuzzy for diet recommendation for diseased person. They focused on the problem of malnutrition and the disease known as marasmus which occurs due to lack of nutrients like, carbohydrates, fats, proteins, lipids, glycogen, etc. [17].

Expert recommendation was proposed by Chen et al. for the optimized nutrition. Sometimes the genetic history of a person is required for the recommendation of personalized diet. People used to get test their tests done for getting detailed information about their genes as it is required for the customized diet. The author has designed a system to correlate the genotypic information of a person and grocery product information [18].

Iwendi et al. proposed IoMT-assisted patient diet recommendation. For this author has used LSTM and proved that it has more accuracy, and random forest classifier was used for analyzing essential features. In this paper, huge dataset was used and that was divided into 70:30, i.e., 70% for training and 30% for testing data. The major concerns of population are health issue, but nutritional needs of a person could not be taken into account always [19]. Toledo et al. proposed a recommendation system for diet considering nutrition and user preferences. Author has used decision table and AHP sort. Their paper includes 600 food items and 20 nutrients food profiles, but in this paper food history of consumers is not present [20].

One more article by Chen et al. [21] illustrated the work based on the usage of tea and similar beverages, and how these affect the health of the user. The model to amp the health issues and the recommendation of the respective diet is shown in the paper [22]. Kardam et al. proposed the diet/food recommendation for diabetic patients using K-Means and Random Forest classifier [23].

In the above text, various existing models and approaches are discussed to encourage the proposed system. As none of the system perfectly define the remedial and preventive plan. The proposed system uses the hybrid model to give better results in terms of recommendations and the suggestions to the user.

3 Healthcare Recommendation for Early Diabetic Patient

In this paper, author has used multiple classification algorithm for early diabetes prediction as manually it was not possible for any medical practitioner or expert for doing that accurately. Classifiers are used to map dataset to specific category are *AdaBoost*, *Gradient Boost*, Random Forest, Support Vector Machine, *Bagging New Ensemble Classifier*.

The strength of the paper is the use of BMI, i.e., Body Mass Index, which measures body fat indirectly and gives height is to weight ratio. BMI is one of the main factors for diabetes prediction as it reflects that if a diabetic person is underweight, normal weight, and overweight but doesn't indicate the changes in muscle mass and body fat. On the basis of BMI author is calculating basal metabolic rate, i.e., BMR for calorie need per day required by a diabetic person according to its health category, as the BMI is not sufficient for measuring the calorie needed by a person to stay healthy.

For implementation results and data analysis, Anaconda 2.1.4 is used in this paper. Anaconda is a package management system, and it manages versions and packages for predictive analysis and data management [1]. The PIMA India diabetes data is for the input data for predictive analysis of the life-threatening diabetic disease. It has been checked in the dataset if any of the input parameters among the pregnancy, insulin, body mass index, skin thickness, glucose, blood pressure, diabetes pedigree function, and age is checked to see if some missing values are there or not and replaced

Table 1 Health status and BMI levels

Health status	BMI_Level
Under weight	15–19.9
Normal weight	20–24.9
Over weight	25–29.9
Level-1_Obesity	30–34.9
Level-2_Obesity	35–39.9
Level-3_Obesity	≥ 40

with null values. Then, it was also checked for columns dataset have correlation with other data is there or not. 1 is replaced with true and 0 is replaced with false value. The original data was split into training and testing into 70 and 30%. The outcome 1 shows the person is diabetic and 0 shows it's not diabetic considering the values of all the parameters. The author has trained the dataset on various classification algorithms, i.e., Ada Boost, Gradient Boost, Random Forest, Support Vector Machine, Bagging, and New proposed ensemble classifier. The result of these models showed us that whether the disease is identified or not. It's been identified that all the trained models give accurate results. The efficiency of the proposed model is also checked and compared for predicting diabetes, and various other parameters are also compared as given in Table 4. The new proposed classifier has 82% accuracy as compared to support vector machine mentioned by Nguyen et al. [7]. The performance report below also shows the performance matrix of all the classifiers.

The health status is checked according to the BMI of the diabetic patient as given in Table 1. BMI is one of the parameters in the training dataset. The BMI is the ratio of weight and height. It does not indicate changes in the body fat and muscle mass.

$$\text{BMI} = (\text{weight in kilograms})/\text{Height in meters} \tag{1}$$

For finding the weight of a particular person:

$$W = \text{BMI}_{\text{normal}} * \text{height}^2 \tag{2}$$

Based on the above equation, any person can be categorized as underweight, normal, and overweight. As mentioned, BMI measures the weight status of your body in relation to fat. A person can figure out its excessive body fat and associated risk and diseases of carrying that extra body fat. Only BMI is not sufficient to maintain a healthy body weight and control diabetes. There is strong correlation between BMI and BMR.

Basal metabolic rate is the calorie amount burn by a person when body is resting. BMR is characterized depending upon individual age, weight, and height. BMR gives the count of calorie that a person needs to remain in order to control diabetes according to its health status. According to Harris–Benedict equation [22], BMR is shown both for men and women is

Table 2 Calorie need per day calculation based on BMR

Daily physical activity	Calorie need per day
Sedentary (very light or no exercise)	1.2 * BMR
Somewhat active (light exercise/physical activity 1–3 days/week)	1.375 * BMR
Temperately active (moderate exercise/physical activity 3–5 days/week)	1.55 * BMR
Always active (hard exercise/physical activity 6–7 days a week)	1.725 * BMR
Extra active (very hard exercise/physical activity and aerobics)	1.9 * BMR

$$\text{For Men: BMR} = 66.5 + (13.75 * \text{weight in kg}) + (5.003 * \text{height in cm}) - (6.75 * \text{age})$$

$$\text{For Women: BMR} = 655.1 + (9.563 * \text{weight in kg}) + (1.850 * \text{height in cm}) - (4.676 * \text{age})$$

$$\text{For Women: BMR} = 655.1 + (9.563 * \text{weight in kg}) + (1.850 * \text{height in cm}) - (4.676 * \text{age})$$

The Harris–Benedict equation uses your BMR and then according to an activity factor to determine your total daily energy expenditure while doing nothing. For this BMR of a parson is multiplied by the appropriate activity factor, as follows:

$$\text{Daily Needed Calorie} = \text{BMR} * \text{daily physical activity factor} \tag{3}$$

Table 2 shows calorie estimation for diabetic patient based on the regular physical activity and its BMR. If a person is diabetic and has BMI that falls into particular category, its BMR is calculated first for calculating daily needed calories, and then the list of food items is recommended according to health status for breakfast, lunch, and dinner. The three-meals of the day are recommended based on the calorie a diabetic person can take. The author has divided the total recommended calorie into 50, 30, and 20% in breakfast, lunch, and dinner. If a person has BMI > 20, has standard weight and diabetic in nature, the amount of calorie intake required is 1000–1200 cal in a day. Then, the breakfast contains 500–600, lunch 300–400 and dinner 100–200 cal in a day, and food items will be suggested accordingly. Similarly, the calories can also be calculated for other BMI criteria.

4 Proposed Framework and Implementation Results

In this paper, author has used machine learning algorithm and classifiers for diet recommendation according to the requirement and need of the diabetic patients. Figure 1 shows the proposed framework for diabetes prediction and control.

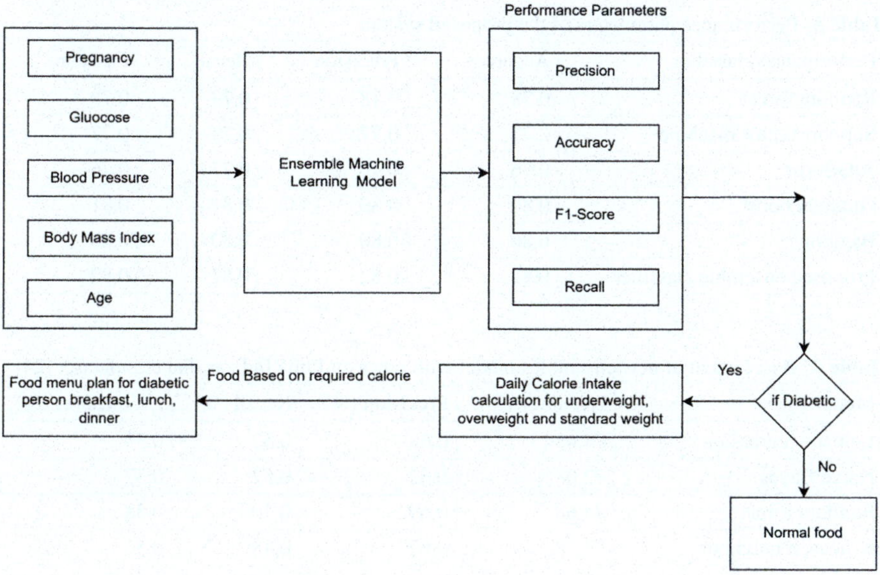

Fig. 1 Proposed framework for diabetes prediction and control

The proposed framework shows the working of the recommendation model. Python is used for implementation of various classification and clustering algorithms in machine learning for diet recommendation to a user. In the above framework, author has used machine learning classifiers for the prediction of diabetes. The dataset used for PIMA India diabetes dataset that is available on Kaggle and UCL libraries, the dataset carries 100,000 entries of the patient. The author used random forest, support vector machine, AdaBoost, Gradient Boost, Bagging, and New ensemble model on the above-mentioned dataset. The mentioned models were trained on various parameters like pregnancy, Glucose, BMI, Blood Pressure, and age for predicting whether a person is diabetic or not. For the prediction, outcome is set to 0 and 1. If a person is diabetic, three-meal course is recommended along with list of food items which suits for appropriate health status. For food recommendation, food dataset is taken from Kaggle. If a person is diabetic, its BMI is checked to find the health status whether a diabetic person is under weight, over weight, and healthy. Then according to health status and BMI score, the calorie count required by a person per day is calculated, and on the basis of daily needed calorie, the list of food items is recommended that a person can eat in breakfast, lunch, and dinner. The list of food recommended is for the control of diabetes at early stage.

Performance Matrix

The calculation of performance matrix and accuracy of training data is measured using various machine learning parameters like: Accuracy, precision, recall, and $F1$-score are calculated and compared with other algorithms. The author has found

Table 3 Performance parameters of the proposed system

Performance/classifier	Accuracy	Precision	Recall	$F1$-score
Random forest	0.78	0.78	0.79	0.78
Support vector machine	0.77	0.77	0.78	0.77
AdaBoost	0.76	0.76	0.77	0.76
Gradient boost	0.80	0.80	0.81	0.81
Bagging	0.80	0.80	0.80	0.80
Proposed ensemble classifier	0.82	0.82	0.82	0.82

Table 4 Assessment of new ensembles model with other on Pima Indians diabetes dataset [24]

Model	Accuracy (%)	Precision (%)	Recall (%)	$F1$-meaures (%)
Logistic regression	74.68	0.68	0.52	0.59
Naïve Bayes	72.08	0.62	0.52	0.57
Random forest	74.68	0.69	0.50	0.58
K-nearest neighbor	73.38	0.67	0.48	0.56
Decision tree	74.03	0.63	0.63	0.63
Support vector machine	74.68	0.70	0.48	0.57
Proposed ensemble model	82.00	0.82	0.82	0.82

performance parameters using Random Forest, support vector machine, AdaBoost, gradient boost, and bagging. The data given in Table 3 mentions that accuracy of proposed classifier is high that when it is compared with other parameters.

The summary of performance metrics, given in Table 3, accuracy, precision, recall, $F1$-score of selected datasets individually is presented.

Table 4 shows the comparison of performance metrics parameters of work done by various others authors. Among all the proposed ensemble classifier has shown the better accuracy. The below mentioned are the results of food recommendation for standard weight category person who is diabetic in nature. In Fig. 4a–c, the list of food items along with the images of the food is shown.

In the same manner, author has calculated for overweight and underweight diabetic person and gives the list of food items to control diabetes.

5 Conclusion

Disease prediction can be effectively work in medical and healthcare recommendation. The proposed work shows early diabetes prediction and control through managing the diet recommendation system based on the user's input values. As in today's life among the busy schedule of everyone's life, it is difficult to plan or take diet according to health status and calorific requirement. The dataset is chosen from

	Food_items	Calories
14	Cashew Nuts	553
20	Protein Powder	411
33	Pop Corn	429
40	Brownie	407
45	Almonds	579
55	Chia seeds	486
80	Beef sticks	550
81	Banana Chips	519
87	Chocolate milk	535

(a)

Your Recommended Lunch contains 300-400 Calories

	Food_items	Calories
6	Brown Rice	362
50	White Rice	360
76	Nachos	350
77	Chicken Popcorn	351

(b)

Your Recommended Dinner contains 100-200 Calories

	Food_items	Calories
22	Tuna Salad	187
23	Tuna Fish	184
34	Dosa	168
35	Idli	156
41	Noodles	108
42	Uttapam	188
44	Dal Makhani	109
57	Salmon	127
60	Lentils	101
61	Pasta with corn homemade	126
66	Goat meat	109
67	Rabbit meat	114
70	Mexican Rice	195
72	Spaghetti and meatballs	170
73	Macroni n Cheese	194
75	Bacon cooked	146
79	Oyster cooked	159
85	Strawberry Icecream	192

(c)

Fig. 4 **a** Recommended food items for breakfast. **b** Recommended food items for lunch. **c** Recommended food items for dinner online repository Pima India Diabetes Dataset (PIMA) and food data. The classifiers are applied to diabetes dataset and checked for the prediction. The accuracy and various other parameters were calculated and analyzed. The performance measures F-measures, Recall, Accuracy, and precision were used for checking the accuracy. The proposed model helps to improve the accuracy whether a person is diabetic or not. The accuracy of new proposed ensemble model is 85% which is found to better than other existing models. The proposed models work well against prediction and recommendation of food items. Finally based on the prediction of the disease, a list of food is recommended to user according to its health status and calorific needs. The calorie intake of a diabetic person is calculated using BMR formula, and list of food items along with images of the food is given to user. Our proposed system helps to take the prediction of diabetes at initial stage and suggested the food items a person should have to remove diabetes and maintaining health.

References

1. Jackins V, Vimal S, Kaliappan M, Lee MY (2020) AI based smart prediction of clinical disease using random forest classifier and Naïve Bayes. J Supercomput 5199–5219
2. Aminah R, Saputro AH (2019) Diabetes prediction system based on iridology using machine learning. In: Proceedings of the 2019 6th international conference on information technology, computer and electrical engineering (ICITACEE), Semarang, Indonesia, Sept 2019
3. Mohebbi A, Aradottir TB, Johansen AR, Bengtsson H, Fraccaro M, Mørup M (2017) A deep learning approach to adherence detection for type 2 diabetics. In: Proceedings of the 39th annual international conference of the IEEE Engineering in Medicine and Biology Society (EMBC), Jeju, Korea, July 2017
4. Sneha N, Gangil T (2019) Analysis of diabetes mellitus for early prediction using optimal features selection. J Big Data 6
5. Karthikeyan S, Sanjay Kumar P, Madhusudan RJ, Sundaramoorthy S, Namboori P-K-K (2019) Detection of multi-class retinal diseases using artificial intelligence: an expeditious learning using deep CNN with minimal data. Biomed Pharmacol J 12
6. Dinh A, Miertschin S, Young A, Mohanty SD (2019) A data driven approach to predicting diabetes and cardiovascular disease with machine learning. BMC Med Inform Decis Mak 19
7. Nguyen BP, Pham HN, Tran H et al (2019) Predicting the onset of type 2 diabetes using wide and deep learning with electronic health records. Comput Methods Prog Biomed 182
8. Rajalakshmi R, Subashini R, Anjana RM, Mohan V (2018) Automated diabetic retinopathy detection in smartphone-based fundus photography using artificial intelligence. Eye 32
9. Chen M, Yang J, Zhou J, Hao Y, Zhang J, Youn C-H (2018) 5G-smart diabetes: toward personalized diabetes diagnosis with healthcare big data clouds. IEEE Commun Mag 56(4):16–23
10. Choudhary P, De Portu S, Arrieta A, Castañeda J, Campbell FM (2019) Use of sensor-integrated pump therapy to reduce hypoglycemia in people with type 1 diabetes: a real world study in the UK. Diabetic Med 36
11. Osadchiy T, Poliakov I, Olivier P, Rowland M, Foster E (2018) Recommender system based on pairwise association rules. Expert Syst Appl. https://doi.org/10.1016/j.eswa.2018.07.077
12. Yuan Z, Luo F (2019) Personalized diet recommendation based on K-means and collaborative filtering algorithm. J Phys. https://doi.org/10.1088/1742-6596/1213/3/032013
13. Teng CY, Lin Y-R, Adamic LA (2012) Recipe recommendation using ingredient networks. In: 4th annual ACM web science conference, June 2012, pp 298–307. https://doi.org/10.1145/2380718.2380757

14. Sookrah R, Devee Dhowtal J, Devi Nagowah S (2019) A DASH diet recommendation system for hypertensive patients using machine learning, pp 1–6. https://doi.org/10.1109/ICoICT.2019.8835323
15. David DS et al (2022) Enhanced detection of glaucoma on ensemble convolutional neural network for clinical informatics. CMC-Comput Mater Contin 70(2):2563–2579
16. Rehman F, Khalid O, Haq N, Khan A, Bilal K, Madani S (2017) Diet-right: a smart food recommendation system. KSII Trans Internet Inf Syst 11(6). https://doi.org/10.3837/tiis.2017.06.006
17. Mahrishi et al (ed) (2020) Machine learning and deep learning in real-time applications. IGI Global. https://doi.org/10.4018/978-1-7998-3095-5
18. Iwendi C, Khan S, Anajemba JH, Bashir AK, Noor F (2020) Realizing an efficient IoMT-assisted patient diet recommendation system through machine learning model. IEEE Access 8:28462–28474. https://doi.org/10.1109/ACCESS.2020.2968537
19. Yera Toledo R, Alzahrani AA, Martínez L (2019) A food recommender system considering nutritional information and user preferences. IEEE Access 7:96695–96711
20. Chen YS, Cheng CH, Hung WL (2021) A systematic review to identify the effects of tea by integrating an intelligence-based hybrid text mining and topic model. Soft Comput 25:3291–3315
21. Kim JC, Chun K (2019) Knowledge based hybrid decision model using neural network for nutrition management. Inf Technol Manag 29–30
22. Kardam SS, Yadav P, Thakkar R, Ingle A (2021) Website on diet recommendation using machine learning. Int Res J Eng Technol (IRJET) 2021:3708–3711
23. Jain A, Singhal A (2022) Personalized food recommendation—state of art and review. In: Ambient communications and computer systems: proceedings of RACCCS 2021, July 2022, pp 153–164
24. Griffith R, Shean R, Petersen CL, Al-Nimr RI, Gooding T, Roderka MN, Batsis JA (2022) Validation of resting energy expenditure equations in older adults with obesity. J Nutr Gerontol Geriatr 1–14

Chapter 9
Enhancing Image Caption with LSTMs and CNN

Vishakha Gaikwad, Pallavi Sapkale, Manoj Dongre, Sujata Kadam, Somnath Tandale, Jitendra Sonawane, and Uttam Waghmode

1 Introduction

In the last few years, the advancements that are taking place in computer vision with applications such as image classification and object detection along with natural language processing being extensively used for chat bots and smart assistants give rise to the possibility of more advanced applications of artificial intelligence in a way that gives computers the ability to not only classify images but also draw meaning from them. This application of AI is known as image captioning [1, 2]. Image captioning is a process where a computer learns to generate one or more sentences to understand the visual content of an image which includes not only objects in an image but also understanding the relationship between these objects and the activities that are being performed. Creating captions is a need for computer vision and natural language

---

V. Gaikwad, P. Sapkale, M. Dongre, S. Kadam, S. Tandale, J. Sonawane, U. Waghmode—authors contributed equally to this work.

V. Gaikwad (✉) · P. Sapkale · M. Dongre · S. Kadam · S. Tandale · J. Sonawane · U. Waghmode
Department of Electronics and Telecommunication, Ramrao Adik Institute of Technology,
Dr. D.Y. Patil Deemed to be University, Navi Mumbai, Maharashtra 400706, India
e-mail: vishakha.gaikwad@rait.ac.in

P. Sapkale
e-mail: pallavi.sapkale@rait.ac.in

M. Dongre
e-mail: manoj.dongre@rait.ac.in

S. Kadam
e-mail: sujata.kadam@rait.ac.in

S. Tandale
e-mail: somnath.tandale@rait.ac.in

J. Sonawane
e-mail: jitendra.sonawane@rait.ac.in

U. Waghmode
e-mail: uttam.waghmode@rait.ac.in

V. Devedzic et al. (eds.), *Proceedings of the International Conference on Intelligent Computing, Communication and Information Security*, Algorithms for Intelligent Systems, https://doi.org/10.1007/978-981-99-1373-2_9

Fig. 1 This is a widefig. This is an example of image captioning

processing. The motivation of our work is to capture how objects in the image are related to each other and express them in the English language. Existing image captioning techniques are used to convert raw image files. Such techniques are not able to give the exact context of the image and therefore cannot be used to gain insight into the image files. To overcome such problems, advanced techniques are used by [3]. The goal of this work is to generate captions using neural language models. Image captioning has significant potential applications such as generating titles for news images, providing information for blind users, the interaction between humans and robots, and generating descriptions for medical images. Image captioning is a complex and complicated task that uses advanced forms of artificial intelligence such as deep neural networks. Due to the importance that images captioning plays in future of AI, it is open for thorough and high-powered theoretical and practical research. Figure 1 shows the example of image captioning.

As a consequence of the recent rise in interest for image captioning due to the availability of large data sets that are easily accessible that can be used to train complex networks for this specific task . The recently introduced sequence-to-sequence learning, based on recurrent neural networks, provides a solid framework to generate proper meaningful sentences with the help of new deep learning networks such as auto-encoders and LSTMs.

2 Related Work

There are various works related to image captioning developed. Image captioning was firstly developed under constrained conditions in [4] an image captioning method that uses a dictionary of objects and language templates. Two new methods of image captioning were described in [5, 6]. In [7], a hierarchical LSTM using an adaptive attention-based encoder-decoder model with visual captioning was focused. The existing multi-model neural network method [8], including the object detection and localization models, is the same as the human sensory system which learns the way to develop the content of images by default and the LSTM units are complex and inherently sequential across time.

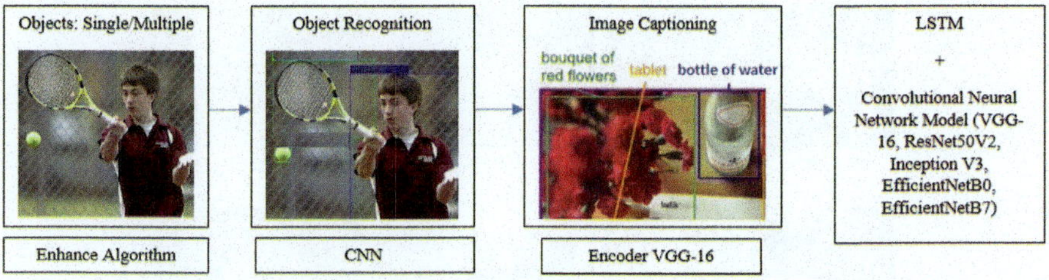

Fig. 2 Proposed image captioning technique

2.1 Challenges

Image captioning is a task that comes easily to humans as we can understand the contents of an image and express it in the form of natural language, i.e., however, we require, but for computers, this requires an integrated effort of various advanced technologies like image processing computer vision and natural language processing. The goal of image captioning is to design a model that can completely describe the information conveyed in an image in the form of a human-like description. It is not enough for the model to identify the objects present in the image or recognize the scene; The model needs to analyze the states and relationships between these objects and generate a semantically and syntactically correct caption. For instance, in an image that contains a person riding the waves on a surfboard, it is not enough to just identify the person and the surfboard individually, but also to understand that the person is riding the surfboard. Due to such peculiarities, the model needs to understand and identify the image's salient parts and the image as a whole.

3 Methodology

In this section, we discussed various existing techniques as well as algorithms for image captioning. Figure 2 shows the proposed image captioning method. In this proposed method, object recognition, classification, image caption, and analysis are consider.

3.1 Image Classification

This involves classifying an image into predefined categories using specific features from the image to differentiate it from other classes. Image classification is extensively used to automate image organization and improve product discover ability with visual search. Image classification is shown in Fig. 3

Fig. 3 Image classification

Fig. 4 Image annotation

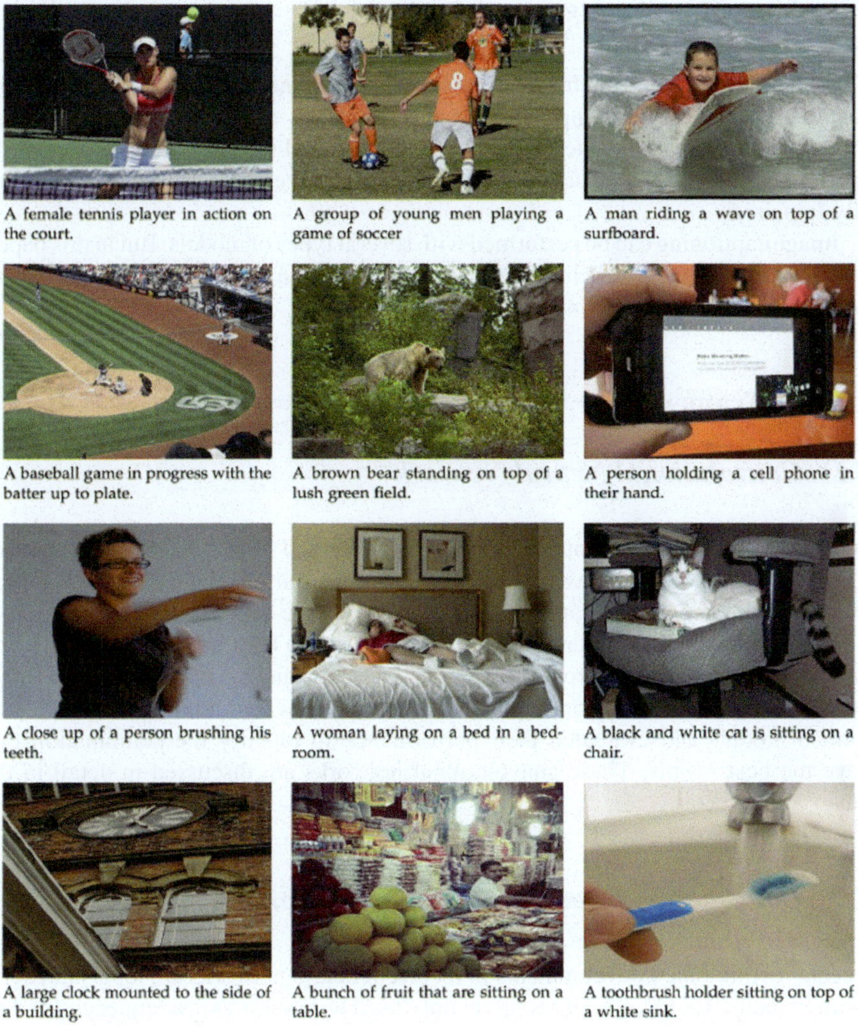

Fig. 5 Image description

3.2 *Image Annotation*

This involves identifying objects and pinpointing their location in the image. This is widely used in computer vision tasks such as activity recognition, face detection, face recognition, and video object segmentation, for instance, tracking the movement of a ball during a cricket match. From Fig. 4, annotation is displayed.

3.3 Image Captioning

This is the process in which a small description is generated to describe an image accurately. Image captioning is shown in Fig. 5. It is extensively used to caption images available on Websites and improve search ability. Potential uses for image captioning include assistance in self-driving cars and scene descriptions for the visually impaired.

Image captioning can be performed with several types of models. But in this paper, we will are implementing a hybrid model consisting of a convolution neural network and a recurrent neural network in the form of LSTMs.

4 The Framework

4.1 Convolutional Neural Network

With the rapid improvements in image processing and object detection, a plethora of pre-trained models are available for use in the form of transfer learning. These models are highly complex and are trained on very large data sets for a long period of time. These models are extensively optimized to give a high accuracy and are being constantly updated to improve prediction time and quality. For the convolutional neural network in our hybrid model, we have implemented a number of such pre-trained models and evaluated their performance to identify the combination that gave the best results. These convolutional networks are discussed in detail in the next section.

4.2 Convolutional Recurrent Neural Network

Recent revolutionary developments in recurrent neural networks are long short-term memory cells (LSTMs). LSTMs are mainly used to process entire sequences of data such as picture or video and also time series prediction. The advantage that LSTM cells possess compared to a common recurrent unit is its cell memory unit. LSTMs have a memory pipeline that possesses the ability to store the previous results and use them in combination with the current inputs to give much more accurate predictions (Fig. 6).

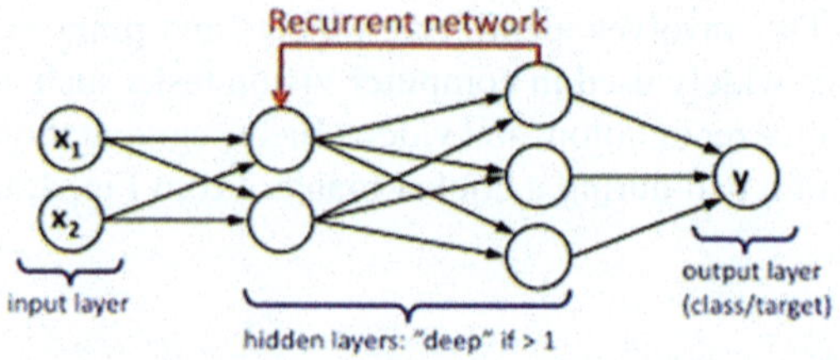

Fig. 6 Recurrent neural network

5 Hybrid Model

The proposed hybrid model that will be used for image captioning is a combination of convolutional neural network and LSTMs, where the convolutional neural network is used to extract features from the input images and LSTM simultaneously used to generate semantically correct sentences using them to finally obtain an output in the form of a sentence that accurately describes the input image. It consists of two parts such as encoder and decoder. The convolutional neural network acts as the encoder and extracts image features into a vector representation of the image. In the decoder, the RNN takes the word from the vector representation of the captions as an input (Fig. 7). Architecture of the model: For the encoder, various pre-trained models can be used. Hence, the architecture is divided into multiple sections with each section representing a different CNN with the same decoder. The decoder consists of an embedding layer followed by multiple stacked LSTMs. The input to this encoder is the tokenized sequence-to-sequence representation of the captions associated with the image in a progressive word-to-word manner (Fig. 8). The encoder in the form of convolution neural networks takes an image as input, and the image is passed through multiple convolutions depending on the framework of each model. All of these models are trained to return an output in the form of a list of 1000 elements with containing the probability of every class that the model is trained to predict. However, we are interested in the way in which the model identifies features and not classification. Hence, we remove the last layer of the model and use the output of the penultimate layer to identify features and pass it on to the decoder. The input of

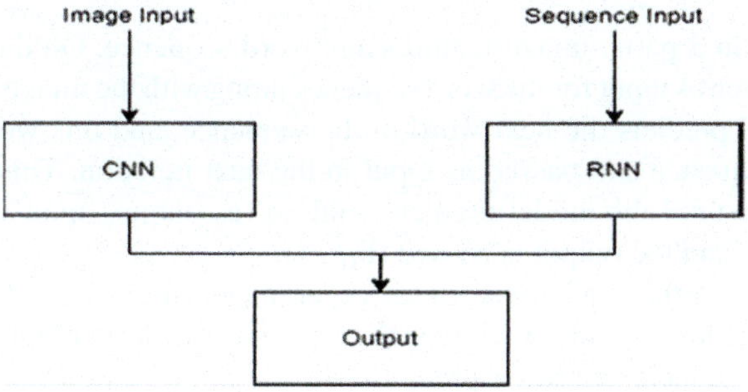

Fig. 7 Hybrid neural network

Fig. 8 Structure of the model

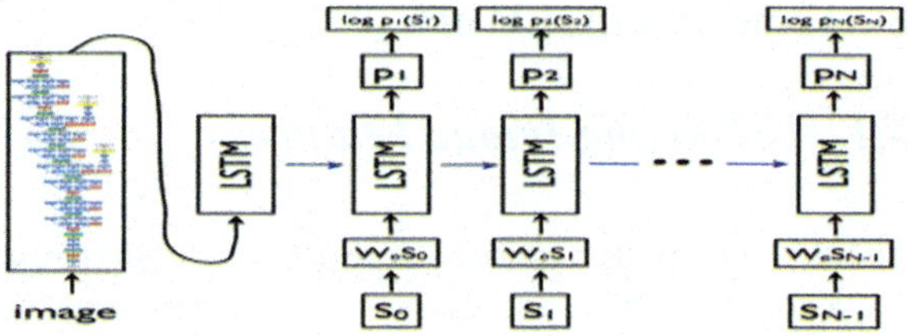

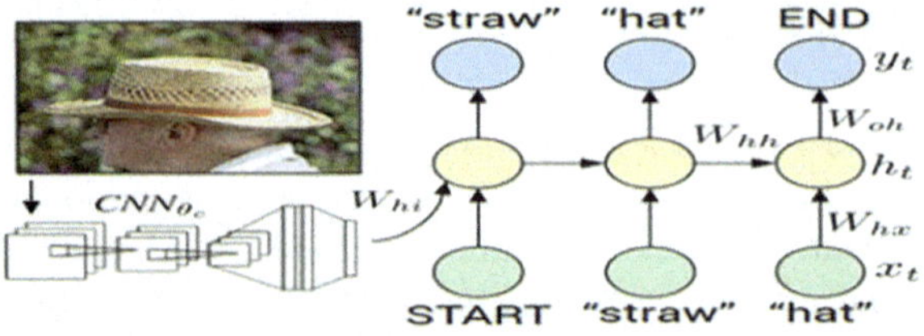

Fig. 9 Word by word predictions

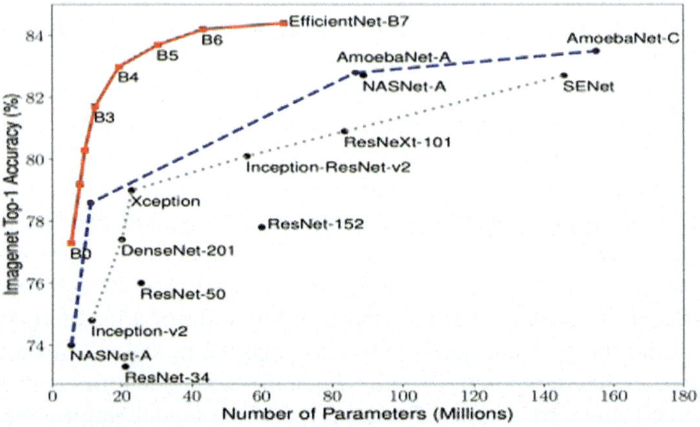

Fig. 10 Accuracy versus parameters for efficient net

the decoder is in 2 parts: image features and word sequence. On the first iteration, 'startseq' is given as input for the word sequence along with the image features. Using this, the model predicts the next word in the sequence, and this word is appended to the word sequence and passed as input in the next iteration. This continues in a similar manner until the model predicts 'endseq' as output, upon which the word sequence ends, and the output is passed (Fig. 9).

Figure 10 shows the graph for accuracy vs parameters for EfficientNet. Advantages of LSTM are highest accuracy on ImageNet, scalable, much smaller in size compared to the previous models and faster predictions.

6 Experimental Work

6.1 Extracting Image Features

The first step of the experimentation process was to extract features from every image. To accomplish this, a directory was created, and the images were loaded into this directory. All the images are looped over one by one, and the extracted image vector is stored in a dictionary where the key is the file name, and the value is the feature array. These features are picked and stored for future use.

6.2 Preparation of Text Data

The labels are in the form of five sentences that describe each image. In order to use these image labels for predictions, the labels need to be cleaned. All the labels are loaded into a nested list where every list contains five strings. Then, each string is passed through pre-processing functions where the strings are converted to lowercase, punctuation's are removed, and stopwords are discarded. Once a clean description data is obtained, a vocabulary is created.

6.3 Model Design

Once the training data is ready, the model is to be defined and trained. We use the tensorow library to create a sequential model with two different inputs. The first input is the array of image features, and the second input is the padded sequences. The feature vector is passed through to a dense layer, and the padded sequences are passed through to an embedding layer and an LSTM layer. The outputs of both these layers are added and passed through to another dense layer followed by a softmax layer with size equal to the size of vocabulary. Since the training data size for every iteration is very large, the entire training data set cannot be passed at once. To combat this challenge, we use a generator function which yields one instance of data at a time, and the model is trained with this instance. This leads to faster execution and a much better memory management system (Fig. 11).

6.4 Execution

Since the models and the data itself are very large in nature, it is not feasible to train these models on local machines as it would require a large amount of computing time and high energy requirements. Hence, we decide to train them using GPUs provided by the Google Colab environment. Details of said GPU are mentioned below (Figs. 12 and 13; Tables 1 and 2):

- GPU Name: Tesla T4
- Compute Capability: 7.5
- Memory Limit: 14674281152 bytes 25

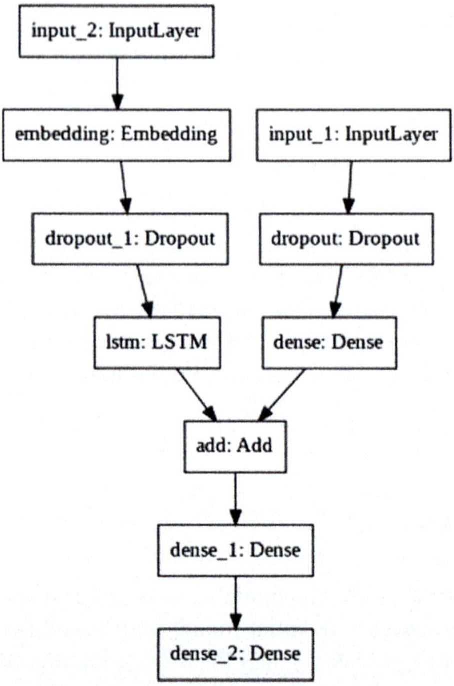

Fig. 11 Encoder-decoder

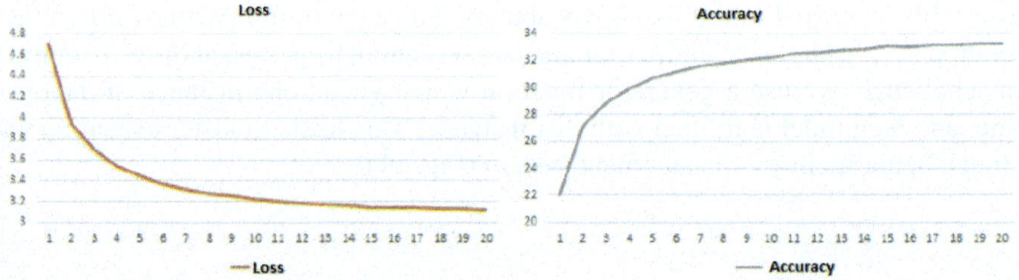

Fig. 12 Loss and accuracy for InceptionV3

7 Results

The following are the BLEU scores for the trained models:

Image Captions:

The following are the predictions for the sample images when given as input to the trained model:

From these results, we observe that EfficeintNetB7 has the highest and most consistent scores, and VGG16 has the lowest scores. The outputs from EfficientNetB7 give the most meaningful sentences and accurate predictions (Fig. 14).

Table 1 Inception-LSTM

Epoch	Time token	Loss	Accuracy
1	664	5.0192	0.1915
2	655	3.7273	0.2877
3	651	3.4418	0.3057
4	648	3.2762	0.3159
5	652	3.1585	0.3240
6	659	3.0703	0.3291
7	663	3.0021	0.3346
8	655	2.9466	0.3402
9	656	2.9046	0.3439
10	657	2.8634	0.3479
11	614	2.8149	0.3527
12	604	2.7898	0.3558
13	607	2.7703	0.3588
14	625	2.7568	0.3596
15	619	2.7369	0.3620
16	614	2.7287	0.3633
17	616	2.7199	0.3651
18	612	2.7135	0.3653
19	619	2.7028	0.3671
20	619	2.6997	0.3686

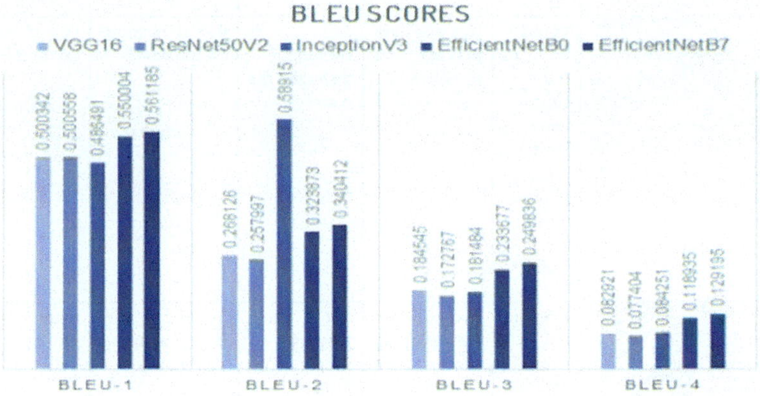

Fig. 13 BLEU scores for trained models

Table 2 BLEU scores for trained models

Model name	BLEU-1	BLEU-2	BLEU-3	BLEU-4
VGG16	50.03	26.86	18.45	8.29
ResNet50V2	50.05	25.79	17.27	7.74
InceptionV3	48.64	58.91	18.14	8.42
EfficientNetB0	55	32.38	23.36	11.89
EfficientNetB7	56.11	34.04	24.98	12.91

Image	VGG16	ResNet50V2	InceptionV3	EfficeintNetB0	EfficeintNetB7
	man in red gear riding a bike down a dirt hill	man in the red helmet is riding a bike down a dirt trail	man on a motorcycle is riding on the dirt road	man in red shirt and helmet is riding bike on dirt hill.	-
	man in red shirt and helmet is riding bike down hill	man is climbing up rock face	man in red shirt is standing on top of mountain	man in red shirt and red pants is standing on top of rocks	man in red shirt is standing on top of mountain
	young boy in red shirt and blue jeans is playing with the water	two dogs are playing in the snow	dog is running through the water	dog is running on the beach	dog is running on the beach
	man in red shirt is walking down the street	two children are playing soccer on the grass	two men playing soccer on field	boy in red shirt is jumping over hurdle	tennis player is hitting tennis ball

Fig. 14 Comparison of images with caption

8 Conclusion

The objective of this research was to design and implement an efficient method for generating image captions using computer vision technologies and also to use this method to design multiple deep learning models with different architectures and perform a comprehensive comparative study that demonstrates which models tend to outperform others using specific evaluation parameters. Based on the conveyed analysis, a system was designed that can extract and store image features and train hybrid models efficiently. On Google Colab GPUs, the average training time for one model was 11 min per epoch. During the entirety of the project, extensive use of cloud computing and storage was made so as to minimize the load on local machines. In this project, we designed and implemented a total of 5 deep learning models using 5 different computer vision architectures, namely VGG16, ResNet50V2, InceptionV3, EfficeintNetB0, and EfficeintNetB7. All 5 of these models were trained on the same data set for the same number of epochs, and the results were compared elaborately. It was concluded with firm evidence that the EfficientNetB7-LSTM Framework had the best performance (highest prediction accuracy and high training and prediction speeds). This model can be successfully used for making real-life predictions in image captioning applications.

References

1. Gu IJ, Wang G, Cai J, Chen T (2017) An empirical study of language CNN for image captioning. In: Proceedings of the international conference on computer vision (ICCV)
2. Gan RC, Gan Z, He X, Gao J, Deng L (2017) Stylenet: generating attractive visual captions with styles. In: Proceedings of the IEEE conference on computer vision and pattern recognition
3. Bourgeoys AJ, Joint M, Thomas C (2004) Automatic generation of natural language descriptions for images. In: Proceedings of the Recherche Dinformation Assistee Par Ordinateur, Avignon, France, 26–28 Apr 2004, pp 1–8
4. Hejrati M, Sadeghi MA, Young P, Rashtchian C, Hockenmaier J, Forsyth D (2010) Every picture tells a story: generating sentences from images. In: Proceedings of the European conference on computer vision, Heraklion, Crete, Greece, vol 5–11, pp 15–29
5. Premraj V, Dhar S, Li S, Choi Y, Berg AC, Berg TL (2011) Baby talk: understanding and generating simple image descriptions. In: Proceedings of the computer vision and pattern recognition, Colorado Springs, CO, USA, vol 20–25, pp 1601–1608
6. Li X, Song J, Shen HT (2019) Hierarchical LSTMs with adaptive attention for visual captioning. IEEE Trans Pattern Anal Mach Intell 42:1112–1131
7. Sucharita V, Jyothi S, Mamatha DM (2013) A comparative study on various edge detection techniques used for the identification of Penaeid Prawn species. Int J Comput Appl (0975-8887) 78(6)
8. Farhadi A, Hejrati M, Sadeghi MA, Young P, Rashtchian C, Hockenmaier J, Forsyth D (2010) Every picture tells a story: generating sentences from images. In: Proceedings of the 11th European conference on computer vision: Part IV, ECCV'10. Springer, Heidelberg, p 1529
9. Li S, Kulkarni G, Berg TL, Berg AC, Choi Y (2011) Composing simple image descriptions using web-scale grams. In: Proceedings of the fifteenth conference on computational natural language learning, CoNLL '11 pages 220228,

10. Hodosh M, Young P, Hockenmaier J (2013) Framing image description as a ranking task: data, models and evaluation metrics. J Artif Intell Res 47: 853–899 (Association for Computational Linguistics, Stroudsburg, PA, USA)
11. Nankani H et al (2021) A formal study of shot boundary detection approaches—comparative analysis. In: Sharma TK, Ahn CW, Verma OP, Panigrahi BK (eds) Soft computing: theories and applications. Advances in intelligent systems and computing. Springer, Heidelberg.https://doi.org/10.1007/978-981-16-1740-9
12. Karpathy A, Fei-Fei L (2015) Deep visual-semantic alignments for generating image descriptions [Online]. Available: https://cs.stanford.edu/people/karpathy/cvpr2015.pdf

Chapter 10
Design and Implementation of S-Box Using Galois Field Approach Based on LUT and Logic Gates for AES-256

K. Janshi Lakshmi and G. Sreenivasulu

1 Introduction

Cryptography is described as the science of secure communication over a public channel, and the security of data is greatly aided by the use of cryptography in network security systems. Sensitive data can be stored and transmitted using cryptography over insecure networks like the Internet, so that the intended recipient is the only one who can read it.

See Fig. 1, Alice is a sender, Bob is a recipient, and Charlie is a hacker. For example, Alice is sending her credit card details (plaintext) through channel like internet to Bob without secure key. Between Alice and Bob, the hacker Charlie broke into the channel and grabbed information including credit card details. He may edit, modified, and duplicated of original data. Bob received stolen or modified data. This process is not secure.

See Fig. 2, Alice is sender, Bob is recipient, and Charlie is hacker. For example, Alice is sending her credit card details (plain text) and secure key. It became cypher text through channel like Internet to Bob. The Charlie cannot hack the data from channel. He cannot edit, modified, and duplicated of original data because secure key added to plaintext. Bob received secure original data. This process is secure, and it is called cryptography technique. Types of cryptographic functions: the same key is used for encryption and decryption, i.e. secret or symmetric key. Asymmetric or public key: various keys are used for encryption and decryption.

K. Janshi Lakshmi (✉) · G. Sreenivasulu
Department of Electronics and Communication Engineering, Sri Venkateswara University College of Engineering, Sri Venkateswara University, Tirupati, Andhra Pradesh, India
e-mail: jansikaramala@gmail.com

G. Sreenivasulu
e-mail: gunapatieee@rediffmail.com

V. Devedzic et al. (eds.), *Proceedings of the International Conference on Intelligent Computing, Communication and Information Security*, Algorithms for Intelligent Systems, https://doi.org/10.1007/978-981-99-1373-2_10

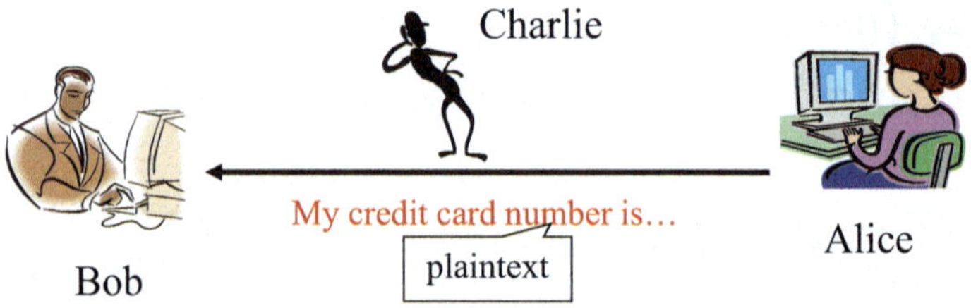

Fig. 1 Without secret key

Fig. 2 With secret key

The 'National Institute of Standards and Technology' (NIST) standardised information security in 1997 by using the data encryption standard (DES) algorithm (NIST 1999). Because the key size (56 bits) was insufficient, DES data (64 bits) was required. Increasing computing power is considered vulnerable to a complete key search attack. Triple DES was designed to overcome this type of disadvantage, but it proved to be a time-consuming process. The disadvantages are that the key is too small, there is less security, and the process is slow [1, 2].

'The National Institute of Standards and Technology (NIST) standardised the AES encryption and decryption algorithm in 2001, which became the 'Federal Information Processing Standard' (FIPS) (NIST 2001). The AES algorithm was developed by cryptographers names which are Joan Daemen and Vincent Rijmen' [3]. Uses of cryptography AES are the Intel Core processor families, security tokens, solid-state devices, wireless networks with the IEEE 802.11i WPA2 standard for secure encryption and streaming video systems, security tokens, magnetisation cards, automated teller machines (ATMs), cellphones, servers for World Wide Web [2].

2 The AES-256 Algorithm

AES means advanced encryption standard. Encryption algorithms make use of the encryption and decryption concepts

Plaintext (P) + Cipher key (K) $\rightarrow$ Cipher text (C) $\rightarrow$ (Encryption Process)
Cipher text (C) + Cipher key (K) $\rightarrow$ Plaintext (P) $\rightarrow$ (Decryption Process)

Table 1 Characteristics

Characteristics	AES-128	AES-192	AES-256
Key size (Word;Bytes;Bits)	4;16;128	6;24;192	8;32;256
Plaintext block size (Word;Bytes;Bits)	4;16;128	4;16;128	4;16;128
Number of rounds	10	12	14
Round key size (Word;Bytes;Bits)	4;16;128	4;16;128	4;16;128
Expanded key size (Word;Bytes)	44;176	52;208	60;240

The entire algorithm is contained in the creation of these cipher text by adding cipher key to plaintext. AES's main operations are S-box, Inverse S-box, Mixcolumns, and Inverse Mixcolumns [4]. A substitution-permutation network is the foundational design idea for AES. The advanced encryption sandard-256 algorithm is mainly consists of two major parts: (i) cipher, (ii) key expansion. The part that performs encryption as well as decryption on blocks of input data is called a cipher. In encryption process, it converts plain text to cipher text using key. Using a key, the decryption process transforms ciphertext into plaintext. A key scheduled created by key expansion is utilised in the cipher process [5].

(i) The transformation known as AddRoundKey comes before the first round.
(ii) There are four transformations in the first Nr-1 round (see Table 1), as follows:

(a) SubBytes Transformation: This transforms the state in a nonlinear way and is represent by the S-box. (b) ShiftRows Transformation: This transforms the state by shifting its rows in a circular manner. (c) MixColumns Transformation: This transforms the state in a linear way. (d) AddRoundKey Transformation: This transforms by XORing a 128 bit key to the state. (iii) The three transformations that make up the final round are as follows: transformations include (a) SubBytes, (b) ShiftRows, (c) AddRoundKey, (d) SubBytes which is an S-box-based transformation. The design of the Substitute Box has a significant impact on AES security [6].

In order to create SubBytes, the multiplicative inverse (inv) of the input element over Galios field $(2^4)^2$ was used, followed by affine transformation. This stage of the AES is the only nonlinear transformation. As opposed to that, Inv SubBytes is made to decrypt data by first applying an inv Affine transformation. The final output of the inv S-box is the multiplicative inv over $GF(2^4)^2$ of the output of the inverse affine transformation [7].

2.1 *AES-256 Encryption*

The encryption operation includes a plaintext of 128 bits or 16 bytes, a cipher key of 256 bits or 32 bytes, and '14' rounds in AES 256 (see Table 1) (see Fig. 3). The AES-256 encryption process has 4 kinds of functions that are repeatedly in a number of rounds: SubByte operation, ShiftRow operation, MixColumn operation, and

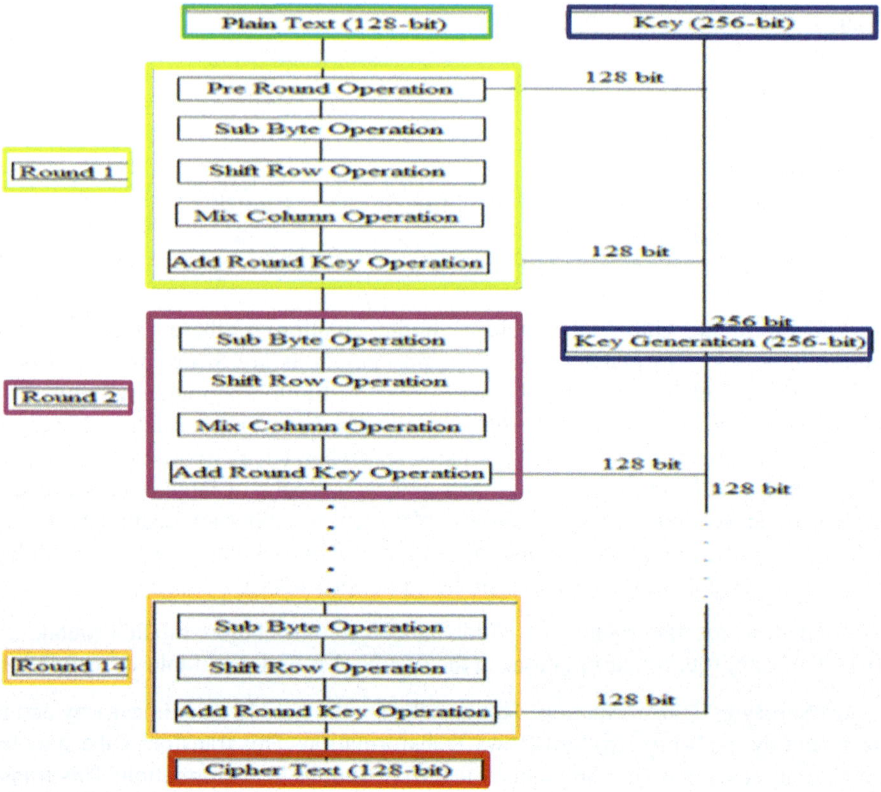

Fig. 3 AES-256 encryption [8]

AddRoundKey operation. In the first round have five operations: Pre_Round operation, Subbyte operation, Shiftrows operation, Mixcolumns operation operation, and Addround key operation. SubByte operations, ShiftRows operations, MixColumns, and AddRoundKey operations are available from the second to the thirteenth round. The final 14th round consists of only three operations. SubByte operation, ShiftRows, and AddRoundKey operations [8].

2.1.1 Input and Output

Plain Text is sequences of 128 bits block such as binary numbers '0' or '1'. Secret Key is 256 bits such as binary numbers '0' or '1'. The Cipher Text means combination of input data and secret key, i.e. 128 bits.

2.1.2 Different Key Length in Bytes

Block length or input data $\rightarrow$ 128 bits or 16 Bytes
Length of a key $\rightarrow$ 128 bits or 16 bytes
Length of a key $\rightarrow$ 192 bits or 24 bytes
Length of a key $\rightarrow$ 256 bits or 32 bytes.

2.1.3 Hexadecimal Bit Patterns

$0000 \rightarrow 0, 0001 \rightarrow 1, 0010 \rightarrow 2,$
$0011 \rightarrow 3, 0100 \rightarrow 4, 0101 \rightarrow 5,$
$0110 \rightarrow 6, 0111 \rightarrow 7, 1000 \rightarrow 8,$
$1001 \rightarrow 9, 1010 \rightarrow A, 1011 \rightarrow B,$
$1100 \rightarrow C, 1101 \rightarrow D, 1110 \rightarrow E, 1111 \rightarrow F$

2.2 AES-256 Decryption

Cipher Text is used as input 128 bits or 16 bytes in the decryption process (see Fig. 4). The Input key is passed to the Key Generation module, which generated new and fresh keys, which are then passed to the Inverse mixColumns, which provide the key for the fourteen rounds of decryption. The Inverse SubByte operation, Inverse ShiftRow operation, Inverse MixColumn, and Inverse AddRoundKey operations are all repeated in a number of rounds during the AES-256 decryption process. The first round includes all five operations: Pre_round operation, Inverse SubByte operation, Inverse ShiftRows operation, Inverse MixColumns operation, and Inverse AddRoundKey operation. From the second to the thirteenth round, there are four operations, such as the Inverse SubByte operation and the Inverse ShiftRows operation, Inverse AddRoundKey operations, and Inverse MixColumns. The final 14th round consists of only three operations, namely Inverse SubByte operations, Inverse ShiftRows, and Inverse AddRoundKey operations [5, 8].

2.3 AES-256 Key Expansion

The 256 bit input key is divided into eight parts during the key expansion process (see Fig. 5). Each component is made up of 32 bits that are organised in a 4×8 matrix. The last column is treated as a row matrix 1×4, and the shift row operation is performed. S-box receives the output of the shiftRows operation and performs a subByte operation. 8 bits of subByte operation output from XOR with the round constant on the MSB side (i.e. round constant value changes for every round). The 0th column of the new and fresh key is EX-ORed with the 1st column as to input key,

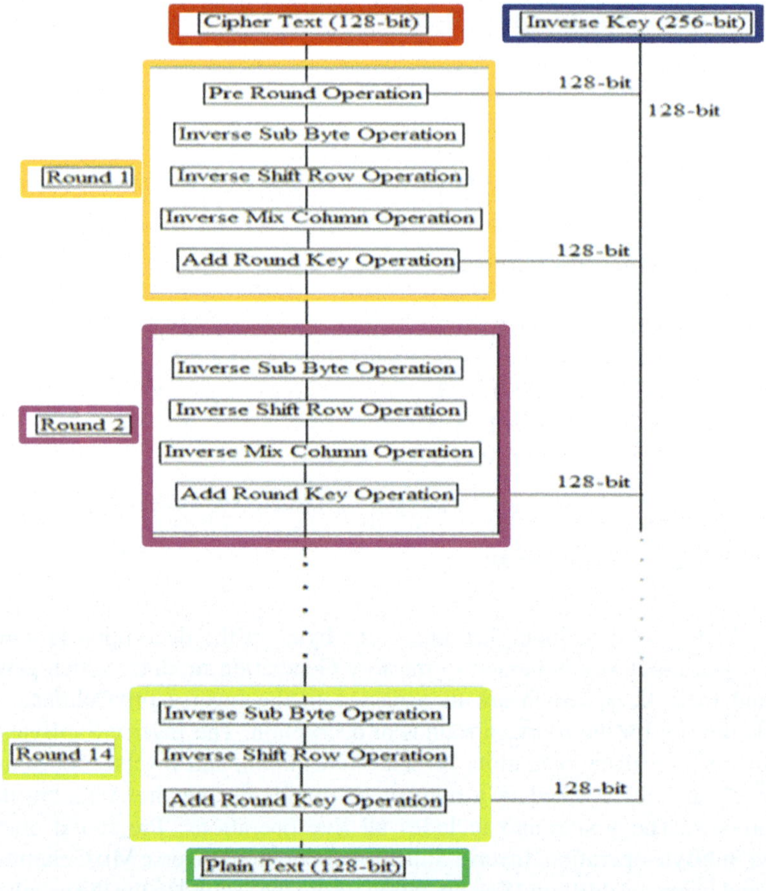

Fig. 4 AES-256 decryption [8]

and the output is taken as the created new key of the 1st column as to new or fresh key. This produces the XOR round constant output, which is XORing with the 0th column of the input key. Similar to how the 1st column of the new key is generated by XORing the first or 1st column of the input key with the 2nd column, the 2nd column of the new key is created by XORing 3rd column of the input key with 2nd column, and so on. To perform an XOR operation, the S-box is given the newly created key from the third column. The output of S-box is XOR with the 4th column of the input key, it gives to the 4th column of new or fresh key, XOR with the fifth or 5th column of input key, which gives the fifth or 5th column of new key, XOR with the sixth or 6th column of input key, which gives the sixth or 6th column of new key, and finally XORed with the seventh or 7th column of input key, which gives the seventh or 7th column of new key. This process is depicted in Fig. 5 AES-256 Key expansion; in

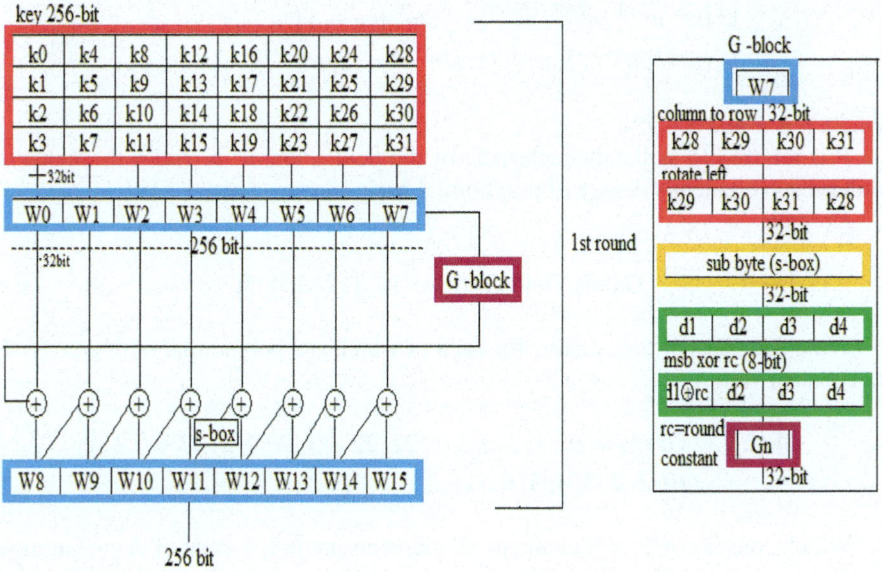

Fig. 5 AES-256 key expansion

this manner, to generate new 256 bits or 32 bytes keys in the AES 256's algorithm by attaching the obtained 8th columns of fresh key.

3 Galois Field (GF) Approach for AES-256

Galois field, also called finite field and named after Evariste Galois, is a field with a constrained number of components. It is especially effective for converting computer data, which is encoded in binary form. In other words, computer data consists of two numbers, 0 and 1, which are the components of a GF with two elements [9]. 'Adding, multiplying, inversion of multiplication, and exponential operations are the four kinds of arithmetic operations such S-box in the AES circuit which performs as the extension field of binary GF(2^m). The elements in the finite field have the following expression form: typically used to determine the space and time complexity of the planned multiplier in the extended field of binary GF(2^m). The basis that is chosen determines the finite field's elements expression form. PolynomialBasis (PB), DualBasis (DB), and NormalBasis (NB) are the three most broadly used substrates' [10]. Mathematical procedures can quickly and efficiently jumble data when it is represented as a vector in a Galois field [9].

The following is a definition of the elements of the Galois field GF(p^n):

$$\text{GF}(u^n) = (0, 1, 2, 3, \ldots, u-1) \cup (u, u+1, u+2, u+3, \ldots, u+u-1)$$

$$\cup \left(u^2, u^{2+1}, u^{2+2}, u^{2+3}, \ldots, u^2 + u - 1\right) \cup \ldots$$
$$\cup \left(u^{n-1}, u^{n-1+1}, u^{n-1+2}, u^{n-1+4}, \ldots, u^{n-1+p-1}\right)$$

where $u \in P$ and $n \in Z^+$.

P is referred to as the characteristic of the field's, and u^n provides the fields of order. Each element's degree of polynomial basis is capped at $n - 1$ [9].

Example:

$$\text{Galois Field GF}(6) = [0, 1, 2, 3, 4, 5]$$

Which consists of 6 elements, for each of which is a polynomial of degree is '0' (a constant value).

$$\begin{aligned} \text{GF}\left(2^3\right) &= \text{GF}(8) = (0, 1, 2, 2+1, 22, 22+1, 22+2, 22+2+1) \\ &= (0, 1, 2, 3, 4, 5, 6, 7) \end{aligned}$$

Which consists of $2^3 = 8$ elements all elements are polynomial of degree at most two evaluate at two [9].

3.1 GF(2^4) Addition Using Logic Gate

The added of these 2 elements in the Galois field (GF) can be relocated to simple bitwise EX-OR operation between these 2 element.

3.2 GF(2^4) Squaring Using Logic Gate

Assume that $k = q^2$ and that each of the two, as indicated by the binary values, is a component of GF(2^4) are $(k3\ k2\ k1\ k0)_2$ and $(q3\ q2\ q1\ q0)_2$ correspondingly [9].

3.3 Multiplying by a Constant (λ and φ) Using Logic Gate

Assume $k = q\lambda$

Where $k = (k3\ k2\ k1\ k0)_2$,

$q = (q3\ q2\ q1\ q0)_2$

$\lambda = (1100)_2$ are GF(2^4) elements

Assume $k = q\varphi$

Where $k = (k1\ k0)_2$,

$q = (q1\ q0)_2$
$\varphi = \{10\}_2$ are GF(2^2) elements

The formulas for multiplying by a constant (λ and φ) are presented see Fig. 7.

4 Galois Field (GF) Approach Based on LUT for AES

A traditional approach is to implement the advanced encryption standard (AES) algorithm using LUTs. It is quite easy to implement the needed functionality. As a result, the following LUTs are used for this purpose. LUTs for implemented SubBytes operation, InvSubBytes transformations are shown in Figs. 6 and 7, respectively. See Figs. 8, 9, 10, 11, 12, 13, 14 and 15 shows LUTs for obtaining the results of Galois Field—GF ((2^{28}) multiplications [2].

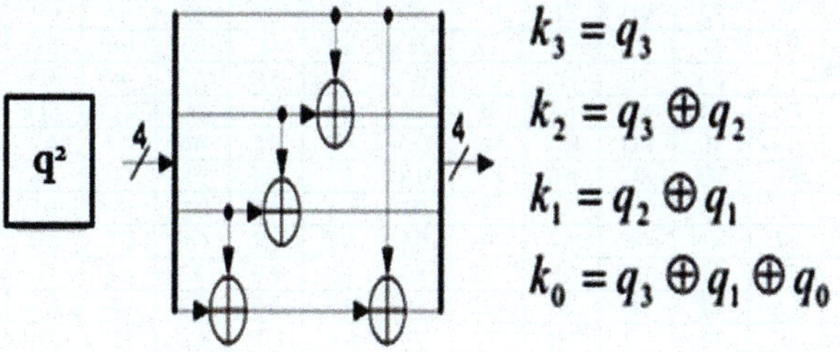

Fig. 6 GF(2^4) squaring hardware diagram

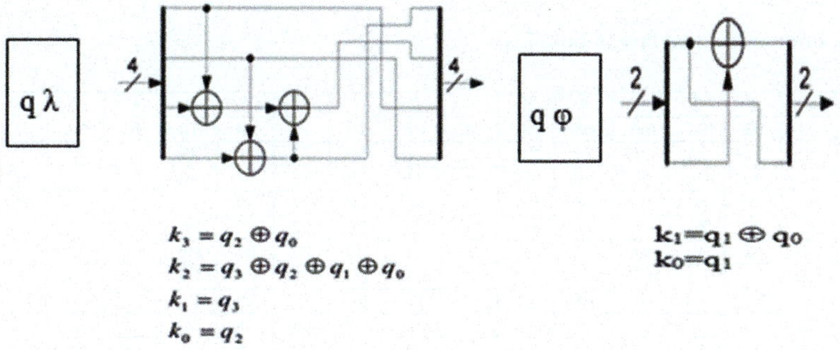

Fig. 7 Multiplying by a constant (λ and φ) hardware diagram

	0	1	2	3	4	5	6	7	8	9	A	B	C	D	E	F
0	63	7C	77	7B	F2	6B	6F	C5	30	01	67	2B	FE	D7	AB	76
1	CA	82	C9	7D	FA	59	47	F0	AD	D4	A2	AF	9C	A4	72	C0
2	B7	FD	93	26	36	3F	F7	CC	34	A5	E5	F1	71	D8	31	15
3	04	C7	23	C3	18	96	05	9A	07	12	80	E2	EB	27	B2	75
4	09	83	2C	1A	1B	6E	5A	A0	52	3B	D6	B3	29	E3	2F	84
5	53	D1	00	ED	20	FC	B1	5B	6A	CB	BE	39	4A	4C	58	CF
6	D0	EF	AA	FB	43	4D	33	85	45	F9	02	7F	50	3C	9F	A8
7	51	A3	40	8F	92	9D	38	F5	BC	B6	DA	21	10	FF	F3	D2
8	CD	0C	13	EC	5F	97	44	17	C4	A7	7E	3D	64	5D	19	73
9	60	81	4F	DC	22	2A	90	88	46	EE	B8	14	DE	5E	0B	DB
A	E0	32	3A	0A	49	06	24	5C	C2	D3	AC	62	91	95	E4	79
B	E7	C8	37	6D	8D	D5	4E	A9	6C	56	F4	EA	65	7A	AE	08
C	BA	78	25	2E	1C	A6	B4	C6	E8	DD	74	1F	4B	BD	8B	8A
D	70	3E	B5	66	48	03	F6	0E	61	35	57	B9	86	C1	1D	9E
E	E1	F8	98	11	69	D9	8E	94	9B	1E	87	E9	CE	55	28	DF
F	8C	A1	89	0D	BF	E6	42	68	41	99	2D	0F	B0	54	BB	16

Fig. 8 S-box (standard box) [2]

	0	1	2	3	4	5	6	7	8	9	A	B	C	D	E	F
0	52	09	6A	D5	30	36	A5	38	BF	40	A3	9E	81	F3	D7	FB
1	7C	E3	39	82	9B	2F	FF	87	34	8E	43	44	C4	DE	E9	CB
2	54	7B	94	32	A6	C2	23	3D	EE	4C	95	0B	42	FA	C3	4E
3	08	2E	A1	66	28	D9	24	B2	76	5B	A2	49	6D	8B	D1	25
4	72	F8	F6	64	86	68	98	16	D4	A4	5C	CC	5D	65	B6	92
5	6C	70	48	50	FD	ED	B9	DA	5E	15	46	57	A7	8D	9D	84
6	90	D8	AB	00	8C	BC	D3	0A	F7	E4	58	05	B8	B3	45	06
7	D0	2C	1E	8F	CA	3F	0F	02	C1	AF	BD	03	01	13	8A	6B
8	3A	91	11	41	4F	67	DC	EA	97	F2	CF	CE	F0	B4	E6	73
9	96	AC	74	22	E7	AD	35	85	E2	F9	37	E8	1C	75	DF	6E
A	47	F1	1A	71	1D	29	C5	89	6F	B7	62	0E	AA	18	BE	1B
B	FC	56	3E	4B	C6	D2	79	20	9A	DB	C0	FE	78	CD	5A	F4
C	1F	DD	A8	33	88	07	C7	31	B1	12	10	59	27	80	EC	5F
D	60	51	7F	A9	19	B5	4A	0D	2D	E5	7A	9F	93	C9	9C	EF
E	A0	E0	3B	4D	AE	2A	F5	B0	C8	EB	BB	3C	83	53	99	61
F	17	2B	04	7E	BA	77	D6	26	E1	69	14	63	55	21	0C	7D

Fig. 9 Inverse standard box (s-box) [2]

	0	1	2	3	4	5	6	7	8	9	A	B	C	D	E	F
0	00	02	04	06	08	0A	0C	0E	10	12	14	16	18	1A	1C	1E
1	20	22	24	26	28	2A	2C	2E	30	32	34	36	38	3A	3C	3E
2	40	42	44	46	48	4A	4C	4E	50	52	54	56	58	5A	5C	5E
3	60	62	64	66	68	6A	6C	6E	70	72	74	76	78	7A	7C	7E
4	80	82	84	86	88	8A	8C	8E	90	92	94	96	98	9A	9C	9E
5	A0	A2	A4	A6	A8	AA	AC	AE	B0	B2	B4	B6	B8	BA	BC	BE
6	C0	C2	C4	C6	C8	CA	CC	CE	D0	D2	D4	D6	D8	DA	DC	DE
7	E0	E2	E4	E6	E8	EA	EC	EE	F0	F2	F4	F6	F8	FA	FC	FE
8	1B	19	1F	1D	13	11	17	15	0B	09	0F	0D	03	01	07	05
9	3B	39	3F	3D	33	31	37	35	2B	29	2F	2D	23	21	27	25
A	5B	59	5F	5D	53	51	57	55	4B	49	4F	4D	43	41	47	45
B	7B	79	7F	7D	73	71	77	75	6B	69	6F	6D	63	61	67	65
C	9B	99	9F	9D	93	91	97	95	8B	89	8F	8D	83	81	87	85
D	BB	B9	BF	BD	B3	B1	B7	B5	AB	A9	AF	AD	A3	A1	A7	A5
E	DB	D9	DF	DD	D3	D1	D7	D5	CB	C9	CF	CD	C3	C1	C7	C5
F	FB	F9	FF	FD	F3	F1	F7	F5	EB	E9	EF	ED	E3	E1	E7	E5

Fig. 10 Multiply by 2 Galois multiplications LUT [2]

	0	1	2	3	4	5	6	7	8	9	A	B	C	D	E	F
0	00	03	06	05	0C	0F	0A	09	18	1B	1E	1D	14	17	12	11
1	30	33	36	35	3C	3F	3A	39	28	2B	2E	2D	24	27	22	21
2	60	63	66	65	6C	6F	6A	69	78	7B	7E	7D	74	77	72	71
3	50	53	56	55	5C	5F	5A	59	48	4B	4E	4D	44	47	42	41
4	C0	C3	C6	C5	CC	CF	CA	C9	D8	DB	DE	DD	D4	D7	D2	D1
5	F0	F3	F6	F5	FC	FF	FA	F9	E8	EB	EE	ED	E4	E7	E2	E1
6	A0	A3	A6	A5	AC	AF	AA	A9	B8	BB	BE	BD	B4	B7	B2	B1
7	90	93	96	95	9C	9F	9A	99	88	8B	8E	8D	84	87	82	81
8	9B	98	9D	9E	97	94	91	92	83	80	85	86	8F	8C	89	8A
9	AB	A8	AD	AE	A7	A4	A1	A2	B3	B0	B5	B6	BF	BC	B9	BA
A	FB	F8	FD	FE	F7	F4	F1	F2	E3	E0	E5	E6	EF	EC	E9	EA
B	CB	C8	CD	CE	C7	C4	C1	C2	D3	D0	D5	D6	DF	DC	D9	DA
C	5B	58	5D	5E	57	54	51	52	43	40	45	46	4F	4C	49	4A
D	6B	68	6D	6E	67	64	61	62	73	70	75	76	7F	7C	79	7A
E	3B	38	3D	3E	37	34	31	32	23	20	25	26	2F	2C	29	2A
F	0B	08	0D	0E	07	04	01	02	13	10	15	16	1F	1C	19	1A

Fig. 11 Multiply by 3 Galois multiplications LUT [2]

	0	1	2	3	4	5	6	7	8	9	A	B	C	D	E	F
0	00	09	12	1B	24	2D	36	3F	48	41	5A	53	6C	65	7E	77
1	90	99	82	8B	B4	BD	A6	AF	D8	D1	CA	C3	FC	F5	EE	E7
2	3B	32	29	20	1F	16	0D	04	73	7A	61	68	57	5E	45	4C
3	AB	A2	B9	B0	8F	86	9D	94	E3	EA	F1	F8	C7	CE	D5	DC
4	76	7F	64	6D	52	5B	40	49	3E	37	2C	25	1A	13	08	01
5	E6	EF	F4	FD	C2	CB	D0	D9	AE	A7	BC	B5	8A	83	98	91
6	4D	44	5F	56	69	60	7B	72	05	0C	17	1E	21	28	33	3A
7	DD	D4	CF	C6	F9	F0	EB	E2	95	9C	87	8E	B1	B8	A3	AA
8	EC	E5	FE	F7	C8	C1	DA	D3	A4	AD	B6	BF	80	89	92	9B
9	7C	75	6E	67	58	51	4A	43	34	3D	26	2F	10	19	02	0B
A	D7	DE	C5	CC	F3	FA	E1	E8	9F	96	8D	84	BB	B2	A9	A0
B	47	4E	55	5C	63	6A	71	78	0F	06	1D	14	2B	22	39	30
C	9A	93	88	81	BE	B7	AC	A5	D2	DB	C0	C9	F6	FF	E4	ED
D	0A	03	18	11	2E	27	3C	35	42	4B	50	59	66	6F	74	7D
E	A1	A8	B3	BA	85	8C	97	9E	E9	E0	FB	F2	CD	C4	DF	D6
F	31	38	23	2A	15	1C	07	0E	79	70	6B	62	5D	54	4F	46

Fig. 12 Multiply by 9 Galois multiplications LUT [2]

	0	1	2	3	4	5	6	7	8	9	A	B	C	D	E	F
0	00	0B	16	1D	2C	27	3A	31	58	53	4E	45	74	7F	62	69
1	B0	BB	A6	AD	9C	97	8A	81	E8	E3	FE	F5	C4	CF	D2	D9
2	7B	70	6D	66	57	5C	41	4A	23	28	35	3E	0F	04	19	12
3	CB	C0	DD	D6	E7	EC	F1	FA	93	98	85	8E	BF	B4	A9	A2
4	F6	FD	E0	EB	DA	D1	CC	C7	AE	A5	B8	B3	82	89	94	9F
5	46	4D	50	5B	6A	61	7C	77	1E	15	08	03	32	39	24	2F
6	8D	86	9B	90	A1	AA	B7	BC	D5	DE	C3	C8	F9	F2	EF	E4
7	3D	36	2B	20	11	1A	07	0C	65	6E	73	78	49	42	5F	54
8	F7	FC	E1	EA	DB	D0	CD	C6	AF	A4	B9	B2	83	88	95	9E
9	47	4C	51	5A	6B	60	7D	76	1F	14	09	02	33	38	25	2E
A	8C	87	9A	91	A0	AB	B6	BD	D4	DF	C2	C9	F8	F3	EE	E5
B	3C	37	2A	21	10	1B	06	0D	64	6F	72	79	48	43	5E	55
C	01	0A	17	1C	2D	26	3B	30	59	52	4F	44	75	7E	63	68
D	B1	BA	A7	AC	9D	96	8B	80	E9	E2	FF	F4	C5	CE	D3	D8
E	7A	71	6C	67	56	5D	40	4B	22	29	34	3F	0E	05	18	13
F	CA	C1	DC	D7	E6	ED	F0	FB	92	99	84	8F	BE	B5	A8	A3

Fig. 13 Multiply by 11 Galois multiplications LUT [2]

	0	1	2	3	4	5	6	7	8	9	A	B	C	D	E	F
0	00	0D	1A	17	34	39	2E	23	68	65	72	7F	5C	51	46	4B
1	D0	DD	CA	C7	E4	E9	FE	F3	B8	B5	A2	AF	8C	81	96	9B
2	BB	B6	A1	AC	8F	82	95	98	D3	DE	C9	C4	E7	EA	FD	F0
3	6B	66	71	7C	5F	52	45	48	03	0E	19	14	37	3A	2D	20
4	6D	60	77	7A	59	54	43	4E	05	08	1F	12	31	3C	2B	26
5	BD	B0	A7	AA	89	84	93	9E	D5	D8	CF	C2	E1	EC	FB	F6
6	D6	DB	CC	C1	E2	EF	F8	F5	BE	B3	A4	A9	8A	87	90	9D
7	06	0B	1C	11	32	3F	28	25	6E	63	74	79	5A	57	40	4D
8	DA	D7	C0	CD	EE	E3	F4	F9	B2	BF	A8	A5	86	8B	9C	91
9	0A	07	10	1D	3E	33	24	29	62	6F	78	75	56	5B	4C	41
A	61	6C	7B	76	55	58	4F	42	09	04	13	1E	3D	30	27	2A
B	B1	BC	AB	A6	85	88	9F	92	D9	D4	C3	CE	ED	E0	F7	FA
C	B7	BA	AD	A0	83	8E	99	94	DF	D2	C5	C8	EB	E6	F1	FC
D	67	6A	7D	70	53	5E	49	44	0F	02	15	18	3B	36	21	2C
E	0C	01	16	1B	38	35	22	2F	64	69	7E	73	50	5D	4A	47
F	DC	D1	C6	CB	E8	E5	F2	FF	B4	B9	AE	A3	80	8D	9A	97

Fig. 14 Multiply by 13 Galois multiplications LUT [2]

	0	1	2	3	4	5	6	7	8	9	A	B	C	D	E	F
0	00	0E	1C	12	38	36	24	2A	70	7E	6C	62	48	46	54	5A
1	E0	EE	FC	F2	D8	D6	C4	CA	90	9E	8C	82	A8	A6	B4	BA
2	DB	D5	C7	C9	E3	ED	FF	F1	AB	A5	B7	B9	93	9D	8F	81
3	3B	35	27	29	03	0D	1F	11	4B	45	57	59	73	7D	6F	61
4	AD	A3	B1	BF	95	9B	89	87	DD	D3	C1	CF	E5	EB	F9	F7
5	4D	43	51	5F	75	7B	69	67	3D	33	21	2F	05	0B	19	17
6	76	78	6A	64	4E	40	52	5C	06	08	1A	14	3E	30	22	2C
7	96	98	8A	84	AE	A0	B2	BC	E6	E8	FA	F4	DE	D0	C2	CC
8	41	4F	5D	53	79	77	65	6B	31	3F	2D	23	09	07	15	1B
9	A1	AF	BD	B3	99	97	85	8B	D1	DF	CD	C3	E9	E7	F5	FB
A	9A	94	86	88	A2	AC	BE	B0	EA	E4	F6	F8	D2	DC	CE	C0
B	7A	74	66	68	42	4C	5E	50	0A	04	16	18	32	3C	2E	20
C	EC	E2	F0	FE	D4	DA	C8	C6	9C	92	80	8E	A4	AA	B8	B6
D	0C	02	10	1E	34	3A	28	26	7C	72	60	6E	44	4A	58	56
E	37	39	2B	25	0F	01	13	1D	47	49	5B	55	7F	71	63	6D
F	D7	D9	CB	C5	EF	E1	F3	FD	A7	A9	BB	B5	9F	91	83	8D

Fig. 15 Multiply by 14 Galois multiplications LUT [2]

5 Simulation Results of Galois Field (GF) Approach Based on LUT for AES-256

The below figures are from Figs. 16, 17, 18, 19, 20, 21, 22, and 23. The inputs are '*a*' 8 bit, and outputs are '*y*' 8 bit. In S-box (standard box) (see Fig. 16), input '*a*' is '55' means row is 5, column is 5. The result of s-box (5, 5), *y* is 'fa' (see Fig. 8). In inverse standard box (s-box) (see Fig. 17), input '*a*' is '55' means row is 5, column is 5.

The result of inverse s-box (5, 5), *y* is 'ed' (see Fig. 9). In Galois, multiply by 2 Galois multiplications LUT (see Fig. 18), input '*a*' is '5a' means row is 5, column is a. The result is (5, a), *y* is 'b4' (see Fig. 10). Multiply by 3 Galois multiplications LUT (see Fig. 19), input '*a*' is '3f' means row is 3, column is f. The result is (3, f), *y* is '41' (see Fig. 11). In multiply by 9 Galois multiplications LUT (see Fig. 20),

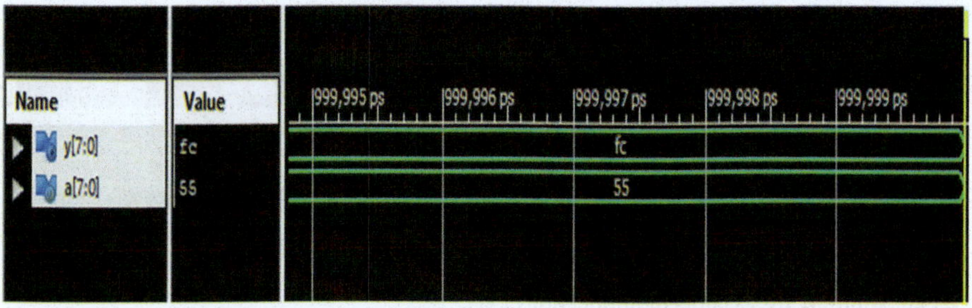

Fig. 16 S-box (standard box)

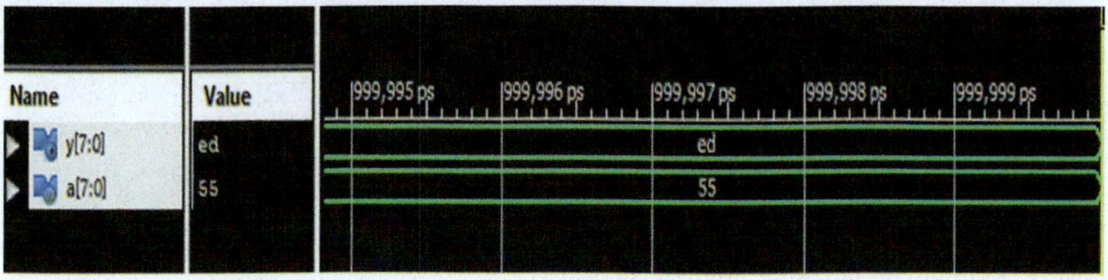

Fig. 17 Inverse standard box (s-box)

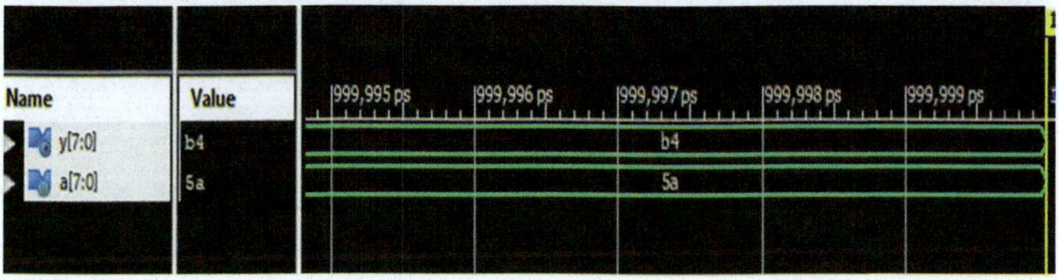

Fig. 18 Multiply by 2 Galois multiplications LUT

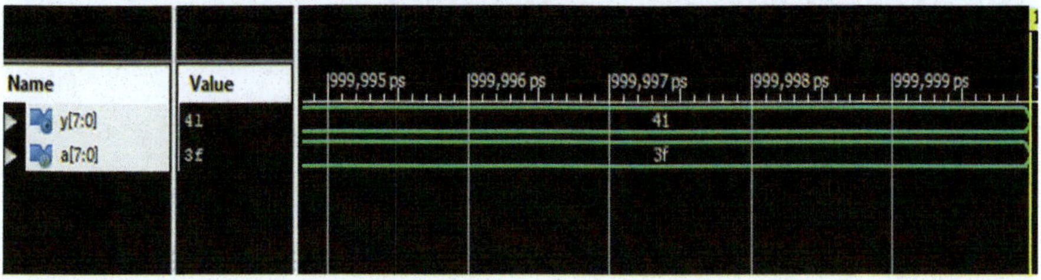

Fig. 19 Multiply by 3 Galois multiplications LUT

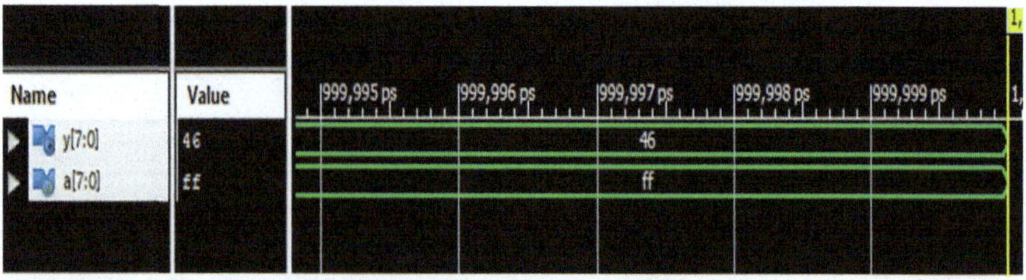

Fig. 20 Multiply by 9 Galois multiplications LUT

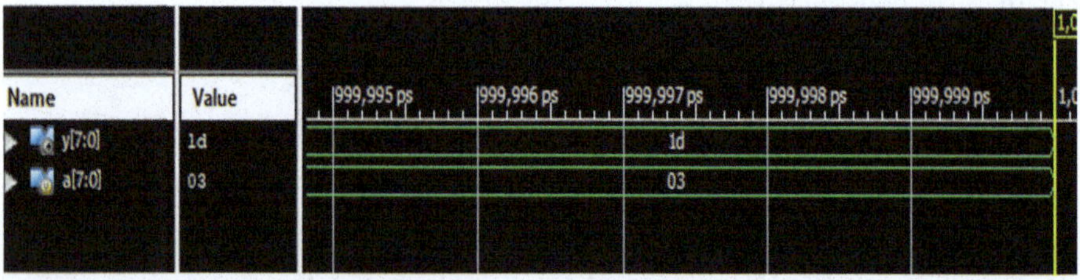

Fig. 21 Multiply by 11 Galois multiplications LUT

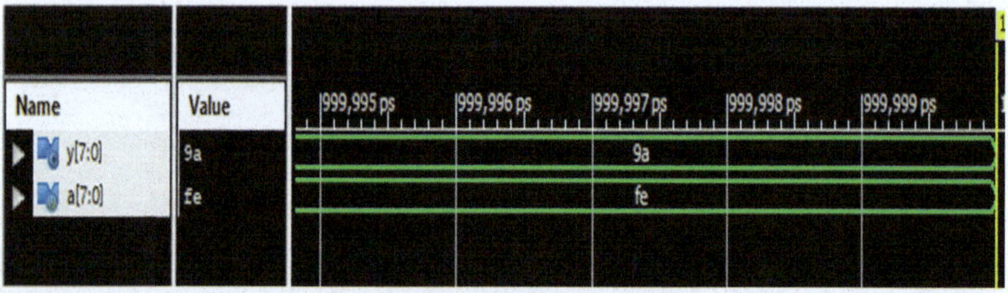

Fig. 22 Multiply by 13 Galois multiplications LUT

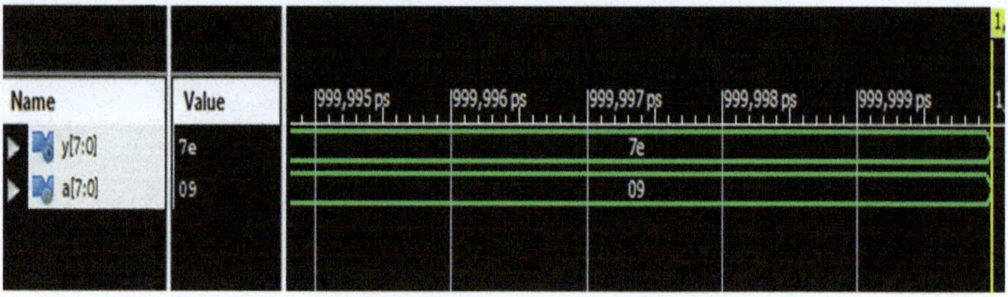

Fig. 23 Multiply by 14 Galois multiplications

Fig. 24 AES RTL schematic diagram

input '*a*' is 'ff'. The result is (f, f), *y* is '46' (see Fig. 12). Multiply by 11 Galois multiplications LUT (see Fig. 21), input '*a*' is '03'. The result is (0, 3), *y* is '1d' (see Fig. 13). In multiply by 13 Galois multiplications LUT (see Fig. 22), input '*a*' is 'fe'. The result is (f, e), *y* is '9a' (see Fig. 14). In multiply by 14 Galois multiplications LUT (see Fig. 23), input '*a*' is '09'. The result is (0, 9), *y* is '7e' (see Fig. 15).

6 AES-256 RTL Schematic Diagram

AES-256 algorithm encryption and decryption schematic diagram are depicted in Fig. 24. Here, taking clk, rst, and enc_dec are input active low it act as encryption. Later, clk, rst, and enc_dec are input active high it act as decryption. And aesin is 128 bit input, keyin is 256 bit, and aesout is 128 bit.

7 Simulation Results of Galois Field (GF) Approach Based on Logic Gate for AES-256

All of these above waveforms were implemented in Verilog-test bench and later simulated in the Xilinx ISE design suite 14.7 tool. AES-256 algorithm encryption and decryption simulation results hexadecimal and ASCII code format as shown in figures are from Figs. 25, 26, 27, and 28. When clk, rst, and enc_dec are input active low it act as encryption. Input is aesin 128 bit, and key is keyin 256 bit. It generates aesout also called cipherkey 128 bit. This cyperkey is input of decryption and give key same as encryption keyin 256 bit. When clk, rst, and enc_dec are input active high it act as decryption. Later, it generates original data that is aesin 128 bit. All inputs and outputs shown in above table in form of HEX code and ASCİİ code forms (see Table 2).

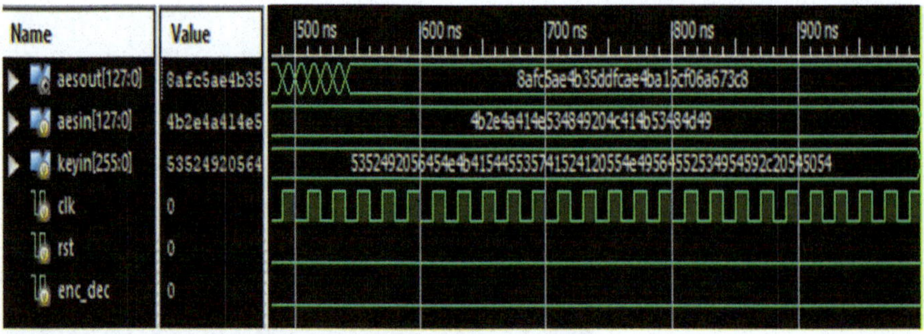

Fig. 25 AES-256 algorithm encryption (Hex code)

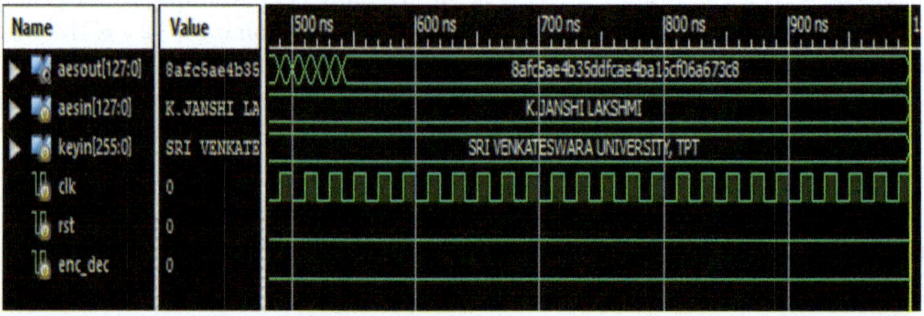

Fig. 26 AES-256 algorithm encryption (ASCII code)

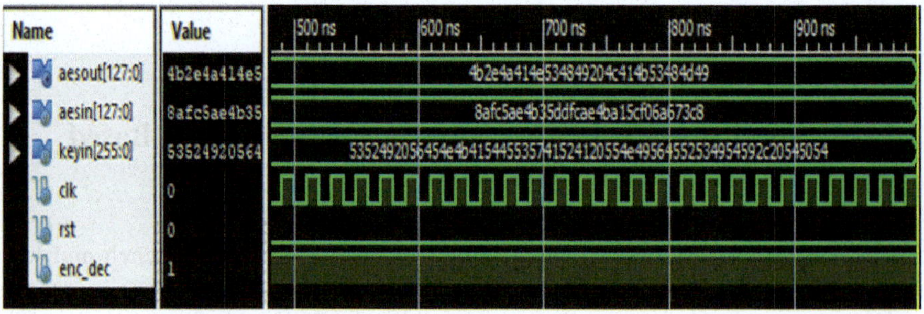

Fig. 27 AES-256 algorithm decryption (Hex code)

8 Conclusion

Blocks of input are 128 bits or 16 bytes can be encrypted and decrypted using the AES algorithm method employing cryptographic keys of 128 bits or 16 bytes, 192 bits or 24 bytes, and 256 bits or 32 bytes. AES uses a configurable number of rounds that are determined by the key length. '10' rounds, '12' rounds, and '14' rounds are used by

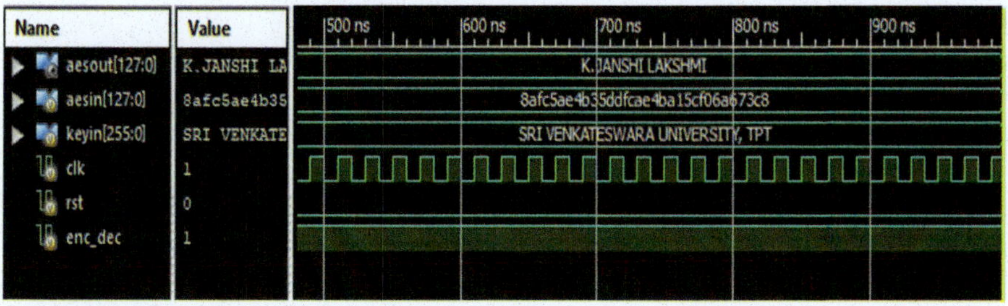

Fig. 28 AES-256 algorithm decryption (ASCII code)

Table 2 AES-256 input, cipher key, and output data

AES-256 encryption data	
Hexadecimal code	
aesin	4B2E4A414E534849204C414B53484D49
keyin	5352492056454E4B415445535741524120554E495645525349545592C20545054
aesout (cipher key)	8afc5aedb35ddfcae4balscf06a673c8
ASCII code	
aesin	K.JANSHI LAKSHMI
keyin	SRI VENKATESWARA UNIVERSITY, TPT
aesout (cipher key)	8afc5aedb35ddfcaedbal5cf06a673c8 (Hex code)
AES-256 decryption data	
Hexadecimal code	
aesin (cipher key)	8afc5aedb35ddfcaedbalscf06a673c8
keyin	5352492056454E4B415445535741524120554E495645525349545592C20545054
aesout	4B2E4A414E534845204C414B53484D49
ASCII code	
aesin (cipher key)	8afc5aedb35ddfcaedbal5cf06a673c8 (Hex code)
keyin	SRI VENKATESWARA UNIVERSITY, TPT
aesout	K.JANSHI LAKSHMI

AES for keys with 128 bits or 16 bytes, 192 bits or 24 bytes, and 256 bits or 32 bytes, respectively. The number of rounds repeated when a key with a size of 256 bits or 32 bytes is utilised equals 14. Searching through a fixed table (S-box) provided in the design replaces the 16 input bytes. Implemented and simulated using the XILINX ISE design suite 14.7 tool for the S-box is to use GaloisFieldGF(2^8) approach based on lookup table and logic gates for calculating Addition, squaring and multiply with constants of GaloisFieldGF(2^8).

References

1. Nation Institute of Standards and Technology (NIST), Data Encryption Standard (DES) (1999) National Technical Information Service, Springfield
2. Sai Srinivas NS, Akramuddin MD (2016) FPGA based hardware implementation of AES Rijndael algorithm for encryption and decryption. In: International conference on electrical, electronics, and optimization techniques (ICEEOT)
3. FIPS PUB 197 Advanced Encryption Standard (AES) (2001)
4. Balupala HK, Rahul K, Yachareni S (2021) Galois field arithmetic operations using Xilinx FPGAs in cryptography. In: IEEE international IOT, electronics and mechatronics conference (IEMTRONICS), 21–24 April 2021
5. Stallings W. Cryptography and network security-principles and practice, 5th edn. Prentice Hall, Pearson. ISBN: 978-0-13-609704-4
6. Nitaj A, Tonien WSJ (2020) A new improved AES S-box with enhanced properties. Part of the Lecture notes in computer science book series (LNSC), vol 12248
7. Teng Y-T, Chin W-L (2022) VLSI architecture of S-box with high area efficiency based on composite field arithmetic. IEEE Access 10
8. Daemen J, Rijmen V (1999) AES proposal: Rijndael. AES algorithm submission, Sept 3, 1999
9. Benvenuto CJ (2012) Galois field in cryptography
10. Qin P, Zhou F, Wu N, Xian F (2021) A compact implementation of AES S-box based on dual basis. In: IEEE 4th international conference on electronics technology

Chapter 11
Load Balancing for MEC in 5G-Enabled IoT Networks and Malicious Data Validation Using Blockchain

Jayalakshmi G. Nargund, Chandrashekhar V. Yediurmath, M. Vijayalaxmi, and Vishwanath P. Baligar

1 Introduction

Real-time multimedia applications are playing an important role in communication system between the users. To fulfill the demand of users in these applications, 5th generation (5G) network is more suitable. Low-latency and high-quality of services and the low computing power are requirements of real-time multimedia applications such as smart agricultural, campus, and city. Mobile edge computing (MEC) enabled with 5G is the preferred solution [1]. Small dense cells in 5G communicate with MEC through a base station (BS). This leads to effective usage of resources and computing power [2].

The new architecture of MEC with 5G also introduces multiple choices of computations such as offloading data. Authors of this article are used ACMLO strategy to balance and optimize the load between the cells, which are connected to servers. The proposed LB_MECs algorithm balances the load between the MEC servers based on average load of edge network.

ACMLO makes decisions on the division of offloading tasks and the allocation of channel bandwidth, so as to complete the requested task with minimal system cost [3]. ALN is based on size of waiting and service queues of each server in it. If the load of the MEC server is more than the average load, then it is considered as overloaded server. Similarly if the server load is less than the average load, then it is listed as under-loaded. The LB_MECs algorithm transfers overloaded server load to under loaded servers.

The combination of cloud and edge computing leads to a new paradigm in IoT-based 5G networks [4]. OMECC algorithm is presented in this paper to offload data from the MEC server to cloud. It leads to an effective computations between MEC and cloud. Hence, it helps to meet the QoS requirements of diverse set of users.

J. G. Nargund (✉) · C. V. Yediurmath · M. Vijayalaxmi · V. P. Baligar
KLE Technological University, Hubballi 580031, India
e-mail: jayagn@kletech.ac.in

V. Devedzic et al. (eds.), *Proceedings of the International Conference on Intelligent Computing, Communication and Information Security*, Algorithms for Intelligent Systems, https://doi.org/10.1007/978-981-99-1373-2_11

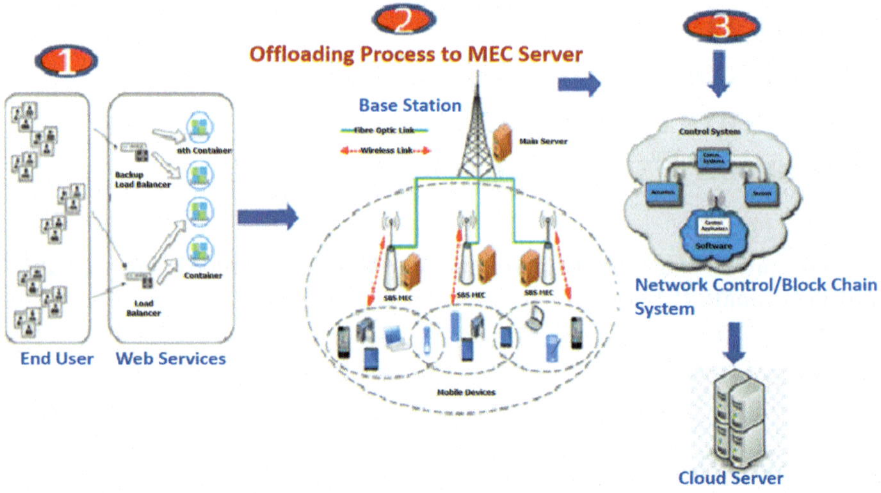

Fig. 1 MEC with blockchain-based cloud system

Blockchain is emerging technology, which promises to tackle security issues in IoT-enabled 5G networks [5] . Smart contract tool is adopted to offer access control solutions to manage and protect cloud resources among nodes [6]. Authors of this article are using blockchain to detect malicious data records of MEC.

Figure 1 represents the proposed model, and it includes three phases. The services of MEC servers are accessed by end users through Web interfaces via wireless network environment. It is depicted as Phase 1 of the research work. 5G HetNet includes set of small dense cells which are connected MEC server through base station (BS) of that corresponding cells. Thus, hierarchical structure is established between MEC servers. Top level MEC server is connected to cloud via base station. This is considered as Phase 2 of the proposed work. The data generated by edge devices is offloaded to MEC servers through BS using optical wireless communication [7]. The computations required to provided services to MDs are performed by MECs. Since huge data from edge devices is dumped in to MEC servers, balancing the load is necessary. Authors of this paper propose load balancing algorithm based on ALN for MECs. Thus, the proposed model drastically improves the response time and throughput of the network. In Phase 3, data offloaded to cloud is verified by blockchain to detect malicious data records. Hence, useful and authorized data is stored on the cloud.

The rest of the paper is organized as follows: Sect. 2 illustrates the survey on load balancing algorithms for MECs and secure offloading in blockchain-based IoT networks. The proposed work is described in Sect. 3. The result discussion is presented in Sect. 4. The article is concluded in Sect. 5.

2 Related Work:

This section explores the research work carried out about load balancing for MEC, ant colony, and offloading data in blockchain-based IoT networks.

Yaser Khamayseh et al. in [1] propose a load balancing algorithm for a distributed environment. The modified particle swarm optimization (MPSO) algorithm is used to avoid falling into local optimum in the iterative process. The task assignment model is proposed by using different strategies like software-defined fog networking (SDFN) strategy. According to that classification results, first strategy MPSO minimizes the completion time. Second strategy SDFN balances the load between nodes, and it improves the quality of service which is a solution for mobile edge computing.

Yijie Wang et al. in [2] present an analysis of the number of ants (i.e., edge nodes) in ant colony system algorithm. The study focuses on the effect of changing the number of ants in the algorithm behavior rather than find the optimum number. The factors which investigate in this study are algorithm execution time, best solution, pheromones accumulative, pheromone dispersion, and the number of new solutions found by the ants. The results show that the number of ants changes the algorithm behavior intensely. Therefore, tuning the parameter number of ants in ant colony system could be easier by applying the min and max number of ants recommended in this study.

Dinh C. Nguyen et al. in [3] propose a blockchain-empowered framework for implementing resource trading and task assignment in edge computing. The smart contracts tool can provide reliable resource transactions and immutable records in the blockchain. This tool can also be adopted for facilitating a vehicular edge consortium blockchain. It enables secure communication to share their computation resources with service requester. To support security and privacy for cognitive edge computing, spatial-temporal smart contracts are used with intensive mechanisms to accelerate the economy sharing in smart campuses.

Ming Ding et al. in [4] present the offloading strategies utilized for edge or cloud services to tackle blockchain-based IoT computation issues. The objective of research is minimizing offloading costs and cloud resources by leveraging conventional convex optimization methods. But such conventional offloading optimization algorithms only work well for low-complexity online models and usually require prior knowledge of system statistics that is difficult to acquire in practical scenarios.

Jaesung Park et al. in [8] improve the resource efficiency of MEC systems. It is important to distribute the imposed workload evenly among MECs. To address this issue, authors propose a task redirection method to balance loads among servers in a distributed manner. Thus, the method enables a congested MECs to distribute tasks to a set of under-loaded servers.

The survey motivates the authors of this article to propose the load balancing algorithm for MECs in 5G network. Authors also extend their work to detect malicious records using blockchain strategy.

3 Proposed Model

In this section, authors explore the proposed system model, LB_MECs load balancing algorithm and access control of OMECC system.

The smart campus application has developed by authors of this paper to monitor vehicle/people movement, air and sound pollution, etc. This application is generating data and provides interaction between user and network. IoT aware 5G network test bed in NS3 [9] uses collected data, which is generated by the smart campus application. Different sensors such as pH, temperature, moisture sensors, and more are used to collect the data. The generated data by the application is stored in the local server.

Authors consider a two-tier computation offloading model presented in Fig. 2, and it is performed in Phase 1, and Phase 2 considers a number of users equipment (UE), which are connected to base station. Each base station (BS) is connected to a MEC server. ACMLO algorithm is used to optimize the load on MEC server. The proposed LB_MECs algorithm is used to balance the load between the MEC servers. The cluster of MEC servers is connected to the cloud through hierarchically top level base station.

The Phase 2 proposes an access control model on a blockchain network for the OMECC system as shown in Fig. 2. Since offloading data from BS to MEC is considered in Phase 1, access control scheme is included in Phase 2. Key components of this phase are OMECC, admin, smart contracts, and miners. The steps in workflow of access control on blockchain are depicted in Fig. 2, and they are as follows:

1. A data from MEC is offloaded to blockchain-based cloud following the initialization process of a cloud.
2. OMECC manager which is implemented using smart contracts tool will verify the malicious data in cloud.
3. OMECC manager processes the requests of MEC in first come first serve (FCFS) manner.
4. Verification of requests is done by the OMECC according to the secured control policy [7]. If it is successful, then data is considered as authenticated data.
5. Offloading transactions are grouped into data blocks, which are inserted into the transaction pool for confirmation by miners (mining process unit).
6. The miners validate the data blocks and sign them with digital signature to append to the blockchain.
7. The offloading transaction is added to the blockchain network and broadcasted to all MECs from the OMECC system.
8. The offloading transaction is updated for tracking via the blockchain client.

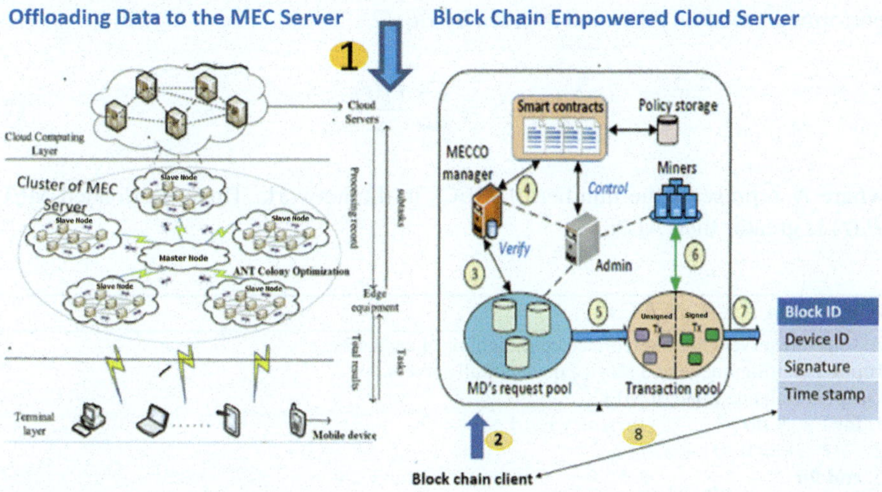

Fig. 2 Proposed system architecture

4 Implimentation

The proposed work primarily takes into account an orthogonal frequency division multiple access (OFDMA), multi-cell, and multi-server edge computing situation [10]. The further part of this section describes the LB_MECs and OMECC algorithms.

Initialize the ants (offloding tasks) based on meta-heuristics that can be used to find approximate solutions for offloading task requests n ACMLO [5]. It sorts the tasks in increasing order according to their load. ACMLO generates the optimization path to calculate their fitness (multiplier penalty factor).

4.1 Load Balancing Between MEC Servers

The proposed algorithm LB_MECs balances the load among MECS based on average load of the network ALN (Table 1).

The load of each MEC is based on length of waiting and service queues. Table 2 shows descriptions of the symbols which are used in Algorithm 1 LB_MECs . Hence, the load of the MEC is given by (1)

$$P_i(t) = W_i(t) + S_i(t) \tag{1}$$

Consider two MECs for comparing their load. MEC_i is having load higher than MEC_j, so MEC_i wants to transfer the task or load to MEC_j. This process is performed based on ALN and is given by Eq. (2).

$$\mathbf{ALN} = \sum_{i=1}^{N} \frac{P_i(t)}{N} \tag{2}$$

where N represents the number of MECs in the network. Thus, overloaded MEC $P_i(t)$ is greater than ALN.

Algorithm 1 LB_MECs

Assume all servers are interconnected (neighbor of each other)
Input : Waiting queue and service queue of all servers.
Output: Balance Load of MECs
1. **for** i < N **do** $//initially$ i = 1
2. $P_i(t) = W_i(t) + S_i(t)$
3. **end for**
4. **ALN** = $\sum_{i=1}^{N} \frac{P_i(t)}{N}$
5. **for** $i < N$ **do** $//initially\ i = 1$
6. **if** $P_i(t) < ALN$**then**
7. *Include* MEC_i *in UL*
8. **else**
9. *Include* MEC_i *in OL*
10.**end for**
11.**if** $|OL| \le |UL|$ **then**
12. **for** i *in OL* **do**
13. **for** j *in UL* **do**
14. **if** $(ALN - P_i(t)) \le (ALN - P_j(t))$
15. $C = (ALN - P_j(t)) - (ALN - P_i(t))$
16. *Transfer C amount load to* $P_j(t)$
17. **end if**
18. **end for** // *UL*
19. **end for** // *OL*
20.**else**
21. *Not possible to Balance all MECs in OL*

In Algorithm 1, the for loop between Line numbers 1 and 3 calculates the load of each MEC servers. Average load ALN is computed on Line number 4. The for loop between Line numbers 5 and 10 separates the MECs as over loaded and under-loaded servers. Overloaded servers are included in set OL, and under-loaded servers are included in set UL. The Line number 11–21 balances the load of MECs. If the over loaded servers are more than under-loaded, then balancing is not possible. Excess load of overloaded server is less than or equal to remaining load of under-loaded servers, then overload server transfers the load to under-loaded server. Otherwise overloaded server checks for other under-loaded server.

Table 1 Variable description of LB_MECs algorithm

Sl. No.	Variable	Meaning
1	P_i (t)	Load of i-th MEC at time t
2	W_i(t)	Load of i-th MEC imposed by the tasks in waiting queue at t
3	$S_i(t)$	Load of MECS i imposed by the tasks in service Queue at t
4	ALN	Average load of a MEC system at t
5	N	Number of MECs
6	UL	The set of under-loaded MEC servers
7	OL	The set of over loaded MEC servers

4.2 Secure Computation Offloading for OMECC System with Access Control

OMECC algorithm is depicted in Algorithm 2 and its is developed as an access control protocal which is executed when MD requests for offloading data to edge cloud. The access control protocol includes two phases: transaction pre-processing (executed by the OMECC manager) and verification (executed by the admin). In the first phase, the OMECC manager receives a new transaction Tx from a MD. The OMECC manager will obtain the public key PK of the requester by using the command Tx.getSenderPublicKey(), and it sends to the contract for validation. After receiving a transaction with a PK of MD from OMECC manager (msg.sender = OMECC manager), the admin will verify access rights of the requester based on its PK in the policy list of the smart contract. If the PK is available in the list, the request is accepted, and now, a task offloading permission is granted to the requester. Otherwise, the smart contract will issue a penalty to this request through the Penalty() function. In this case, all offloading activities are denied, and the request is discarded from the blockchain network. Algorithm 2 illustrates the whole processes of transaction pre-processing and verification.

5 Simulation Experiments and Result Analysis

The research work is carried out in three different phases with 5G environment. In Phase 1, data collected from end devices through IoT network, and it is transferred to MEC servers through base stations. In Phase 2, ACMOL algorithm is implemented for balancing and optimizing the load between the small cells, which are connected

Algorithm 2 Access control for computation offloading

```
Input: Tx : offloading request to the block chain system
Output: Access control result for offloading request Tr
Initializationby OMECC Manager
Recive a new transaction Tx from a MD
Get the public key of the requester
Send the public key to admin    [message : sender = OMECC manager]
Preprocessing the request by the system.
  if public key is available in the policy list then
    policy List(PK) ← true
endIf
Decode the transaction    [decoded Tr ← Abi Decoder.decode Method(Tx)]
Addr ← web3. eth get Data (decoded Tr ([Data Index])
Specify DeviceID : Addr (Index[DID]);
Verification by the smart contract
while true do
  if policy List (PK) → true Then
    if policy List (DeviceID) → true Then
      Result Penalty (PK; "Successful!")
      break;
    else
          Result ← Penalty (PK, :"Failed")
          break;
    endif
  else
          Result ← Penalty (PK, :"Failed")
          break;
  endif
endwhile
```

Table 2 Variable description of OMECC algorithm

Sl. No.	Variable	Meaning
1	T_X	Transaction
2	T_r	Transaction request
3	PL	Policy list
4	T_n	N-th transaction request
5	PK	Public key
6	PN	Penalty (detecting an unauthorized offloading request to cloud)
7	W_n	The allocated bandwidth to the MD n

Table 3 Simulation environment for LB_MECs

Sl No	Variable	Meaning
1	System bandwidth	20 GHZ
2	Transmission power	24 dBm
3	Number of small cells	50
4	Antenna mode	Isotropic
5	Number of users	80
6	Path loss	146.4+43.4log10(R)
7	Fading	Log-normal

to MEC. The proposed LB_MECs algorithm is implemented for balancing the load between MECs. In Phase 3, blockchain-based cloud is implemented to detect the malicious data records. All three phases are simulated using NS3 [7].

5G network environment is simulated in NS3 (version 3.27) [10] using "5G new radio access configuration modules". The collected data in Phase 1 is trained using linear regression machine learning model. The processed data is in the form of excel file, and it is exported to NS3. The excel file is copied to ns3 base folder. The tracePath in build.py Python script is used to access IOT-enabled real-time data as input by the simulator NS3. The configuration of MEC server is also performed by build.py.

5.1 Analysis of ACMLO

As the part of Phase 2 of the research work, ACMLO and LB_MECs are included in test.py and build.py. NS3 is reconfigured to simulate these two algorithms.

ACMLO algorithm allocates the tasks to cells based on efficient channel bandwidth. Random task allocation(RA) method allocates the tasks to cells randomly. Average task allocation (AA) method allocates the tasks based on average load of cells connected to the MEC server.

The system cost represents the allocation of offloading tasks to the small cells in 5G network. Figure 3 shows the comparative study of system cost ACMLO, RA, and AA against number of cells. RA does not result in effective system cost. However, ACMLO reduces system cost 14.28%, when it is compared with AA.

End-to-end delay comparison of ACMLO, RA, and AA is depicted in Fig. 4. RA shows large delay compared to other two methods. ACMLO shows 38% less delay than AA. The result discussion illustrates that ACMLO is better load balancing and optimizing method between cells of MEC, when it is compared with RA and AA methods.

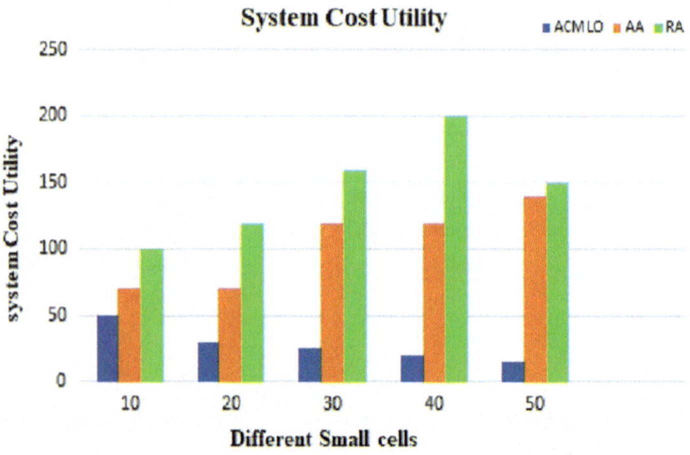

Fig. 3 Analysis of system cost

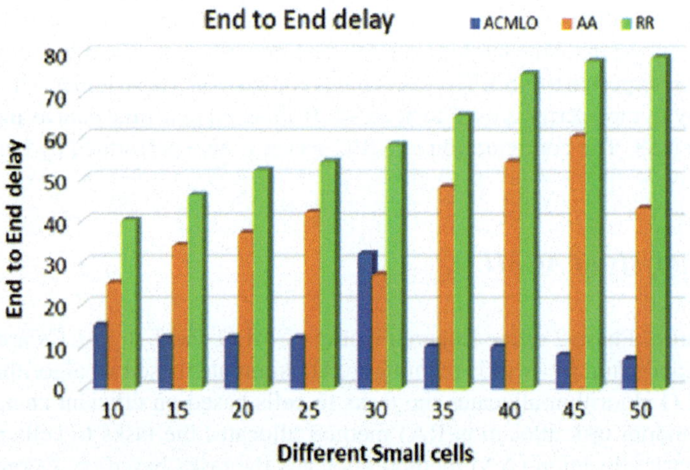

Fig. 4 Analysis of energy consumption

5.2 *Analysis of LB_MECs*

The proposed load balancing algorithm between MEC server LB_MECs is analyzed using delay and throughput parameters. Table 3 depicts the parameter descriptions used in the simulation. Figure 5 shows the delay analysis of LB-MECs algorithm against offloading rate. At the beginning of the simulation due network set up, even though offloading rate is less, delay is more. LB_MECs shows linear decrease in delay as offloading rate increases. Figure 6 shows the analysis of throughput of LB_MECs algorithm. As transferring or offload rate increases, throughput also increases linearly. Thus, LB_MECs distributes the load uniformly between the servers.

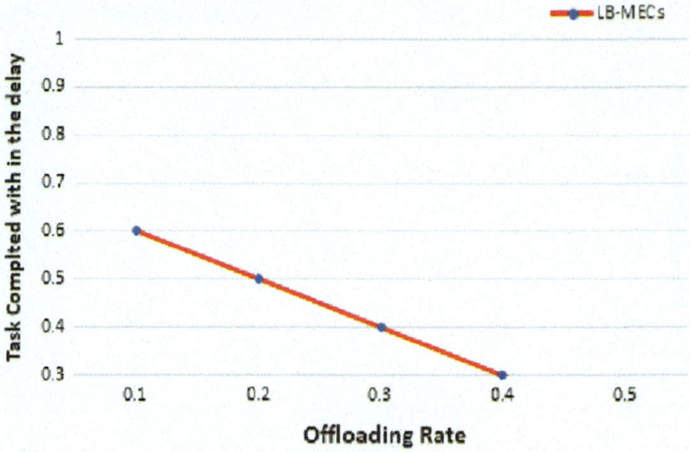

Fig. 5 Delay analysis of LB_MECs

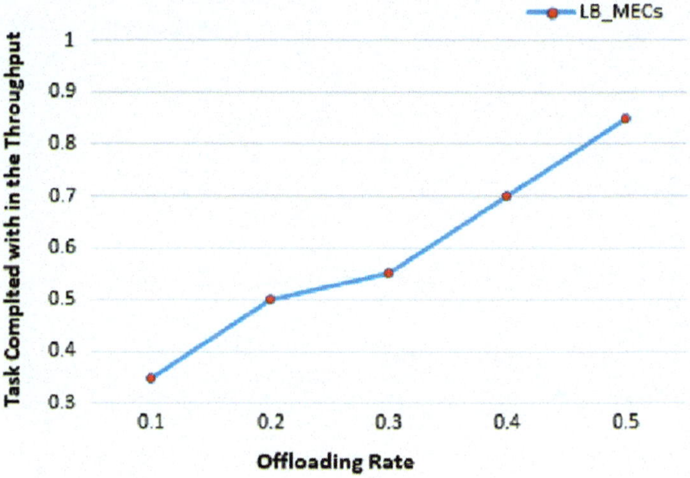

Fig. 6 Throughput analysis of LB_MECs

5.3 Analysis of OMECC

Blockchain is implemented using consensuses such as proof of work (PoW) and proof of authority (PoA). PoW is used for mining the data, and PoA is used for validating the data. It is implemented using Ethereum [11]. Offloading data from MEC to cloud and malicious data record detection is performed using OMECC algorithm. This algorithm is implemented with smart contract tool [12]. Python scripting is also used

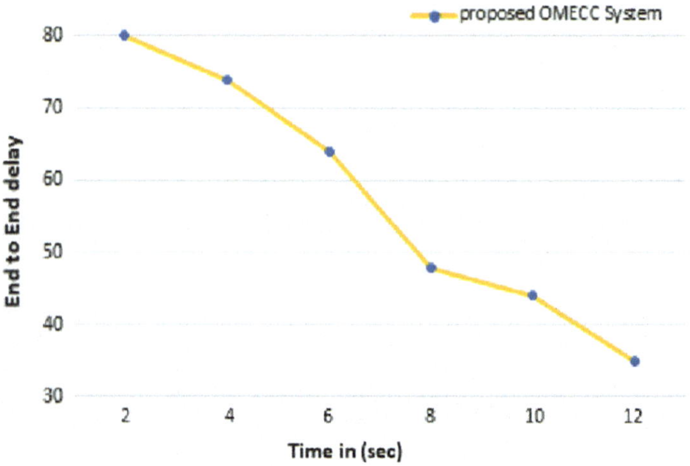

Fig. 7 Delay analysis of OMECC

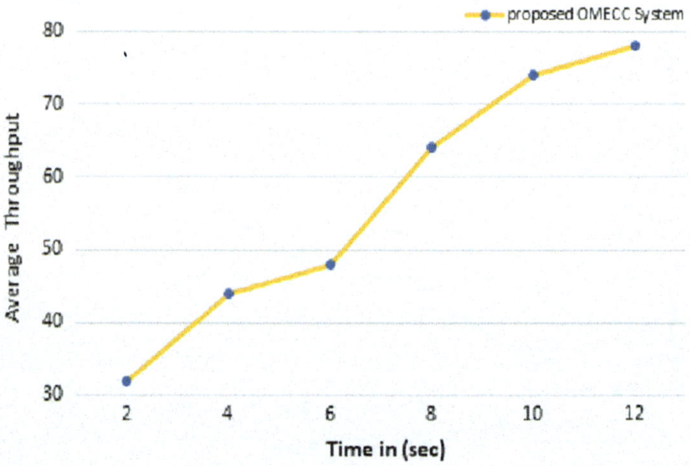

Fig. 8 Throughput analysis of OMECC

to simulate this process in NS3. End-to-end delay and throughput of OMECC are shown in Figs. 7 and 8, respectively. As simulation data increases, delay of OMECC decreases. The throughput of OMECC increases as simulation proceeds.

6 Conclusion

The load balancing and optimizing the load in dense network like 5G is essential to give expected performance. Authors of this article carried out their research work to balance the load between the MEC servers in 5G environment. Advance technologies like IoT-based 5G network, MEC with 5G, and blockchain-based cloud are used in the research. The proposed research is carried out in three different phases. The Phase 1 performs collection of data through IoT network and transfers data to the MEC through BS of the cell in 5G. In Phase 2, ACMLO algorithm is applied to balance and optimize the load between the cells of MEC. The proposed LB_MECs algorithm balances the load between MEC servers. In Phase 3, MEC data is loaded to cloud. The proposed OMECC algorithm based on block chain is used to detect malicious data records in the cloud. ACMLO is compared with RA and AA method. Results show that ACMLO is better than RA and AA. The analysis of LB_MECs algorithm shows the load distribution among MECs are uniform. Similarly, OMECC shows improved delay and throughput.

References

1. Addali KM, Bani Melhem SY, Khamayseh Y, Zhang Z, Kadoch M (2019) Dynamic mobility load balancing for 5G small-cell networks based on utility functions. IEEE Access 7:126998–127011. https://doi.org/10.1109/ACCESS.2019.2939936
2. Addali K, Kadoch M (2019) Enhanced mobility load balancing algorithm for 5G small cell networks. In: IEEE Canadian conference of electrical and computer engineering (CCECE), pp 1–5. https://doi.org/10.1109/CCECE.2019.8861598
3. Singh H, Rai V et al (2022) An enhanced Whale optimization algorithm for clustering. In: Multimedia tools and applications, pp 1–20. https://doi.org/10.1007/s11042-022-13453-3
4. Shahid SM, Seyoum YT, Won SH, Kwon S (2020) Load balancing for 5G integrated satellite-terrestrial networks. IEEE Access 8:132144–132156. https://doi.org/10.1109/ACCESS.2020.3010059
5. Zhang H, Song L, Zhang YJ (2018) Load balancing for 5G ultra-dense networks using device-to-device communications. IEEE Trans Wireless Commun 17(6):4039–4050. https://doi.org/10.1109/TWC.2018.2819648
6. Zhang Q, Xu X, Zhang J, Tao X, Liu C (2020) Dynamic load adjustments for small cells in heterogeneous ultra-dense networks. IEEE Wireless Commun Netw Conf (WCNC) 2020:1–6. https://doi.org/10.1109/WCNC45663.2020.9120688
7. Mei C, Xia X, Liu J, Yang H (2020) Load balancing oriented deployment policy for 5G core network slices. In: IEEE international symposium on broadband multimedia systems and broadcasting (BMSB), pp 1–6. https://doi.org/10.1109/BMSB49480.2020.9379563
8. Drozdova V, Akhpashev R (2020) The usage of load intensity balance approach for 5G MAC protocol characteristics investigation. In: Ural symposium on biomedical engineering. Radio-electronics and information technology (USBEREIT), pp 292–294. https://doi.org/10.1109/USBEREIT48449.2020.9117725
9. Giuseppi A, Maaz Shahid S, De Santis E, Ho Won S, Kwon S, Choi T (2020) Design and simulation of the Multi-RAT load-balancing algorithms for 5G-ALLSTAR systems. In: 2020 international conference on information and communication technology convergence (ICTC), pp 594–596. https://doi.org/10.1109/ICTC49870.2020.9289485

10. Zhang H, Song L, Zhang YJ (2018) Load balancing for 5G ultra-dense networks using device-to-device communications. IEEE Trans Wireless Commun 17(6):4039–4050. https://doi.org/10.1109/TWC.2018.2819648
11. Hatipoğlu A, Başaran M, Yazici MA, Durak-Ata L (202) Handover-based load balancing algorithm for 5G and beyond heterogeneous networks. In: 2020 12th international congress on ultra modern telecommunications and control systems and workshops (ICUMT), pp 7–12. https://doi.org/10.1109/ICUMT51630.2020.9222456
12. Jain P et al (2014) Impact analysis and detection method of malicious node misbehavior over mobile Ad Hoc networks. Int J Comput Sci Inf Technol (IJCSIT) 5(6):7467–7470
13. Yanzhi S, Yuming L, Quan W, Cheng F, Liqiong J (2021) Multi-factor load balancing algorithm for 5G power network data center. In: 2021 3rd international conference on advances in computer technology, information science and communication (CTISC), pp 188–193. https://doi.org/10.1109/CTISC52352.2021.00042

Chapter 12
Path Exploration Using Hect-Mediated Evolutionary Algorithm (HectEA) for PTP Mobile Agent

Rapti Chaudhuri, Suman Deb, and Partha Pratim Das

1 Introduction

UGV finds its usage in the field of various industrial applications like grip and throw, pick and place, assimilation of goods as well as other official services including transport of substances from one place to another reducing the manual labor [1]. For accomplishment of such purposes, a point-to-point collision-free goal achiever strategy is definitely needed to be executed in the domain of service robotics. Nature inspired path planning strategies have proved their superiority with respect to performance, time of execution, iteration, and perfection over other graph theoretic-based path searching techniques [2]. Bio-intelligent path finding techniques basically work on the principle of Darwin's theory of evolution. One of the most applicable example of evolutionary algorithm is GA (Genetic Algorithm) [3], which has been constructed on the basis of best-suited chromosome selection. The work described in this paper proposes a specialized technique inspired from the existing working architecture of Genetic Algorithm for finding a valid feasible optimized goal-oriented path by a customized mobile robot platform. VSLAM (Visual Simultaneous Localization and Mapping) guarantees near-perfect solution for robust real-time visual mapping of the traversed path by the mobile robot [4]. The created map cites as significant reference to be followed to get the idea of next step in a non-deterministic dynamic surrounding. In this context, a combined concept of run-time simultaneous mapping followed by evolutionary path planning approach has been proposed for obtaining optimal-time navigation. Geometrical description of the proposed work has been presented

R. Chaudhuri (✉) · S. Deb · P. P. Das
NIT Agartala, Agartala, India
e-mail: rapti.ai@nita.ac.in

S. Deb
e-mail: sumandeb.cse@nita.ac.in

V. Devedzic et al. (eds.), *Proceedings of the International Conference on Intelligent Computing, Communication and Information Security*, Algorithms for Intelligent Systems, https://doi.org/10.1007/978-981-99-1373-2_12

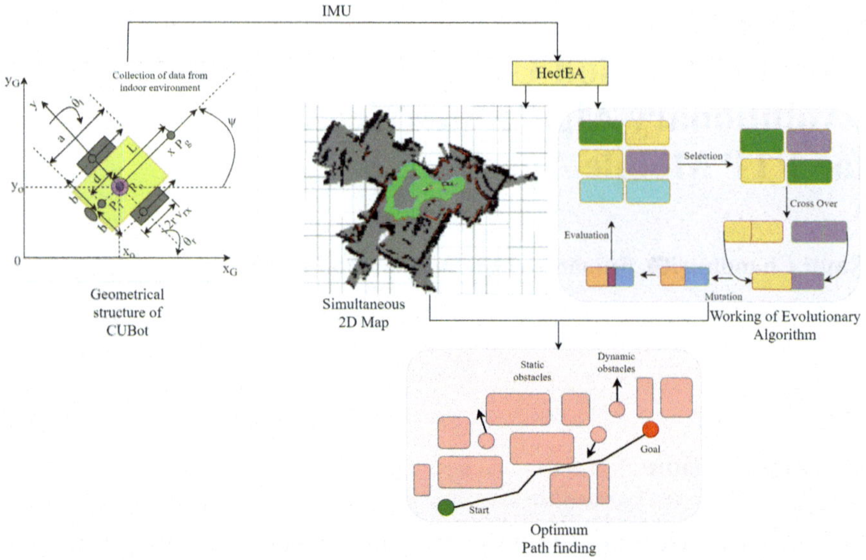

Fig. 1 Geometrical architecture of the entire research procedure resulting in optimized path navigation briefly presented as visual reference

in Fig. 1. Primary contributions of this work have been divided into following points described below:

- Customization of a differential drive designed mobile robot platform (CUBot) for carrying out the experimental applications.
- Filtered out RANSAC (Random Sample Consensus)-smooth path explored by the mobile robot (CUBot) with perfect detection of each possible on-road obstructions.
- Proposal of 2D Hect-mediated evolutionary algorithm (HectEA) for executing optimal-time path navigation.

The outline of the paper includes brief description about related works, methodology of the whole working procedure adapted, and comparative analysis of the obtained experimental results with other existing conventional evaluation processes.

2 Related Works

Extensive studies on Simultaneous Localization and Mapping and Path planning by a mobile robot has resulted into extraction of certain related works mentioned in this section. The description of respective research works enabled to identify the research gaps and accordingly develop the particular mentioned idea to be executed specifically in a considered indoor surrounding.

Prieto et al. presents a brief description on autonomous exploration in [5] required for mobile robot by the use of frontier-based exploration strategy along with SLAM and reactive navigation. Javier et al. in paper [6] portrayed a special distributed technique for mobile agent navigation in 2D as well as 3D environments equipped with different class of obstacles. The same work has also incorporated Sequential convex programming to execute target collision-free navigation plan based on local optimal parameters. The research work described in the paper [7] simulates Hector SLAM in ROS environment in a customized *L*-shaped environment. In the paper [8], real-time mapping and localization results are presented with respect to acquired measured data with precise accuracy by comparing estimated landmark position. The work in [9] has cited a bio-inspired neural network-based approach in case of solving coverage planning problems which is specially applicable in UAV (Unmanned Aerial Vehicle). Chaudhuri et al. [10] proposed a unique solution for combined 3D point cloud reconstruction of indoor environment with intelligent object identification by machine learning approach for tracking the memory of already explored path by customized mobile robot. Bio-intelligent path searching based on sparrow search algorithm is computed in the paper [11] to find highly convergent and qualitative optimized route. The presented research work separates itself from the aforesaid reviewed works by fusing the idea of real-time simultaneous mapping with evolutionary path searching approach for achieving comparative optimal-time path navigation by the considered UGV.

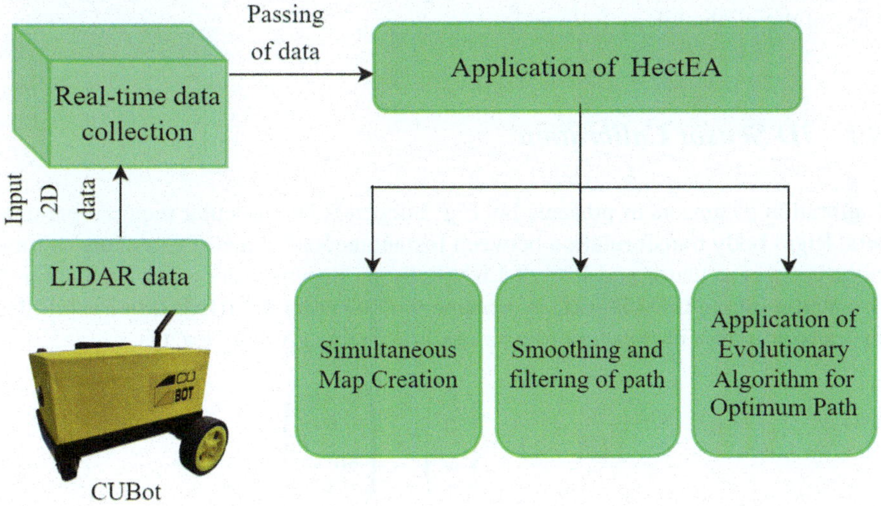

Fig. 2 Schematic representation of whole working architecture of the method proposed in this concerned research work

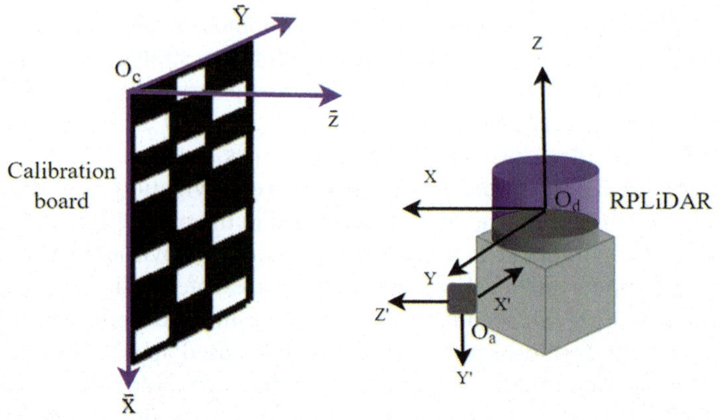

Fig. 3 LiDAR coordinate calibration with along with referential calibration board

3 Methodology

This section portrays the entire working architecture with visual and schematic presentation of the proposed idea. This has been categorized into subsequent subsections to depict the work with good clearance and precision. Figure 2 presents the schematic view of the methodology followed.

3.1 2D Sensor Calibration

Calibration of sensors as presented in Fig. 3 depends on particular required parameters. Rigid body transformation between Instantaneous Center of Curvature or rotation (ICC) of actuator and ICC of LiDAR frame depends on proper estimation of actuated spinning LiDAR [12]. Achievement of 2D point in LiDAR frame is fulfilled by transformation of polar coordinate to Cartesian as given by Eq. (1).

$$^{\psi}x(\theta_i, w_i) = \begin{pmatrix} x_i \\ 0 \\ z_i \\ 1 \end{pmatrix} = \begin{pmatrix} \cos\theta_i \\ 0 \\ \sin\theta_i \\ 1 \end{pmatrix} w_i \tag{1}$$

where ψ denotes LiDAR frame and w denotes range measurement.

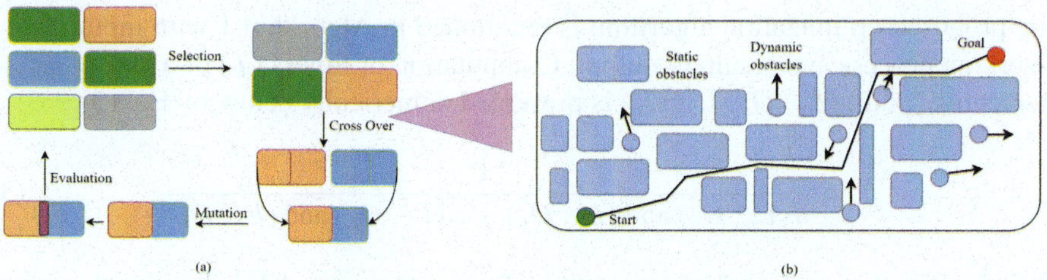

Fig. 4 **a** Portray of internal working procedure of Genetic Algorithm-inspired proposed evolutionary technique. **b** Evolutionary technique resulting in optimized path

3.2 Execution of Proposed HectEA (2D Hect-Mediated Evolutionary Algorithm)

Conceptual 2D simultaneous mapping. The idea of 2D simultaneous mapping has been inspired from Hector SLAM (Hector Simultaneous Localization and Mapping) [13]. The inbuilt laser triangulation system of LiDAR has been developed by SLAM-TEC [14] equipped with high-quality overall performance. 2D Hect is applied for scanning cross section of single plane with 2D LiDAR sensor. Instead of odometry particle filter-based mapping is preferred in case of 2D Hect. Roll or pitch motion of the sensor is executed by 2D Hect module of the proposed algorithm. Primary concern of this technique is estimation of optimum pose of the robot represented by rigid body transformation.
$(\epsilon) = [P_x, P_y, \rho]$ from the robot to the prior map [15]. $S(R_i(\epsilon))$ is the value of the map at $R_i(\epsilon)$, world coordinate of scan end points $R_i = (Z_{i,x}, X_{i,y})^T$. $S(R_i(\epsilon))$ obeys the function mentioned in Eq. (2).

$$R_i(\epsilon) = \begin{bmatrix} \cos\rho & -\sin\rho \\ \sin\rho & \cos\rho \end{bmatrix} \begin{bmatrix} Z_{i,x} \\ Z_{i,y} \end{bmatrix} + \begin{bmatrix} P_x \\ P_y \end{bmatrix} \tag{2}$$

Optimal pose estimation is achieved as a result after adding minimum error to the original pose estimation.

The second module of the proposed algorithm is inspired from Genetic Algorithm (GA), an evolutionary heuristic search technique based on natural evolution theory of Darwin [16]. The primary working principle depends on adaptation of group of animals to the environment. Subsequent parts in evolution involves selection, reproduction, crossover, mutation, and finally production of new population as portrayed in Fig. 4.

It follows the bio-inspired procedure to find global minimum solution. Each chromosome of an individual acts as the possible path to be traversed which has been mentioned as selective_path in the algorithm. After undergoing of multiple steps of exploration, evaluation, and integration, the final chromosome giving the best fit result is taken as optimized path, mentioned as new_p in algorithm. The stochastic proposed optimization algorithm is mentioned in Algorithm 1 with input, entire working process and resultant output. Computation of fitness (f) of a chromosome (selective_path) (L_u, O_u, L_v, O_v) is presented numerically below by Eq. (3).

$$f = \frac{1}{d(L_u, O_u) \text{ and } (L_v, O_v) + d(L_v, O_v) \text{ and } (L_g, O_g)} \tag{3}$$

'v' denotes the generated chromosome after crossover and 'g' denotes the target chromosome. 'd' depicts the distance between the respective chromosomes, i.e., respective selective paths.

4 Experimental Analysis and Results

This section describes the experimental procedure and analyzes the computed results. This has also been divided into subcategories expressing the type of input sensor data followed by filtering and smoothing of traversed noisy path by the robot platform.

4.1 Collection of 2D Data

This paper makes a profound priority of A1M8 series RPLiDAR (RoboPeak LiDAR) over other 2D optical sensors for capture of indoor structure [17]. Boundary of the on-road obstacles situated in same plane as the sensor are scanned according to its limited triangular measurement. Maximum scanning range of A1M8 series LiDAR is approximately 12 m by radius. Internal architecture of RPLiDAR has been shown in Fig. 5. Precise detection and perfection of this series distinguish it from other existing sensors and establish the importance to be used in execution of concerned experiments.

4.2 Filtered Out and Smooth Results of 2D Hect

Filtered out path map captured by LiDAR. 2D linear map has been created from simultaneous visualization of traversed trajectory in ROS (Robot Operating System) environment. The algorithmic structure gets simulated in backened section of ROS, and the front end visualization is obtained from ROS visualizer like FoxGlove, Rviz, and RTABmap [18]. HectEA divides the work into different section. First, it forms run-time visualization of path and produces the map by filtering and smoothing the respective map structure. Second, it passes the map into next module where evolutionary algorithm has been performed for decision-making in case of obtaining optimal-time navigation by the mobile agent.

Algorithm 1 HectEA Algorithm for Optimized Navigation

Input: - Laser data, $map_{a-priori}$, $many\_paths$[p],target[p], $selective\_path$[p], fitness(f), output[p], $path\_function$(n(t)), smoothing parameter (ϕ)

Output: - Optimized estimated pose ϵ^*, new_p

1: Grid map division.
2: **for** $P_s \in \quad map_{a-priori}$ **do**
3: $\quad \nabla\ S(P_s) = \left(\frac{\delta S}{\delta x}(P_s), \frac{\delta S}{\delta y}(P_s)\right)$
4: Approximation of gradient and derivatives using four closest integer coordinates P_{00}, P_{01}, P_{10} and P_{11}.
5: $S(P_s) \approx \frac{y-y_0}{y_1-y_0}\left(\frac{x-x_0}{x_1-x_0}S(P_{11}) - \frac{x_1-x}{x_1-x_0}S(P_{01})\right) \quad + \frac{y1-y}{y1-y0}\left(\frac{x-x_0}{x_1-x_0}S(P_{10}) - \frac{x1-x}{x1-x0}M(P_{10}) - \frac{x1-x}{x1-x0}M(P_{00})\right)$
6: $\frac{\partial M}{\partial x}(P_s) \approx \frac{y-y_0}{y_1-y_0}S\Big((P_{11}) - S(P_{01})\Big) + \frac{y_1-y}{y_1-y_0}S\Big((P_{10}) - S(P_{00})\Big)$
7: $\frac{\partial S}{\partial y}(P_s) \approx \frac{x-x_0}{x_1-x_0}M\Big((P_{11}) - S(P_{10})\Big) + \frac{x_1-x}{x_1-x_0}S\Big((P_{01}) - S(P_{00})\Big)$
8: Application of RANSAC
9: Minimization of Penalized Square Error (PSE)
10: $PSE_{\phi}(n)$ = [y-n(t)]^T [y-n(t)]+$\phi\ J_2[n]$
11: Pose expression in 2D Environment.
12: $\in= (P_x, P_y, \chi)^T$.
13: Optimization of the pose expression by matching the laser data and the map.
14: **for** $\in= (Px, Py, \chi)^T, map_{output} \leftarrow S(R_i(\in))$ at $R_i(\in)$ **do**
15: $\quad \in^* = \arg\min_{\in} \sum_{i=1}^{N}[1 - S(R_i(\in))]^2$
16: $\quad$ **for** $\epsilon \leftarrow \epsilon + \Delta\epsilon$ **do**
17: $\quad\quad \sum_{i=1}^{N}[1 - S(R_i(\epsilon))]^2 \rightarrow 0$
18: Use iterative and gradient descent for solving minimization problem of $\Delta \in$.
19: $\in^* \leftarrow \in$
20: p = $many\_paths$;
21: f = 0;
22: **for** q in range p **do**
23:     **if** selective_path[q] $\neq$ target[q] **then**
24:         f = f + 1;
25:         return f;
26: **if then** $output[q] < target[q]$
27:     **for** q in range p **do**
28:         parents[q];
29:         Computation and placement of crossover result in variable *crossed_p*
30:         new_p = crossed_p;
31:         repeat step 3
return new_p

Smoothness operation on perceived data inputs. For perfect detection of all possible situated obstacles, Luna LiDAR has also been used whose radial detection range is 0.2–8 m approximately with 90% reflectivity. It makes actual perception of all obstacles required to be avoided in the movement resulting in a smooth filtered out explored path. Moreover, overshooted areas are also get rejected in this process of smoothing. The actual graphically presented smooth path is obtained by the applica-

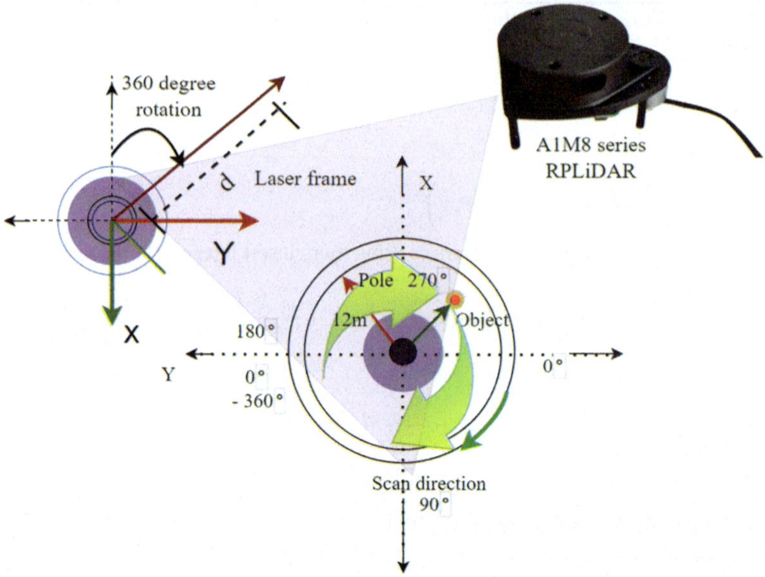

Fig. 5 Geometrical description of RPLiDAR internal working architecture along with the model used to carry out the experiment

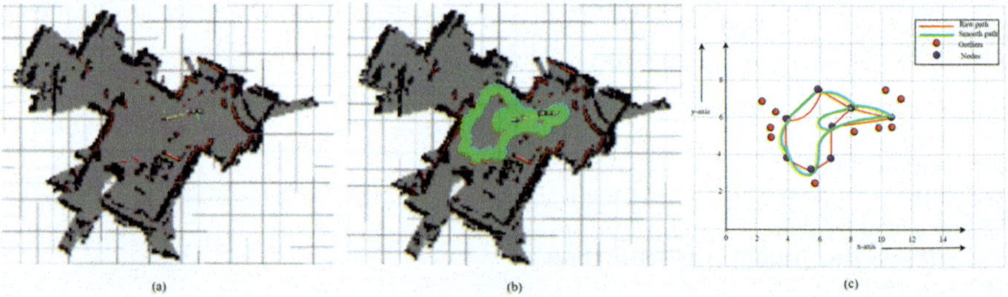

Fig. 6 **a** Boundary detection of obstacles in considered surrounding with sensor head. **b** Path obtained after filtration and smoothing of raw line. **c** Graphical representation of smoothing algorithm with raw straight and obtained spline path from start to goal point

tion of RANSAC (Random Sample Consensus) algorithm [19]. Elimination of little wriggles are depicted and accommodated with spline lines replacing the straight path. RANSAC algorithm is extensively studied and applied for obtaining best fit line from scattered noisy data. Numerical analysis of smoothing operation applied on rough penalties is done below. Considering a function $n(t)$, instantaneous slope of $n(t)$ is given as Eq. (4).

$$Dn(t) = \frac{\mathrm{d}}{\mathrm{d}t} n(t) \tag{4}$$

The curvature $D^2 n(t)$ is presented in Eq. (5):

Table 1 Experimental table presenting comparison between the performance of respective algorithms

Sample candidate function	Algorithm	x_{value}	y_{value}	Optimized value of f
$f = 9x - 4y + 5$	HectEA	**– 1349.8**	**753.4**	**– 15,156**
	CEA	– 1291	871	– 15,103
$f = 5x^2 + 7y^3$	HectEA	**– 0.1**	**– 1247**	**– 1.3573e+10**
	CEA	– 1.5	– 1169	– 1.1182e+10
$f = 2x^4 - 8y^5 + 9$	HectEA	**– 0.6**	**1348.6**	**– 3.5689e+16**
	CEA	– 1	1290	– 2.8578e+16

$$D^2 n(t) = \frac{d^2}{dt^2} n(t), \ldots, D^k n(t) = \frac{d^k}{dt^k} n(t), \ldots \tag{5}$$

The size of the curvature for all n is described and presented in the form of Eq. (6):

$$J_2(n) = \int [D^2 n(t)]^2 dt \tag{6}$$

The minimization of the penalized squared error (PSE) is illustrated numerically in Eq. (7):

$$\text{PSE}_\phi(n) = [y - n(t)]^T [y - n(t)] + \phi J_2[n] \tag{7}$$

where ϕ presents the smoothing parameter which denotes the measurement of compromise between data fit and smoothness. Increased ϕ is directly proportional to roughness penalizing which results $n(t)$ to be linear. Continuous decreased ϕ reduces the penalty allowing $n(t)$ to better fit data. Figure 6 portrays the visualization of the optimized path using real-time simultaneous map creation procedure.

4.3 Comparative Performance Analysis of HectEA with Other Conventionally Executed Technique

This subsection portrays the comparison between the performances of proposed technique and the existing bio-inspired evolutionary technique for obtaining near-optimum path by the considered customized mobile robot. The comparative numerical analysis presented in Table 1 proves the superiority of the proposed technique over other existing CEA (Conventional Evolutionary Algorithm). The respective performances are captured on certain sample candidate functions taken on concerned indoor environment with respect to the structure of the customized wheeled mobile agent.

5 Conclusion

The experimental context presents more optimized value of the considered function in case of proposed algorithm as compared to the existing conventional technique. The result verifies the proposed strategy to be relatively efficient for optimized path planning accompanied with near-perfect on-road obstacles detection. The simulation results prove the successful combined algorithmic construction of simultaneous 2D map creation and evolutionary path search by customized mobile robot (CUBot). Benchmark of performance by the respective algorithm presents the superiority of the presented algorithmic architecture. An estimation of smooth and filtered out path serve as the real-time visual reference for probabilistic obstacle detection followed by perfect path search in near-optimal amount of time. The presented work would be proved as a definite instance of citation for executing future research in the domain of indoor mobile robot navigation.

References

1. Abu-Dakka FJ, Valero F, Mata V (2012) Evolutionary path planning algorithm for industrial robots. Adv Robot 26(11–12):1369–1392
2. Chaudhuri R, Deb S, Shubham S (2022) Bio inspired approaches for indoor path navigation and spatial map formation by analysing depth data. In: 2022 IEEE international conference on distributed computing and electrical circuits and electronics (ICDCECE), pp 1–6
3. Lamini C, Benhlima S, Elbekri A (2018) Genetic algorithm based approach for autonomous mobile robot path planning. Proc Comput Sci 127:180–189
4. Karlsson N, Di Bernardo E, Ostrowski J, Goncalves L, Pirjanian P, Munich ME (2005) The vSLAM algorithm for robust localization and mapping. In: Proceedings of the 2005 IEEE international conference on robotics and automation. IEEE, pp 24–29
5. Prieto RA, Cuadra-Troncoso JM, Álvarez-Sánchez JR, Santosjuanes IN (2013) Reactive navigation and online SLAM in autonomous frontier-based exploration. In: International work-conference on the interplay between natural and artificial computation. Springer, pp 45–55
6. Alonso-Mora J, Montijano E, Schwager M, Rus D (2016) Distributed multi-robot formation control among obstacles: a geometric and optimization approach with consensus. In: 2016 IEEE international conference on robotics and automation (ICRA). IEEE, pp 5356–5363
7. Nagla S (2020) 2D hector SLAM of indoor mobile robot using 2D lidar. In: 2020 international conference on power, energy, control and transmission systems (ICPECTS), pp 1–4
8. Beinschob P, Reinke C (2015) Graph SLAM based mapping for AGV localization in large-scale warehouses. In: 2015 IEEE international conference on intelligent computer communication and processing (ICCP). IEEE, pp 245–248
9. Godio S, Primatesta S, Guglieri G, Dovis F (2021) A bioinspired neural network-based approach for cooperative coverage planning of UAVs. Information 12(2):51
10. Chaudhuri R, Deb S (2022) Adversarial surround localization and robust obstacle detection with point cloud mapping. In: Das AK, Nayak J, Naik B, Vimal S, Pelusi D (eds) Computational intelligence in pattern recognition. Springer Nature, Singapore, pp 100–109
11. Zhang Z, He R, Yang K (2021) A bioinspired path planning approach for mobile robots based on improved sparrow search algorithm. In: Advances in manufacturing, pp 1–17
12. Alismail H, Browning B (2015) Automatic calibration of spinning actuated lidar internal parameters. J Field Robot 32(5):723–747

13. Saat S, Abd Rashid WN, Tumari MZM, Saealal MS (2020) Hectorslam 2D mapping for simultaneous localization and mapping (SLAM). J Phys Conf Ser 1529:042032
14. Nitta Y, Bogale DY, Kuba Y, Tian Z (2020) Evaluating SLAM 2D and 3D mappings of indoor structures. In: ISARC. Proceedings of the international symposium on automation and robotics in construction, vol 37. IAARC Publications, pp 821–828
15. Zhang X, Lai J, Xu D, Li H, Fu M (2020) 2D lidar-based SLAM and path planning for indoor rescue using mobile robots. J Adv Transp 2020:1–4
16. Randeep S et al (2022) Analysis of network slicing for management of 5G networks using machine learning techniques. Wirel Commun Mob Comput 2022:9169568
17. Kumar T et al (2022) A comprehensive review of recent automatic speech summarization and keyword identification techniques. Springer International Publishing, Cham, pp 111–126
18. Das S (2018) Simultaneous localization and mapping (SLAM) using RTAB-map. arXiv preprint arXiv:1809.02989
19. Raguram R, Frahm J-M, Pollefeys M (2008) A comparative analysis of RANSAC techniques leading to adaptive real-time random sample consensus. In: European conference on computer vision. Springer, pp 500–513

Chapter 13
Experimental Analysis on Fault Detection in Induction Machines via IoT and Machine Learning

Om Prakash Singh, V. Shanmugasundaram, Ayaz Ahmad, and Subash Ranjan Kabat

1 Introduction

Induction motors, in particular squirrel cage motors, power many electric drives. Examples include machinery use, electric automobiles, and rail systems. In the industrial sector, they are known for their simplicity, strength, and ease of maintenance [1]. However, just like any other component, they are not indestructible and may fail. In the last several decades, researchers have studied these devices' failure processes and probabilities and the approaches for fault detection and diagnostics (FDD). Several subsystems make up a general electric drive, but electric machines are among the most important [2]. It is essential to examine the many failures that might occur in these subsystems because of their influence on the induction machine's behavior. Failure mechanisms for the inverter's MOSFET, IGBT, and diode power semiconductors are among the most prevalent (MOSFET, IGBT). In semiconductors, short and open-circuit failures are the most prevalent.

O. P. Singh (✉)
Department of Computer Science and Engineering, Vidya Vihar Institute of Technology, Purnia, Bihar, India
e-mail: opsingh6612@gmail.com

V. Shanmugasundaram
Department of Electrical and Electronics Engineering, Sona College of Technology, Salem, Tamilnadu 636005, India

A. Ahmad
National Institute of Technology Patna, Patna 800005, India
e-mail: ayaz@nitp.ac.in

S. R. Kabat
Department of Electrical Engineering, Radhakrishna Institute of Technology and Engineering, Bhubaneswar, Khordha, India

V. Devedzic et al. (eds.), *Proceedings of the International Conference on Intelligent Computing, Communication and Information Security*, Algorithms for Intelligent Systems, https://doi.org/10.1007/978-981-99-1373-2_13

For this reason, it is a serious fault that requires immediate action to shut down the drive [3]. A catastrophic failure occurs when the inverter's output loses all of its currents, which is not always the case. These faults may lie unnoticed in the system for a lengthy period, making it necessary to develop health management approaches to spot irregularities before their occurrence. It is not uncommon for DC-link components to fail, such as electrolytic capacitors, due to the demanding operating circumstances. It may not be possible to maintain a constant DC voltage when the capacitance and corresponding series resistance are changed.

On the other hand, sensors control electric drives' safety functions [4]. If they fail, they might lessen the traction force or cause an emergency stop; depending on the situation, they may cause electrical machines to malfunction, diminish their efficiency, or possibly stop working altogether.

As demonstrated, two of the most common causes of failure are the measurement's gain or offset or the device's direct disconnect. Stator and rotor failures may be distinguished when looking at the subsystem of an induction motor. Short-circuits in the stator winding (of various sorts), vibration, and failures in the phase connection are the most common stator problems. Often caused rotor failures are broken bars, misaligned rotors, or issues with bearings. As a result, numerous developed FDD solutions are for various applications, including electric automobiles, rail traction motors, and renewable energy systems [5, 6].

2 Data Collection Process

Developing data-driven strategies for FDD is challenging due to a lack of accessible data. It is not easy to create, verify, and test machine learning algorithms because of how electrical equipment is built [7]. Because of this, there is a lot of data on healthy machine operations but tiny on usual problems. As a consequence, many applications have datasets with inequalities.

Only a small number of real-world industrial datasets are still available. The cost and time required for a successful data gathering strategy are prohibitive. Thus, while developing data-driven condition monitoring systems based on machine learning and deep learning, the problem of data shortage and imbalance has become a significant negative.

The training dataset uses simulations to get around these restrictions. It cannot be replaced in different operating circumstances and reasons for failure in real-world data; hence, simulations are employed in a digital environment [8]. This research uses a MATLAB/Simulink framework to model the effects of inadequate induction motor couplings. An electric traction drive's 160 kW power output is affected by two induction motors coupled in series. Our industrial partner has proved that this platform can be used to construct railway traction systems. When compared to laboratory results in healthy persons, the platform's findings. In the previous articles, our study team has also looked at the system's normal behavior and the fault insertion block.

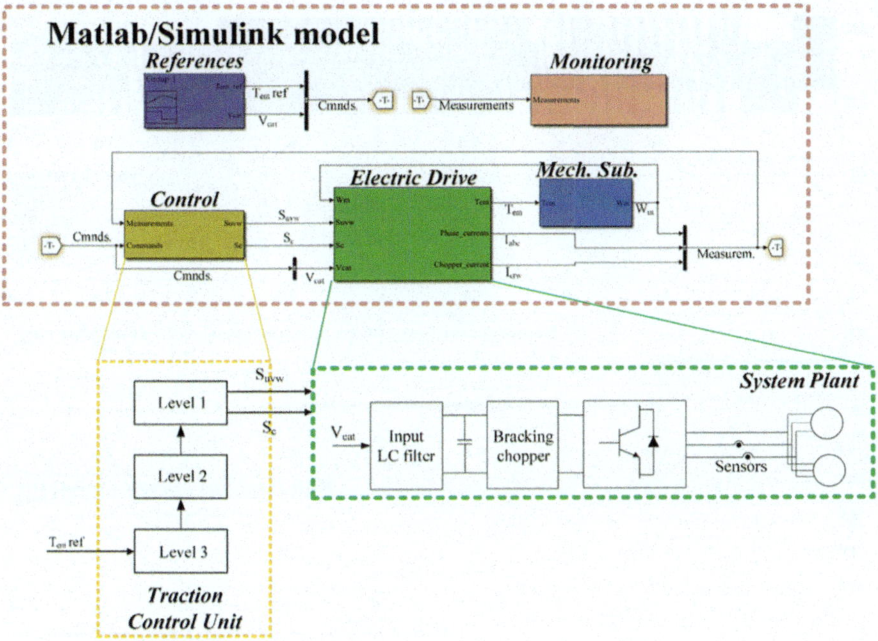

Fig. 1 MATLAB/Simulink platform replica

The MATLAB/Simulink platform replicates the operation of an electric motor using many blocks (see Fig. 1). The input stage contains a filter and a crowbar with a 3-phase inverter, two parallel induction motors, and contractors. It is a two-step process: 1. static and inertia. Simulink or Simscape blocks may be used to simulate power electronics and traction motors, depending on the accuracy and simulation speed required for a particular application. *Sample Heading (Forth Level).*The contribution should contain no more than four levels of headings.

The software-in-the-loop approach is used to incorporate the control features.

Because of the device's control software, this simulation is accurate. TCU has three control levels. Level 3 control determines level 2's reference points. First, level 3 gathers torque and flux measurements for vector control. This level may activate bus voltage control or torque reference limitations.

After level 2 sends inverter voltage references to level 1, modulation algorithms calculate inverter and crowbar switching states. Two motors parallel to one inverter and one current sensor per inverter phase. Controlling IM vectors, the total inverter current is measured.

Using this platform, the control method may be tested in various situations. Consequently, a set of healthy baseline simulations have been created to examine the consequences of power connection problems [9, 10]. In order to achieve the appropriate acceleration and deceleration rates and goal speed, torque is adjusted according to a specified profile. There were three different speed profiles to replicate in this example.

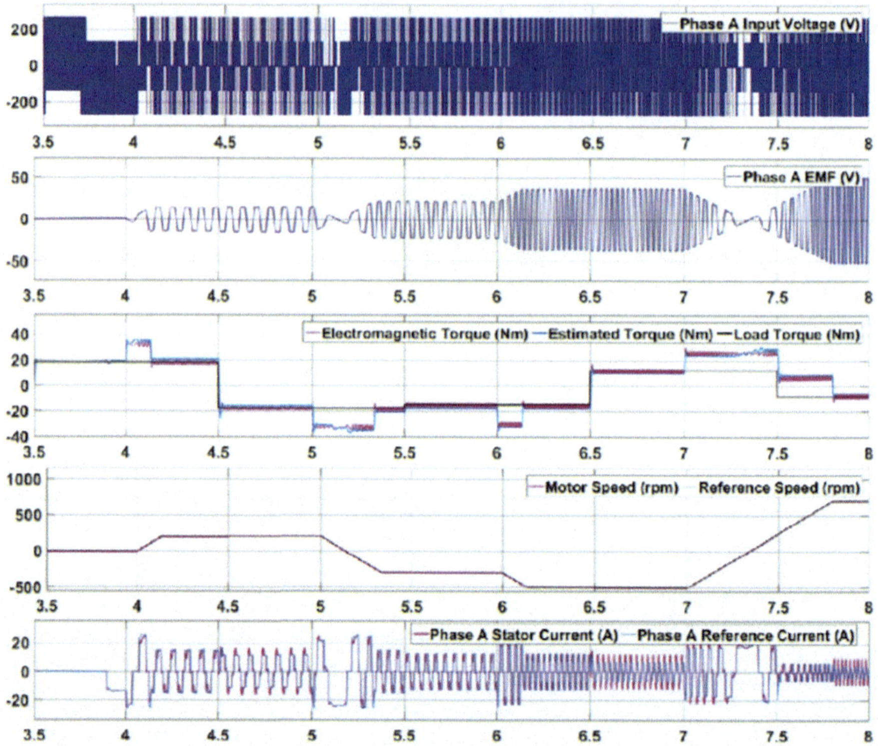

Fig. 2 Input data from the DC-link voltage meter at 2200 rpm and 60 N m load are used to calculate the following values

For a 2200 rpm goal speed and a 60 N m load torque simulation setting, the torque, phase currents, speed, and DC-link voltage are shown in Fig. 2.

Based on the previous baseline simulations, power connection faults have been incorporated into the plant model [11]. The problems simulated were HRC faults, open-phase faults, and opposite-phase wire connections. Because of the Simscape toolkit in Simulink, these errors may be inserted simply into the simulation. According to Fig. 3, a series resistor was connected to one motor phase, while the other two failures were caused by directly modifying the motor connections.

This updated model and criteria might simulate poor operation scenarios. Run 70 simulations to produce erroneous data (5 methods, with three fault modes injected at four different instants). This technique had 355 million 60-ns samples.

The following are some simulation results. Figure 4 shows the simulated indications for an open-phase failure in induction motor no. 1. The IM-1 phase currents demonstrate that phase separated, causing the rest to rise. Vector control is average because two motors are paralleled with one inverter. One motor failure impacts the other. Due to current measurement feedback and vector control mechanism, torque oscillates. High load inertia reduces speed-range torque variations. Note this (bottom

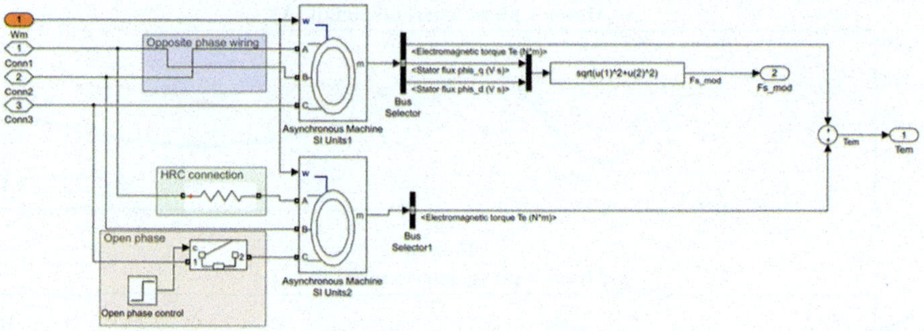

Fig. 3 Modeling of IM power connection failures in the SiL platform

right graph). Since the simulation has no speed control loop, accurate torque and speed fall when a malfunction occurs.

It would be possible to compensate for this loss of speed by raising the torque instruction from the user [12].

Since there is a closed torque control loop, it also sets the current in the opposite-phase wiring mode (shown in Fig. 5). Due to a wiring error, one of the motors' torques does not match its reference, resulting in an inaccurate target speed.

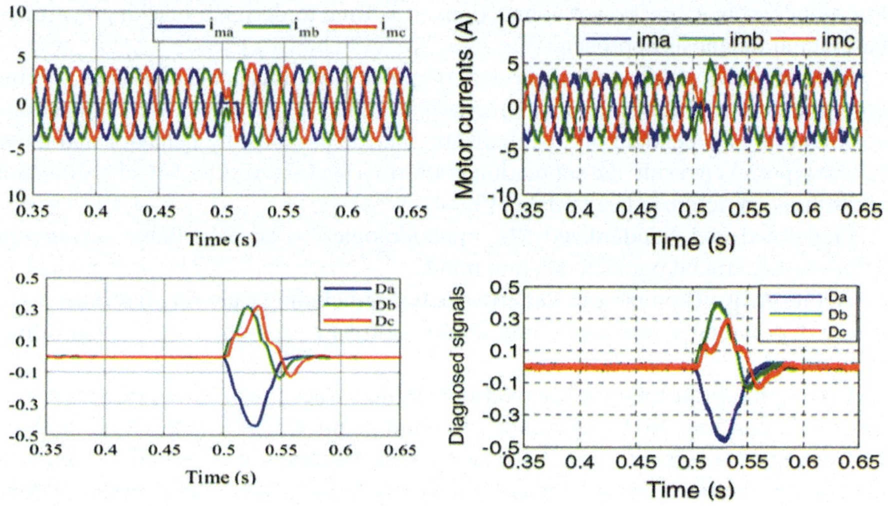

Fig. 4 Phase currents of motor 1 and motor 2 at 2200 rpm and 60 N m load

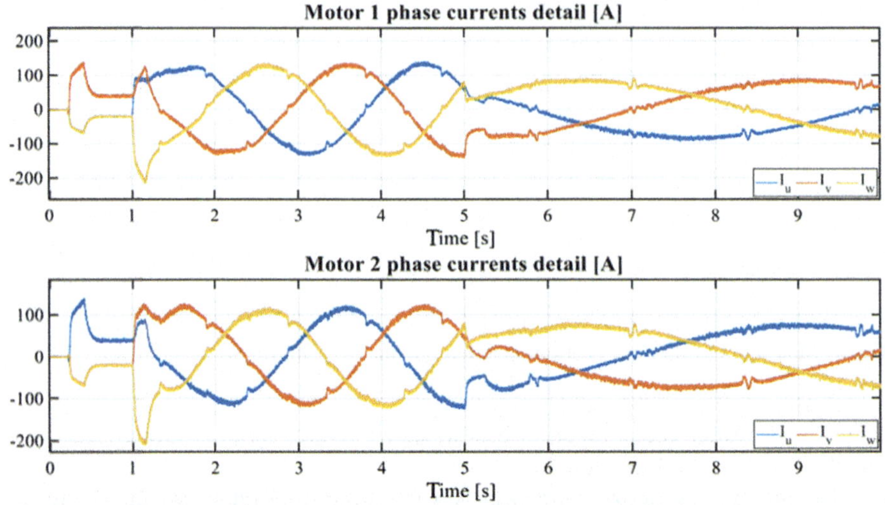

Fig. 5 Phase currents of motor 1 and motor 2 at 2200 rpm and 100 N m load

3 Diagnostic Strategy for Machine Learning Process

For induction machine power connections, it is time to design a machine learning-based fault diagnosis approach.

Data-driven strategies strive to create computer systems capable of doing activities that generally need human brains. This study's induction machine power connections examination is dependent on distinguishing between unhealthy behavior and these failure types. As a result, the simulation platform trains and verifies ML classification algorithms on fictitious health data [13, 14].

Predefined and standardized ML methodologies to develop these data-driven solutions are crucial points to keep in mind.

Additionally, it comprises a variety of actions, including gathering raw data, organizing it, prepping it, and integrating it into a software program, among other things (Fig. 6).

Critical stages are therefore defined and optimized according to this area's application needs. It is here. To be successful, the solution must be able to tell the difference between normal behavior and aberrant conduct. However, it is essential to prevent generating false positives. Equipment shutdowns due to false positives might substantially impact the availability and profitability of applications such as the railway [15].

The notion that simulations may create unpleasant scenarios is as important as the capacity to include various failure modes. This classification problem can be solved using supervised machine learning. In these cases, tagging data samples is much easier since the exact failed injection time and associated attributes are known. The

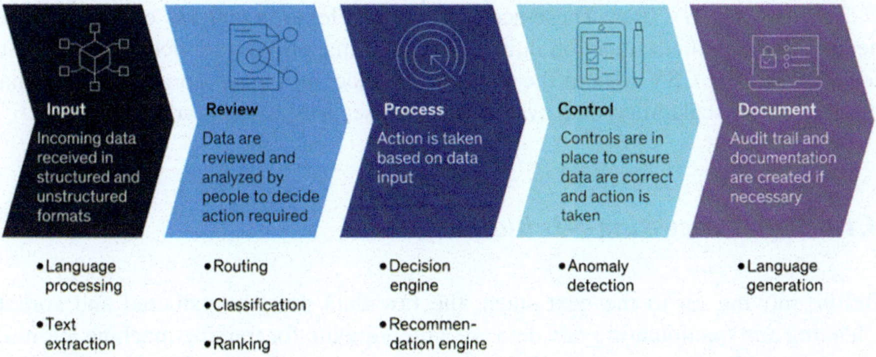

Fig. 6 Machine learning may be used more successfully if there is a standardized procedure

labeling task in a real-world application scenario requires expertise. The technique of supervised ML [16].

Training dataset samples (Xin) information about their present health condition (Yin). The new dataset (X0in) can easily interpret the ML model's output predictions (Yout). An ML method for three-class classification is part of this research project. Even though more than three health states, the open-phase fault and the HRC fault have equivalent effects on phase current imbalance and therefore, must teach an algorithm to discriminate between healthy (H), CI, and OPWF classifications (OPW).

Amazon Web services (AWSs) are the cloud service platform utilized to build ML's defect detection algorithm. Commercial cloud-based solutions have been a secondary priority when designing the FDD strategy. The proposed method is more likely to be employed in business to make handling massive datasets easier than utilizing tools like MATLAB. For the FDD approach based on data.

4 Data Collection Process

Data collection will need to be created with the same recorded variables, sample frequency, and collection modality as possible, with actual application data as nearly comparable as possible in the simulation (average, RMS, etc.). As a consequence, it must alter to correctly depict a real-world context. Outcomes of the simulation [17].

At a pace of 50 s, for example, the simulation described previously is running. Consequently, the sampling rate is 20 kHz, and many other variables are available. Every variable is not available in real-world applications, nor is there a high acquisition frequency. It is possible to emulate real-world sensor constraints by taking 64 ms samples of 11 real-world variables. Here is a quick look at all the data collected throughout the experiment. There are about 26,800 samples for each variable in the raw dataset. After that, I did a full data export.

The data as CSV files. These acceleration/deceleration patterns capture various health states and fault injection intervals. Used Amazon Web services (AWSs) platform for storing CSV files in the AWS S3 service, an object storage solution that offers the best scalability, data availability, and security in the market today.

4.1 Data Processing—Initial Stage

Before moving on to the next stage, the raw data must be gathered and sorted. Cleaning and manipulating raw data are a prerequisite for training machine learning algorithms. It is necessary to divide the general preprocessing and feature engineering processes [18].

Information derived from the raw dataset via AWS SageMaker service. Filtering, anomaly detection, normalization, and even segmentation are just a few of the first steps in data processing (as discussed below). Because this research's raw data come from a simulation platform rather than a cleaning crew, cleaning is not as important.

After cleansing the raw data, feature engineering delivers as much helpful information as possible to machine learning algorithms. Feature selection (FS) and feature extraction (FE) are two essential parts of feature engineering (FS).

Feature extraction attempts to keep as much of the original data as possible while extracting numerical characteristics from raw data. Manually or automatically utilize PCA and LDA. Extracted five time-domain features for each of 11 recorded variables using a hopping window. New statistics for standard deviation, maximum, and minimum. An 11-variable raw dataset matrix provided 55 time-domain parts totaling 26,800 samples.

Feature selection involves assessing the obtained attributes and removing those that do not meet assessment requirements. Python's SelectKBest implements F-test filtering. ML algorithms should use the full 12 statistical details. Figure 7 shows feature ratings in a bar graph.

ML algorithms may be trained and tested on this final dataset. For example, there is a blue, red, and green sample with opposite-phase faults (yellow). Both models with failures caused by imbalance and opposing connection phasing do not exhibit these characteristics physically, even though parts with nutritional status (blue) show positive speed values and stable torque and phase current ranges. It is easy to see that torque and current are very variable when looking at the first failure scenario (red). In other words, torque vibrations by phase current ripples develop when the control approach. When a phasing defect (yellow) occurs, it affects speed and mechanical torque. Samples with mechanical torques and rates that are primarily negative may never reach the reference.

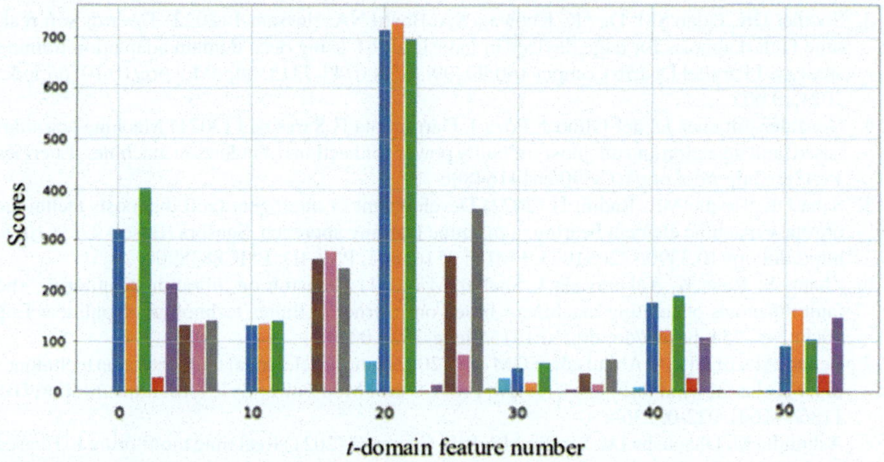

Fig. 7 T-domain features and scores from the feature selection process

5 Conclusions

E-motor: This article employs ML-based FDD to tackle high-resistance, open-phase, and opposite-phase wiring difficulties. This approach uses a software-in-the-loop simulation in MATLAB/Simulink due to the lack of field data. Training and assessing the algorithms used SVM, RF, KNN, and logistic regression (LR). Feature extraction and raw data preprocessing may improve ML workflow efficiency.

The precision and accuracy of the random forest ML approach improved to 98.5% and 96.6%, respectively. Because industrial shutdowns may be costly, the false positive rate is the first measure to be evaluated when assessing ML algorithms.

May recognize an unbalanced motor from opposite wiring issues by applying the proposed method, as demonstrated in this example. The algorithm's advice in identifying and isolating faults will be helpful for future maintenance tasks and may help avert more damage. A data-driven approach previously model- or signal-based failure types. As a result, future industrial applications will have less time to adapt to the new system since it is implemented using online marketing services cloud service.

References

1. Tran M-Q, Elsisi M, Mahmoud K, Liu M-K, Lehtonen M, Darwish MMF (2021) Experimental setup for online fault diagnosis of induction machines via promising IoT and machine learning: towards industry 4.0 empowerment. IEEE Access 1. http://doi.org/10.1109/ACCESS.2021.3105297
2. Huang M, Liu Z, Tao Y (2020) Mechanical fault diagnosis and prediction in IoT based on multi-source sensing data fusion. Simul Model Pract Theor 102:101981. ISSN 1569-190X. http://doi.org/10.1016/j.simpat.2019.101981

3. Soother DK, Ujjan SM, Dev K, Khowaja SA, Bhatti NA, Hussain T (2022) Towards soft real-time fault diagnosis for edge devices in industrial IoT using deep domain adaptation training strategy. J Parallel Distrib Comput 160:90–99. ISSN 0743-7315. http://doi.org/10.1016/j.jpdc.2021.10.005
4. Gonzalez-Jimenez D, del-Olmo J, Poza J, Garramiola F, Sarasola I (2021) Machine learning-based fault detection and diagnosis of faulty power connections of induction machines. Energies 14:4886. http://doi.org/10.3390/en14164886
5. Saha DK, Hoque ME, Badihi H (2022) Development of intelligent fault diagnosis technique of rotary machine element bearing: a machine learning approach. Sensors (Basel) 22(3):1073. http://doi.org/10.3390/s22031073. PMID: 35161814; PMCID: PMC8838900
6. Zhang X, Rane K, Kakaravada I, Shabaz M (2021) Research on vibration monitoring and fault diagnosis of rotating machinery based on internet of things technology. Nonlinear Eng 10(1):245–254. https://doi.org/10.1515/nleng-2021-0019
7. Bebars AD, Eladl AA, Abdulsalam GM et al (2022) Internal electrical fault detection techniques in DFIG-based wind turbines: a review. Prot Control Mod Power Syst 7:18. https://doi.org/10.1186/s41601-022-00236-z
8. Ashmitha M, Dhanusha DJ, Vijitlin MS, Biju George G (2021) Real time monitoring IoT based methodology for fault detection in induction motor. Ir Interdisc J Sci Res (IIJSR). Available at SSRN: https://ssrn.com/abstract=3849600
9. Choudhary D, Malasri S (2020) Machine learning techniques for estimating amount of coolant required in shipping of temperature sensitive products. Int J Emerg Technol Adv Eng 10(10):67–70. http://doi.org/10.46338/ijetae1020_12
10. Muqodas AU, Kusuma GP (2021) Promotion scenario based sales prediction on E-retail groceries using data mining. Int J Emerg Technol Adv Eng 11(6):9–18. http://doi.org/10.46338/IJETAE0621_02
11. Hymavathi J, Kumar TR, Kavitha S, Deepa D, Lalar S, Karunakaran P (2022) Machine learning: supervised algorithms to determine the defect in high-precision foundry operation. J Nanomaterials 2022
12. Singh C, Rao MSS, Mahaboobjohn YM, Kotaiah B, Kumar TR (2022) Applied machine tool data condition to predictive smart maintenance by using artificial intelligence. In: Balas VE, Sinha GR, Agarwal B, Sharma TK, Dadheech P, Mahrishi M (eds) Emerging technologies in computer engineering: cognitive computing and intelligent IoT. ICETCE 2022. Communications in computer and information science, vol 1591. Springer, Cham. http://doi.org/10.1007/978-3-031-07012-9_49
13. Chouhan A, Gangsar P, Porwal R, Mechefske CK (2020) Artificial neural network based fault diagnostics for three phase induction motors under similar operating conditions. Vibroengineering Procedia 30:55–60. http://doi.org/10.21595/vp.2020.21334
14. Nabanita D et al (2019) IOP Conf Ser Mater Sci Eng 623:012016
15. Zhang X, Rane KP, Kakaravada I, Shabaz M (2021) Research on vibration monitoring and fault diagnosis of rotating machinery based on internet of things technology. Nonlinear Eng 10(1):245–254. http://doi.org/10.1515/nleng-2021-0019
16. Sousa PHF, Nascimento NMM, Almeida JS, Reboucas Filho PP, Albuquerque VHC (2019) Intelligent incipient fault detection in wind turbines based on industrial IoT environment. J Artif Intell Syst 1:1–19. http://doi.org/10.33969/AIS.2019.11001
17. Gascón A, Casas R, Buldain D, Marco Á (2022) Providing fault detection from sensor data in complex machines that build the smart city. Sensors 22:586. https://doi.org/10.3390/s22020586
18. Jalayer M, Kaboli A, Orsenigo C, Vercellis C (2022) Fault detection and diagnosis with imbalanced and noisy data: a hybrid framework for rotating machinery. Machines 10:237. https://doi.org/10.3390/machines10040237

Chapter 14
AENTO: A Note-Taking Application for Comprehensive Learning

Kanika, Pritam Kumar Dutta, Arshdeep Kaur, Manish Kumar, and Abhishek Verma

1 Introduction

Note-taking is an integral activity in the process of learning [1]. Note-taking seems to be prominently present at all levels of education. Whether it is online learning or typical classroom learning, students believe that note-taking affects their retention abilities [2]. Many students have a habit of noting down everything taught by the teacher using memos or recordings. Notes are a great help during exam season for quick revisions. According to students' perception, making notes helps one improve their academic performance [3]. Colorful and well-decorated notes are a great way to develop an interest in a certain topic, rather than repeatedly reading dull black and white thick books [2]. Note-taking hence has its own set of advantages. For instance, it improves writing speed, vocabulary, and fluency in language [4]. The activity helps to improve the memory of students, quick-wits, and decision-making and teaches them the importance of organization and presentation in several aspects [4, 5].

Note-taking has been with us for more than a century now. Today, connectivity and technology have digitized the entire process of note-taking. There are mobile and web applications that assist the students in various ways to make their note-taking efficient. Applications like Samsung notes [6]. Evernote [1, 7, 8]. NotTaker [6, 9, 10] focuses on acting like a scratchpad. A few other mobile applications help by reading physical notes and converting them into digital form. Moving a step ahead, applications like LiveNote, MicroNote [6, 7], CoScribe [6, 11] also have collaboration features to improve the student's experiences [6, 12, 13].

Although taking of notes is an important part of the process of teaching–learning, not much literature is available. The studies available mostly focus on aspects like

Kanika · P. K. Dutta (✉) · A. Kaur · M. Kumar · A. Verma
Department of Computer Science and Engineering, Chandigarh University, Mohali, Punjab, India
e-mail: pritamkumar11.pd@gmail.com

V. Devedzic et al. (eds.), *Proceedings of the International Conference on Intelligent Computing, Communication and Information Security*, Algorithms for Intelligent Systems, https://doi.org/10.1007/978-981-99-1373-2_14

security, collaboration, platform of the application and ease of access. The applications for note-taking available provides features of adding and storing class notes and other features like memo and annotations [1, 14, 15]. However, certain features like creating handwritten notes using a stylus or fingers are considered to be unnecessarily time-consuming and less efficient [16–18]. Rather than working on providing merely storage, one can think of processing and extracting the meaningful information available in the class notes. If the raw data present can be processed, its information can further be used to improve the learning experiences of students [19]. As far as collaborations are concerned, one can work on collaborating with students as well as students and teachers [6, 9]. Presumably, no extensive analysis of the informative data class notes produces is available.

We aim to develop a note-taking application that not only stores students' notes topic-wise but also processes the notes to extract meaningful phrases. These important phrases are then compared with the teachers' notes. A student is recommended the core concept discussed in class. Also, we propose to compare the content of the student's notes with that present in the teachers' notes. After comparison, the system identifies the concepts the students missed while learning in the class. This helps in providing a comprehensive learning experience. Additionally, the application also recommends some interesting real-life applications relevant to the class notes uploaded by the students.

The following section, Sect. 2 presents the review of literature on note-taking systems and applications available. Section 3 expounds on the technical details of the improved note-taking application and its system architecture. In the next section, Sect. 4, we elaborate on the results obtained. This section enlists the reported feedback on the use of the note-taking system we propose. In the end, we conclude with some important findings.

2 Prior Work

Several authors have acknowledged the value and importance of class notes and note-taking activities [19]. A few have also listed the challenges in the note-taking process when it is digitalized. It is believed that the progress in the transition from physical to digital notes is a bit slow. The attributed reasons are a dilemma in learning with technology, complexity, inefficiency, and integrity [6].

Some prefer traditional note-taking with pen and paper, while some are willing to give note-taking the touch of the digital era. As in, it was surveyed whether it's preferable to digitalize note-taking. It is concluded that traditional note-taking encourages students to be active listeners and the practice can help them be better learners. However, sometimes handwritten notes become clumsy or may get lost, and so the digitalization of note-taking is believed to be able to provide much ease [5].

Nowadays, smartphones are so ubiquitous that every student either owns or has access to one. So, in [13], a study was conducted to develop a smartphone-based collaborative learning app. However, smartphones can also serve as a distraction during activities. Although it enables ease of note-taking, it does not possess features of storage or sorting the notes [20].

The study indicates that note-taking activity and reviewing the content of notes have a positive influence on learning [21]. The activity also improves the recall capabilities of students. However, it is also believed that the devices used for digitization of class notes can also serve as a distraction inside the class [13].

Recently, in [12] a mobile app was proposed for note-taking. The unique feature of this application was its security features. These features were added to protect sensitive data. Similarly, Stephan Pachikov developed the Evernote app in 2008 for note-taking and task management. This app provides the feature to instantly record texts or voices, which can be later viewed and edited as necessary. This app was mainly developed with the aim of capturing meeting key points effectively so that meetings can be more fruitful and organized. Though it has not yet been tested in the field of education, the approach it has taken has great scope for implementation in the field of note-taking [6].

Classroom Presenter is another application developed to enable collaboration between students and teachers on presentations [22]. With this app, the presented material can also be shared among students and they can simultaneously take notes on it or capture parts for reviewing and editing in the future. It is a very helpful app for collaborative learning and noting down student feedback developed by the University of Washington (2005). Amazon Kindle is another app which supports reading uploaded material from the database. Developed in late 2007, it uses an electronic medium to display paperback sheets. It provides features of easy editing and is environment-friendly [6].

Various other note-taking applications, both desktop as well as mobile applications are available. Several other studies highlight the importance of class notes [23]. However, none of them provides the features that can assist in complete and comprehensive learning. The students get help in various forms from all these note-taking applications. Despite developments in the field, to the best of our knowledge, no application exhaustively utilizes the content of class notes to improve the learning of students.

3 Working of AENTO

AENTO consists of five modules as it is shown in the architectural diagram in Fig. 1. At first, the notes are stored in a particular order so that retrieval is easy and efficient. Since the note-taking application is more than mere storage space, the system proposed also has a recommendation module. The editing feature is scratchpad. The scratchpad has some unique features, discussed in the coming subsections.

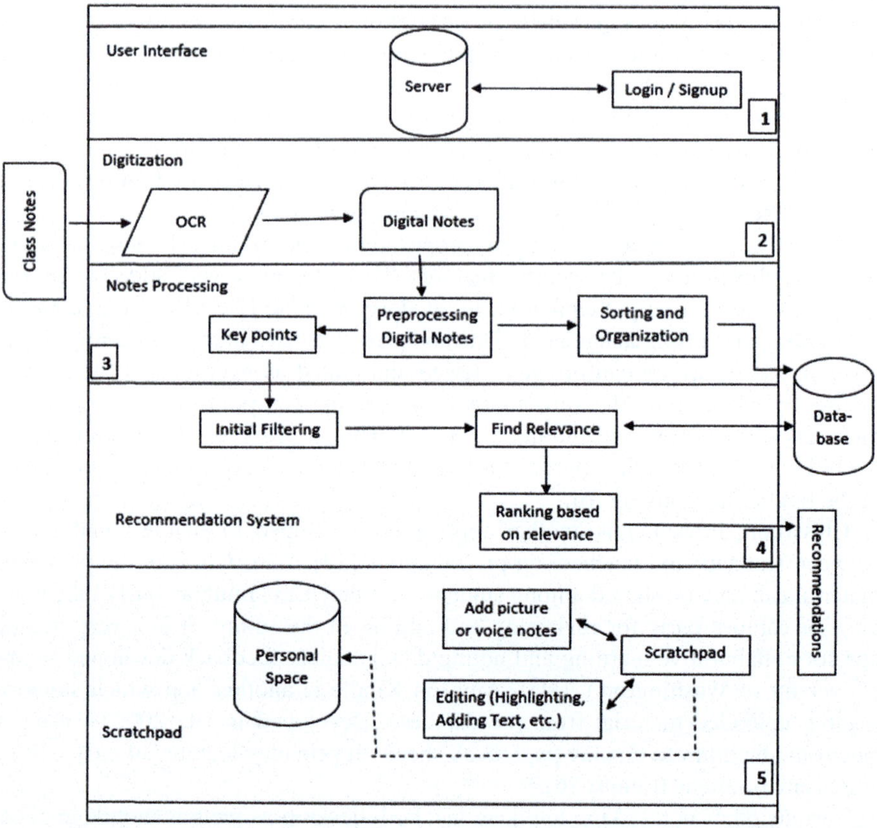

Fig. 1 Architectural diagram of AENTO

3.1 User Interface

This is the interface the users interact with on opening AENTO. It is the login page where the user can create an account or simply sign in if already exist as a user on the application. The system uses two databases in this project. One is for storing sign-in credentials and the other one is for saving all the digital notes which can be later accessed by the user. When a user tries entering the app, a login prompt comes up. If the user is not registered, the sign-up pop-up shows up. After registering, the login credentials are securely stored in the server via encryption [12, 18, 24].

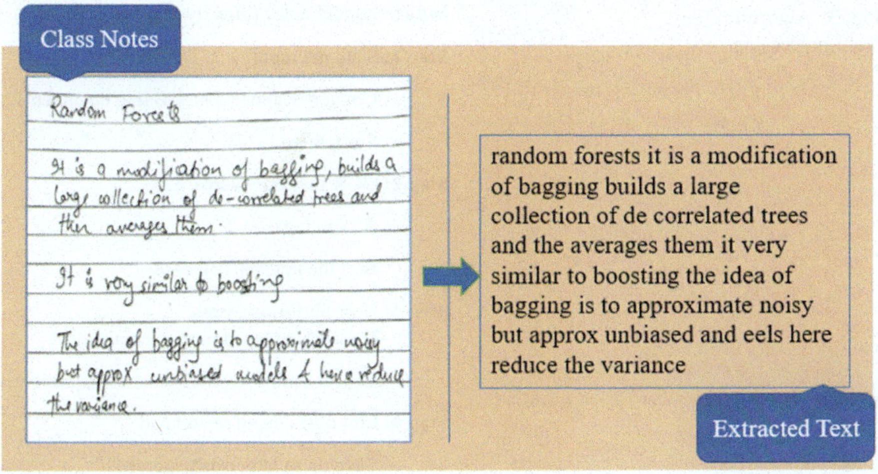

Fig. 2 Scanning notes

3.2 Digitization

It is one of the components of our project which will help to convert all the handwritten notes into digital notes. Optical character recognition (OCR) is used in the app to scan the alphabet from handwritten or typed notes. The string of texts thus generated is processed to extract the key points, as it is demonstrated in Fig. 2. The key points present in the notes are identified by filtering out all the nouns and noun adjective pairs [7, 25].

3.3 Note-Processing

The digital notes are then processed for keywords according to the algorithm shown in Fig. 3. In the database, the keywords are used as tags which makes organization and searching much easier. These keywords are mainly technical terminologies used in the notes. The uploaded notes are grouped under these keywords, in the database. So, if one wants to find notes on a certain topic, they can just search the topic name and all the relevant materials available having that topic name as a tag will appear. It makes the organization much easier and more efficient. In Fig. 4, the keywords are identified by the app using the training dataset and some are learnt through improvised learning from the notes or Wikipedia. Upon identification of key points from the class notes, it also recommends the missing concepts. Those recommendations can either be existing notes or Wikipedia links. The system can store both types of notes–the ones coming from the scanner as well as direct digital notes submitted by the student in text form. After storing it will perform a sorting operation and sort all the data

Fig. 3 Algorithm for Note-processing

Input: Student notes (directly scanned or images)

Step 1: Scan the input.

Convert scanned material into documents.

Store them.

Step 2: OCR finds keywords.

Assign tags.

Sort the notes in database.

Step 3: User performs search:

If matching tags found: fetch the data.

Repeat until the whole database is searched.

Step 4: Edit notes using scratchpad

Save them in personal storage.

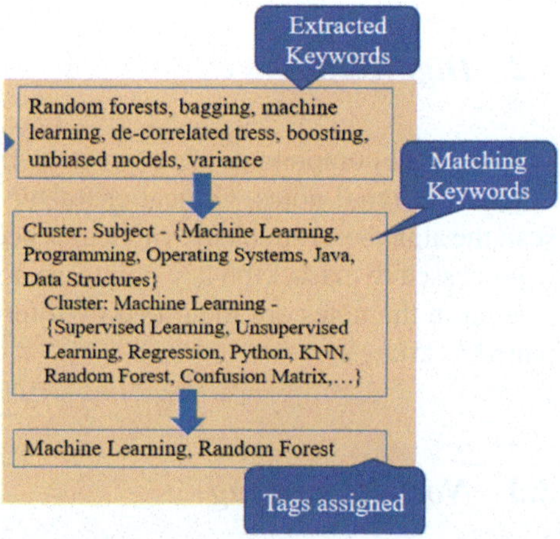

Fig. 4 Note-processing

based on the key points. Then stores it the data in the database under appropriate tags, so that they can be easily found later using keywords [10].

3.4 *Recommendation System*

AENTO is unique because it will provide what you have missed out on in class. It is assumed that the teacher as well as students of a class submit notes to the system. By comparing the teachers' notes with that of students' text, the system identifies the keyword student has missed noting down. The application recommends a list of keywords to ensure comprehensive learning following the algorithm in

Fig. 5 Algorithm for recommendation system

Input: Keyword (topic/ chapter name)

Step 1: Search for matching tags.

Step 2: Match the keywords with existing tags in database:
If match found: count number of tags matching

Repeat until whole database is scanned.

List the materials in order of count value.

Step 3: Search for materials on internet using the keywords:
for matches found: sort them based on relevance

List them in order of relevance.

Step 4: Display output.

Output: Recommendation List.

Fig. 5. Additionally, using the similarity equation proposed by Kanika et al. [26], the application also recommends relevant content to the students. To make the content similar only to the keyword, we chose the balance factor, alpha as 1 [20, 26–28].

AENTO scans for matching expressions of <adjective> * <noun> in the saved notes or websites and sorts them according to the frequency of overlapping with the keywords. It creates a list in order of relevance and presents it to the user.

To eliminate redundant technical words from the recommendation, a specific algorithm is used by the system. By using Eq. 1, it measures the similarity between two sequences in a range of 0–1.

$$r = \frac{2.0 \times l}{n} \tag{1}$$

Here, the total length of the two sequences combined is represented by *n*, the length of the continuous longest part of them is represented by *l* and *r* represents the similarity score. The technical phrase with shorter length is dismissed in case the value of *r* is greater than 0.7.

AENTO then analyzes the semantic relativeness of each article. A pre-trained word embedding based on the Stanford's GloVe algorithm is employed [5]. Words are represented by vectors. These vectors express the semantic relevance of the words. The text is then converted into a form of pairs of word and frequency, $A = \{(w_1, f_1), \ldots (w_i, f_i), \ldots (w_n, f_n)\}$, which means word w_i has occurred f_i times.

Suppose a phrase with multiple words is $X = \{l_1, l_2, l_3, \ldots, l_n\}$, which has the possibility of being an image label *m* or a key phrase in *L*. The below Eqs. 2 and 3 is used to derive how similar article *A* and *X* are semantically:

$$\text{Similarity}(A, X) = \frac{\sum_{(w_i, f_i) \in A} \sum_{l_j \in X} f_i \times \text{Similarity}_{ij}}{|A||X|} \tag{2}$$

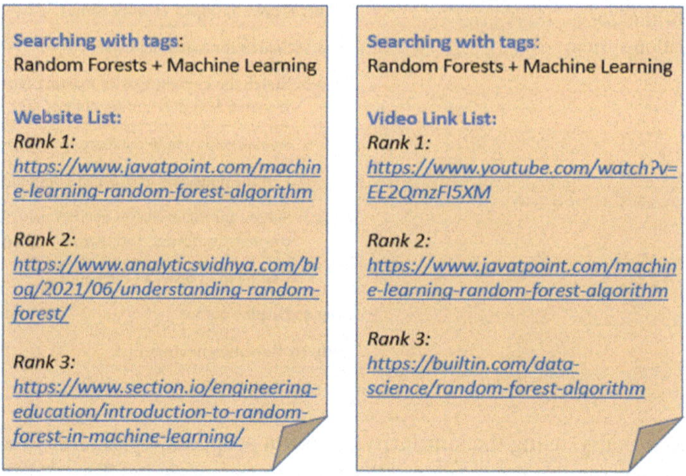

Fig. 6 Recommendation system

$$\text{Similarity}_{ij} = \frac{\vec{w}_i \cdot \vec{l}_j}{|\vec{w}_i|\left|\vec{l}_j\right|} \tag{3}$$

Here $\vec{w}_i$ and $\vec{l}_i$ are the vector representing w_i and l_j, respectively, are the respective. The lengths of the article A and phrase X [Kowser, Sultana, Shuchita Rahman] are represented by $|A|$, $|X|$, respectively. Semantic similarity for phrase m in any image label M and a key phrase p in L, is given by the Eq. 4, which is represented as a linear combination of a prefixed weighting factor α:

$$\text{Similarity}_{\text{score}} = \alpha \times \text{Similarity}(A, p) + (1 - \alpha) \times \text{Similarity}(A, m) \tag{4}$$

Whether the site recommendation is closer to the image or to the class notes semantically, is determined by the weighting factor α.

Thus, in Fig. 6, the finalized recommendation list is formed in order of relevance and AENTO displays that output to the user.

3.5 Scratchpad

This feature enables users to make quick memos whenever necessary. It has been improved by adding some simple features of font-changing, font styles, and text colors. These memos are saved for reviewing and editing later. Besides, it also offers the feature of editing digitized notes from the database. A copy of the data from the database can be fetched to customize them according to the user's necessity by highlighting certain parts or adding sticky notes at parts or attaching voice notes or attaching some pictures and memos for reference. This edited copy of data will be stored in the personal space provided to users. It is virtual storage provided to registered users of the app for storing their memos and customized documents. This space is personal storage for the user, where neither administrator nor other users have any access unless explicitly provided by the user. However, the storage capacity is limited and hence, users need to smartly manage their personal space [14, 29, 30].

4 Experimental Results

Our workflow revolved around the plan depicted in Fig. 7. In order to assess the usage of the note-taking application. We conducted a survey among 250 students. The respondents were engineering students in their third year. A pre-test assessing their knowledge of software engineering courses was taken. The test consisted of 10 questions of average to complex difficulty levels. At the end of 4 weeks, after the pre-test was taken, a post-test was taken to assess the present knowledge of students. The post-test also consisted of 10 questions. Each question was for 2 marks and there was no negative marking for the wrong answer. The scores of the pre-test and post-test were compared using a t-test. The descriptive statistics for the test are given in Table 1. As available in Table 1, the mean for pre-test scores is relatively lower than that of the post-test. The same is indicated by a low p-value. With the p-value of 0.000, a notable difference in pre-test score and post-test score can be observed. With this, we can establish that the scores of students or their academic performance have improved significantly with the use of the note-taking application.

Survey Questionnaire: In order to assess the performance of AENTO, a survey-based field evaluation was done. We mainly aimed to evaluate the relevance of the generated recommendations and how the students perceive the app. For the survey, we prepared one qualitative question and four feedback questions, as given below:

Q 1: *How was the impact of the app on your note-taking speed?*

Students were to state how impactful the app was on their note-taking speed on a scale of 1–5. The scales denote 1: extremely slow 2: a bit slow 3: neutral 4: a bit fast 5: very fast.

Q 2: *How efficient is the app for note-taking?*

Students were to state how helpful the app was for note-taking on a scale of 1–5. The scales denote 1: extremely inefficient 2: slightly inefficient 3: quite efficient 4: very efficient 5: extremely efficient.

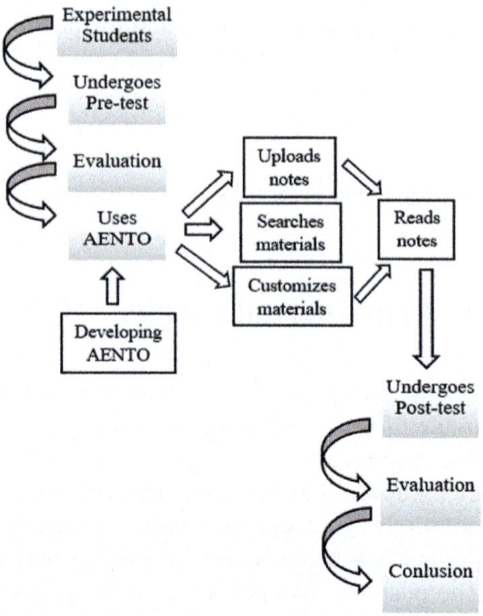

Fig. 7 Flowchart showing research workflow

Table 1 Survey data count

Group	Pre-test	Post-test
Mean	4.6	7.17
SD	1.97	1.61
SEM	0.12	0.1
N	250	250

Q 3: *How comfortable do you find the usage of the app?*

Students were to state the difficulty in using the app on a scale of 1–5. The scales denote 1: extremely difficult 2: a bit difficult 3: feels alright 4: moderately easy 5: extremely easy.

Q 4: *How likely it is for you to use this app in the future?*

Students were to state their willingness of using this app in the future on scale of 1–5. The scales denote 1: extremely unlikely 2: unlikely 3: can't tell 4: likely 5: extremely likely.

Q 5: *Write your experience while using this app.*

The first four questions evaluated the performance of AENTO for the sample datasets, and the last one evaluated how the app was perceived by the students for their own datasets.

Survey Response Analysis: Although the scores indicate a significant improvement in the academic performance of students, we also conducted a survey to find out

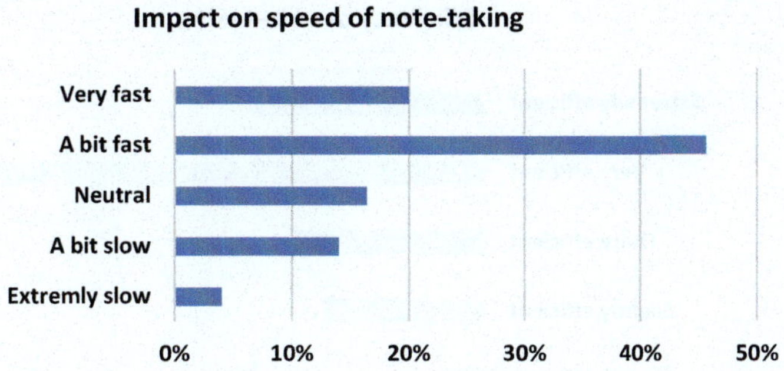

	Extremly slow	A bit slow	Neutral	A bit fast	Very fast
Impact on speed of note-taking	4%	14%	16.40%	45.60%	20%

Fig. 8 Distribution of responses for impact on speed of note-taking

the experience of users. The survey consisted of four questions. Question 1 assessed the impact of the AENTO on the speed of taking notes. The students were asked to assess the impact of the note-taking application on their speed of taking down notes on a scale of 1–5. Here 1 represented a negative impact and 5 indicated a very positive impact. The responses are given in Fig. 8. As depicted in the graph, no central tendency is obtained. The "extremely slow" received minimal responses. The second question assessed the efficiency while taking down notes.

Figure 9 shows the distribution of responses for question 2 of the survey. Maximum students found the use of the note-taking application "a bit efficient". 20% of the students have responded to the category "very efficient", which is a positive indication.

Similarly, the distribution of responses to question 3 are picturized in Fig. 10. This question asks the students about the ease with which they were able to use the application. Here, as depicted in Fig. 10, more than 40% of the respondents found it easy enough to use.

At last, we wanted to know whether the students would use the application in future or not. This question was answered by 4% of the students in a negative response, as it is shown in Fig. 11. All others found it useful and they responded in other categories indicating willingness to use the application in near future.

5 Conclusion

We have developed a note-taking application. The note-taking application not only helps in storing the class notes of students meaningfully, but it also makes classroom learning comprehensive. This is achieved by preprocessing the notes and converting

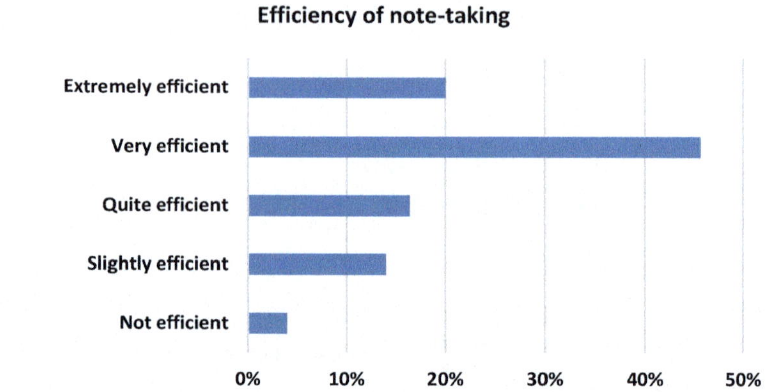

	Not efficient	Slightly efficient	Quite efficient	Very efficient	Extremely efficient
Efficiency of note-taking	4%	14%	16.40%	45.60%	20%

Fig. 9 Distribution of responses for efficiency of note-taking

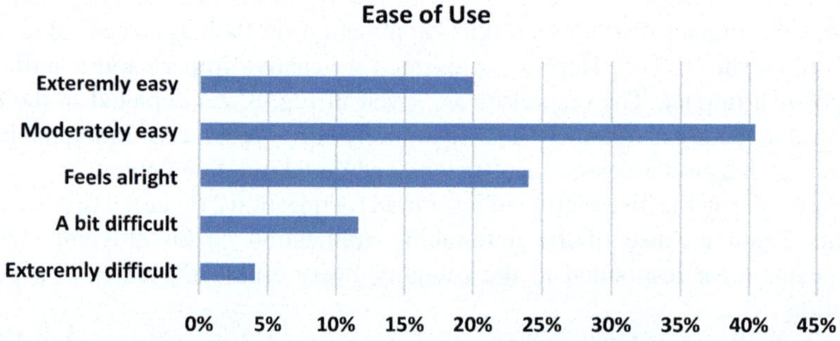

	Exteremly difficult	A bit difficult	Feels alright	Moderately easy	Exteremly easy
Ease of Use	4%	11.60%	24%	40.40%	20%

Fig. 10 Distribution of responses for ease of use

them into important keywords. A recommendation system module helps the students by recommending various content to students. We gave 250 engineering students the application for four weeks. The effectiveness of note-taking applications is assessed with statistical tests on the scores of students in pre-test and post-test. To assess the usability of the application, we also conducted a survey. The survey consisted of four questions. The questions assessed the ease of use and students' experiences with the application. The results of the statistical test indicated significant improvement in the post-test scores. This gives us an insight into the impact of the note-taking application on students' learning. The responses to survey questions are also quite positive. Overall, the results are encouraging enough to work on the system.

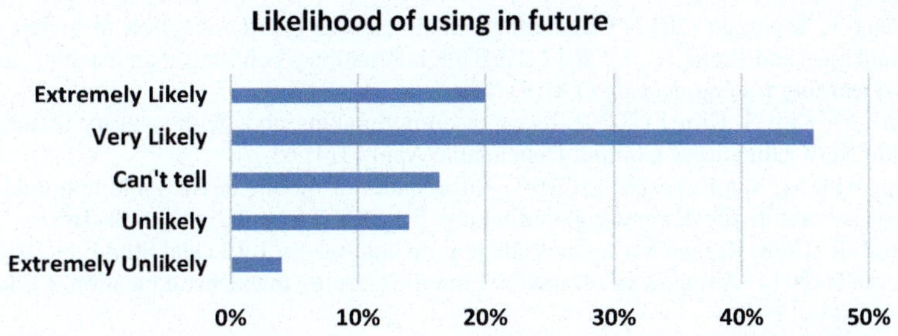

	Extremely Unlikely	Unlikely	Can't tell	Very Likely	Extremely Likely
Likelihood of using in future	4%	14%	16.40%	45.60%	20%

Fig. 11 Distribution of responses for likelihood of using the app in future

Although the system received positive feedback, we have added only basic features. In future, we would work on improving the recommendations by making them relevant yet serendipitous. Also, we aim to provide a summarization feature in the improved version of the application. In the present study, we have given the application to students only for 4 weeks. Hence, in future, we would analyze the usage impact for the entire semester.

References

1. Ganske L (1981) Note-taking: a significant and integral part of learning environments. ECTJ 29(3):155–175
2. Hartley J, Davies IK (1978) Note-taking: a critical review. Programmed Learn Educ Technol 15(3):207–224
3. Hartley J, Marshall S (1974) On notes and note-taking. High Educ Q 28(2):225–235
4. Rahmani M, Sadeghi K (2011) Effects of note-taking training on reading comprehension and recall. Reading 11(2):116–128
5. Pennington J, Socher R, Manning CD (2014) Glove: global vectors for word representation. In: Proceedings of the 2014 conference on empirical methods in natural language processing (EMNLP), pp 1532–1543
6. Mosleh MAA, Baba MS, Malek S, Alhussein MA (2016) Challenges of digital note taking. In: Advanced computer and communication engineering technology. Springer, Cham, pp 211–231
7. Havryliuk N, Osaulchyk O, Dovhan L, Bondar N (2020) Implementation of E-learning as an integral part of the educational process. In: Society. Integration. Education. Proceedings of the international scientific conference, vol 4, pp 449–459
8. Al-Zaidi MS, Joy M, Jane S (2013) Exploring the use of micro note-taking with social interaction features for education. In: Proceedings Edulearn13, pp 6098–6106
9. Lusenaka E (2011) Challenges of note-taking in consecutive interpreting training at the University of Nairobi (Doctoral dissertation, University of Nairobi, Kenya)
10. Esposito F, Ferilli S, Basile T, Di Mauro N (2008) Machine learning for digital document processing: from layout analysis to metadata extraction. In: Machine learning in document analysis and recognition. Springer, Berlin, Heidelberg, pp 105–138

11. Wang T, Towey D (2017) Open educational resource (OER) adoption in higher education: challenges and strategies. In: 2017 IEEE 6th international conference on teaching, assessment, and learning for engineering (TALE). IEEE, pp 317–319
12. Park M, Kim S, Kim J (2020) Research on note-taking apps with security features. J Wirel Mob Netw Ubiquitous Comput Dependable Appl 11(4):63–76
13. Van Wyk M, Van Ryneveld L (2018) Affordances of mobile devices and note-taking apps to support cognitively demanding note-taking. Educ Inf Technol 23(4):1639–1653
14. Tesch R (1988) Researcher memos and pop-up note-takers. Int J Qual Stud Educ 1(3):281–285
15. Baran E (2014) A review of research on mobile learning in teacher education. J Educ Technol Soc 17(4):17–32
16. Chen Y-T, Hsu C-H, Chung C-H, Wang Y-S, Babu SV (2019) Ivrnote: design, creation and evaluation of an interactive note-taking interface for study and reflection in vr learning environments. In: 2019 IEEE conference on virtual reality and 3D User interfaces (VR). IEEE
17. Boye A (2012) Note-taking in the 21st century: tips for instructors and students. Teaching, Learning, and Professional Development Center
18. Mahrishi M, Morwal S, Muzaffar AW, Bhatia S, Dadheech P, Rahmani MKI (2021) Video index point detection and extraction framework using custom YoloV4 Darknet object detection model. IEEE Access 9:143378–143391. https://doi.org/10.1109/ACCESS.2021.3118048
19. Erkens M, Bodemer D, Hoppe HU (2016) Improving collaborative learning in the classroom: text mining-based grouping and representing. Int J Comput-Support Collaborative Learn 11(4):387–415
20. Reilly M, Shen H (2011) The design and implementation of the smartphone-based groupnotes App for ubiquitous and collaborative learning. In: iUBICOM'11: the 6th international workshop on ubiquitous and collaborative computing, vol 6, pp 46–55
21. DeZure D, Kaplan M, Deerman MA (2001) Research on student notetaking: implications for faculty and graduate student instructors. CRLT Occas Pap 16:1–7
22. Anderson R, Anderson R, Davis P, Linnell N, Prince C, Razmov V, Videon F (2007) Classroom presenter: enhancing interactive education with digital ink. Computer 40(9):56–61
23. Chakraverty S, Chakraborty P, Aggarwal A, Madan M, Gupta G (2022) Enriching WordNet with subject specific out of vocabulary terms using existing ontology. In: Data engineering for smart systems: proceedings of SSIC 2021, Springer, Singapore, pp 205–212
24. Kiewra KA (1987) Note taking and review: the research and its implications. J Instr Sci 16:233–249
25. Memon J, Sami M, Khan RA, Uddin M (2020) Handwritten optical character recognition (OCR): a comprehensive systematic literature review (SLR). IEEE Access 8:142642–142668
26. Kanika SC, Chakraborty P, Agnihotri S, Mohapatra S, Bansal P (2019) KELDEC: a recommendation system for extending classroom learning with visual environmental cues. In: Proceedings of the 2019 3rd international conference on natural language processing and information retrieval, pp 99–103
27. Rajeshkumar G, Vinoth Kumar M et al (2022) An improved multi-objective particle swarm optimization routing on MANET. Comput Syst Sci Eng 44(2):1187–1200. https://doi.org/10.32604/csse.2023.026137
28. Van Es, EA, Sherin MG (2002) Learning to notice: scaffolding new teachers' interpretations of classroom interactions. J Technol Teach Educ 10(4):571–596
29. Atrash A, Abel M-H, Moulin C (2015) Notes and annotations as information resources in a social networking platform. Comput Hum Behav 51:1261–1267
30. Ferilli S (2011) Automatic digital document processing and management: problems, algorithms and techniques

Chapter 15
Predicting Power Consumption Using Tree-Based Model

Dhruvraj Singh Rawat and Dev Mithunisvar Premraj

1 Introduction

Over the last few years, the availability of energy has altered the course of human civilization. Not only have new energy sources been discovered, but power consumption (PC) is also rising as nations develop around the world. If advancements in energy efficiency are unable to offset this rise in demand, our global PC will keep expanding every year. As the PC increases, it becomes more challenging to switch from fossil fuels to low-carbon energy sources; new low-carbon energy must be developed to meet this increased demand while also attempting to replace the existing fossil fuels in the energy mix. Energy plays an important role not only in the everyday lives of humans but also in the activities of the economy. Therefore, the PC per capita of a country is regarded as an essential indicator of economic development; hence, it is vital to manage our PC. PC prediction in the early stages can be done by building a machine learning (ML) model which can infer patterns from a vast amount of historical energy data. This will be an essential tool for energy producers and managers by providing accurate results for PC prediction.

Over the past few decades, PC prediction has gained the spotlight in academia [1–5]. Several ML models based on artificial neural network (ANN) have been shown to work well-predicting PC [6–8]. In [9], five ML models, such as genetic programming (GP), deep neural network (DNN), ANN, support vector machine (SVM), and multiple regression (MR), are compared for predicting PC in a building. The ANN model obtained the best results for PC prediction. The prediction of PC for a hotel in Madrid, Spain, was compared using ANN and random forest (RF) in [10]. The results demonstrated the parallel validity of both models for PC forecasting. The [11]

D. S. Rawat
LNM Institute of Information Technology, Jaipur, India

D. M. Premraj (✉)
St. Joseph's Institute of Technology, Chennai, India
e-mail: devmithunisvar@gmail.com

V. Devedzic et al. (eds.), *Proceedings of the International Conference on Intelligent Computing, Communication and Information Security*, Algorithms for Intelligent Systems, https://doi.org/10.1007/978-981-99-1373-2_15

compares the performance of many ML models for predicting PC, including decision tree (DT), RF, ANN, and SVM regression that used a radial basis kernel function. Based on the data collected in the experiments, RF proved to be the most effective method. Using PC data from three separate Tetouan city's power distribution networks, the authors of [12] compare and contrast three ML models-SVM, ANN, and RF to see which one has the highest prediction capabilities. Based on experimental data, the RF model provides the lowest mean absolute deviation (MAD) and the best correlation (R^2) between observed and projected PC throughout all stages of testing. Due to the success that the author of this paper got from using the random forest algorithm (bagging-based tree algorithm) in predicting solar power generation, in his work [13], we have tried to find the best as well as the most accurate model to predict PC by comparing the results from four different tree-based ML models that comprises decision tree, AdaBoost (boosting), XGBoost (boosting), and random forest (bagging). The open-source dataset [14], used in the study, contains PC for every ten minutes, from three different regions of Tetouan city from January 1, 2017, till December 31, 2017. The effectiveness of these regression models is measured by statistics like the coefficient of determination (R^2), the R-squared value, the Pearson correlation coefficient, and the mean absolute error (MAE). The rest of this study can be structured as follows. The literature review is described in Sect. 2. Section 3 describes the exploratory analysis done on the dataset. Section 4 describes the models and evaluation metrics used, Sect. 5 gives the experimental results and findings, and future research is discussed and summarized in Sect. 6.

2 Literature Review

PC refers to the electrical energy consumed or utilized per unit of time to operate something, such as a home appliance. Machine learning (ML) is an area of artificial intelligence (AI) that helps programs improve their predictive abilities without further human input. We give historical data to ML algorithms to predict new output values. In [15], RF, regression tree, and support vector regression are employed to predict PC for buildings. The authors concluded that the RF model can effectively predict the building's PC with high accuracy. In [16], the XGBoost and the RF model, respectively, introduce an advanced data denoising technique—complete ensemble empirical mode decomposition with adaptive noise, the two models were contrasted to find an accurate model for predicting water quality. The findings of the study confirmed the superiority of predicting power of both the models, and the XGBoost model outperforms the RF model in finer margins.

In [11], PC prediction is made to compare different machine learning models: feedforward neural network with an RF, back-propagation algorithm, DT, and SVR with radial basis function kernel. The study's goal was to predict the PC every 10 min and/or every hour and ascertain which approach was the most successful. Tetouan is a city in Northern Morocco, and the dataset [14] utilized in the research is generated from three separate power distribution networks of the city. The SCADA system was used to deliver the energy and data captured every 10 min, between January 1, 2017, and December 31, 2017. Results showed that the RF model had lower prediction errors compared to its rivals. SVM, ANN, and RF are the three ML models used to forecast PC in [17]. The models' efficacy is verified using the same publicly available dataset from the previous research. The experimental findings demonstrate that the results from the RF model have the strongest correlation between the measured and projected PC. In [18], when the proposed optimization-based generation rescheduling was implemented together with the XGBoost-based oscillation damping estimation model, the XGBoost model improved the accuracy in predicting the damping ratios of an inter-area oscillation by 58.10%. In the RF method, there is a high possibility that most of the trees could have made random predictions with some random chances since each of the trees had its circumstances like class imbalance, over-fitting, sample duplication, inappropriate node splitting, etc. Hence in our study, we decided to compare the RF method with three other ML models, namely DT, AdaBoost (boosting), and XGBoost(boosting), on the same dataset [14] used in the above two studies. One of the most critical differences between RF and XGBoost is that the former gives more preference to hyper-parameters to optimize the model. In contrast, the latter offers more preference for functional space when reducing the cost of a model. While using the RF model, we need to be cautious; a slight change in the hyper-parameter can alter the prediction because it will affect almost all trees in the forest.

3 Exploratory Analysis

3.1 Dataset

The research dataset [14] containing the power consumption of Tetouan city was used for analysis. The dataset is publicly accessible at the UCI Machine Learning Repository and has the records at every 10 min intervals. The data is present for dates between January 1, 2017, and December 31, 2017. The city had three separate power distribution networks, and data for each zone is present. There were no missing values in the dataset, and it contains 52416 records. The dataset contained six different independent features, namely date and time, diffuse flows, general diffuse flows, wind speed, humidity, and temperature for Tetouan city. To gain more insight, the following features were additionally created using the DateTime column—day, month, hour, minute, day of week, quarter of year, and day of year.

3.2 Case Study

A line graph is used in the study to depict how the PC varied for the three zones with respect to the hours of a day and months of the year 2017. Figure 1 shows the variation of PC for all three regions with respect to the hours of the day. The line

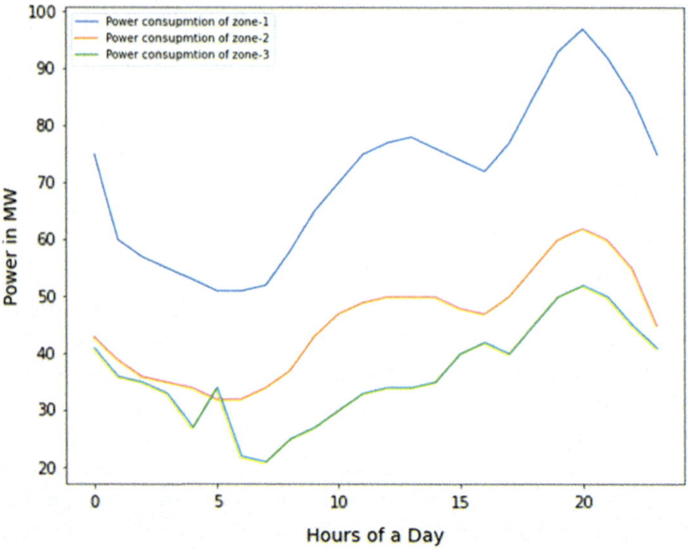

Fig. 1 Variation of PC for the three zones with respect to hours of a day aggregated for the entire year

graph is plotted with the hours of the day in the X-axis and the PC in Megawatts (MW) in the Y-axis. It could be inferred from the graph that the PC during various hours of the day is almost the same for zone-2 and zone-3, but zone-1 had higher PC when compared to them. On a general overview, PC occurred at its peak during the 20th hour of the day for all the three zones, whereas the lowest figures were observed during the 7th hour of the day for all three zones.

Figure 2 is a line graph, and it depicts the variation of PC for all three regions with respect to the months of the year 2017. The line graph is plotted with the twelve months of the year 2017 in the X-axis, and the PC in Megawatts (MW) in the Y-axis. It is observed that the PC during the first nine months was almost the same for zone-2 and zone-3, but zone-1 had higher PC when compared to them for all the twelve months. On a general overview, PC occurred at its peak during the month of August for zone-2, during September for zone-1 and for zone-3 during July. The lowest figures of PC were observed during the month of February for zone-1 and zone-2, where as zone-3 had the lowest value at December.

A heat map (HM) is a data visualization technique used to show a phenomenon's magnitude as a color in two dimensions. HM is used to visualize density; the darker shades of the chart represent higher-value metrics, whereas the lighter colors indicate low-value metrics. In our research, we used Pearson correlation HM to determine the degree of similarity between pairs of characteristics. Figure 3 is the correlation matrix which depicts how well the different attributes of the dataset are correlated with the PC for all the three zones. It is observed that only temperature has a moderate link with PC across zones, whereas all other parameters have a weaker relationship.

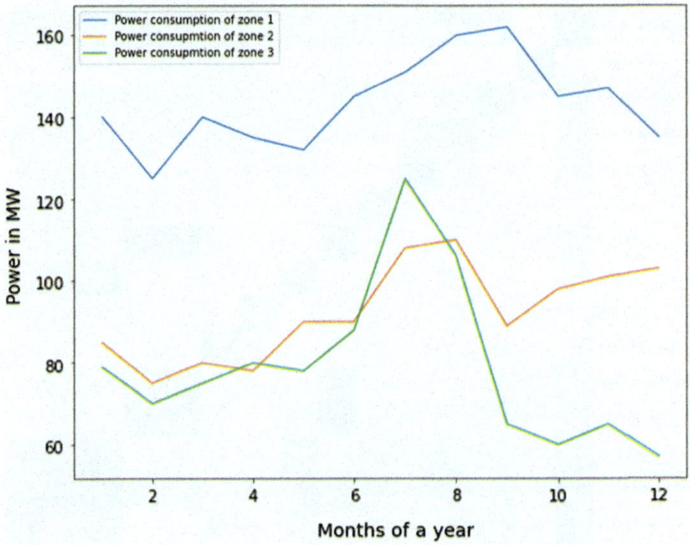

Fig. 2 Variation of PC for the three zones with respect to months of the entire year

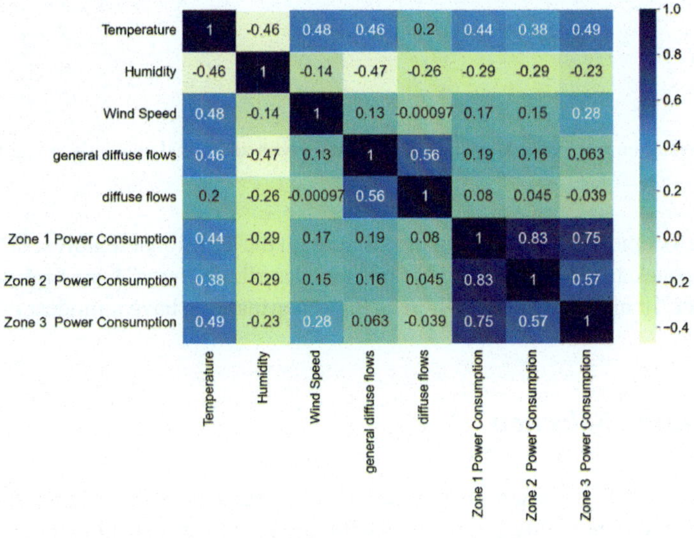

Fig. 3 Correlation matrix for attributes present in dataset

Since only one attribute showed good correlation with PC for three zones, we created six more attributes, namely day, month, hour, year, quarter of the year, and minute for our study. These attributes were created using the "DateTime" attribute already present in the dataset.

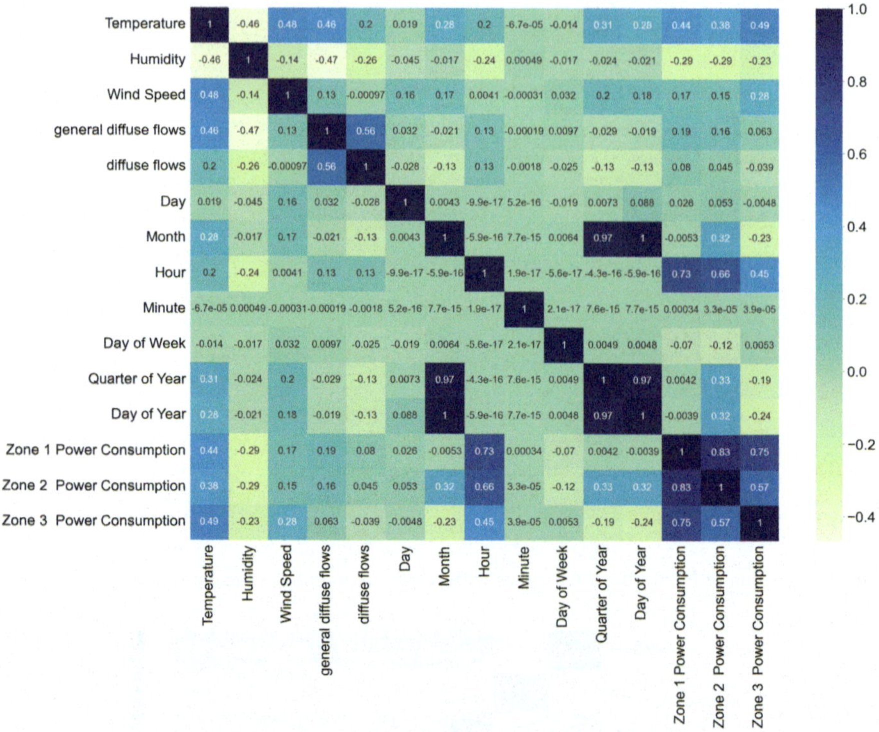

Fig. 4 Correlation matrix for attributes dataset along with additional manually created features

Figure 4 depicts the correlation heat map with the additional features. On a general overview, one can see that the newly introduced attribute "hour" has a high correlation with the PC in all three zones, whereas the other features have a modest correlation.

3.3 *Feature Selection*

When a lot of data is present in a model, feature selection becomes inevitable due to computational constraints and for removing the noisy variables. Feature selection is also an important step made for making a good model.

In SelectKBest method [19], which is a part of the sklearn, "feature_selection module" is used in our study to find the relative importance of the independent variables. It is a univariate feature selection process that works by selecting the best-performing features based on the univariate statistical test.

Table 1 gives us the score for the feature importance of each feature in the dataset; the important features have higher scores. Figure 5 contains the mean accuracy decrease that is found after using the permutation feature method. In permutation

Table 1 Attribute and Z-score

Column	Score
Temperature	12,599.188076
Humidity	4719.862820
Wind speed	1511.949826
General diffuse flows	1919.645156
Day	36.205083
Month	1.498047
Hour	59,084.947978
Minute	0.006156
Day of week	255.934980
Quarter of year	0.925660
Day of year	0.800526

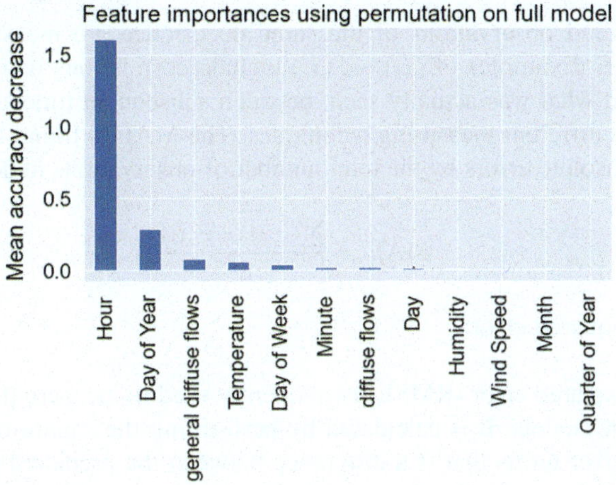

Fig. 5 Important features

feature selection, the value of a particular column is shuffled and its effect on model predictive power is observed. The feature that has the highest effect will have the highest mean accuracy decrease in model performance.

Based on the above analysis, the following features were used for training the model—hour, day of year, general diffuse flows, temperature, day of week, minute, diffuse flows, and day.

4 Evaluation Metrics and ML Models

Evaluation metrics (EM) are used to determine how well a statistical or machine learning model performs. The ML models and EM used to evaluate the efficacy of the ML modes in our study are discussed in this section.

4.1 Evaluation Metrics

In this research, we evaluate the ML models using four distinct statistical metrics. They are chosen based on their usefulness in PC prediction in previous studies [20–22]. The following are the evaluation metrics used in our study.

- Mean absolute error:

  Errors in paired observations of the same occurrence are measured using the method MAE. Examples of $(\hat{x_i})$ vs. (x_i) include correlations between what was expected and what was actually seen, between subsequent time and initial time, and between different measuring techniques. The MAE is found by dividing the sum of all absolute errors by the total number of observations in the study.

$$\text{MAE} = \frac{\sum_{i=1}^{N} |x_i - \hat{x_i}|}{N} \tag{1}$$

- Root mean squared error:

  Root mean squared error (RMSE) is frequently used to measure the performance of regression models. It is calculated by performing the square root of the sum of squared error terms that is a difference between the predicted $(\hat{x_i})$ and actual terms (x_i) whole divided by the number of terms. RMSE gives higher weightage to outlier error terms.

$$\text{RMSE} = \sqrt{\frac{\sum_{i=1}^{N} \left(\sum x_i - \hat{x_i}\right)^2}{N}} \tag{2}$$

- Mean squared error:

  Mean squared error (MSE) is a risk function that is equivalent to the anticipated value of the squared error loss. MSE is typically non-negative and never negative due to random factors. No consideration is given by the estimator to data that would lead to a more precise estimate. Since MSE is based on the square of the Euclidean distance, it is always positive and tends to go down as the error gets closer to zero

$$\text{MSE} = \frac{1}{N}\sum_{i=1}^{N}(x_i - \hat{x}_i)^2 \quad (3)$$

- Coefficient of determination:

R-squared (R^2), often referred to as the coefficient of determination, is a statistical measure in a regression model used to ascertain the amount of variation in the dependent variable that the independent variable can account for. R^2 is used to determine the goodness of fit or the degree to which the data fits the regression model.

$$R^2 = 1 - \frac{\text{RSS}}{\text{TSS}} \quad (4)$$

where N indicates the total number of samples, RSS is the sum of squares of residuals, and TSS is the total sum of squares.

4.2 Machine Learning Models

- Random forest model:

RF, an ensemble classifier, was introduced by Breiman [23]. Given its fast training time, capacity to solve regression, predict accurately, and also the ability to work with large datasets, RF has found widespread usage in the scientific and technical communities [24]. In RF, numerous non-pruned datasets are generated, and their results are aggregated using majority voting; the diversity of trees is increased by using the bootstrap data drawn from training data to construct trees. The samples not used in construction are denoted as "out-of-bag" (OOB) data. This out-of-band data is used internally by the algorithm as validation throughout the training phase. The RF model used for PC prediction has ten ensemble trees, each of which is trained using a minimum of 10 data. The prediction model of RF [23] is represented as

$$f_{\text{rf}}^{N} = \sum_{n=1}^{N} T_{\text{ree}}(y) \quad y = y_1, y_2, \ldots, y_p \quad (5)$$

y is a p-dimensional vector of inputs, N represents the average number of regression trees constructed by RF, and T_{ree} refers to DT

- Decision tree model:

A DT is a flowchart-like tree structure in which internal nodes represent tests (on input data patterns) and leaf nodes depict categories (of these patterns). The design and potential of DT are explained in [12]. DT can be used to replace statistical procedures for finding data, retrieving text, finding missing data in a class, enhancing search engines, and exploring various applications in medical fields. The training dataset is divided into several subsets depending on the values of the splitting attribute and splitting criteria (information gain, Gini index, gain ratio). The DT algorithm loops until all instances of a subset belong to the same class.

- AdaBoost model:

Adaptive boosting, or AdaBoost, is a machine learning approach that functions as an ensemble method. In 1995, Freund and Schapire [25] published a boosting technique that significantly improved upon previous attempts by addressing many of the problems that had plagued previous implementations. AdaBoost's most popular technique is a DT with a single split, also known as a decision stub. The method constructs a model in which each data point is given the same importance and then provides more weight to misclassified data. In the new model, more emphasis is placed on all the points that have larger weights. The same method of training the models will be repeated until or until a lower error is obtained.

- XGBoost model:

Extreme gradient boosting, or XGBoost, is a distributed gradient boosted decision tree (GBDT), ML Library that was developed by Tianqi Chen as part of the Distributed Machine Learning Community (DMLC). In this algorithm, DT is constructed in a sequential fashion, weights are given to each independent variable that is input into the DT, and predictions are made based on the DT's output. If a variable is predicted wrong, its weights are increased and fed to the second DT. A strong and more precise model is obtained by ensembling these individual classifiers/predictors. This model can be used for regression, classification, ranking, and user-defined prediction problems.

5 Experimental Result and Observation

The author performs a comprehensive study to find the capability of DT, AdaBoost, XGBoost, and RF models for PC prediction; statistical measurements and visual representations are used to evaluate the results. Results of models' performance on the training and testing set are given in Tables 2 and 3. Python-based ML environment running on Windows 10 with 2.1 GHz AMD Ryzen 3 and 8 GB RAM is used to build the model. On the general analysis of the statistical results provided in Table 2, it can be easily inferred that the XGBoost model outperformed the other models. The XGBoost model obtained the lowest MSE, MAE, and RMSE and the highest R^2 score for all the three zones as compared to other ML models used in the study. In Table 3, the averaged metrics of all three zones are represented to get the overall results. The XGBoost performed best and had 41.05, 54.97, and 36.53% lower MAE, MSE, and RMSE errors as compared to the worst-performing model. When looking at the R^2 score at the overall level, the XGBoost has a 46.9% higher score as compared to the worst-performing model. When compared to RF, the XGBoost model needs a significantly lower fit time for all three zones, as given in Table 2. These findings demonstrate that the XGBoost provides a reasonable fit time as well.

The scatter plot (SP) plotted between the predicted PC value and the measured PC values is used to visually illustrate the performance of each model. For the ideal best-performing model, all the points should lie on a linear line. Thus, the model which has the most points closer to the linear line has the best performance. To numerically evaluate the strength of the relation, the R^2 score is also embedded in each scatter plot. The SP for zone-1 is shown in Fig. 6, and the XGBoost model has the highest R^2 scores of 0.985 and 0.98 for training and test set. These high R^2 scores can also be visually confirmed by the high concentration of points on the trendline.

Likewise, the SP for zone-2 is presented in Fig. 7 and XGBoost was the best-performing model with R^2 scores of 0.981 and 0.988 for the testing and training set. Similarly, SP for zone-3 is shown in Fig. 8 and the XGBoost again achieves the best performance with R^2 scores of 0.994 and 0.991, respectively, for training and testing sets. The performance of XGBoost for the prediction of PC as represented by SP is in line with data presented in Table 2. For all three regions, XGBoost is the best-performing model followed by RF, while DT and AdaBoost models were not efficient in prediction when compared to them. **RQ1: What are the training time and inference time of the proposed model?** Table 2 shows the fit time and inference time for all the ML models used in the study. On average, the fit time for our proposed model (XGBoost) varies from 5.894 to 5.778 s, with an inference time varying from 0.0257 to 0.0235 s. On average, it took 5.8687 s to fit and 0.02445 s to infer.

6 Conclusion

This research is set out to develop a reliable model for PC prediction using ML methods. Five statistical measures (Pearson's correlation coefficient, R^2, root mean squared error, mean squared error, and mean absolute error) are used to compare four ML models (decision tree, AdaBoost, XGBoost, and random forest) for predicting energy consumption. Based on the consolidated results presented in Table 3, XGBoost came out as the best-performing model for PC prediction, followed by random forest, decision tree, and AdaBoost. XGBoost also had the lowest standard deviation while calculating most of the evaluation metrics. AdaBoost came out as the worst-performing model, and it even performed poorer even when compared with the decision tree model. According to Table 3, the decision tree had the lowest fit time. And among the ensemble model, AdaBoost had the lowest fit time followed by XGBoost and random forest. Random forest took 40,928.64% greater fit time than

Table 2 Statistical performance of PC prediction using fivefold cross-validation

Model	Zone	Metric	MAE	MSE	RMSE	R^2	Fit time	Inference time
Decision tree	Zone-1	Mean	2.118	8.923	2.801	0.809	**0.121**	**0.0029**
		Std.	0.784	6.858	1.038	0.133	**0.004**	**0.0003**
	Zone-2	Mean	1.816	5.7982	2.3408	0.729	**0.122**	**0.0029**
		Std.	0.479	3.0008	0.5643	0.1288	**0.0045**	**0.0005**
	Zone-3	Mean	1.967	8.2025	2.5926	0.67294	**0.1225**	**0.0025**
		Std.	0.944	7.3871	1.2169	0.1667	**0.0031**	**0.0004**
Random forest	Zone1	Mean	1.9832	8.3309	2.6767	0.8233	49.859	0.1378
		Std.	0.7716	6.9687	1.0797	0.13275	0.2965	0.0046
	Zone-2	Mean	1.5492	4.3324	2.0203	0.79448	49.7669	0.1354
		Std.	0.415	2.097	**0.4984**	**0.1105**	0.4815	0.0038
	Zone3	Mean	1.8113	7.2650	2.4497	0.6834	50.7941	0.1344
		Std.	0.8701	6.3586	1.1345	0.2089	1.7379	0.0077
AdaBoost	Zone-1	Mean	2.788	12.924	3.4477	0.7184	0.668	0.0068
		Std.	**0.7504**	6.9901	**0.9332**	0.1352	0.1004	0.0021
	Zone-2	Mean	2.240	8.0506	2.7722	0.62506	0.6625	0.0056
		Std.	0.5201	3.4275	0.6045	0.1619	0.0062	0.0051
	Zone-3	Mean	2.7528	12.7486	3.4473	0.3443	0.6584	0.0055
		Std.	**0.7808**	6.8932	**0.9296**	0.4317	0.0086	0.0004
XGBoost	Zone-1	Mean	**1.7394**	**6.6065**	**2.3404**	**0.8619**	5.8949	0.0235
		Std.	0.7583	**6.6336**	1.0629	**0.1251**	0.3386	0.0019
	Zone-2	Mean	**1.3892**	**3.6588**	**1.8392**	**0.8231**	5.7782	0.0241
		Std.	**0.4138**	**2.1268**	0.5255	0.1172	0.2562	0.0023
	Zone-3	Mean	**1.4579**	**4.9375**	**1.9743**	**0.7961**	5.9342	0.0257
		Std.	0.801	**5.0214**	1.0195	**0.1386**	0.4774	0.0031

The bold values indicate the best value in each column and row

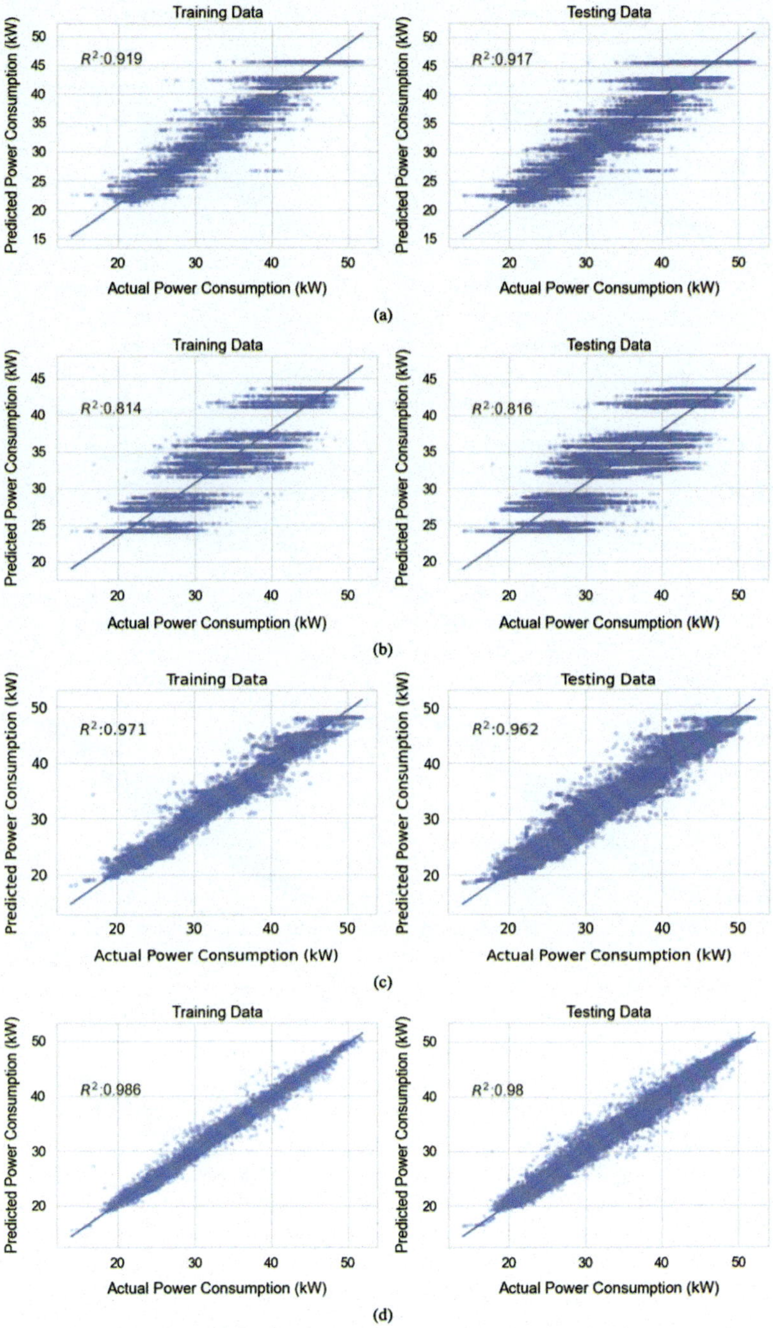

Fig. 6 Scatter plots performance of the zone-1 PC prediction by the DT (**a**), AdaBoost (**b**), RF (**c**), and XGBoost (**d**) for training and testing

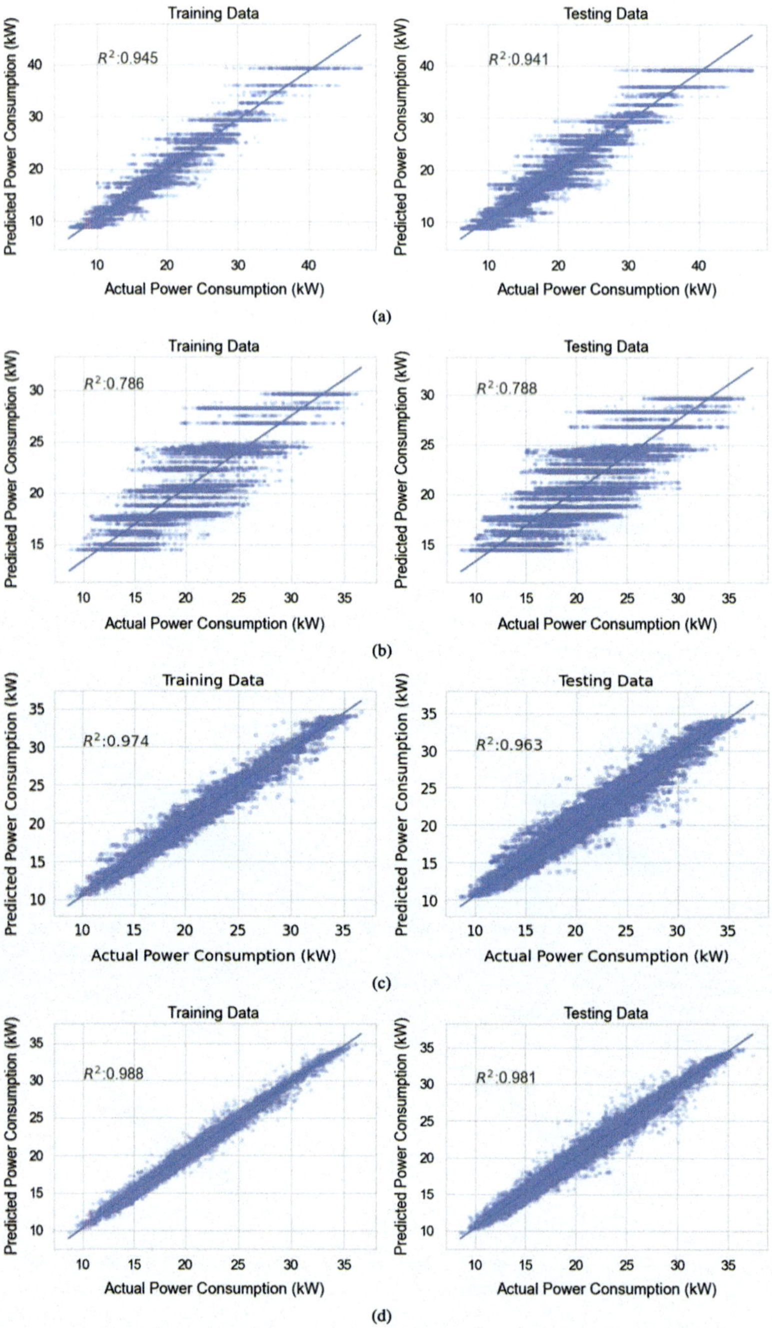

Fig. 7 Scatter plots performance of the zone-2 PC prediction by the DT (**a**), AdaBoost (**b**), RF (**c**), and XGBoost (**d**) for training and testing

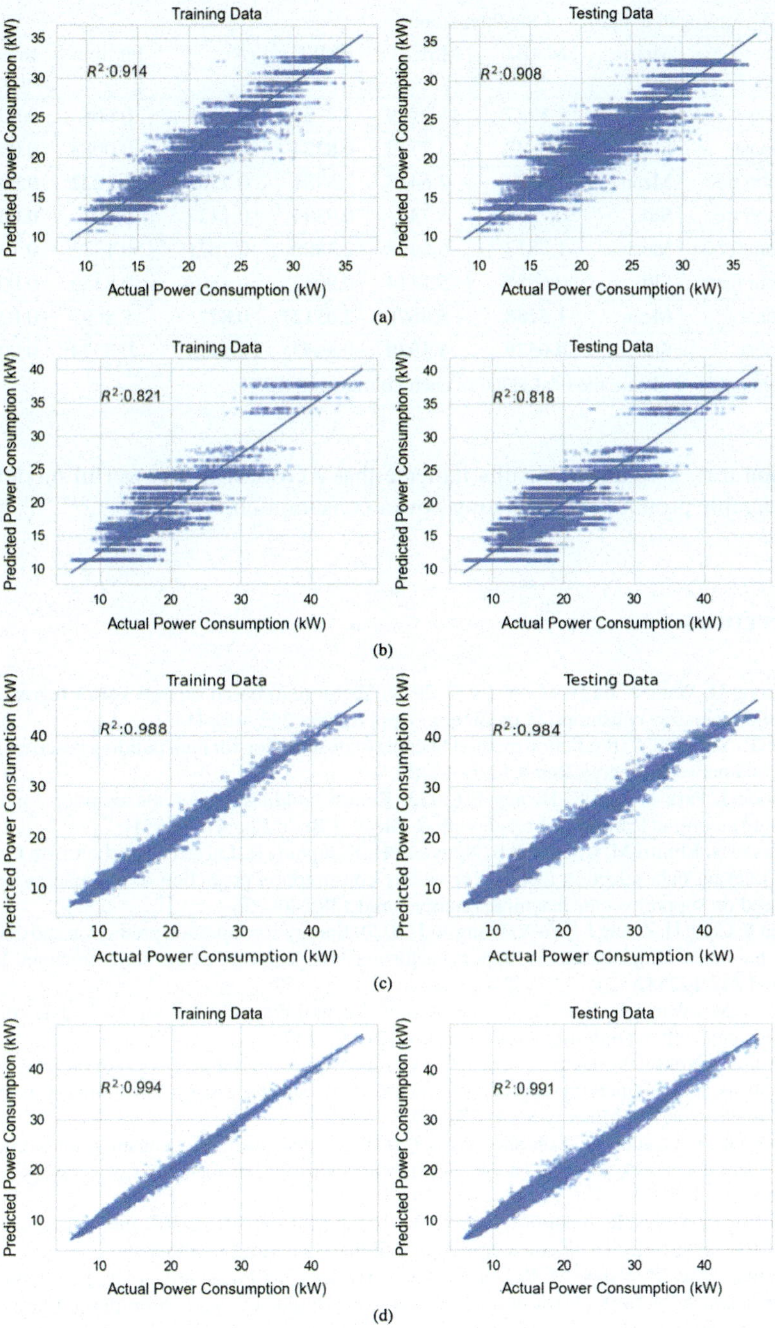

Fig. 8 Scatter plots performance of the zone-3 PC prediction by the DT (**a**), AdaBoost (**b**), RF (**c**), and XGBoost (**d**) for training and testing

Table 3 Aggregated metrics of the three zones

Model	Metric	MAE	MSE	RMSE	R^2	Fit time	Inference time
AdaBoost	Mean	2.5937	11.2539	3.2322	0.5626	0.663	0.00601
AdaBoost	Std.	0.6838	5.7719	**0.8224**	0.2429	0.0083	0.00103
Decision tree	Mean	1.967	7.6413	2.5781	0.7369	**0.1222**	**0.0027**
Decision tree	Std.	0.7359	5.741	0.9399	0.1427	**0.004**	**0.0004**
Random forest	Mean	1.7813	6.6428	2.3808	0.7671	50.1372	0.1355
Random forest	Std.	0.6885	5.1414	0.9042	0.1507	0.8386	0.0054
XGBoost	Mean	**1.5288**	**5.0676**	**2.05127**	**0.827**	5.8691	0.0244
XGBoost	Std.	**0.6579**	**4.5939**	0.8693	**0.127**	0.3574	0.0024

The bold values indicate the best value in each column and row

decision tree. Overall the results indicate that XGBoost is a powerful model with considerable promise for predicting power consumption.

References

1. Zhong H, Wang J, Jia H, Mu Y, Lv S (2019) Vector field-based support vector regression for building energy consumption prediction. Appl Energy 242:403–414
2. Lei R, Yin J (2022) Prediction method of energy consumption for high building based on LMBP neural network. Energy Rep 8:1236–1248
3. Moon J, Park S, Rho S, Hwang E (2022) Robust building energy consumption forecasting using an online learning approach with R ranger. J Build Eng 47:103851
4. Paudel S, Elmitri M, Couturier S, Nguyen PH, Kamphuis R, Lacarrière B, Le Corre O (2017) A relevant data selection method for energy consumption prediction of low energy building based on support vector machine. Energy Build 138:240–256
5. Liu Y, Chen H, Zhang L, Wu X, Wang X-J (2020) Energy consumption prediction and diagnosis of public buildings based on support vector machine learning: a case study in china. J Clean Prod 272:122542
6. Wong SL, Wan KK, Lam TN (2010) Artificial neural networks for energy analysis of office buildings with daylighting. Appl Energy 87(2):551–557
7. Aydinalp-Koksal M, Ugursal VI (2008) Comparison of neural network, conditional demand analysis, and engineering approaches for modeling end-use energy consumption in the residential sector. Appl Energy 85(4):271–296
8. Azadeh A, Ghaderi S, Sohrabkhani S (2008) Annual electricity consumption forecasting by neural network in high energy consuming industrial sectors. Energy Convers Manage 49(8):2272–2278
9. Amber K, Ahmad R, Aslam M, Kousar A, Usman M, Khan MS (2018) Intelligent techniques for forecasting electricity consumption of buildings. Energy 157:886–893
10. Ahmad MW, Mourshed M, Rezgui Y (2017) Trees vs. neurons: comparison between random forest and ANN for high-resolution prediction of building energy consumption. Energy Build 147:77–89
11. Salam A, El Hibaoui A (2018) Comparison of machine learning algorithms for the power consumption prediction:-case study of Tetouan city-. In: 2018 6th international renewable and sustainable energy conference (IRSEC). IEEE 2018, pp 1–5

12. Swain PH, Hauska H (1977) The decision tree classifier: design and potential. IEEE Trans Geosci Electron 15(3):142–147
13. Rawat DS, Padmanabh K (2021) Prediction of solar power in an IoT-enabled solar system in an academic campus of India, pp 419–431
14. Dua D, Graff C (2017) UCI machine learning repository. [Online]. Available: http://archive.ics.uci.edu/ml
15. Wang Z, Wang Y, Zeng R, Srinivasan RS, Ahrentzen S (2018) Random forest based hourly building energy prediction. Energy Build 171:11–25
16. Lu H, Ma X (2020) Hybrid decision tree-based machine learning models for short-term water quality prediction. Chemosphere 249:126169
17. Zogaan WA, Power consumption prediction using random forest model
18. Asvapoositkul S, Preece R (2021) Decision tree-based prediction model for small signal stability and generation-rescheduling preventive control. Electr Power Syst Res 196:107200
19. Pedregosa F, Varoquaux G, Gramfort A, Michel V, Thirion B, Grisel O, Blondel M, Prettenhofer P, Weiss R, Dubourg V, Vanderplas J, Passos A, Cournapeau D, Brucher M, Perrot M, Duchesnay E (2011) Scikit-learn: machine learning in Python. J Mach Learn Res 12:2825–2830
20. Ma Z, Ye C, Li H, Ma W (2018) Applying support vector machines to predict building energy consumption in China. Energy Proc 152:780–786
21. Vinagre E, Pinto T, Ramos S, Vale Z, Corchado JM (2016) Electrical energy consumption forecast using support vector machines. In: 27th international workshop on database and expert systems applications (DEXA). IEEE 2016, pp 171–175
22. Ngo N-T, Truong TTH, Truong N-S, Pham A-D, Huynh N-T, Pham TM, Pham VHS (2022) Proposing a hybrid metaheuristic optimization algorithm and machine learning model for energy use forecast in non-residential buildings. Sci Rep 12(1):1–18
23. Jiang R, Tang W, Wu X, Fu W (2009) A random forest approach to the detection of epistatic interactions in case-control studies. BMC Bioinform 10(1):1–12
24. Genuer R, Poggi J-M, Tuleau-Malot C (2010) Variable selection using random forests. Pattern Recogn Lett 31(14):2225–2236
25. Freund Y, Schapire RE (1997) A decision-theoretic generalization of on-line learning and an application to boosting. J Comput Syst Sci 55(1):119–139

Chapter 16
A Novel Hybrid Approach for Dimensionality Reduction in Microarray Data

Devendra K. Tayal, Neha Srivastava, Neha, and Urshi Singh

1 Introduction

In microbiology, data have increased significantly over time, both in terms of cases and features [1]. Microarray data contain thousands of genetic information that can be used to provide reliable estimates of variance [2]. They are widely used to study gene expression variations under different experimental settings [3]. Microarray technology enables the analysis of a wide variety of samples, including new and previously unrecorded samples. The frequency of specific markers in malignancies can also be determined this way [4]. Microarrays identify the marker genes from high-throughput arrays. Not to build models that predict disease symptoms from genuine samples. The liability of severe overfitting is a fundamental problem with high-dimensional data. Datasets where the number of features is equal to or more than the number of observations are the subject of high-dimensional data. Hundreds of samples may be present in microarrays, which analyse gene expression. Several thousands of genes could be identified in each sample. The representativeness of the samples provided is another issue beyond our control. A worthless feature set can be produced by overfitting unrepresentative samples [5]. Feature selection is often suboptimal in gene expression datasets due to high complexity and limited sample size. The pre-processing phase of machine learning, called feature selection, reduces dimensionality, removes irrelevant inputs, improves learning accuracy and improves understanding. A set of the genuine features [1] were chosen in order to efficiently diminish the feature space w.r.t certain criterion. Aside from general feature selection

---

D. K. Tayal · N. Srivastava · Neha (✉) · U. Singh
Department of Computer Science and Engineering, Indira Gandhi Delhi Technical University for Women, James Church, New Church Rd, Opp. St, Kashmere Gate, New Delhi, Delhi 110006, India
e-mail: neha015mtcse21@igdtuw.ac.in

V. Devedzic et al. (eds.), *Proceedings of the International Conference on Intelligent Computing, Communication and Information Security*, Algorithms for Intelligent Systems, https://doi.org/10.1007/978-981-99-1373-2_16

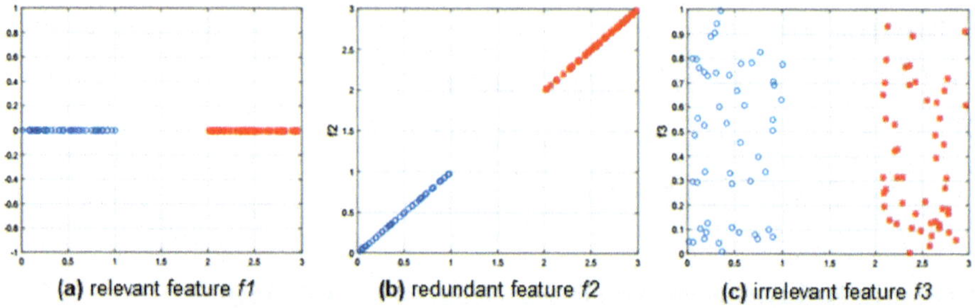

Fig. 1 An illustrative example of **a** redundant, **b** relevant, **c** irrelevant features

processes, domain-specific approaches have been developed in areas such as bioinformatics [6–8], text classification [9] and multimedia [10]. Any machine learning technique can perform classification based on a collection of features [11].

We can eliminate these properties by feature selection without significantly impacting learning performance, reducing memory and computational costs [12]. For example, feature $f1$ in Fig. 1a is a relevant feature that can distinguish two classes (clusters). However, given feature $f1$, feature $f2$ in Fig. 1b is unnecessary because it is highly correlated with $f1$. Feature $f3$ in Fig. 1c is useless because it has no ability to distinguish between two classes (clusters). Therefore, removing $f2$ and $f3$ has no negative impact on training performance. However, recent increases in data dimensionality have had a serious negative impression on the behaviour and accuracy of some existing feature selection algorithms [1]. Feature selection strategies based on classification may be classified into three classes [11].

Filter Method: Variable ranking algorithms are the main criterion for ordering variable selection in filter techniques. Ranking techniques are used because they are simple and have a good track record in practical situations [13].

Wrapper Method: This method evaluates the variable subset by considering the medium as a "black-box" and its performance to be the detached function. Several search techniques may be utilized to identify a subspace of variables that maximize the detached function.

Embedded Method: In order to speed up the calculation time needed in wrapper techniques for reclassifying various subsets, embedded methods [10, 13–15] are designed to do so. The key strategy is to incorporate feature selection during training. Feature addition and removal in the model are based on the findings from earlier training.

Goals of feature selection (FS) [6]: To decrease the size of the problem, which will make our algorithms less difficult in terms of space and time. To enhance classifiers by eliminating irrelevant features and lowering the risk of overfitting to noisy data.

To determine what characteristics may be significant to a certain problem. For example, to demonstrate which gene expressions are relevant in a particular disease.

Kennedy and Eberhart suggested PSO—a novel mutative estimation method [14, 16]. Simulations of social act in living things served as an inspiration [17]. Shen et al. [14] suggested combining Tabu search techniques with PS optimization for selection of genes. On the other side, findings from the suggested hybrid technique are less important since PSO cannot search in all feasible search areas when Tabu methods are used. Then Chuang et al. suggested an enhanced binary PSO [18]. In many datasets, this technique reached 100% classification accuracy, although it did so at the expense of a sizable number of properly selected genes. Since the fitness number of the global best particles stayed constant after three iterations, they were all reset to the same place. For the same goal, Li et al. combined PSO and GA [18]. GW optimizer is the one of the adequate swarm intelligence-based metaheuristic algorithms. Metaheuristic algorithms seek to find the best solution over a broad search space with the least amount of computational overhead [19]. The metaheuristic algorithms have an extremely straightforward concept, are simple to put into practice and don't need data inclination to discover the nearly ideal answer [8]. The primary goal of these algorithms is to deal with any optimization problem. Heuristics and metaheuristics differ primarily in that heuristics seek out the optimal answer to a particular problem, whereas metaheuristics are created to solve a particular problem. On the other hand, metaheuristics are a universal strategy that may be applied to almost any kind of optimization issue. The captaincy structure and the chase of grey wolves serve as motivation for this population-based optimization technique [16, 20].

2 Literature Survey

Among the most often used evolutionary variants in hybrid approaches is particle swarm optimization because of its capacity to find global optimums, speed of convergence and simplicity. Several studies in the text have combined different particle swarm optimization variants with other metaheuristics, including PSO-GA [21], PSO-DE and PSO-ACO [16]. The likelihood of becoming trapped in the local optimum is reduced by these hybrid algorithms. A recently developed optimization technique called GSA [17] was motivated by nature. The many hybrid particle swarm optimized algorithm versions are discussed here.

Kennedy and Eberhart suggested PS optimization for nonlinear function optimization [17]. Applications like nonlinear methods optimization and neural networks training were suggested. The connections between artificial life, genetic algorithm and PS optimization were examined. It looked at how, in comparison with its forerunners, the idea of particle swarm optimization developed from social simulation to optimizer. Mohamad et al. [22] proposed an enhanced B-PSO to choose a limited, almost optimal selection of instructive genes. On two actual microarray datasets, the suggested approach was assessed. According to the experimental findings, IPSO performed better than BPSO and similar prior studies. This is due to the fact that the updated method in IPSO for modifying the positions of particles only chooses a certain number of genes per iteration, ultimately building a subgroup of instructive genes for cancer classification that is very near to being optimal.

Chuang et al. [18] IB-PSO were created for *K*-nearest neighbour and feature selection. As an IB-PSO authority on microarray data categorization, it serves as a resource. The outcomes of our trials show that our method effectively decreased the number of genes (features) required while streamlining the gene selection (features). The recommended approach makes the best accuracy in 9/11 microarray data precedent, and it is comparable to the classification efficiency in the final two test issues. The recommended method is a pre-processing way to aid with the process of feature selection optimization considering it improves classification accuracy while consuming less computer power.

For the arrangement microarray data, Alba et al. [23] provided a correlation of PS optimization and a genetic algorithm, both of which were enhanced using SVM. Both strategies are used to identify small clusters of valuable genes among thousands.

Li et al. Tan introduced a hybrid PS and GA approach for gene selection utilizing the SVM as the classifier [24]. They build a method that was tested on three benchmark microarray datasets. The experimental results show that the proposed technique may improve classification precision, verify the most valuable gene subset and reduce dataset complexity.

Shen et al., created a hybrid PS optimizer and TS algorithm (HPSO-TS) approach for microarray data. By leveraging Tabu search, the HPSO-TS algorithm can effectively overleap local optima. Three separate microarray datasets are used with the suggested methodology.

Ahmed et al. [15] introduced a hybrid PSO form called HPSO-M. HPSO-M's main objective was to combine the GA mutation approach with PS optimization. In relation with quality of solution and the stability of solution, convergence rate and capacity to search the global optimum, the variation outperforms the PS optimization variant, as the authors stated, who evaluate the hybrid variation's performance on a large number of classical functions.

Mirjalili and Hashims [25] propose a unique HPSO-GSA that combines PSO with the gravitational search algorithm. The concept is to optimize the advantages of both strategies by combining the inspection and developing abilities of the gravitational search algorithm with particle swarm optimization.

Mirjalili et al. [26] in this study suggested a brand-new SI optimization strategy that was motivated by grey wolves. The suggested approach was dependent on the civil structure and hunting techniques of grey wolves. The effectiveness of the suggested approach/algorithm was assessed using 29 test functions when it comes to convergence, prevention of localized optima, utilization and exploration.

Chhabra et al. [7] developed the GWA clustering (GWAC), a GWA-based clustering method, via their study.

The GWA's search function was utilized to locate the best cluster cores in the provided feature margin. The novel GWAC method was compared to six known metaheuristic-based clustering ways using both simulated and real datasets. It was exciting to see from the computational results that GWAC gives better accuracy, recall, G-measure and intracluster distance values. A gene expression data collection is used to compare the effectiveness of GWAC with other techniques. Experimental results show that the GWAC is more successful than other approaches.

Zhang et al. [20] in this study propose a unique H-GWO for clustering that joins the particle swarm optimization method with high-level hybridization (HGWOP) for application to clustering optimization. The computational complexity is reduced, and GWO's global search capabilities are enhanced while retaining the advantages of its strong local search functionality. With the use of the poor-for-change approach, the two updated algorithms are organically linked to form a unique hybrid GWO that more effectively balances exploitation and exploration.

Our paper presented a hybrid PS Optimization using GWO approach for selection of gene in three different microarray dataset. The convergence problem in the HPSO-GWO algorithm is resolved with the addition of GWO as a local enhancement method. The concept and associated Pseudocode are described in full in the publication. To assess HPSO-GWO's effectiveness, the suggested technique is used to publically accessible microarray datasets.

3 Methods

3.1 Particle Swarm Optimization Algorithm (PSO)

Kennedy and Eberhart suggested the PSO algorithm [17], underlying justification was largely influenced by models of behaviour of sociable animals like bird flocking and fish schooling. The birds either disperse or flock together in search of food before settling on a spot where they may find it. Even the birds emigrate from one place to other in quest of food, and there is always a bird that has a strong sense of smell, which indicates that it is aware of the spot where the food can be obtained and is carrying the appropriate food resource message. Due to the fact that they're disseminating the message, especially the valuable message, sometime after moving from location to location in quest of food, the birds will swarm to the place, the food can be found.

```
Step 1: For every particle
          Initialize all particle
        END
Step 2: Do
          For every particle
            determine fitness Value
            If  fitness  is better than the p_best
                then set the current value = new p_best
          END
Step 3: Choose the particle with best fitness value as a g_best
Step 4: For every particle
          Update particle velo. using Eq (1)
          Update particle pos. using Eq (2)
        End
Step 5: While Max Itr or Min Criteria is not attained
```

Pseudocode 1 Particle Swarm Optimization (PSO)

Each bird is referred to as a particle in this approach to global optimization functions/problems, which is inspired by animal behaviour.

$$V_i^{k+1} = V_i^k + c_1 r_1 (p_i^k - x_i^k) + c_2 r_2 (g_{\text{best}} - x_i^k) \quad (1)$$

$$x_i^{k+1} = x_i^k + V_i^{k+1} \quad (2)$$

Using the above mathematical equations, the global search margins are updated, to position each crowd partner in the PSO technique.

3.2 *Grey Wolf Optimization Algorithm (GWO)*

Hunting and searching behaviours have influenced a sizable portion of the swarm intelligence techniques developed to date. But there isn't a swarm intelligence approach that mimics the hierarchy of the wolf pack, which is widely known due to its group hunting [27]. The GWO, a novel swarm intelligence approach, was presented by Mirjalili et al. [25] and was encouraged by grey wolves and prompted by several algorithms. The GWO's capabilities in handling both theoretical and actual applications were examined. The GWO variant impersonates the social organization and method of trapping wild wolves. To imitate the leadership structure, the audience in GWO is broken into four groups termed alpha-α, beta-β, delta-δ and omega-ω. They are referred to as the highest predators since they are the top predators in the food chain. They often like to stay in groups. The superiors can be male or female and are referred to as alphas-α. Typically, the alpha-α is responsible for making commitments like when to go to bed, wake up, go hunting and do more things. The top-level wolf (alpha) receives help from the beta wolf in decision-making and other group tasks. The beta-β task to respect the pack's alpha wolf while directing the other lower-level members. The beta wolf, who also gives the alpha input, reinforces the directives of the alpha. The grey wolf is ranked at level 3 as Omega. The scapegoat is the wolf. The third rung of grey wolves is regularly subject to submission to all assertive wolves.

There are three fundamental hunting techniques—seeking a prey for meal, surrounding prey and attacking the prey.

Each technique is calculated by using the below formulas:

$$d = |\, c \,. x_{P(t)} - x(t) \,| \quad (3)$$

$$X(t+1) = XP(t) - a \cdot d \quad (4)$$

The mathematical formulation of the vectors a and c is:

$$\begin{aligned} a &= 2\ell \cdot \Gamma_1 \\ c &= 2 \cdot \Gamma_2 \end{aligned} \quad (5)$$

Since there is no known global optimum for optimization issues, it has been assumed that top-level wolf, second-level wolf and delta have a decent understanding of where they are. This assumption is plausible given that these three solutions are the best for the population as a whole. Other wolves should be required to update their places in the manner described below:

$$\chi(t+1) = 1/3(\chi_1 + \chi_2 + \chi_3) \tag{6}$$

Values of X_1, X_2, X_3 are calculated by:

$$\begin{aligned} \chi_1 &= \chi_\alpha(t) - A_1.d_\alpha \\ \chi_2 &= \chi_\alpha(t) - A_2.d_\alpha \\ \chi_3 &= \chi_\alpha(t) - A_3.d_\alpha \end{aligned} \tag{7}$$

Values of d_α, d_β, $d_\delta$are calculated by:

$$\begin{aligned} d_\alpha &= |C_1 \text{-} X_\alpha \text{-} X| \\ d_\beta &= |C_2 \text{-} X_\beta \text{-} X| \\ d_\delta &= |C_3 \text{-} X_\delta \text{-} X| \end{aligned} \tag{8}$$

Exploration refers to the capacity to locate prey, whereas exploitation refers to the capacity to engage in prey assault. Using random values for A, the search is shifted apart from the prey. The population is compelled to separate from the prey, if When $|A| > 1$.

Step 1 : Ini. the population Xi (randomly{ i=from 1 to n}

Step 2: Set alpha=2 ,A & C = default values using Eq (5)

Step 3: Calculate the fitness of each member of population (using Eq (7))

Step 4: for t=1 & upto Maximum Itr:

Determine the position using Eq (8)

Determin alpha, A,C using Eq (5)

a=2 [1-t/T]

Calculate fitness of all search agents

Update $X_\alpha X_\beta X_\delta$

END

Step 5: Return X_α

Pseudocode 2 Grey Wolf Optimized Algorithm (GWO)

3.3 Hybrid PSO-GWO

Our approach suggested in this paper employs the particle swarm optimized and grey wolf optimized metaheuristic approaches. The PSO approach is popular and acclaimed. Despite being comparably recent in the literature, the grey wolf optimized algorithm is a metaheuristic approach that, like the PSO algorithm, has been demonstrated to yield effective outcomes. A mixed strategy has been used. These two algorithms deliver effective outcomes. Without changing the PSO and GWO algorithms' fundamental workings, we developed the hybrid PSO-GWO algorithm (HPSO-GWO). The PSO algorithm is capable of successfully resolving almost all real-world issues. To lessen the possibility of the PS optimize algorithm entangling in a local minimum, a solution is necessary. In our suggested strategy, the GW optimize algorithm is employed to help the PS optimize algorithm lessens the likelihood of striking local minima.

According to Sect. 3.1's explanation of the PSO algorithm, some particles are sent to arbitrary places with a slim chance of avoiding local minimums. These guidelines may provide certain risks in terms of departing from the global minimum, as discussed in Sect. 3.2. By guiding certain particles to sites that have been somewhat enhanced by the GW optimization algorithm rather than arbitrary positions, the capabilities of the GWO algorithm are employed to reduce these hazards.

INPUT :MAX ITRE
    :Population_Size (no. of pollution size) & Prob.
Step 1: init. the particles
Step 2: for i,j=1 & upto MAX ITRE & Population_Size
Step 3: Run Particle Swarm Optimization Technique
Step 4: Update velocity and position using Eq 9 & 10
    To avoid Local minima { check random(0,1) is smaller then Prob.}
Step 5: Set alpha , A & C Values
Step 4: Run for loop from 1 to 10 small number of iteration & population size
Step 5: Run GWO
    Update Position of alpha , beta & Gamma wolves
    Again Set alpha , A & C Values
Step 6: Export the mean of these wolves
Step 7: Return Idea Particles and Swarm values
Step 8: MAX ITRE reached then end Procedure { if not restart from step 2}

Pseudocode 3 Hybrid PSO-GWO

The GWO method is enforced in accession to the PS optimization algorithm, which lengthens the running time. We merge the potential of exploitation in swarm technique with the exploration capability in grey wolf optimizer to reserve the strength of variants. The proposed mathematical eq. it updates the locations of the first three search agents in the search space in HPSO-GWO by using the inertia constant to control the grey wolf's surveys and search space. The updated equations are as follows:

$$\begin{aligned} d_\alpha &= |c_1.x_\alpha - w * x| \\ d_\beta &= |c_2.x_\beta - w * x| \\ d_\delta &= |c_3.x_\delta - w * x| \end{aligned} \tag{9}$$

By combining PS and GW optimized variants, the velocity or speed and updated eq. for c are:

$$\begin{aligned} V_i^{k+1} &= \omega * (V_i^k + c_1 r_1 (x_1 - x_i^k) + c_2 r_2 (x_2 - x_i^k) + c_3 r_3 (c_3 r_3)(x_3 - x_i^k)) \\ x_i^{k+1} &= x_i^k + V_i^{k+1} \end{aligned} \tag{10}$$

4 Experiment and Results

In this study, three well-known datasets from microarray experiments were used. The colon gene expression dataset is applied for microarray experiments in this study. The dataset is made up of 63 experiments from colon cancer patients, each with 2001 levels of gene expression. There are 22 normal biopsies and 41 tumour biopsies from the same individual's colons. For 1904 breast cancer patients, this dataset contains '31' clinical characteristics, '331' mRNA levels z-scores and '175' gene mutations. Almost all of the PLCO study data for lung cancer screening, incidence and death analysis are accessible in the lung cancer dataset. The colon gene expression dataset is made up of '72' microarray tests from lung cancer patients, each having '3573' gene expression levels. Among them are '24' tumour biopsies and '48' (normal) biopsies from healthy areas of the same patient's lung cancer. These datasets are high-dimensional data because they contain high degree of irrelevant and redundant information. The dimensionality curse is an issue when working with large dimensional datasets. There are several ways to lower the dataset's dimensionality or choose only the best selection of features that typically work the best for a specific issue statement.

We get a solution that outperforms both the algorithms by hybridizing the "particle swarm optimization" algorithm with a lower-performing approach, highlighting the efficacy of our hybridization strategy. We may conclude from this great conclusion that by uniting the PSO with the GWO, and we were able to enhance the PSO's purportedly weak exploration capabilities. These algorithms have two phases which are exploration and exploitation. Exploitation is the capacity of a variation to locate entire regions of a function area, whereas exploration is the convergence to the most fantastic result of the function near a good result. To evaluate the performance of our hybrid particle swarm optimization/grey wolf optimization approach to the PS optimization, GW optimization, cuckoo search algorithm (CSA), genetic algorithm (GA) and flower pollination algorithm (FPA) methods.

The research capacity of the GWO is utilized to mitigate the dangers by guiding some particles to partially improved places rather than arbitrary positions. As GWO is utilized in conjunction with PSO algorithm, its running time is increased. We combine the research ability of PSO with analysis ability of GWO to create the strength of modifications. The population measure was limited to 50 in these experiments, and every procedure was executed for 100 times in every analysis. The parameters used in our hybrid algorithm during implementation are k-value in KNN which is 5, maximum number of iteration $T = 100$, population size $N = 50$, acceleration factor a_min and a_max, velocities $V\max$ and Cognitive Factors $c1$ and $c2$. The entire samples of a dataset are randomly categorized into categories for testing and training (shown in Table 1). We use KNN instead of SVM to evaluate the execution of the suggested algorithm because classifier used will influence the feature selection. Table 2 shows the result analysis of microarray experiments using various methods for selecting shell features.

Examining the results, our HPSO-GWO approach stands out as the only method to achieve the highest precision otherwise achievable for all microarray datasets. HPSO-GWO can identify global minima and reduce overfitting. Our method outperforms the GWO, FSA, CS, PSO and GA methods in all cases. FPA and GA outperform HPSO-GWO. However, when looking at the colon cancer dataset, our method outperformed all wrapper methods, suggesting that our approach improved exploratory power (Figs. 2 and 3).

Table 1 Data format of genes expression classification

Dataset	Sample No.	Categories No.	No. of genes	No. of genes selected
Colon	63	7	2001	1002
Breast Cancer	31	6	693	430
Lung Cancer	72	14	3573	1790

Table 2 Classification accuracies of microarray data

Dataset	PSO	GWO	GA	CS	FPA	HPSO-GWO
Colon	68.303	70.345	65.33	78.947	68.421	73.684
Breast Cancer	73.015	88.88	89.47	92.982	95.906	92.397
Lung Cancer	86.66	86.3636	95.4545	95.454	89.1212	90.9090

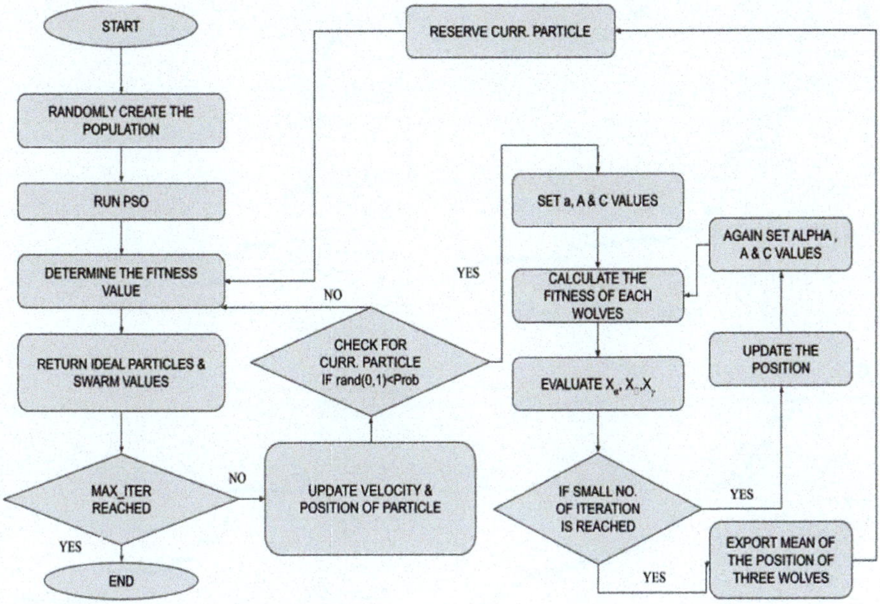

Fig. 2 Flowchart for hybrid PSO-GWO

5 Conclusion

In this work, we provide a unique hybrid algorithm based on PSO and GWO. This Strategy is selected for gene diversity and classification of high-dimensional DNA microarray data that were presented and contrasted. The GWO algorithm's search function is employed to prevent the PSO from obtaining local minima. For this reason, few grains of the PSO algorithm were substituted with a tiny probability by partially modified grains gained by the GWO algorithm. We used PS optimization, grey wolf optimizer (GWO), cuckoo search algorithm (CSA), genetic algorithm (GA) and flower pollination algorithm (FPA) strategies for comparing algorithm performance. Additionally, selected genes are validated using the exact K-neighbours classifier method to classify the data. Hybrid PSO-GWO has higher time complexity than both PSO and GWO because we did not change the underlying equations of the method and used low-probability GWO in the primary flow of PSO. As a result, we can infer that the hybrid technique which was utilized has significantly boosted the exploration capabilities of PSO.

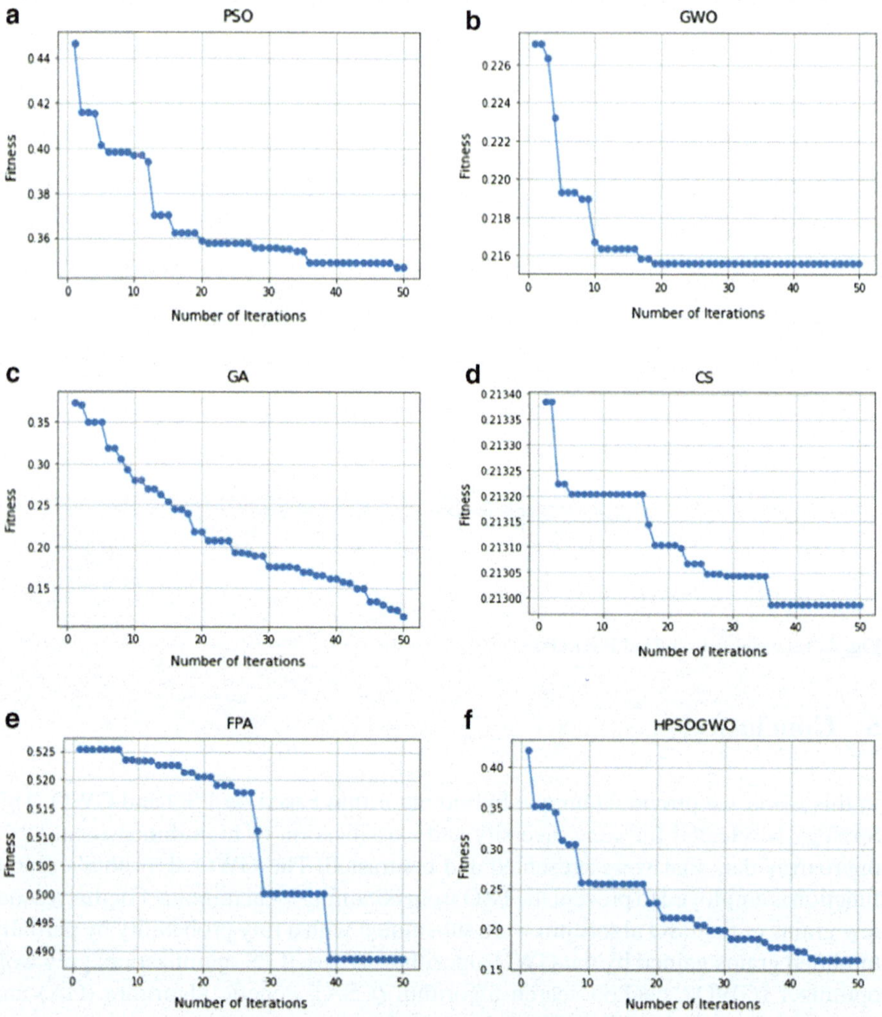

Fig. 3 Graphical representation of feature selection methods: **a** particle swarm optimization (PSO)—fitness versus maximum number of iteration **b**. flower pollination algorithm (FPA)—fitness versus maximum number of iteration **c**. genetic algorithm (GA)—fitness vs maximum number of iteration **d**. cuckoo search (CS)—fitness versus maximum number of iteration **e**. grey wolf optimization (GWO)—fitness versus maximum number of iteration **f**. hybrid PS optimization with GW optimization (HPSO-GWO)—fitness versus maximum number of iteration

Funding This work is funded under the Data Science Research of Interdisciplinary Cyber-Physical Systems (ICPS) Programme of the Department of Science and Technology (DST) [Sanction Number T-54], New Delhi, Government of India, India.

References

1. Yu L, Liu H (2003) Feature selection for high-dimensional data: a fast correlation-based filter solution. In: Proceedings, twentieth international conference on machine learning, vol 2, pp 856–863
2. Yang H, Churchill G (2007) Estimating p-values in small microarray experiments. Bioinformatics 23(1):38–43. https://doi.org/10.1093/bioinformatics/btl548wang
3. Chen JJ, Wang S-J, Tsai C-A, Lin C-J (2006) Selection of differentially expressed genes in microarray data analysis. Pharmacogenomics J 7(3):212–220
4. Govindarajan R, Duraiyan J, Kaliyappan K, Palanisamy M. Microarray and its applications. J Pharm Bioallied Sci. 4(Suppl 2):S310-2
5. Kuncheva LI, Matthews CE, Arnaiz-Gonz´alez A, Rodr´ıguez JJ (2020) Feature selection from high-dimensional data with very low sample size: a cautionary tale. School of Computer Science and Electronic Engineering, Bangor University, Aug 2020
6. Shardlow M (2019) An analysis of feature selection techniques
7. Kumar V, Chhabra JK, KumarD (2016) Grey wolf algorithm-based clustering technique. J Intell Syst
8. Gandomi AH, Yang X-S (2014) Chaotic bat algorithm. J Comput Sci 5(2):224–232
9. Deng X, Li Y, Weng J, Zhang J (2019) Feature selection for text classification: a review. Multimedia Tools Appl 78(3):3797–3816
10. Ahmed A, Esmin A, Matwin S (2013) HPSOM: a hybrid particle swarm optimization algorithm with a genetic mutation. Int J Innovative Comput Inf Control 9(5):1919–1934
11. Saeys Y, Inza I, Larranaga P (2007) A review of feature selection techniques in bioinformatics. Bioinformatics
12. Li J, Cheng K, Wang S, Morstatter F, Trevino RP, Tang J, Liu H (2017) Feature selection. ACM Comput Surv 50(6):1–45
13. Chandrashekar G, Sahin F (2014) A survey on feature selection methods. Comput Electr Eng 40(1):16–28
14. Shen Q, Shi WM, Kong W (2008) Hybrid particle swarm optimization and tabu search approach for selecting genes for tumor classification using gene expression data. Comput Biol Chem 32:53–60
15. Guyon I, Elisseeff A (2003) An introduction to variable and feature selection. J Mach Learn Res 3:1157–1182
16. Wang F, Zhang H, Li K, Lin Z, Yang J, Shen X-L (2018) A hybrid particle swarm optimization algorithm using adaptive learning strategy. Inf Sci 436–437:162–177
17. Kennedy J, Eberhart RC (1995) Particle swarm optimization. In: Proceedings of IEEE international conference on neural networks, vol IV, pp 1942–1948
18. Chuang LY, Chang HW, Tu CJ et al (2008) Improved binary PSO for feature selection using gene expression data. Comput Biol Chem 32:29–38
19. Faris H, Aljarah I, Al-Betar MA, Mirjalili S (2018) Grey wolf optimizer: a review of recent variants and applications. Neural Comput Appl 30(2):413–435
20. Zhang X, Lin Q, Mao W, Liu S, Dou Z, Liu G (2021) Hybrid particle swarm and grey wolf optimizer and its application to clustering optimization. Appl Soft Comput 101:107061
21. Holden N, Freitas AA (2008) A hybrid PSO/ACO algorithm for discovering classification rules in data mining. J Artif Evol Appl 11. Article ID 316145
22. Mohamad MS, Omatu S, Deris S, Yoshioka M (2009) Particle swarm optimization for gene selection in classifying cancer classes
23. Alba E, Garcia-Nieto J, Jourdan L, Talbi E-G (2007) Gene selection in cancer classification using PSO/SVM and GA/SVM hybrid algorithms. In: 2007 IEEE congress on evolutionary computation
24. Li S, Wu X, Tan M (2008) Gene selection using hybrid particle swarm optimization and genetic algorithm. Soft Comput 12:1039–1048
25. Singh N, Hachimi H (2018) A new hybrid whale optimizer algorithm with mean strategy of grey wolf optimizer for global optimization. Math Comput Appl 23(MDPI AG):14

26. Mirjalili S, Mirjalili SM, Lewis A (2014) Grey wolf optimizer. Adv Eng Softw 69:46–61
27. Faris H, Aljarah I, Al-Betar MA, Mirjalili S (2018) Grey wolf optimizer: a review of recent variants and applications. Neural Comput Appl 30:413–435
28. Lee PY, Loh WP, Chin JF (2017) Feature selection in multimedia: the state-of-the-art review. Image Vis Comput 67:29–42
29. Mirjalili S, Hashim SZM (2010) A new hybrid PSOGSA algorithm for function optimization. In: Proceedings of the international conference on computer and information application (ICCIA '10), pp 374–377, Tianjin, China, Nov 2010
30. Xu Y, Fan P, Yuan L (2013) A simple and efficient artificial bee colony algorithm. Math Prob Eng 1–9. Sucharita S, Sahu B, Swarnkar T (2021) A comprehensive study on the application of grey wolf optimization for microarray data
31. Lai X, Zhang M (2009) An efficient ensemble of GA and PSO for real function optimization. In; Proceedings of the 2nd IEEE international conference on computer science and information technology (ICCSIT '09), pp 651–655, Beijing, China, Aug 2009
32. Rashedi E, Nezamabadi-pour H, Saryazdi S (2009) GSA: a gravitational search algorithm. Inf Sci 179(13):2232–2248
33. Zhao W, Wang L, Mirjalili S (2022) Artificial hummingbird algorithm: a new bio-inspired optimizer with its engineering applications. Comput Methods Appl Mech Eng 388:114194. ISSN 0045-7825
34. Djemame S (2021) Cellular automata for edge detection based on twenty-five cells neighborhood. In: 2021 International conference on information systems and advanced technologies (ICISAT), 2021, pp 1–7
35. Hou K, Guo M, Li X, Zhang H (2021) Research on optimization of GWO-BP model for cloud server load prediction. IEEE Access 9:162581–162589. https://doi.org/10.1109/ACCESS.2021.3132052

Chapter 17
An Analytical Approach for Twitter Sarcasm Detection Using LSTM and RNN

Surbhi Sharma and Mani Butwall

1 Introduction

Language is most important part of communication which include words and sentences which is generally filled with nuances and comments that convey the negative or contradicting meaning of original sentence. To express our emotions in the communication person can use our tone and facial expression to convey additional meaning effectively. Contradicting sentiment can be presented as Sarcasm.

The Cambridge [1] word reference characterizes mockery as "comments that mean something contrary to what they say, made to condemn a person or thing in a manner that is interesting to others yet irritating to the individual reprimanded". At the point when utilized in text, the proposed mocking importance is handily lost, because of the absence of a portion of the recently referenced lines, in any case present in discourse.

In this paper author are keen on recognizing sarcasm explicitly in short content. Yet, what's the requirement of differentiate the sarcastic text to normal text? In the area of Natural language processing several resources are required knowledge of sarcasm in the author defined context [2]. As indicated by Poria et al. [3], sarcasm is core for estimation examination or can totally change the context of the fundamental notion.

The wide range of the work already done by authors till the current level of sarcasm has based on lexical resources that are explicit to the language [2]. However, the author has used deep learning in this study, a strategy that has recently demonstrated

S. Sharma
Department of IT, SKIT M&G, Jaipur, India
e-mail: surbhisharma@skit.ac.in

M. Butwall (✉)
Department of CSE, SKIT M&G, Jaipur, India
e-mail: mani.butwall@skit.ac.in

V. Devedzic et al. (eds.), *Proceedings of the International Conference on Intelligent Computing, Communication and Information Security*, Algorithms for Intelligent Systems, https://doi.org/10.1007/978-981-99-1373-2_17

remarkable promise. Generated by Poria et al. [3]. As Already describe that deep learning is a huge area of ML that makes use of deep neural networks. The ability of neural networks to learn summarized presentations of information means that they not only remember information but also learn the basic pattern or notion being discussed. Learn summarized depictions on many levels of contemplation thanks to a deep structure with several layers [4]. For this research, the author has examined three comprehensive theories of brain structure for the undertaking of mockery discovery: Two RNN with various actuation cells and one CNN.

2 Related Work

As indicated by Poria et al., sarcasm detection in idea appraisal is a relatively young topic that has gained widespread acceptance [3]. Over time, a number of works have been created employing various methodologies. Joshi et al. [5] presented a table that summarized previous studies' datasets, methodologies, clarifications, features, and settings. The works of those author've discovered are regarded as top-tier.

Sarcasm detection on Czech and English Twitter, Ptáek et al. [2] addressed the issue by employing greatest entropy and SVM as classifiers. They looked into two languages, English and Czech, and gathered information from Twitter. The English datasets, which include one that has been modified and one that has been skewed, were created by combining all tweets with the hashtag #sarcasm to show inconsiderate tweets. Each English dataset has 100,000 tweets, with the first dataset being imbalanced (50,000 sarcastic tweets vs. 50,000 non-sarcastic tweets) and the second dataset being modified (50,000 sarcastic tweets vs. 50,000 normal tweets) (25,000 sarcastic tweets and 75,000 ordinary tweets). They received the accuracy in terms of F1-score of 0.946 on the sensible dataset and an $F1$-score of 0.924 on the unbalanced dataset.

Poria et al. [3] employed a different approach in their study A Deeper Investigation of Mocking Tweets Using Deep Convolutional Neural Organizations. They exploited neural associations, namely unmistakably important CNNs, for the project, as the title suggests. They took crucial characteristics out of the CNN and used them as dedication to an SVM in the final representation. They employed three distinct datasets containing mean and non-taunting tweets for their investigation. Ptáek et al. [2] provided two of the datasets, while the sarcasm locator provided the third. Using a typical convolution neural network and three pre-trained features: feeling, character, and appraisal, they were able to get an accuracy with $F1$-score of 0.977. Their regular technique, which did not include the pre-trained features, received an $F1$-score of 0.95 on the practical test.

Ghosh and Veale [6] employed a neural association model comprised of a CNN, followed by an LSTM, and finally a deep neural network to get an $F1$-score of 0.92 in their work deep earth drilling sarcasm using neural organization. This neural model was distinguished from a recursive SVM. They also scored the displays of the associations independently, CNN + CNN ($F1$-score: 0.872) and LSTM + LSTM

($F1$-score: 0.879), where LSTM + LSTM indicates that the associations are stacked on top of one other. In addition to testing their models on a dataset they created with 39,000 tweets, they surveyed their system using two open datasets from [7, 8].

In addition to us, two of the three works mentioned above used the neural association approach. Ptácek et al. [2], on the other hand, employed most notable entropy and SVM as classifiers.

The difference between our attempted models and the ones utilized by Ghosh and Veale [6] is that ours are less unpredictable. Ghosh and Veale [6] utilized a model with several associations stacked on top of each other, however author tested CNN and RNN separately with only one mystery layer. Ghosh and Veale [6], on the other hand, evaluated each association's introduction entirely on the basis of what was communicated previously. Another qualification is that [6] do not use LSTM in their RNN, which is unusual for them.

3 Methodology

In the proposed work the datasets used, which contains the checked tweets. Before moving the work on applied model, author have to processed the dataset of approx. 150,000 tweets belongs to each category sarcastic and non-sarcastic, and for data pre-processing demanding cleaning system has used. Author also remembered an opportunity for how author should manage marks, like different hashtag, customer notifications, and web URL's. Its only done to remove unwanted information that would not require by the classifier, while keeping the syntactic significance. To start with the training step, author need to design a specific neural network with their properties, and these properties which are controlled by network called hyperparameters. These hyperparameters imposed the network self-learning ability that means author will searching an appropriate installation of hyperparameter to improve.

Author anticipated to design the associations and their properties before moving on to the next step in the planning process. Hyperparameters are the terms used to describe the various properties author may have controlled. These have an impact on an organization's ability to recognize, implying that developing a credible set of hyperparameters could be a major step forward in terms of increasing execution. To complete the association, author chose Google's open source programming library Tensorflow.

3.1 Description of Used Models

Three distinct models have been used to carry out and test the proposed work. 2 different RNNs, 1 with cells of LSTM and the rest one cells of GRU, both with hidden secret units of number 256, as well as a CNN model with 128 channels and channel widths of 3, 4, and 5. For a complete model representation, see Table 1.

Table 1 Brief description of the defined models

Model	Number of parameters
Representing model LSTM	Number of hidden memory units-256
Representing model GRU	Number of hidden memory units-256
Representing model CNN	Size of the filter = 128 + width of the filter = 3, 4, 5

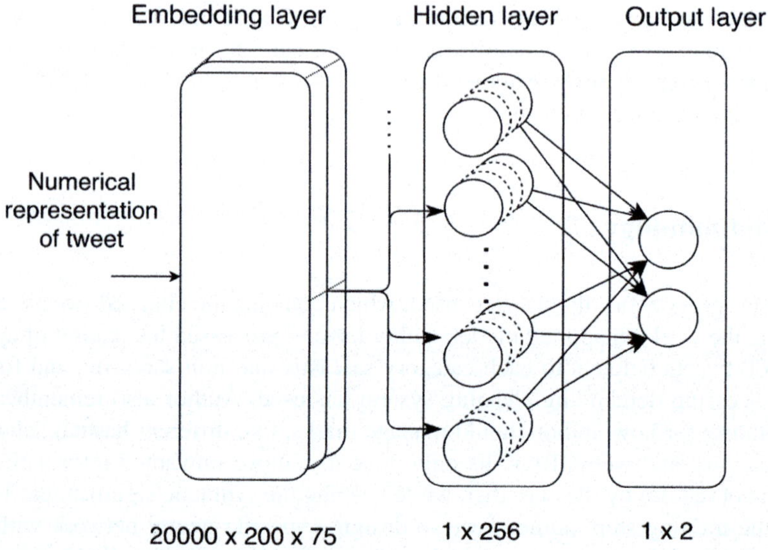

Fig. 1 Implementation diagram of the LSTM/GRU network

Each has an information layer that is linked to an installation layer that is linked to the organization's primary units. Proposed framework with LSTM, GRU, or CNN layer, depending on the type of organization, which is thus associated with an entirely associated yield layer with a softmax actuation work (Fig. 1).

3.2 Hyperparameter Optimization

A piece of enhancing the organization execution here and there lies in tracking down the ideal design of the hyperparameters. A few organizations are delicate to these boundaries and discovering them is essential. For our situation authors were keen on directing examinations with a few unmistakable qualities going from most minimal to most elevated determined worth, see Table 2.

Table 2 Hyperparameters ranging from the minimum to the maximum value were employed to conduct the hyperparameter search

Parameter	Minimum value	Maximum value
Dropout	0.5	1.0
Rate of learning	0.0006	0.0018
Decay by weight factor	0.02	0.08

Table 3 For conducting training experiment in this work, chosen hyperparameters are:

Rate of Dropout	0.73
Learning rate	0.002
Regularization parameter	0.048

Before of creating the organization, scalar estimations of the hyperparameters were picked by an arbitrary uniform dispersion from Table 3 characteristics, and the preparation cycle was iteratively rehashed with carelessly chosen values. Proposed approach will not generate the most optimal hyperparameters, but it should generate some reasonable ones or, better yet, demonstrate that proposed model will not very optimize to change the hyperparameters.

Every cycle was logged alongside the outcomes for further analysis to see if the choice of hyperparameters had a substantial impact on organization exhibits.

3.3 Training and Testing Models

Since a strategy for pre-processing the datasets have been chosen, and tracked down the decision generated by the hyperparameters, all proposed model was prepared and assessed on all described component of the datasets, then we can create a model for the dataset and it is assessed on test cases of the information in the other datasets. It will generate extra outcomes for efficient assessment whether the prepared model generate well.

3.4 Network Functionality Analysis

Author conducted a handful of analyses to assess the organizations' usefulness. Despite the fact that the results may appear to be little at first glance, this is not a minor query. The explanation is that the company may develop distinct patterns for snarky tweets, such as a high recurrence of generally witty terms, tweet length, and excessive hashtag usage. The consequences of the organization directly match with the specific words of the recurrence model to generate characterization because one can get respectable execution merely by looking at word content.

The experiments that followed were designed to produce outcomes that would help us respond to these questions. Every study was limited to just working along the organization and d generated dataset had the highest F1-score on the test segment for the underlying tests. In the examinations of the proposed model, it's considered as a best case. Furthermore, for ADAM augmentation, defined network were created and analysed over the dataset repeatedly to quantify the evaluating parameters like mean, $F1$-score, and standard deviation for suddenly introduced loads, hyperparameters, and groups. A little parameter of standard deviation would shows that the exploratory strategies were significant for the score variations along with the comparison to the base case.

4 Result and Conclusion

In this section, author shows the outcomes of the various organizations' preparation and testing. The search for perfect hyperparameters revealed that the models were typically uninterested in a variety of scenarios. With arbitrarily formed sets of hyperparameters, initial loads, and bunch sizes for prepping, $F1$-scores just changed imperceptibly. As a result, the published scores are all generated by networks with similar hyperparameters that were browsed in a run that generate approximate results depicted in Table 3.

In this paper, author used the dataset that produced the best results, in this case the for RNN with LSTM-cells model and the reasonable dataset from [2].

4.1 Training and Testing the Models

Result generated by the construction and testing of the three separate models: Recurrence neural network with LSTM, recurrence neural network with GRU, and convolution neural network are presented in the accompanying area. Each dataset was divided into three sections: preparation, approval, and testing, with 70, 15, and 15% of the instances used in each.

Note that the table does not include the outcomes obtained by cross-dataset evaluation in the range of DA and DB. It would happen due to datasets contain a large number of tweets. As a result, the organization may focus on a few information points in one set's train parcel that occur in the test segment of the other, resulting in one-sided findings.

Table 4 Data set DA tags having frequency of occurrence in the interval [− 1, 1] and − 1 or 1 represent sarcastic and non-sarcastic tags respectively

	Tags of tweets	Frequency rate
DA	<hashtag>	0.03789
DA	<user>	− 0.1254
DA	<url>	− 0.6733

4.2 Analysis of Network Functionality

The findings of the experiments are presented in the next section, after that next section show the discussion of how these findings are support or fragile the effectiveness of our model. The RNN network with LSTM-cells and dataset DA were defined in the following tests. After 30 sample runs, this base case gave a mean $F1$-score of 0.85 on the test set with a standard deviation of 0.0029.

4.2.1 Bag of Words Model

The demonstration of the sack of words model is introduced in this paper. The tables are ordered in the same order as in segment 4.1, implying that all datasets are used. Because of the neuronal groupings, the best estimate of 0.809 was obtained on DA when labels were included. The presentation shifted when the labels were removed, with a mean decrease of 0.002, which is insignificant for our model. The recurrence esteems in Table 8, as well as the occurrence of individual labels, help to explain this minor variation. The Bag of Words model's layout not affected by the presence of low recurrence generated by the most frequently occurring tags. Presented model has a higher negative score, however the tag is applied infrequently in the dataset (Table 4).

4.2.2 Word Scrambling

In this section author are introduce testing on a mixed tag dataset. The preparation and testing were done completely within the specified dataset only; no inter change dataset execution was used in this section. The dataset concept and shading coding are identical to those in segment 4.a. Mixed information and unique information denotes that the testing was done separately on the mixed and unique forms of the test tags.

4.2.3 Impact of Word Embeddings

The purpose of this test was to see how the addition of the word embeddings affected our results. Both the preparation and testing were carried out on segments from the decent dataset (DA), one in which the embeddings were set freely and the other in which author used word vectors from a GLoVe model. As shown in Table 5, the effect was much smaller than that observed by Poria et al. [3], who found a significant decline in $F1$-score when using arbitrarily generated embeddings (Tables 6 and 7).

These figures suggest that the GLoVe model, unlike the Word2Vec model, may have been mismatched for our task. The results of Poria et al. [3] show that Word2Vec word embeddings were a better fit because the increase in execution was more noticeable than in our model using GLoVe (Table 8).

Table 5 Training of first dataset with recurrence neural network with LSTM and assessed on both the first and word mixed test proportions

	Jumble data	Original data
DA	0.827	0.842
DB	0.683	0.723
DC	0.77	0.796

Table 6 Training of first dataset with recurrence neural network with GRU cells and assessed on both the first and word mixed test proportions

	Jumble data	Original data
DA	0.821	0.836
DB	0.639	0.697

Table 7 Poria et al. examined results for randomly initialized word embeddings versus pre-trained GLoVe or Word2Vec models [3]

Model	$F1$-score
Define model With random embedding	0.8220
Define model With GLoVe embedding	0.8420
Poria et al. [3] with random embeddings	0.8623
Poria et al. [3] with Word2Vec embeddings	0.9771

Table 8 On the same dataset, the average $F1$-score for LSTM network classifications

Evaluating parameter	Network
$F1$-score	0.875
Standard deviation	0.021

The performance of neural network model's standard deviation is evaluated on 30 separate runs

While it's reasonable to expect humans to outperform machines in detecting sarcasm, author can't say why the results were the opposite. However, it's worth noting that a neural organization approaches groups of comparison data from its preparation phase, whereas people were not offered any information prior to the investigation. Characteristics such as sarcasm repetition or the content of specific phrases may be highlighted as a result, and the organization will be more familiar with them.

References

1. Cambridge. Meaning of "sarcasm" in the english dictionary, 2017. http://dictionary.cambridge.org/dictionary/english/sarcasm
2. Ptáček T, Habernal I, Hong J (2014) Sarcasm detection on czech and english twitter. http://pure.qub.ac.uk/portal/files/17977967/Coling2014.pdf
3. Poria S, Cambria E, Hazarika D, Vij P (2016) A deeper look into sarcastic tweets using deep convolutional neural networks. ArXiv e-prints, Oct 2016. https://arxiv.org/pdf/1610.08815.pdf
4. Deng L, Yu D (2014) Deep learning methods and applications. In: Le Q, Mikolov T (ed) Distributed representations of sentences and documents. International conference on machine learning—ICML 2014, vol 32, pp 1188–1196. ISSN 10495258. https://doi.org/10.1145/2740908.2742760. http://arxiv.org/abs/1405.4053
5. Joshi A, Bhattacharyya P, Carman MJ (2016b) Automatic sarcasm detection: a survey. CoRR, abs/1602.03426. http://arxiv.org/abs/1602.03426
6. Ghosh A, Veale T (2016) Fracking sarcasm using neural network. In: Proceedings of NAACL-HLT. http://anthology.aclweb.org/W/W16/W16-0425.pdf
7. Riloff E, Qadir A, Surve P, De Silva L, Gilbert N, Huang R (2013) Sarcasm as contrast between a positive sentiment and negative situation. https://www.cs.utah.edu/{~}riloff/pdfs/official-emnlp13-sarcasm.pdf
8. Davidov D, Tsur O, Rappoport A (2010) Enhanced sentiment learning using twitter hashtags and smileys. In: Proceedings of the 23rd international conference on computational linguistics: posters, no August, pp 241–249. https://doi.org/10.1.1.185.3112. http://dl.acm.org/citation.cfm?id=1944566.1944594
9. Goodfellow I, Bengio Y, Courville A (2016) Deep learning. MIT Press. http://www.deeplearningbook.org
10. Graves A (2008) Supervised sequence labelling with recurrent neural networks. Image Rochester NY, p 124. ISSN 01406736. https://doi.org/10.1007/978-3-642-24797-2. https://arxiv.org/pdf/1308.0850.pdf
11. Chung J, Gülçehre C, Cho KH, Bengio Y (2014) Empirical evaluation of gated recurrent neural networks on sequence modeling. CoRR, abs/1412.3555. http://arxiv.org/abs/1412.3555
12. Joshi A, Sharma V, Bhattacharyya P (2015) Harnessing context incongruity for sarcasm detection. In ACL 2:757–762
13. History World. History of language, 2017. http://www.historyworld.net/wrldhis/PlainTextHistories.asp?historyid=ab13
14. Chowdhury GG (2005) Natural language processing. Ann Rev Inf Sci Technol 37(1):51–89. ISSN 00664200. https://doi.org/10.1002/aris.1440370103. http://doi.wiley.com
15. US Secret Service (2014) Computer based annual social media analytics subscription. https://www.fbo.gov/?s=opportunity&mode=form&id=8aaf9a50dd4558899b0df22abc31dc0e&tab=core&_cview=0
16. Joshi A, Bhattacharyya P, Carman MJ, Saraswati J, Shukla R (2016a) How do cultural differences impact the quality of sarcasm annotation?: A case study of Indian annotators and American text. In: Proceedings of the 10th SIGHUM workshop on language technology for cultural heritage, social sciences, and humanities, W16-2111

17. Will K (2017) The dark secret at the heart of AI. https://www.technologyreview.com/s/604087/the-dark-secret-at-the-heart-of-ai/
18. González-Ibáñez R, Muresan S, Wacholder N (2011) Identifying sarcasm in Twitter: a closer look. In: Proceedings of the 49th annual meeting of the association for computational linguistics: human language technologies: short papers-volume 2, no 2010, pp 581–586. https://doi.org/10.1.1.207.5253. http://www.aclweb.org/anthology/P/P11/P11-2102.pdf
19. Kriesel D (2007) A brief introduction to neural networks. ISSN 14320711. http://linkinghub.elsevier.com/retrieve/pii/0893608094900515
20. Dahl GE, Sainath TN, Hinton GE (2013) Improving deep neural networks for LVCSR using rectified linear units and dropout. In: 2013 IEEE international conference on acoustics, speech and signal processing. IEEE, pp 8609–8613. ISBN 978-1-4799-0356-6. https://doi.org/10.1109/ICASSP.2013. 6639346. http://ieeexplore.ieee.org/document/6639346/
21. LeCun Y, Bottou L, Bengio Y, Haffner P (1998) Gradient-based learning applied to document recognition. Proc IEEE 86(11):2278–2323. ISSN 00189219. https://doi.org/10.1109/5.726791
22. Karpathy A (2015) The unreasonable effectiveness of recurrent neural networks. http://karpathy.github.io/2015/05/21/rnn-effectiveness/
23. Dadheech et al (2020) Implementation of internet of things-based sentiment analysis for farming system. J Comput Theor Nanosci 17(12):5339–5345 (7). https://doi.org/10.1166/jctn.2020.9426
24. Chung J, Gülçehre C, Cho KH, Bengio Y (2015) Gated feedback recurrent neural networks. CoRR, abs/1502.02367. http://arxiv.org/abs/1502.02367
25. Ruder S (2016) An overview of gradient descent optimization algorithms. Web page, pp 1–12. https://arxiv.org/pdf/1609.04747.pdfhttp://arxiv.org/abs/1609.04747
26. White D, Ligomenides P (1993) GANNet: a genetic algorithm for optimizing topology and weights in neural network design. In: New trends in neural computation. Springer, Berlin, Heidelberg, pp 322–327. ISBN 3540567984. https://doi.org/10.1007/3-540-56798-4_167. http://link.springer.com
27. Dauphin Y, Pascanu R, Gulcehre C, Cho K, Ganguli S, Bengio Y (2014) Identifying and attacking the saddle point problem in highdimensional non-convex optimization. arXiv, pp 1–14. ISSN 10495258. http://arxiv.org/abs/1406.2572
28. Kingma D, Ba J (2014) Adam: a method for stochastic optimization. In: International conference on learning representations, pp 1–13, 2014. ISSN 09252312. https://doi.org/10.1145/1830483.1830503 http://arxiv.org/abs/1412.6980
29. Kumar T et al (2022) A review of speech sentiment analysis using machine learning. In: Kaiser MS, Bandyopadhyay A, Ray K, Singh R, Nagar V (eds) Proceedings of trends in electronics and health informatics. Lecture Notes in Networks and Systems, vol 376. Springer, Singapore. https://doi.org/10.1007/978-981-16-8826-3_3
30. Fawcett T (2006) An introduction to ROC analysis. Pattern Recogn Lett 27(8):861–874. ISSN 01678655. https://doi.org/10.1016/j.patrec.2005.10.010
31. Angela P, Tarbox J, Ranick J, Clair MS (2013) Teaching children with autism to detect and respond to sarcasm. Res Autism Spectr Dis 7(1):193–198. ISSN 17509467. https://doi.org/10.1016/j.rasd.2012.08.005. http://www.sciencedirect.com/science/article/pii/S1750946712000980
32. Ranick J, Persicke A, Tarbox J, Kornack JA (2013) Teaching children with autism to detect and respond to deceptive statements. Res Autism Spectr Dis 7(4):503–508. ISSN 17509467. https://doi.org/10.1016/j.rasd.2012.12.001. http://www.sciencedirect.com/science/article/pii/S1750946712001481

Chapter 18
Fuzzy Logic-Based Outlier Detection Technique for Supporting Stock Market Trading Decision

A. M. Rajeswari, Parul Bhatia, and A. Selva Anushiya

1 Introduction

The stock market is a place where the initial amount invested by prospective traders and investors has ample scope to grow and thus grabs attention from humanity. Meanwhile, stock trading is a risky business line where traders may lose a substantial amount of money. 'Volatility' of the stock market is pretty alarming. Imbalance of stock trading in one direction causes volatility. Beyond risk, the volatile stock markets afford the opportunity of getting profit for the traders. Even a good trader cannot take in profit out of zero volatile stock markets. Thus, to profit from the volatile stock market, it is necessary to have a correct business decision system for traders, especially for beginners and intermediaries.

Stock prices are uncertain and fluctuate from time to time. It moves randomly in their own defined patterns and predicting stock prices may be a grim effort [1]. Therefore, to analyze the trend of stocks over time, most researchers approached in accordance with the regression technique, a statistical approach. Anticipation of volatility with a GARCH model [2] has demonstrated that risk management can be improved and portfolio returns can be maximized. Engle and Sheppard [3] had estimated volatility using an innovative dynamic conditional correlation multivariate GARCH model. A factor loading matrix in generalized orthogonal GARCH model

A. M. Rajeswari (✉)
Velammal College of Engineering and Technology, Madurai, India
e-mail: amrtce@gmail.com

P. Bhatia
Shri Vishwakarma Skill University, Palwal, Haryana, India
e-mail: parul.bhatia84@gmail.com

A. Selva Anushiya
AAA College of Engineering and Technology, Sivakasi, India
e-mail: anushiya@aaacet.ac.in

V. Devedzic et al. (eds.), *Proceedings of the International Conference on Intelligent Computing, Communication and Information Security*, Algorithms for Intelligent Systems, https://doi.org/10.1007/978-981-99-1373-2_18

has been used in estimating volatility for European sector reforms [4]. Time-varying correlation with the new MGARCH model [5] has been explored as compared to the traditional Monte Carlo simulations. TGARCH model [6] has been the most preferred model for volatility among GARCH, EGARCH and TGARCH family. BEKK-GARCH [7] model had been used to study asymmetric relationship between return and volatility linkages. The varied risk patterns due to volatile movements in stock prices of the Brazilian sectoral indices had been captured by GARCH-VAR [8] estimation model. Chakrabarty et al. [9] have observed MRA–EDCC GARCH and VAR–EDCC GARCH models to detect spillover effects for sectoral indices of BSE.

Financial data are usually noisy, and time-varying risk traits may persist, leading to a simple regression/GARCH model to be less robust. This may require alterations in the regression models or a superior technique for predicting stock market volatility. The business development and marketing manager of the Australian trade commission strongly recommended in his article [10] that 'making use of fuzzy logic system (FLS) to detect the uncertain trading decisions in the stock market data (SMD) is meaningful and worthy.' To overcome time-related risks in volatile stocks, researchers began applying FLS along with GARCH models and its variants. Maciel et al. [11] applied the evolving fuzzy technique on the GARCH models for stock market volatility forecasting and found that the model's performance improved significantly. A novel fuzzy linear regression-based GARCH model [12] has been proposed to increase the accuracy of time series forecasting. A fuzzy-based decision support system [13] and probabilistic-based fuzzy logic approach [14] have been suggested for the traders to decide when to sell, hold or buy the shares.

Volatility of stock prices causes the target returns to be more deviated and may be deformed in any one of the directions. Because of these variations, volatility is considered as an outlier. Outliers are abnormal and rare happenings/observations which are also semantically correct, meaningful and thus can be supportive in the decision-making process. The shortcomings of the above discussed traditional regression and its variation models have brought forth the need to explore new and groundbreaking methods for forecasting volatility/outliers. In recent days, supervised/unsupervised machine learning (ML) algorithms like support vector machine [15] and clustering [16] have proven and examined by several researchers for predicting the stock market volatility.

The aforesaid literature suggests that the ML algorithms along with FLS may be safer to employ when dispensing with stock market—the time series data. Since stock prices are numerical data, it has to be partitioned into ranges [17] before performing prediction by the ML algorithms. This will aid to cut down the computation time complexity. Besides, while partitioning the stock prices, the misinterpretation of boundary values due to crisp boundaries should also be managed in a better way for accurate prediction. Hence, we propose a FLS-based Semi-Supervised Outlier Detection (SSOD) technique to remedy all the aforementioned issues which have to be factored into account for predicting accurate outliers in SMD. The proposed SSOD adopts associative classification technique to categorize the outliers/volatility in SMD. In addition, SSOD makes use of a sliding window [18] technique of size

2 by considering the previous day and the follow-up day stocks to handle the time-bounded stock price values. Besides, SSOD makes use of the surprising measure [19] 'Lift' to detect the correct outliers for effective trading decisions. Since Lift measures 'how many times more often any two patterns occur together than expected if they were statistically independent' [20], we found it more relevant surprising measure to detect the volatility/outliers. As a result, forecasted outliers made it possible for traders to make accurate trading decisions.

The performance of SSOD is benchmarked against the Conventional Associative Classification (CAC) technique. Indeed, the two techniques adopt the same procedure for detecting outliers, except for the type of learning. CAC makes use of supervised learning while SSOD use semi-supervised learning. Hence, to realize the effectiveness of FLS along the associative classification technique in detecting outliers/volatility from SMD, SSOD is tested against CAC. Based on experimental results, SSOD has been found to outperform CAC by detecting a large set of precise and meaningful outliers.

The rest of this paper is structured as follows. Section 2 describes the literature work of the proposed approach and its aim. Section 3 discusses the framework, specifications, and benefits of the SSOD technique. In Sect. 4, the experimental evaluation of SSOD against CAC has been discussed most. Finally, conclusion and future work are presented in Sect. 5.

2 Related Work

Occasionally volatility in the stock trading may persist for a period of time. To address such situations, there was an inadequacy in variations of regression-based models beyond the adoption of FLS for its performance improvement. As such, the ML research community has started visualizing the sustained volatility as a classification problem. Shi and Zhuang [21] were explored the ML algorithm artificial neural networks for the Hong Kong stock markets and realized that it performed better in comparison with regression models. Alameer et al. [22] explored the adaptive neuro fuzzy inference system and genetic algorithm in a combined approach for forecasting the volatility of copper prices in financial markets. Madan et al. [23] have exploited the ML algorithms like binomial GLM, SVM, and random forest for anticipating the price of the highly volatile Bitcoin market. Yang et al. [16] proved that using the standard deviation to compute a variable margin along with support vector machine regression method yields a good predictive result in the Hang Seng Index forecast. Also, FLS was utilized to identify the unexpected booms and collapses in the stock market by taking on a subtractive clustering method [17]. Therefore, it may be useful to explore fuzzy techniques to forecast stock market volatility.

Outlier prediction of time series data using frequent pattern-based ML algorithms was successful in several fields including stock market. Association rule mining [24] is the first and foremost frequent pattern-based technique introduced to mine the buying behavior of the customers by the association rules (ARs) and to improve the business thereof, these ARs can theoretically span across many intervals. To avoid spending unnecessary resources in generating irrelevant ARs, Bruno et al. [18] used a time-related parameter called sliding window of size '2' to detect the stock splits occurred in the SMD. The ARs can be slightly tuned to have only the class values as the consequence and are called as Class ARs (CARs). This technique of evolving CARs is called as associative classification and can be used for classification purpose. A recent survey proved that associative classification yields better results than conventional classification technique [25], because the patterns identified by CARs are more closely associated than the patterns identified by the conventional classification models. Hence, the proposed SSOD makes use of associative classification technique to detect the outliers from SMD.

SMD is populated with numerical values and have to be categorized into ranges [26] before the mining process to avoid unnecessary computational time. While using crisp boundary for partitioning, the real values are either allowed to acquaint in the set or not and are indicated as either yes/no or 0/1. This method will have a problem in handling the boundary value semantics. Zadeh [27] states that 'A fuzzy set allows its components to have a degree of membership in the set, and thus, an element can present in more than one set with its own degree of presence.' Since fuzzy rule-based systems (FRBS) deal with IF–THEN rules, they become the reference to conventional rule-based systems. The antecedents and consequents part of FRBSs comprises a compilation of fuzzy logic statements as a replacement for real or categorical values and are similar to ARs. Consequently, the work [26] used FLS to generate ARs for the applications under consideration.

Aggarwal [28] states that 'outliers may be Noise or Anomaly which will be classed as weak and strong outliers, respectively, based on its outlier score.' Furthermore, Aggarwal [28] cited that 'data points with a high outlier score are perfect outliers.' Picking up the trustworthy and appropriate abnormal ARs (to spot the outliers) from the infrequent ARs is a challenging issue for the researchers. This is because some ARs might not be interested, and some may still mislead. The ARs, when applied to detect outliers, bring an additional interesting measure to identify the outlier score of it. The interesting measures like dependency degree [18] and residual leverage [29] were utilized as the pruning measure to extract the interesting temporal association rules (TARs) as outlier for decision making in the fields of finance and medicine, respectively. McNicholas et al. [30] have standardized the Lift, one of the pruning measures for ARs. These surveys gave us the confidence to use Lift as the outlier pruning measure.

Altogether, almost all the above-mentioned ML algorithms-based works are transmitted away with static predefined threshold values for the interesting measures used to detect outliers, which need the domain experts' advice. But, Selvi et al. [31] proposed a non-parametric association-based algorithm to find outliers by enhancing FP-Growth algorithm with an automatic score of minimum support and minimum confidence. The proposed SSOD adopts a similar manner of establishing threshold values for the interesting measures to prune the outliers.

3 Methodology of the Proposed SSOD Technique

The aim of this work is to recognize the precise outlier in a volatile stock market for future trading decision. Based on the literature review, we understood that existing supervised and unsupervised ML algorithms have certain limits in detecting sustainable volatility in the stock market. Our previous study [32] has demonstrated this shortfall in various fields such as medicine, environmental science, education, and finance. Understanding the inadequacies of existing ML algorithms in volatility prediction, we thought about offering a hybrid version of supervised (classification) and unsupervised (association) technique. Hence, the fuzzy-based SSOD technique is proposed.

Outliers for volatility prediction that can be used in the trading decision by stock traders have been brought forth in accordance with the steps outlined in Fig. 1. SSOD adopts the fundamental steps of Apriori algorithm [24] which can be used to predict rare items from SBI stock market data [33]. The special features introduced in SSOD to overcome the limitations of the conventional (supervised) classification technique are also presented in Fig. 1. Besides, the benefits of the proposed solutions are detailed in Table 1.

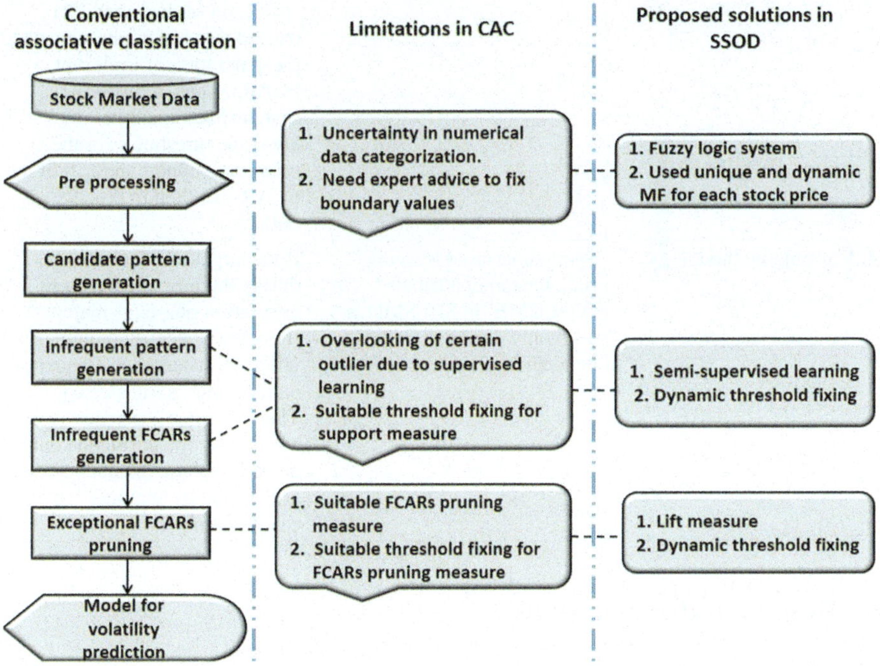

Fig. 1 Framework for volatility prediction using associative classification

Table 1 Special features of SSOD and its benefits

Parameters	Proposed special features	Benefits of the suggested special features of SSOD
1. Categorizing the numerical stock prices	FLS with unique and dynamic MFs for each stock price (OP, CP, HP, LP)	Since the volatility of the target returns depends on multiple, correlated, varying factors of stock prices, the dynamic, unique MF for each price found to be suitable in handling the boundary values of the respective domains in a more precise manner
2. Suitable measure for outliers/volatility pruning	Surprising measure—lift	The relevant FCARs extracted by the Lift are also high confidence by nature Thus, the FCARs extracted by Lift are able to detect the correct and precise outliers which can assist in precise trading decisions
3. Threshold value of the interesting measures	Dynamically computed	Distinct support measure thresholds for infrequent patterns are set dynamically at each level of the infrequent pattern generation phase to avoid the generation of irrelevant FCARs. Thus, reduces the search space and time Dynamic thresholds for the measures support and Lift of the FCARs resulted in precise outliers
4. Learning methodology	Semi-supervised learning. (i.e., the target attribute 'VOLUME' in SBI SMD is used only during the rule generation phase)	If the target attribute is used during the infrequent pattern generation phase, its major (frequent) category 'volume1' will be considered as frequent patterns and got neglected Thus, 'volume1' category will not have scope during the outlier detection phase (actual outliers are present in 'volume1' category only)

4 Results and Discussions

To assess SSOD, we performed a number of experiments on SBI SMD. The performance of SSOD is compared to that of CAC, the standard associative classification technique. Both CAC and SSOD are assessed based on the heap space used by the algorithm, the time required to generate the outliers, the number of outliers generated, and the nature of the instances covered by the outliers. Likewise, the outliers detected by both CAC and SSOD are analyzed and evaluated in terms of precise trading decisions. For the experiments, the data set is randomly partitioned into training data with 70% instances and the remaining 30% instances as testing data with a follow-up of tenfold cross validation. The experimental results are depicted in Table 2. Similarly, the scalability of both the methods is tested by gradually increasing the database size as 25, 50, 75, 100%, and the results are reported in Figs. 2, 3, 4, and 5.

Table 2 CAC versus SSOD techniques in outlier detection

Parameters	Conventional method—CAC	Proposed method—SSOD
1. Algorithm used	Apriori Rare-based associative classification	Apriori Rare-based associative classification
2. Learning methodology	Supervised	Semi-supervised
3. Target attribute (volume) partitioning	Crisp method	Crisp method
4. Stock price partitioning	Crisp boundary	Fuzzy boundary
5. Surprising measure used	Lift	Lift
6. Threshold values of the interesting measures	Min-confidence = 0.95 *Dynamically computed* Max-support = 0.311 $\text{Lift}_{\text{threshold}} = 1.041$	Min-confidence = 0.95 *Dynamically computed* Max-support = 0.254 $\text{Lift}_{\text{threshold}} = 1.196$
7. No. of infrequent rules generated	26 (CARs)	133(FCARs)
8. No. outlier generated	12	20
9. Average coverage (%)	12.86	19.83
10. Memory space used (bytes)	67,71,230	73,30,616
11. Time used to generate outliers (seconds)	10	13

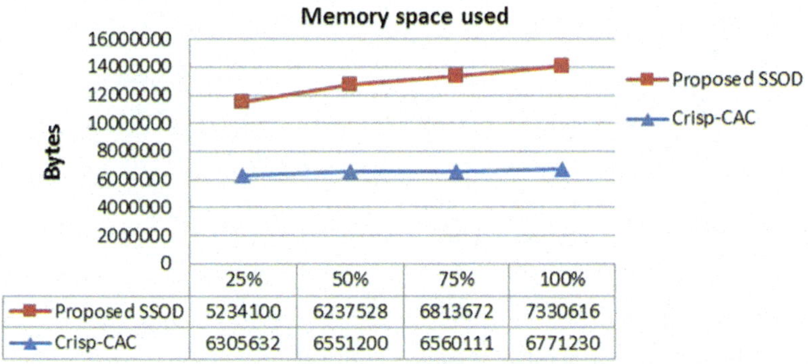

Fig. 2 Scalability test based on heap space used to generate outliers

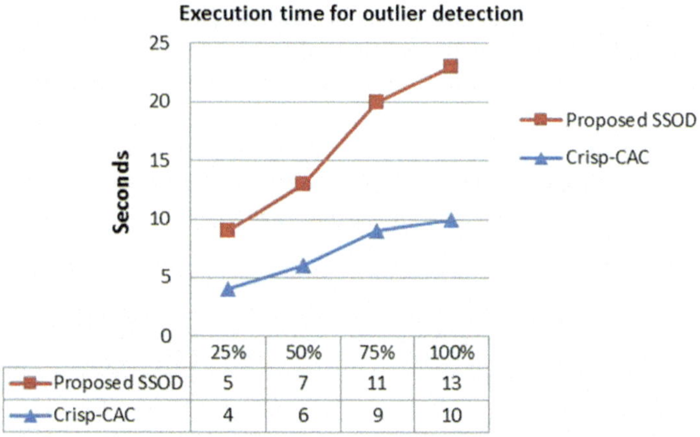

Fig. 3 Scalability test based on time used to generate outliers

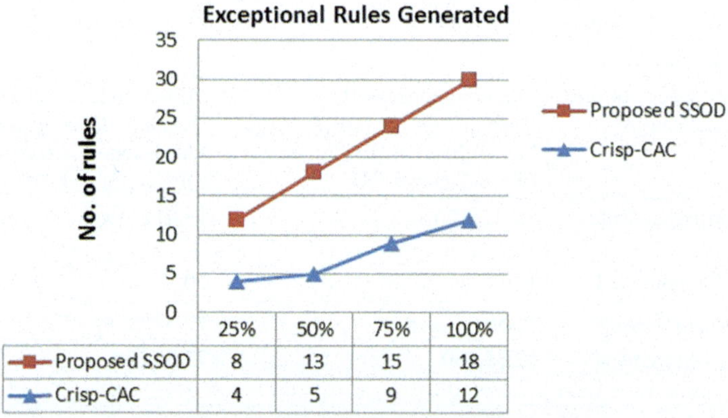

Fig. 4 Scalability test based on the number of outliers generated

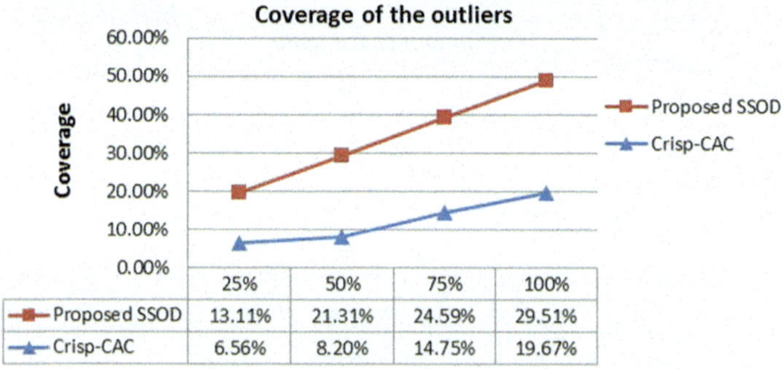

Fig. 5 Scalability test based on nature of outliers predicted

According to the results of Table 2 and Figs. 2 and 3, it is understood that SSOD consumes more time and more memory space to detect outliers. This is because FLS will produce massive patterns in infrequent patterns generation phase. Table 2 depicts that even though the number of infrequent rules generated by SSOD is huge, the number of outliers pruned is proportionately low when compared to CAC. This is because of the natural fuzzy boundaries, dynamic and appropriate threshold values for the interesting measures like support and Lift. According to Table 2, it can be observed that, for CAC, the threshold value of the support measure is high and that of the Lift is low. The reverse is true for SSOD, where the threshold value for the support measure is low (rare occurrence), and the threshold value for the Lift is high (perfect outlier). That is why SSOD generates a wide set of precise outliers.

5 Conclusion and Future Enhancement

An overview of the need to identify outliers and use of fuzzy logic to process volatile stock market data is covered. The use of existing GARCH models and its variants in forecasting volatility and its limitations are surveyed. Accordingly, a semi-supervised hybrid SSOD technique is proposed to detect outliers/volatility in the SBI stock market. The infrequent rules—FCARs are evolved from the SBI stock market using the objective measures like support and confidence. Then the exceptional/abnormal FCARs which can predict the outliers/volatility from stock market are pruned using a surprising measure 'Lift.' The threshold values of these measures are determined dynamically (without expert's advice) based on the distribution of domain values within the database. In addition, the boundary values for the interval fixing of stock prices during preprocessing are handled by FLS in a more naturalistic manner.

As the proposed SSOD technique is based on associative classification, we have measured its performance against Conventional Associative Classification (CAC) based on execution time, heap space used by the algorithm to evolve outliers and the number of outliers detected. The coverage values of the detected outliers are assessed to understand its type, correctness, and completeness. From the observational results, we found that the proposed SSOD uses up more time and heap space to detect the outliers than CAC. This is because SSOD make use of fuzzy logic system, which results in huge set of infrequent patterns. But, SSOD is able to get a perfect lot of outliers which are able to correctly recognize the volatility of SBI stock market. The proposed SSOD outperforms the CAC technique with 19% more accuracy. Also, SSOD has drastically cut the irrelevant patterns during the outlier generation phase. This success is possible because of the right choice of the surprising measure—Lift and dynamic computation of the threshold values of the interesting measures used to detect the outliers.

The proposed SSOD can be applied on a broad set of different stocks with volatility. The surprising measure—Lift—is a common pruning measure and not specifically planned for market trading. As such, Lift measure can also be tested in a wide variety of other areas. Likewise, the proposed SSOD make use of the FLS system with the triangular MF to spot the outliers. From the observational solutions, we found that few of the 'individual outliers' spotted by the SSOD are eliminated as noise. This can be determined by replacing the MF and alternate outlier pruning measure. Also, the proposed SSOD generated an enormous of infrequent patterns, which resulted in more heap space and time consuming. This can also be overcome by an alternate method of generating the infrequent patterns.

References

1. Kendall MG, Hill AB (1953) The analysis of economic time-series-part i: prices. J R Stat Soc Ser A (General), 116(1):11–34
2. Malik F, Hassan SA (2004) Modeling volatility in sector index returns with GARCH models using an iterated algorithm. J Econ Finance 28(2):211–225
3. Engle RF, Sheppard K (2001) Theoretical and empirical properties of dynamic conditional correlation multivariate GARCH. National Bureau of Economic Research. http://www.nber.org/papers/w8554
4. Boswijk HP, Van der Weide R (2011) Method of moments estimation of GO-GARCH models. J Econometrics 163(1):118–126
5. Tse YK, Tsui Albert KC (2001) A multivariate GARCH model with time-varying correlations. Science Direct Working Paper No S1574-0358(04)71166-1. https://ssrn.com/abstract=3162726
6. Ezzat H (2013) The application of GARCH methods in modeling volatility using sector indices from the Egyptian Exchange. J Money, Investment Bank. https://mpra.ub.uni-muenchen.de/51584/
7. Ng SL, Chin WC, Chong LL (2017) Multivariate market risk evaluation between Malaysian Islamic stock index and sectoral indices. Borsa Istanbul Rev 17(1):49–61
8. Bernardino W, Brito L, Ospina R, Melo S (2019) A GARCH-VaR investigation on the Brazilian sectoral stock indices. Braz Rev Finance 16(4):573–610
9. Chakrabarty A, De A, Bandyopadhyay G (2005) A Wavelet-based MRA-EDCC-GARCH methodology for the detection of news and volatility spillover across sectoral indices—evidence from the Indian Financial Market. Glob Bus Rev 16(1):35–49
10. Sergey T (2004) Fuzzy logic in market trader. http://www.timingsolution.com/TS/Articles/Fuzzy/
11. Maciel L, Gomide F, Ballini R (2016) Evolving fuzzy-GARCH approach for financial volatility modeling and forecasting. Comput Econ 48(3):379–398
12. Mohamad Hanapi AL, Othman M, Sokkalingam R, Ramli N, Husin A, Vasant P (2020) A novel fuzzy linear regression sliding window GARCH model for time-series forecasting. Appl Sci. https://doi.org/10.3390/app10061949
13. Ijegwa AD, Vincent OR, Folorunso O, Isaac OO (2014) A predictive stock market technical analysis using fuzzy logic. Comput Inf Sci 7(3):1–17
14. Govindasamy V, Thambidurai P (2013) Probabilistic fuzzy logic based stock price prediction. Int J Comput Appl 71(5):28–32
15. Yang H, Chan L, King I (2002) Support vector machine regression for volatile stock market prediction. In: International conference on intelligent data engineering and automated learning, Springer, Berlin, Heidelberg, pp 391–396
16. Boushehri AG (2000) Applying fuzzy logic to stock price prediction (Doctoral dissertation, Concordia University). https://spectrum.library.concordia.ca/1116/
17. Kumar T et al (2022) A review of speech sentiment analysis using machine learning. In: Kaiser MS, Bandyopadhyay A, Ray K, Singh R, Nagar V (eds) Proceedings of trends in electronics and health informatics. Lecture Notes in Networks and Systems, vol 376. Springer, Singapore. https://doi.org/10.1007/978-981-16-8826-3_3

18. Bruno G, Garza P (2010) TOD: Temporal outlier detection by using quasi-functional temporal dependencies. Data Knowl Eng 69(6):619–639
19. Geng L, Hamilton HJ (2006) Interestingness measures for data mining: a survey. ACM Comput Surv. https://doi.org/10.1145/1132960.1132963
20. Hahsler M (2015) A probabilistic comparison of commonly used interest measures for association rules. https://michael.hahsler.net/research/association_rules/measures.html. Accessed 12 Dec 2020
21. Shi C, Zhuang X (2019) A study concerning soft computing approaches for stock price forecasting. Axioms. https://doi.org/10.3390/axioms8040116
22. Alameer Z, Abd Elaziz M, Ewees AA, Ye H, Jianhua Z (2019) Forecasting copper prices using hybrid adaptive neuro-fuzzy inference system and genetic algorithms. Nat Resour Res 28(4):1385–1401
23. Madan I, Saluja S, Zhao A (2015) Automated bitcoin trading via machine learning algorithms. http://cs229.stanford.edu/proj2014/Isaac%20Madan
24. Agrawal R, Imieliński T, Swami A (1993) Mining association rules between sets of items in large databases. In: Proceedings of the 1993 ACM SIGMOD international conference on management of data, pp 207–216
25. Nguyen LT, Vo B, Hong TP, Thanh HC (2013) CAR-Miner: an efficient algorithm for mining class-association rules. Expert Syst Appl 40(6):2305–2311
26. Rajeswari AM, Deisy C (2019) Fuzzy logic based associative classifier for slow learners' prediction. J Intell Fuzzy Syst 36(3):2691–2704
27. Zadeh LA (1996) Fuzzy sets. In: Fuzzy sets, fuzzy logic, and fuzzy systems: selected papers by Lotfi A Zadeh. https://doi.org/10.1142/9789814261302_0021
28. Aggarwal CC (2017) An introduction to outlier analysis. In: Outlier analysis, Springer, Cham
29. Kshirsagar PR et al (2022) Covid heuristic analysis using machine learning. In: 4th RSRI International Conference on Recent Trends in Science and Engineering, REST Labs, Krishnagiri, Tamil Nadu, India, 27–28, February, 2021, AIP conference proceedings 2393, 020077, pp 020077-1–020077-5. https://doi.org/10.1063/5.0074120
30. McNicholas PD, Murphy TB, O'Regan M (2008) Standardising the lift of an association rule. Comput Stat Data Anal 52(10):4712–4721
31. Selvi CK, Tamilarasi A (2009) Mining association rules with dynamic and collective support thresholds. Int J Eng Technol 1(3):236–240
32. Rajeswari AM, Yalini SK, Janani R, Rajeswari N, Deisy C (2018) A comparative evaluation of supervised and unsupervised methods for detecting outliers. In: 2018 Second international conference on inventive communication and computational technologies (ICICCT). IEEE, pp 1068–1073
33. State Bank of India (2013–2017) Available online: http://economictimes.indiatimes.com/prices.cms?companyid=11984. Last accessed 2022/11/05

Chapter 19
Lung Cancer Detection (LCD) from Histopathological Images Using Fine-Tuned Deep Neural Network

Swati Mishra and Utcarsh Agarwal

1 Introduction

It has been observed from history that lung cancer is the utmost public type of disease found in different age groups of people. The causes of this can be smoking, due to exposure to air pollution, radon gas, certain other chemicals, etc. Lung cancer can be categorized into two kinds, small-cell lung cancer and non-small-cell lung cancer [1]. Approximately, 81–85% of lung cancers belong to NSCLC [1]. There are some main subtypes of lung cancer such as squamous cell carcinoma, adenocarcinoma, and large cell carcinoma of NSCLC. These all subtypes derived from different types of lung cells are grouped as NSCLC.

Artificial intelligence is now more popular in the field of health sectors, and it has value in all aspects of primary care. Computer vision plays a vital role in the early-stage detection of any illness using medical images. So that treatment can be given to the patient in time and the illness can be cured more simply. There have been eminent research on the usage of artificial intelligence in the medical domain such as brain tumor detection, cancer detection from X-Rays, CT scans, pathology images, etc. In this paper, we detect lung cancer from histopathology images with greater accuracy using the transfer learning technique which is applied by fine-tuning a pre-trained model. The flow of the paper is as follows. Previous work done in the field of lung cancer detection is described in preliminaries. After this, the dataset used in this paper is discussed. The proposed methodology, pre-processing, training, testing, and validation have been discussed. Finally, the results are discussed in the next section followed by the conclusion.

S. Mishra (✉) · U. Agarwal
JSS Academy of Technical Education, Noida, India
e-mail: swatimishra@jssaten.ac.in

V. Devedzic et al. (eds.), *Proceedings of the International Conference on Intelligent Computing, Communication and Information Security*, Algorithms for Intelligent Systems, https://doi.org/10.1007/978-981-99-1373-2_19

2 Preliminaries

Machine learning algorithms are used to detect lung cancer using different datasets available. Research has been done using different datasets by researchers to detect lung cancer. A brief description of the work done by researchers in their research based on lung cancer detection using different datasets with different techniques is described. For the detection of lung cancer, UCI Machine Learning Repository and Data World are used to produce different results for each classifier. The results were implemented using DT, LR, NB, and SVM classifiers. Out of all the classifiers, it has been shown that SVM performed well on the dataset [2]. In this paper, the author experimented with using deep neural networks and auto-encoders for lung cancer detection. They have used LIDC-IDRI [3] dataset for this purpose. They were able to achieve accuracy of 79%, 81%, and 79%, respectively [4]. A method of ensembling different combinations of deep residual networks and a classifier on the LIDC-IDRI dataset [3] was proposed, and they were able to achieve an accuracy of 84% using an ensemble of UNet + XGBoost and Resnet + XGBoost which independently have accuracies 74 and 76% [5]. The authors of this paper proposed CNN architectures for lung cancer detection using whole-slide histopathology images. They used VGG and ResNet. The outputs of VGG and ResNet were compared using the receiver operating characteristic (ROC) plot. VGG16 model achieved 75.41% patch-level accuracy and ResNet obtained 72.05% of patch-level accuracy [6].

In this paper, the authors implemented a CNN network on the LC25000 [7]. The authors used CNN model to classify lung cancer into three different classes. They were able to achieve 96.11% and 97.2% of training and validation accuracy with the developed CNN model [8]. The author of this research suggested a system that combines CNN and AlexNet models and achieved 96% accuracy for classifying lung tumors seen in CT scan images as benign or malignant [9]. Aside from that, several findings indicate that EfficientNet Models have benefited them. The authors of this paper used EfficientNet-B2 architecture to classify breast cancer histology images into four classes with an accuracy of 98.33% and 96.67%, using Reinhard and Macenko stain normalization methods, respectively [10]. The paper introduces a rapid recognition approach for diagnosing 20 skin disorders using EfficientNet, achieving an accuracy of 97.10% with EfficientNet-B7, which has a long-trained time, while EfficientNet-B0 achieves an accuracy of 93.35% with the shortest training time [11].

3 Description of Dataset

In this research work, we have used the LC25000 dataset [7] for implementation. This dataset consists of 25,000 color images and 5000 images in each different five classes. This dataset has histopathological images which belong to three types of lung cancer images and two types of colon cancer images. This dataset has histopathological

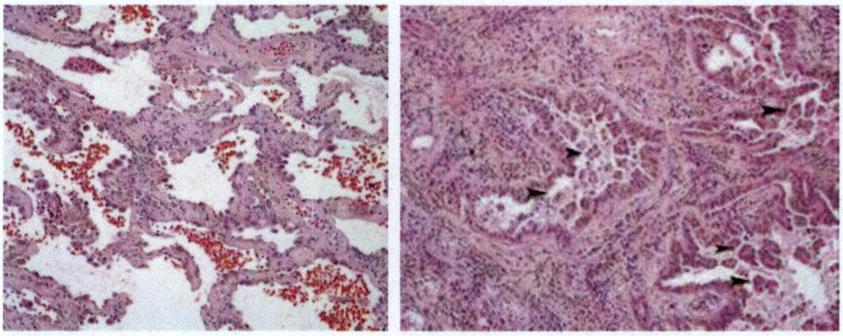

Fig. 1 Histopathological images

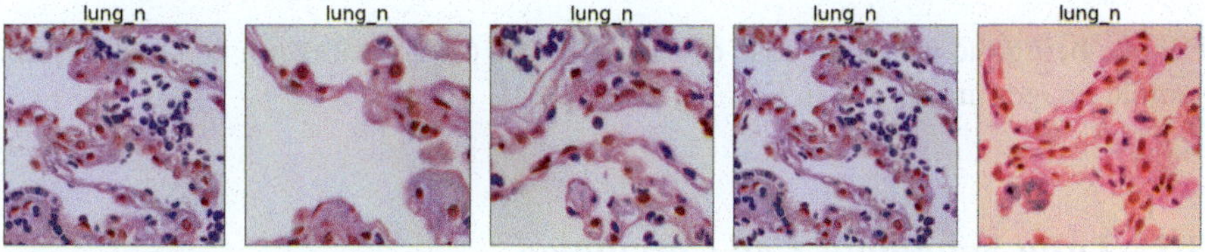

Fig. 2 Lung benign tissue sample images

images. All images are 768 × 768 pixels in size in JPEG file format. Histopathological images are taken from a microscope; therefore, these images are called microscopic images. Histopathology prefers the examination of tissue on microscopic images for the classification of the disease as shown in Fig. 1. Lung benign tissue, lung adenocarcinoma, and lung squamous cell carcinoma are some classes in this dataset having 5000 images in each.

3.1 Lung Benign Tissue

It is an abnormal growth of tissue found not to be cancerous. Early-stage detection of cancer is very important to treat the patient in time. This type of tissue grows slowly and automatically stops after some time. It does not expand to other parts of the body. Figure 2 shows some of the sample images of benign lung tissues.

3.2 Lung Adenocarcinoma

It is the most commonly found cancer. This kind of cancer mostly affects smokers or former smokers. More often than males, women are diagnosed with this kind of cancer. Younger societies are more prone to experience it. Adenocarcinoma often develops in the outer regions of the lung and is more likely to be discovered early on.

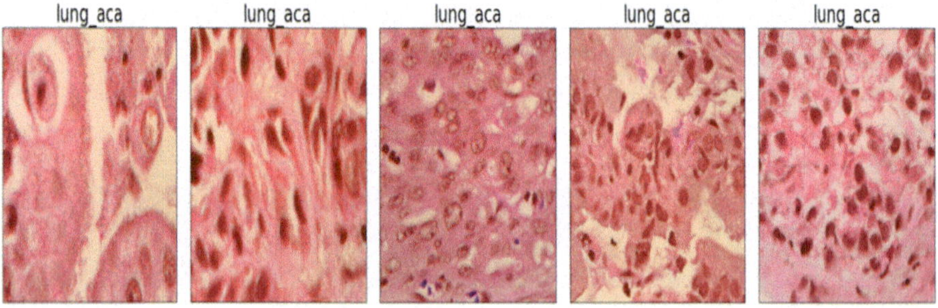

Fig. 3 Lung adenocarcinoma sample images

Despite being the primary cause of cancer death, the risk of death is modest. Some sample images are shown in Fig. 3.

3.3 Lung Squamous Cell Carcinoma

This type of cancer belongs to a malignant type of lung cancer. It starts rising in the cells lining the bronchi. It can spread to nearby lymph nodes and organs. It can also travel through the blood to other parts of the body. It has a strong connection with smoking history. There are some other risk factors which include family history and exposure to smoke. Some sample images are shown in Fig. 4.

4 Methodology

In a brief, we used EfficientNet-B0 [12] pre-trained on ImageNet [13] as our base model and fine-tuned it by replacing the last output layer with a stack of a few

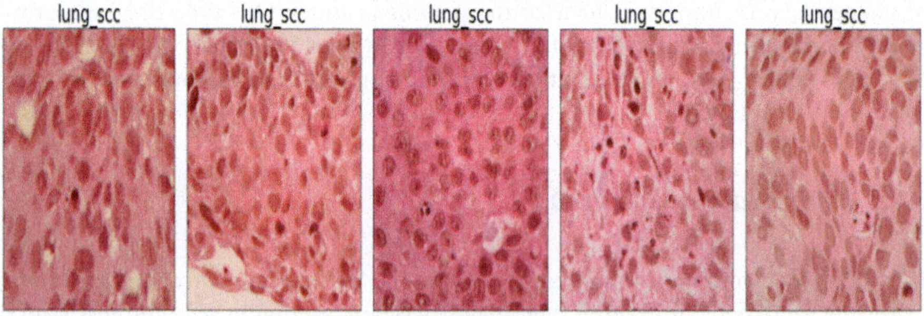

Fig. 4 Lung squamous cell carcinoma sample images

Fig. 5 Structure of the proposed method

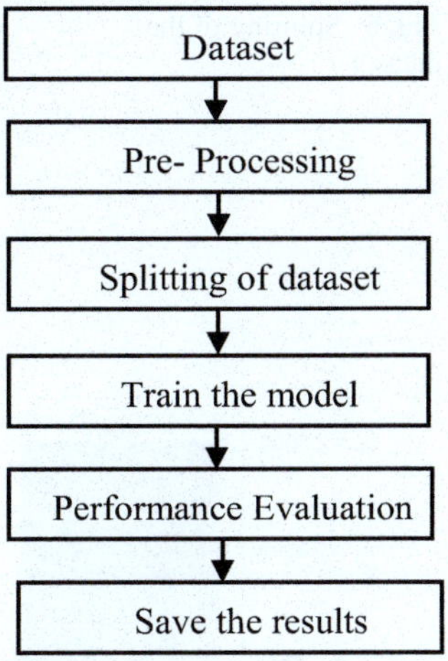

linear and dropout layers and trained the model by freezing all the layers except the additional layers with the 70% of total data and rest is used for evaluation purpose. The model was able to produce more efficiency and accuracy than previous ConvNets.

The step-by-step procedure of this work is shown below (Fig. 5):

Step 1: Input the images from the LC25000 dataset [8].
Step 2: Pre-processing is to be done on the images.
Step 3: Split the dataset into training, validation, and testing sets.
Step 4: Train EfficientNet B0 model using transfer learning through fine-tuning.
Step 5: Evaluate the model's performance.
Step 6: Save results.

4.1 Data Pre-processing

It is pivotal for the classification of histopathological images. The dataset contains quite a large image, while the convolution neural network is usually designed to accept much smaller inputs. Therefore, the images were resized from 768×768 to 224×224. Dataset is divided for training, validation, and testing in the ratio of 7:2:1. The images in all the sets were stratified, meaning that the classes were represented equally. Data splitting into training, validation, and testing are shown in Fig. 6. Dataset distribution for each class is shown in Fig. 7.

The following pre-processing techniques were applied to the data:

Image Resizing: Any machine learning model will be trained faster with the smaller images. The data can be in raw form. It may have images of different sizes and colors etc. Therefore, it is required to make all the images existing in the dataset of the same desired image size before giving it to the model. Resizing is a part of pre-processing of a dataset. The image size is 768×768 in the LC25000 dataset

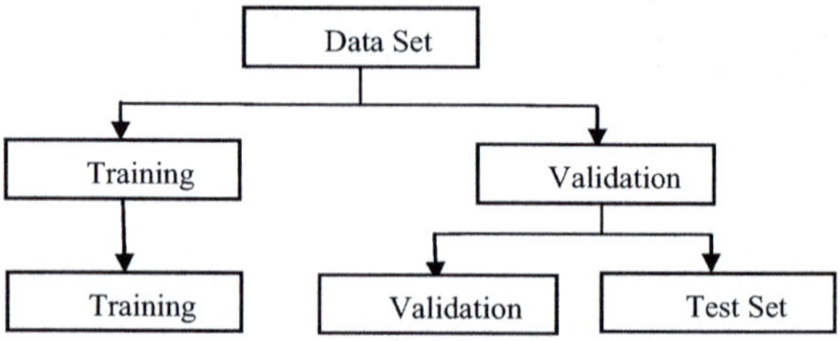

Fig. 6 Splitting of the dataset

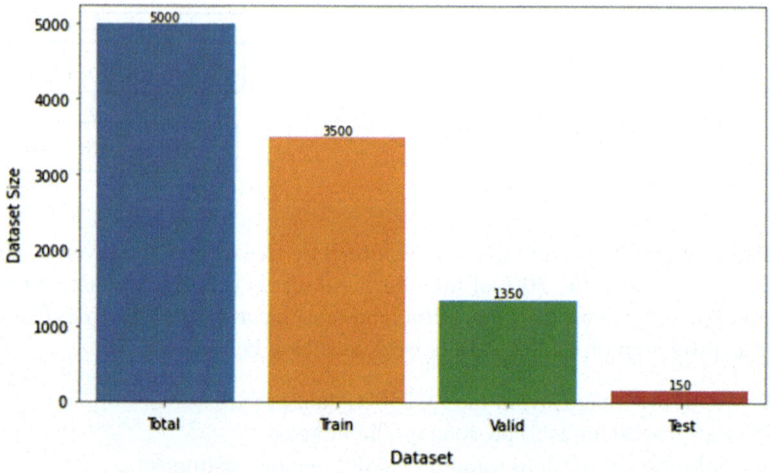

Fig. 7 Dataset distribution for each class

which is very large. For implementation purposes, these images are resized into 224 × 224 so that the model can be trained faster and computationally also it is less expensive.

Dataset Splitting: The model should not be judged on the same set of data that it was trained on, but rather on alternative data. The same dataset would be used to evaluate the model if the dataset was not separated into different sets. The dataset should be split into the train, validation, and test parts to prevent overfitting. We choose to use a tiny portion of the data (in this case 10%) as a test set because we do not have any concealed test set data. The remaining 90% of the data was divided so that the majority of it gets fed during training and we also have a sizable quantity of data to validate the model.

4.2 Fine Tune

The EfficientNet model has produced state-of-the-art results on well-known datasets like ImageNet and has demonstrated numerous outstanding outcomes in the medical

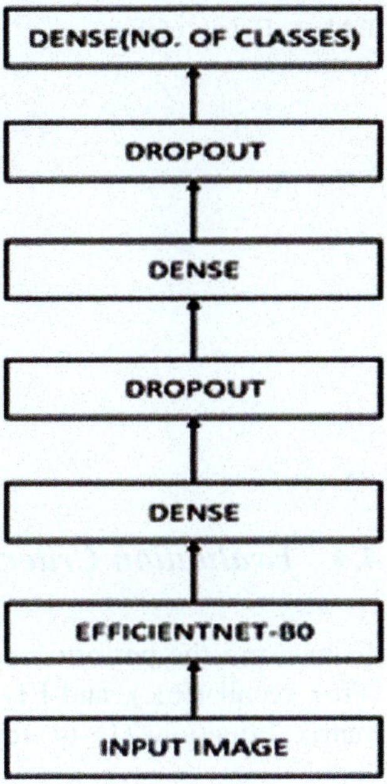

Fig. 8 Block diagram of model

images dataset. We used the EfficientNet-B0 model pre-trained on ImageNet as our base model, and fine-tuning is also done by replacing the last layer with a sequence of a few linear and dropout layers. A final output layer was attached after the addition of two dense layers with an activation function ReLu ($f(x) = \max(0, x)$) as represented in Fig. 8. For the implementation, the dropout rate is set to 0.2. Dropout layers avoid overfitting, and dense layers at the end assist the network to classify the retrieved features more accurately.

4.3 Model Training and Testing Strategy

The model was trained using the Adam optimizer and a categorical cross-entropy loss function was used to calculate the loss. It was trained with a batch size of 32 for 15 epochs. After every epoch, the model is validated on the validation set, and the model weights were saved if there was an improvement in the validation loss so that the model with the lowest validation loss was saved. The PyTorch package is used for implementation. A publicly available pre-trained EfficientNet-B0 model was utilized [14]. The model was trained and tested on Kaggle kernels which provide 12 GB RAM and a 16 GB Nvidia P100 graphics processing unit. Specifications used for implementation are shown in Table 1.

Table 1 Details of the specification

Model	EfficientNet-B0
Used software	Python
Image size	224 × 224
Batch size	32
Learning rate	$1e^{-3}$
Epochs	15
Additional layers	5
Dropout rate	0.2
Optimizer	ADAM
Loss function	Cross-entropy loss

4.4 Evaluation Criteria

To measure the performance of the developed model, accuracy (ACC), precision (Pre), recall (Rec), and F1-score metrics are computed along with the confusion matrix. Equations (1)–(4) represent the formulations of the metrics.

$$ACC = \frac{(TP + TN)}{(TP + FP + TN + FN)} \tag{1}$$

$$Pre = \frac{TP}{(TP + FP)} \tag{2}$$

$$Rec = \frac{TP}{(TP + FN)} \tag{3}$$

$$F1score = 2 * \frac{(Pre * Rec)}{(Pre + Rec)} \tag{4}$$

Confusion matrix (Fig. 9).
Where:
True positive (TP)—number of images classified as cancerous.

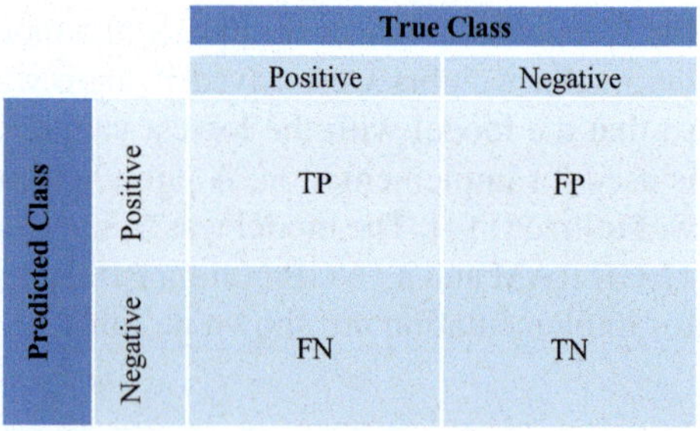

Fig. 9 Confusion matrix of true and predicted classes

Table 2 Model accuracy

Model	Training Acc (%)	Validation Acc (%)	Test Acc (%)
EfficientNet-B0	99.15	99.14	98.67

True negative (TN)—number of normal lung images classified as cancerous.

False positive (FP)—number of normal lung images incorrectly classified as cancerous.

False negative (FN)—number of cancerous lung images misclassified as cancerous.

5 Result and Discussion

The best model weights were saved at the 14th epoch as the model achieved the least validation loss at that epoch. The model attained 99.15%, 99.14%, and 98.67% accuracy on train, validation, and test sets, respectively (Table 2).

Figure 10 (a) shows the plot between training accuracy and validation accuracy to the number of epochs. Figure 10 (b) shows the plot of training loss and validation loss for the number of epochs.

Tables 3, 4, and 5 display the accuracy, precision, recall, and F1-score for different categories of histopathological images for the custom CNN model and our fine-tuned EfficientNet-B0 Model on different sets.

The confusion matrix shown in Fig. 11a, b depicts the actual label vs. the predicted label of the images for the validation and test data, respectively, for a particular labeled category.

6 Conclusion

In this paper, we attempt to detect lung cancer from histopathological images using the transfer learning technique which was employed in the form of fine-tuning. The EfficientNet-B0 model was used as a feature extractor which was then extended with a few additional layers to classify the image into three different classes of lung cancer. The model achieved an accuracy (Acc) of 99.15% on the training set, 99.14% on the validation, and 98.67% on the test. Precision (Pre), recall (Rec), and F1-score were calculated, and a confusion matrix plot was plotted to measure the performance of the model.

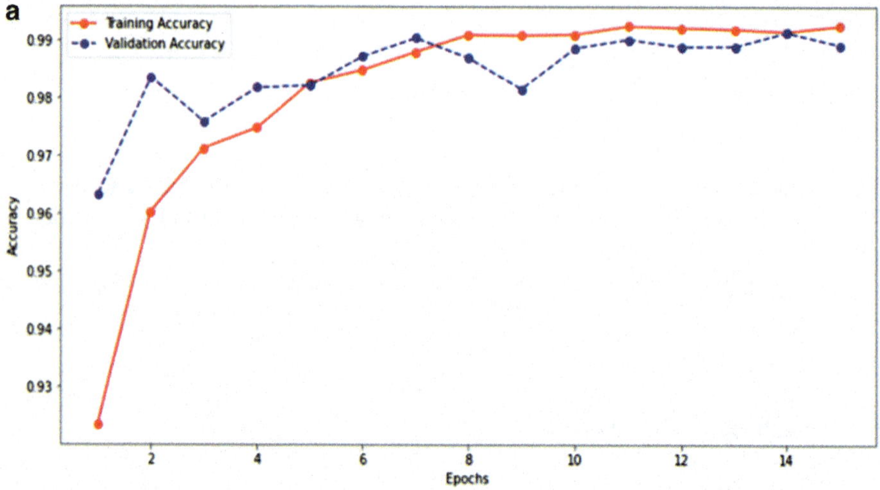

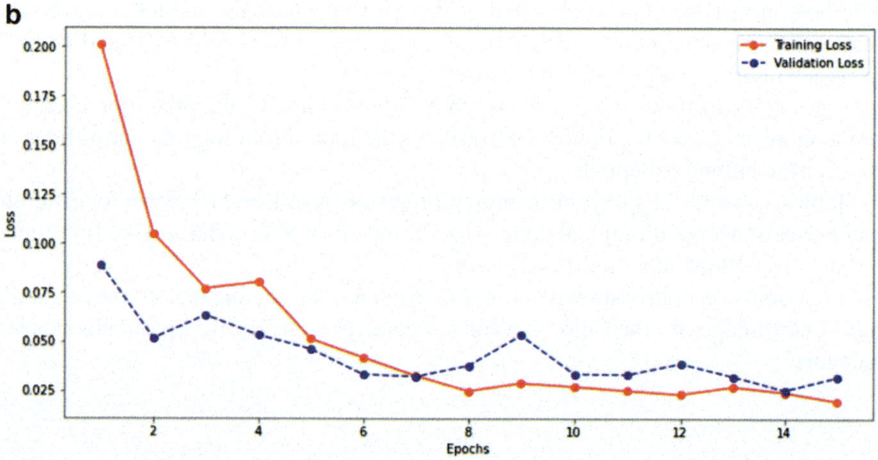

Fig. 10 **a** Training and validation accuracy versus epoch, **b** training and validation loss versus epochs

Table 3 Evaluation metrics result of custom CNN model [8] for different categories on the validation set

Category	Precision	Recall	F1-score
Adenocarcinoma	0.95	0.97	0.96
Benign tissue	1.00	1.00	1.00
Squamous cell carcinoma	0.97	0.95	0.96

Table 4 Evaluation metrics result of fine-tuned EfficientNet-B0 model for different categories on the validation set

Category	Precision	Recall	F1-score
Adenocarcinoma	0.99	0.98	0.99
Benign tissue	1.00	1.00	1.00
Squamous cell carcinoma	0.98	1.00	0.99

Table 5 Evaluation metrics result of fine-tuned EfficientNet-B0 model for different categories on the test set

Category	Precision	Recall	F1-score
Adenocarcinoma	0.99	0.97	0.98
Benign tissue	1.00	1.00	1.00
Squamous cell carcinoma	0.97	0.99	0.98

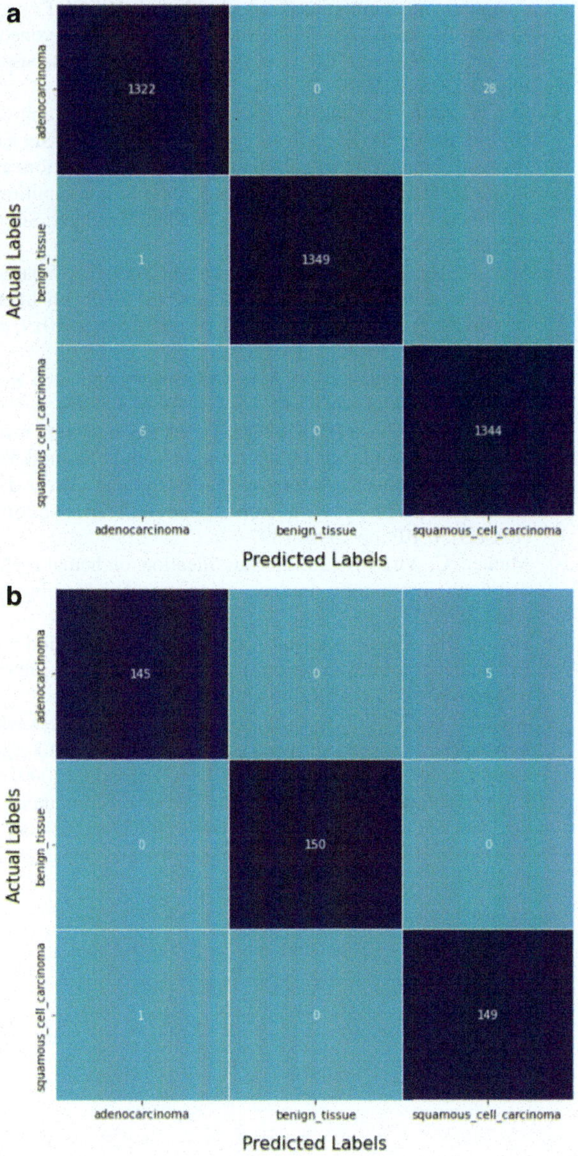

Fig. 11 **a, b** Confusion matrix for validation and test dataset

References

1. American Cancer Society, Lung Cancer. [Online]. Available: https://www.cancer.org/cancer/lung-cancer/about/what-is.html
2. Radhika PR, Rakhi AS, Nair, Veena G (2019) A comparative study of lung cancer detection using machine learning algorithms. In: 2019 IEEE International conference on electrical, computer and communication technologies (ICECCT)
3. Armato SG et al (2011) The lung image database consortium (LIDC) and image database resource initiative (IDRI): a completed reference database of lung nodules on CT scans. Med Phys 38(2):915–931
4. Sun W, Zheng B, Qian W (2016) Computer-aided lung cancer diagnosis with deep learning algorithms. In: SPIE medical imaging. International Society for Optics and Photonics
5. Bhatia S, Sinha Y, Goel L (2019) Lung cancer detection: a deep learning approach. In: Bansal J, Das K, Nagar A, Deep K, Ojha A (eds) Soft computing for problem solving. Advances in intelligent systems and computing, vol 817. Springer, Singapore. https://doi.org/10.1007/978-981-13-1595-4_55
6. Šarić M, Russo M, Stella M, Sikora M (2019) CNN-based method for lung cancer detection in whole slide histopathology images. In: 2019 4th International conference on smart and sustainable technologies (SpliTech), Split, Croatia, 2019, pp 1–4. https://doi.org/10.23919/SpliTech.2019.8783041
7. Mahrishi et al (eds) (2020) Machine learning and deep learning in real-time applications. IGI Global. https://doi.org/10.4018/978-1-7998-3095-5
8. Hatuwal BK, Thapa HC (2020) Lung cancer detection using convolutional neural network on histopathological images. Int J Comput Trends Technol 68(10):21–24
9. Agarwal A, Patni K, Rajeswari D (2021) Lung cancer detection and classification based on AlexNet CNN. In: 2021 6th International conference on communication and electronics systems (ICCES). IEEE, pp 1390–1397
10. Munien C, Viriri S (2021) Classification of hematoxylin and eosin-stained breast cancer histology microscopy images using transfer learning with EfficientNets. Comput Intell Neurosci 2021
11. Hridoy RH, Akter F, Rakshit A (2021) Computer vision-based skin disorder recognition using EfficientNet: a transfer learning approach. In: 2021 International conference on information technology (ICIT). IEEE, pp 482–487
12. Anand R et al (2022) Hybrid convolutional neural network (CNN) for Kennedy Space Center hyperspectral image. Aerosp Syst 5(3):1–8. https://doi.org/10.1007/s42401-022-00168-4
13. Deng J, Dong W, Socher R, Li L-J, Li K, Fei-Fei L (2009) ImageNet: a large-scale hierarchical image database. In: 2009 IEEE Conference on computer vision and pattern recognition, 2009, pp 248–255. https://doi.org/10.1109/CVPR.2009.5206848
14. Wightman R (2019) PyTorch Image Models. https://github.com/rwightman/pytorch-image-models

Chapter 20
Smart Office: Happy Employees in Enhanced and Energy-Efficient (EEE) Workplace

C. M. Naga Sudha, J. Jesu Vedha Nayahi, S. Saravanan, and Subish Daniel

1 Introduction

The Internet is becoming the most influential and powerful tool in our day-to-day life. It has revolutionized the way of living with its development in speed, bandwidth, and functionality. From sending and receiving emails to Skype calling a person in another country, the Internet is reaching new heights. The ongoing development of the Internet will cause fundamental changes in the world to come.

The Internet is engulfing all the advanced technologies, resulting in a world where every other thing is connected to the Internet. It is not just limited to computers or smartphones but has extended to many of the physical devices around us, thus redefining our lifestyle and transforming the way we interact with things via technology. IoT has made a breakthrough on the Internet, and the evolution and advancement of its applications are outpouring. IoT provides opportunities to interconnect any number of devices. Nowadays, the fitness bands used to measure calorie expenditure and heartbeats are almost owned by everyone, which is also an IoT device. The smart devices include lights, fans, washing machines, vacuum cleaners, and many more. IoT provides opportunities to interconnect any number of devices and control them through end devices like mobile phones and computers. IoT has developed existing connections between people and computers to include digitally connected things. This technology provides an easy and efficient way to perform day-to-day tasks. IoT consists of devices which are more intelligent, informative, understanding,

C. M. Naga Sudha (✉) · S. Daniel
Anna University-MIT Campus, Chennai, India

J. Jesu Vedha Nayahi
Anna University, Regional Campus, Tirunelveli, India

S. Saravanan
K. Ramakrishna College of Technology, Trichy, India

V. Devedzic et al. (eds.), *Proceedings of the International Conference on Intelligent Computing, Communication and Information Security*, Algorithms for Intelligent Systems, https://doi.org/10.1007/978-981-99-1373-2_20

and thus, day-to- day communication becomes much easier. IoT is also expected to make advances in the automobile industry like connected cars.

In today's world, IoT is deployed everywhere and has a wide spectrum of applications, including health, automobile, farming sector, and automation of real-life existing systems, thus shaping human life in a way better and more digitized. Its applications are not just concerned only with devices and things; it has grown to provide a smart way of living. Due to physical things connected and controlled digitally in a wireless medium, there is a huge amount of automation done. Machine to machine interaction can lead to faster and timely outputs in many cases where human intervention might cause a delay.

With the accelerating advances in technology and industrialization, the software sector has evolved into a huge infrastructure with several components and devices, making the management of this infrastructure tough. The circumstances of the workplace have a direct impact on the efficiency of the employees. Hence, the contentment of the employees becomes a basic necessity to extract work from them. In the vision of improving the day-to-day lives of breadwinners, the workplace environment is equipped with IoT devices. Such a system intends to create a comfortable and safe working environment that boosts people to collaborate and to be productive. With the automation of devices, the tasks that are done daily can be controlled; thus, no human attention is required.

Smart IoT-driven workplaces are scalable and modular that mainly focus on providing more time for business and spend less on office-keeping chores and bureaucratic procedures like ordering maintenance, and thus, in a nutshell, IoT is used extensively to lessen the burden on humans. Smart office solutions can drastically boost the productivity of the firm. The maintenance of several equipment and devices installed in the firm can be reported by the devices themselves, and thus, the necessity of the employee—maintenance department interaction becomes almost zero. This makes the repairing process effortless and quick. The predictions made by machines are more accurate and faster, hence making them more reliable.

When the staff need not sweat out for the small stuff, they get more time to spend on the bigger and important tasks. Working smart is more profitable than working hard. It makes work done with lesser stress and lesser effort. Also, the people need not remain stuck with the daily chores but instead involve in more creative jobs. Smart office solutions make it possible for many imprompt collaborations of the employees. Any firm that encourages collaboration is likely to tap on the creativity of its workforce, leading to the development of the firm. So, we propose this smart workplace project to help the maintenance of the workplace which has many cabins equipped with several electrical appliances by fully automating them using the concept of IoT to make smart usage of energy and to provide a user-friendly workplace environment.

2 Related Work

MobeenShahroz et al. had proposed the idea of using RFID sensors in the supermarket. The main objective is to reduce the waiting time for the billing process (barcode-based). RFID sensors consist of RFID tag and RFID reader. The tag will be attached to every product, and the reader is used to read the product information. In this system, each customer is provided with a unique customer ID so that they can login and easily manage the products via a mobile application.

It sends the data to the server and generates the bill automatically. With this system, the customers can get the best quality products in a short time [1]. Dan-IoanGota et al. had emphasized the concept of garage door automation and temperature monitoring. The actuators are used for controlling the operation of the door. The temperature of the rooms is also monitored. This system is useful for people with disabilities as they can monitor and control the entire system through smartphones [2].

Rajesh Kannan et al. had propounded the concept of automatic working of smart windows based on the circumstances. This system describes a smart window that automatically opens/closes itself by detecting the amount of gas leakage in its surrounding. It also detects the presence of hazardous gases and responds according to their amount of presence in the surroundings. The data collected through sensors are used for triggering the smart window. The smart window coordinates with some other IoT devices like an exhaust fan and siren which are automatically switched on/off based on the state of the smart window. It also alerts the user about the detected gas leakage and also makes the collected data available to the user through the end device. The smart window can be operated in either manual mode or automated mode according to the user's comfort [3].

Recent works had insisted the importance to solar energy as it plays an important role in saving energy. An idea to maximize the utilization of solar energy is proposed for the lighting system in homes. Whenever there is sufficient sunlight in the environment, the solar panels receive solar energy and charge the solar power battery. The stored energy of the solar power battery is used for the operation of the entire lighting system. The electrical devices in the system are operated using solar power when the solar power battery has sufficient energy. Based on the percentage of charge in the battery, the IoT devices connected with the solar power battery work automatically. Kannapiran et al. had put forward a methodology to save electricity when it is not in use. The motion of the employees is detected with the help of a motion detector, and thus, electronic appliances are controlled. The main advantage of this system is that it requires less maintenance and saves energy. It provides both manual and automatic modes to avoid disruption in the working of the electrical devices [4].

Emakpor et al. had emphasized the RFID system that provides additional security to the users through the reader deployed at the door and RFID tag provided to the users. The main advantage of this system is that it denies the entry of an unauthorized person thus providing greater security [5].

Sinara et al. had proposed the idea of a smart window that holds good for all smart home scenarios. The smart window functions according to the pre-defined parameters and consistently checks for their threshold values. In this system, the smart window is connected to the solar battery which also contributes to efficient energy management. The data collected through sensors decide the status of the smart window. This automated window is designed in a way to reduce human involvement in simple works. The user can check the data and also the state of the smart window through their end device. Thus, the smart window can be operated either in automated mode with the pre-defined parameters or in manual mode by the user through the connected end device [6].

Ilkyu Ha had given the concept of strengthening security through the digital door lock system. It restricts the invalid user's entry and gives a caution to the user's mobile. So, the door opens only when the authorized user approaches the door. Else, the door remains closed, thus ensuring safety [7, 8].

RenukaBhuyar et al. came up with the concept of a fire alarm system that alerts the users by warning them through the end devices whenever the threshold value is crossed. The sensors are used to extract real-time data from the environment. Based on this, the alarm, ensure the safety of the people in the office. It can be extended to the whole building that results in the deduction of energy consumption to a greater extent [9, 10]. MusalaVenkateswara Rao et al. had proposed the idea of achieving energy efficiency by making use of the Node MCU connected to electrical appliances of the workplace. It mainly focuses on security and controlling electrical devices. For the security purpose, it makes use of RFID sensors. When there is an increase in the count of the employees entering the workplace than the usual count, the LED light starts glowing which indicates that the electrical devices should be switched to the cubicle [11]. In [12], security aspects on securing the communication in smart office are discussed.

3 Proposed Work

The IoT-based smart workplace is designed to provide a comfortable and energy-efficient environment for the employees. It automatically works with the help of IoT devices and sensors, which drastically reduces the manpower involved in the daily simple tasks of the office. The automation of the entire system is based on the ON and OFF conditions of various subsystems. Various sensors like a rain sensor, trip sensor, motion detector, photosensor, smoke detector, and fire sensors are used to build an energy-efficient and smart workplace. RFID is used on the front door to provide improved security as shown in Fig. 1.

3.1 Imperative Aspects

The door works on the concept of radio frequency identification which will be unlocked only when the RFID reader detects a valid RFID card of the employee.

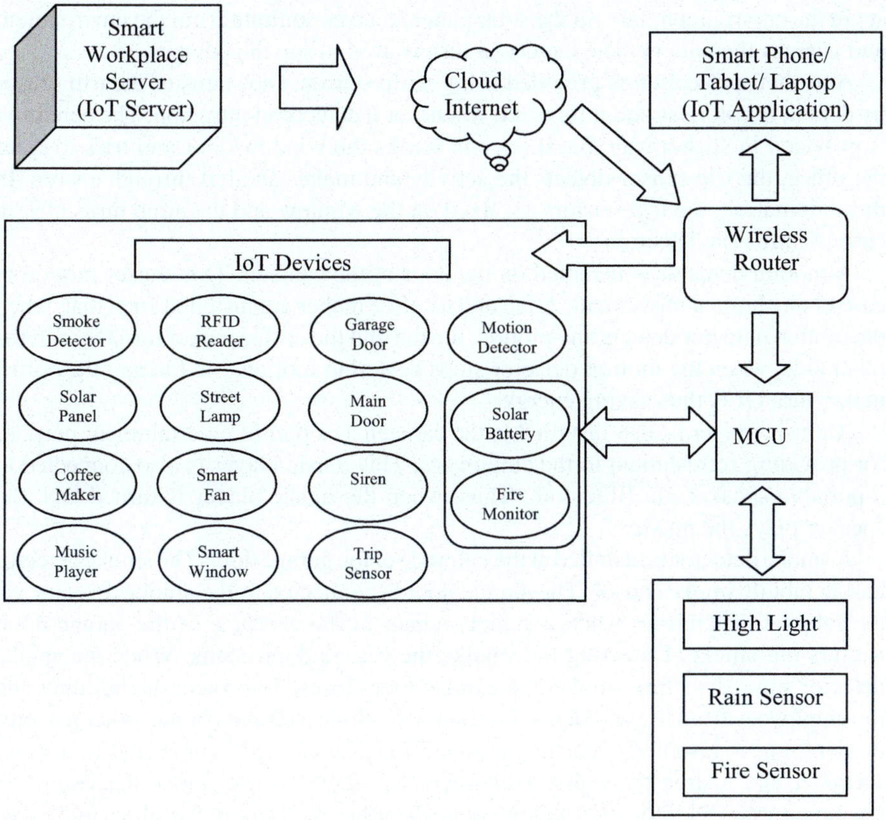

Fig. 1 Block diagram of the smart workplace system

When the employee enters with an invalid RFID card, then the door remains locked; thus, it helps in providing a safer workplace. The reader works based on the radio waves emitted from the RFID card which must be in the particular frequency range for a successful validation by the RFID reader. An antenna that is inbuilt in the RFID card transmits radio waves of a certain frequency. The same frequency range is also set on the reader that is fixed on the main door. The reader sends the data to the cloud and checks whether the frequency is valid or not, and based on the information, the door opens else remains in the closed state. Using solar energy, the LED bulbs, lamp, and fan will function automatically. The battery makes use of solar energy from the panel based on the environmental conditions and gets charged. When the battery power drops to zero, the LED bulbs, lamp, and fan will stop running as these components purely depend on the solar battery for their power.

The LED bulbs, lamp, and fan will automatically start functioning once the battery gets charged again. The battery charges automatically from the solar panel when there is sufficient sunlight in the environment. The environmental conditions are already set in the environment tab. So the solar panel receives sunlight from the environment and charges the battery. The battery in turn is used to run the devices.

Anti-theft protection is provided using a trip sensor. The status of the trip sensor by default is colorless and it turns red whenever it detects an intrusion. The condition is provided in such a way that if anyone breaks the window/door and tries to enter the office, the trip sensor detects the activity and makes an alert through a siren. In this automation, the trip sensors are fixed on the window and the siren turns ON, in case of intrusion detection.

A motion detector is installed on the door of the canteen. This senses motion in case of any human movement. A fan and a coffee maker are installed such that, when the motion detector detects any motion, the fan and the coffee maker turn ON. On the other side, when the motion detector stops detecting motion, the fan and the coffee maker turn OFF, thus saving energy.

A music player is also installed in the canteen as a part of entertainment purpose for providing refreshment to the employees. This music player is also connected to a portable speaker via Bluetooth. Thus, when the music player is turned ON, the speaker plays the music.

A smoke detector is installed at the entrance of the garage door. The smoke detector has an inbuilt smoke sensor. The garage door is connected to the smoke detector via the Internet. Therefore, when a vehicle comes at the entrance of the garage door, it gives out smoke. Detecting the smoke, the garage door opens. When the smoke detector stops detecting smoke, the garage door closes. To protect the building and the employees from fire accidents due to electrical short circuits or any other reasons, the building is installed with fire sensors. This fire sensor is connected to a water sprinkler and a siren through the Internet. So, when the fire sensor detects a fire, the siren turns ON indicating that there is an outbreak of fire in the office and alerts the people. On the other hand, the water sprinkler also starts sprinkling the water as a measure to extinguish the fire. A street lamp is installed in the surroundings of the office. The street lamp has an inbuilt photosensor. Thus, the photosensor detects sunlight from the environment and decides whether the street lamp should be in ON or OFF condition. During the day when there is enough sunlight the street lamp is OFF and, in the evening, when the amount of sunlight drops, the street lamp is ON.

The smart window in our proposed work works based on the combined action of a photosensor and rain sensor that does not require a manual action to close/open it. The photosensor from the street lamp is employed here, and a rain sensor has been installed on the window. The basic states and conditions of windows are that they should be closed during the night and when it rains. Otherwise, the window remains open. The threshold is set on the conditions of the IoT server, such that the windows decide themselves when to open and when to close, and there is no manual attention needed.

4 Flow Chart

In Fig. 2, the process flow of the proposed system is described.

4.1 Algorithm

The step-by-step procedure (algorithm) for the proposed work in this system is as follows:

(1) Start (Window open).
(2) Microcontroller initialization.
(3) Receive input from the rain sensor and photosensor.
(4) The sensors send signals to Node MCU.
(5) Node MCU connects to the cloud via intermediary devices.

Fig. 2 Flowchart of the proposed work

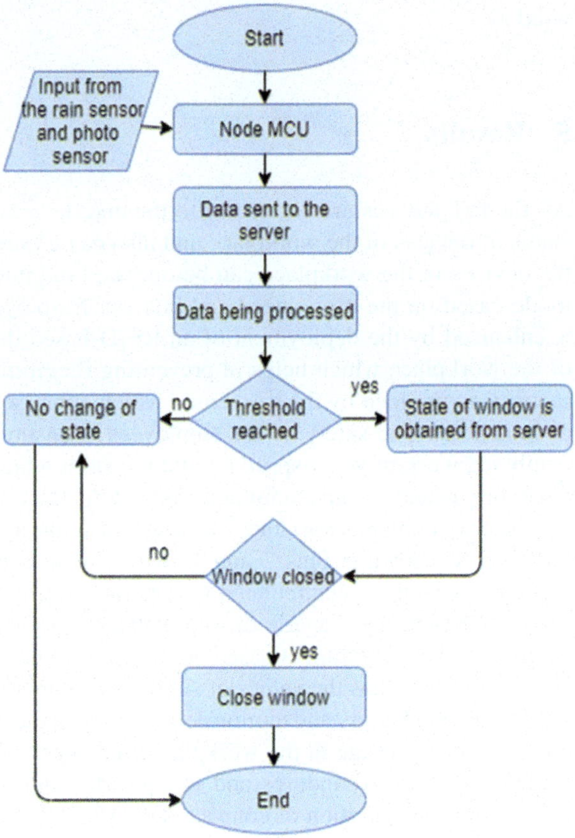

(6) The cloud verifies the signal from Node MCU.
(7) If the threshold is reached for any of the sensors, then close the window.
(8) Stop.

4.2 Description

The conditions for the working of smart window are:

- The window remains open when it is daytime and when it is not raining.
- The window remains close when it is nighttime (or) when it is raining.

We can implement the smart window concerning the above-mentioned conditions by using two sensors, namely photosensors and rain sensors.

When it rains, the rain sensor sends the information to the microcontroller which is then passed to the server. The server has a connection with the street lamp which receives the data about the amount of sunlight in the environment from the photosensor. The combined information from these two sensors controls the status of the window.

5 Results

All the IoT devices are connected to the end device so that the users can view the status of devices in the workplace and also can control them from anywhere. Thus, the devices in the workplace can be operated in either automated mode or manual mode based on the user's need and interest. Employee's security in the workplace is enhanced by the deployment of an RFID-based door lock on the main entrance of the workplace which helps in preventing illegitimate approaches and additional security is provided by the trip sensor which appraises the employees about the risk of intrusions. The safety of the employees is ensured by the fire detector and the combined work of water sprinkler and the siren which takes immediate action over the extinguished fire and simultaneously alerts the users.

The environmental variables like rain and sunlight are changed for the simulation purpose. As shown in Figs. 3 and 4, two separate containers have been created and variables have been manipulated to generate sunlight and rain in the desired time. Figure 5 depicts the scenario of solar battery charging which shows the presence of sunlight in the environment based on the graph (Fig. 3 x-axis values are the hours of a day, and y-axis has the values of sunlight percentage).

Solar power battery and motion detectors make the main contributions in achieving efficient energy usage in the workplace. Battery works according to the amount of sunlight in the surroundings and is converted into electrical power, ensuring the uninterruptible operation of connected devices.

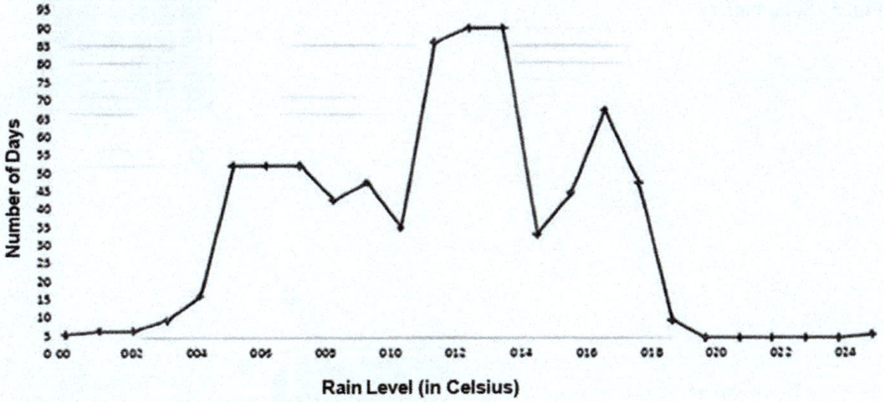

Fig. 3 Sunlight environment

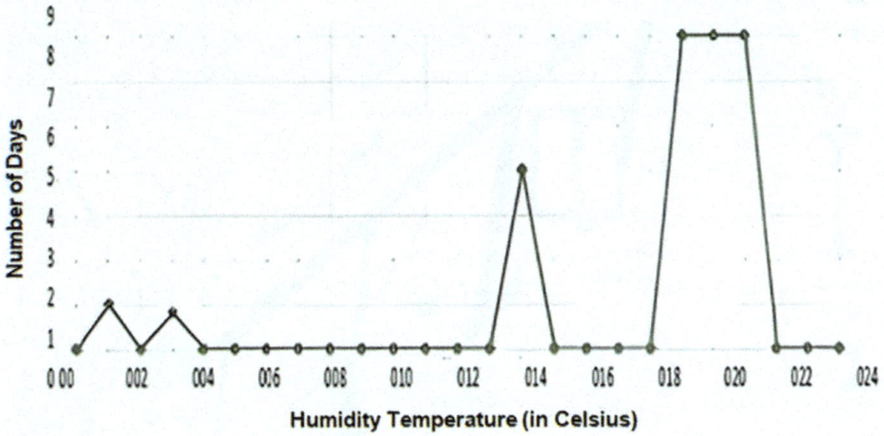

Fig. 4 Rain environment

The presence of sunlight can also be detected using the street lamp which automatically turns on, whenever there is no sufficient sunlight in the surrounding. Motion detectors prevent the wastage of energy to a huge extent, as it controls the connected IoT devices to work based on human movement detection.

Figures 6 and 7 are based on the graph (Fig. 4 x-axis values are the hours of a day, and y-axis has the values of centimeters of rain in the environment). In Fig. 6, the scenario of the smart window closed at the time of raining above 6cm, and in Fig. 7, the scenario of the smart window which automatically opens when there is a drizzle or no rain in the environment is illustrated. The smart window and smoke detectors are deployed in the workplace to provide extra comfort to employees. Smoke detectors reduce the manpower by automatically opening/closing the garage door based on the parking vehicle's positions.

Fig. 5 Solar battery

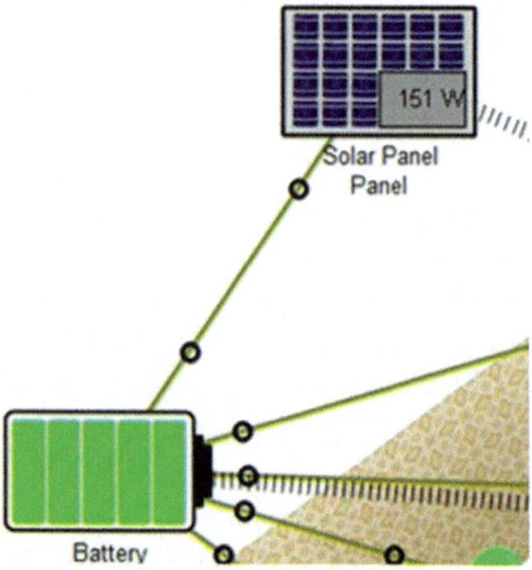

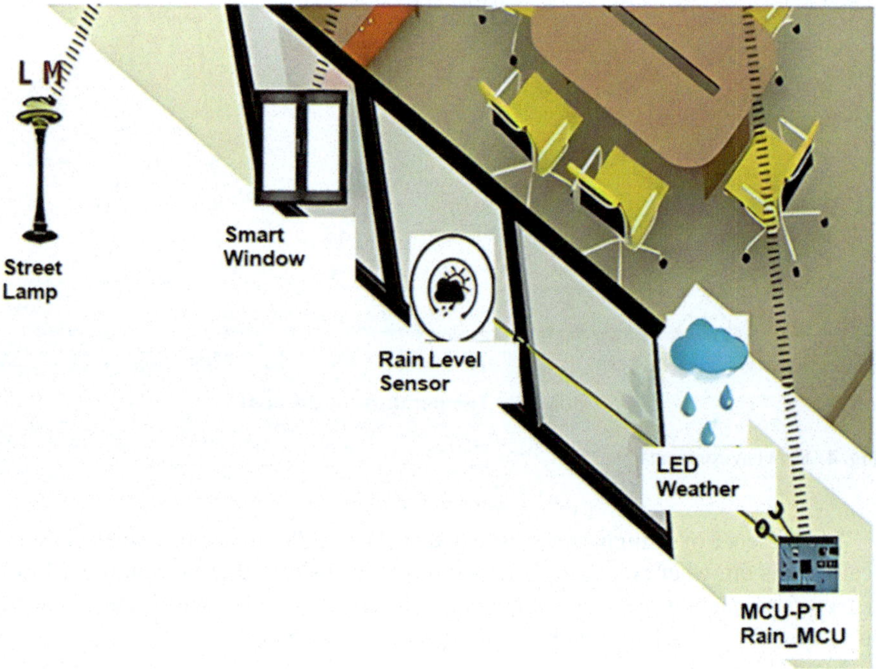

Fig. 6 Presence of rain

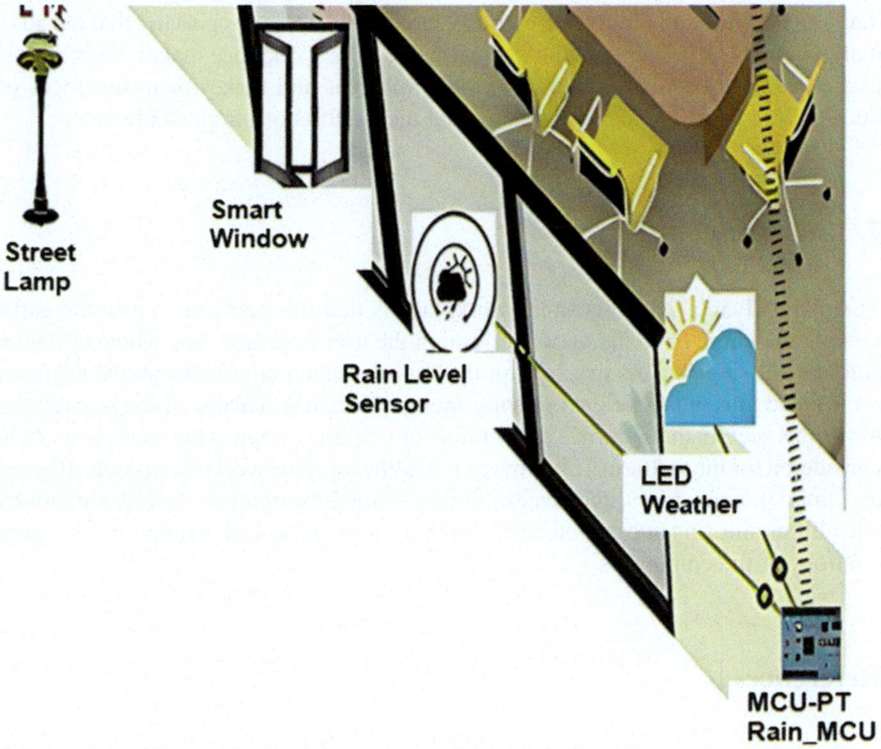

Fig. 7 Absence of rain

The smart window works based upon the main environmental entities viz. sunlight and rain, and thus, it independently changes its state without any need for human intervention.

6 Conclusion

In the proposed system, the user's comfort and satisfaction are the crucial factors kept in mind. The smart street lamp, fire alarm, fire sensor, a smart garage, music player, motion detector, trip sensor, and RFID-based door lock are used in this system, and a smart window based on photosensor and rain sensor has been designed that automatically works ensuring the complete automation of working place which allows the employees to work with more comfort and ease. The improved security of the system involves the RFID-based door lock which allows only those with the valid RFID card inside the workplace, then the fire sensors and trip sensors which give a siren alert making the workplace safer and smarter. The smart workplace system is developed to have an independent and leisure workplace, thus proving that the automatic mode is far better than the manual mode. The workplace makes smart energy usage and provides extra comfort to the employees and makes them full focus on boosting productivity and thus minimizing their efforts on simple daily tasks.

7 Future Scope

The main advantage of this smart workplace is that the user can control the entire system anywhere from the network through the user interface. The whole system is automated using sensors that rely on the environmental conditions and atmosphere events, and thus, it is efficient in saving energy. Since it is scalable, it can be extended to various sectors that improve the comfort of working people. One such sector to be considered for the welfare of employees is healthcare. This workplace is both efficient and smart in many ways, and besides, for assisting the employee's health conditions, a health monitoring subsystem added to this smart workplace would render a great comfort for the employees.

References

1. Shahroz M, Mushtaq MF, Ahmad M, Ullah S, Mehmood A, Choi GS (2020) IoT based smart shopping cart using radio frequency identification. IEEE Access 8
2. Gota D-I, Puscasiu A, Fanca A, Miclea L, Valean H (2020) Smart home automation system using Arduino micro controllers. In: 2020 IEEE International conference on automation, quality and testing, robotics (AQTR)
3. Megalingam RK, Hegde A, Vijaya Krishna Tejaswi P, Doriginti M (2020) IoT based controller for smart windows. In: 2020 IEEE international students' conference on electrical, electronics and computer science (SCEECS)
4. Selvaraj K, Chakrapani A (2017) Smart office automation system for energy saving. Int J Adv Comput Electron Eng 2(9):8–12
5. Emakpor S, Esekhaigbe E (2020) Development of an RFID-based security door system. J Electr, Control Telecommun Res 1:9–16
6. Medeiros SP, Rodrigues JJPC, da Cruz MAA, Ricardo Rabelo AL, Saleem K, Solic P (2019) Windows monitoring and control for smart homes based on Internet of Things. In: 2019 4th International conference on smart and sustainable technologies (SpliTech)
7. Ha I (2015) Security and usability improvement on a digital door lock system based on the Internet of Things. Int J Secur Its Appl 9(8):45–54
8. Bhuyar R, Ansari S (2016) Design, and implementation of smart office automation system. Int J Comput Appl 151(3), 0975-8887
9. Musala VR, Rama Krishna TV, Ganduri R, Roohi A (2018) An effective energy management system for smart office cubicles using IoT. J Adv Res Dyn Control Syst 10(02-Special Issue)
10. Abhyankar KR, Wandhekar AS, Khan SM (2019) IoT based thief security alert system. Int J Res Eng, Sci Manag 2(6)
11. Arun S, Likith AR, Dharshan K, Srinivasa N (2019) Smart office monitoring system using IoT. Int Res J Eng Technol (IRJET) 06(04)
12. Safa H, SakthiPriyanka N, Vikkashini Gokul Priya S, Vishnupriya S, Boobalan T (2016) IOT based theft preemption and security system. Int J Innov Res Sci, Eng Technol 5(3)

Chapter 21
Aquila Optimization-Based Cluster Head Selection and Honey Badger-Based Energy Efficient Routing Protocol in WSN

S. Venkatasubramanian and S. Hariprasath

1 Introduction

It is now widely accepted that WSN is the common and popular evolving network because of its low development costs, inexpensive sensor devices, and the ability to sense numerous physical and ecological considerations [1]. WSN is also capable of transmitting, processing, and sensing information. WSN has a wide range of applications in a variation of industries, including industrial, commercial, medical, defense, and weather forecasting, among others. Base station (BS) [2] uses the collected data to make decisions for a variety of purposes. Limited energy and power resources limit the nodes in WSNs. The base station consumes more energy as it receives data packets, so an energy-efficient routing method is essential.

There are a number of ways in which WSN energy can be used, including data transfer via nodes and environmental sensors. Processing and sensing of data useless energy than the transmission of the data itself did. BS and sensor nodes (SN) both benefit from these routing algorithms for constructing communication paths. These sensor nodes must, however, be able to lead data from the nodes to the base stations securely and effectively. Noded is appointment, low network reliability, generation, hardware restrictions, and soon all have an impact on WSN architecture[3]. Several routing approaches have been developed during the last few decades to address these issues.

A wireless communication medium is used for routing packets between the SNs and BS of WSN. Single and multiple path routing protocols are the two main sorts of protocols for routing. The sink node in WSN receives data packets from the source nodes via intermediary nodes. Depending on the network's energy capacity

S. Venkatasubramanian (✉) · S. Hariprasath
Saranathan College of Engineering, Trichy, India

S. Hariprasath
e-mail: hariprasath-ece@saranathan.ac.in

V. Devedzic et al. (eds.), *Proceedings of the International Conference on Intelligent Computing, Communication and Information Security*, Algorithms for Intelligent Systems, https://doi.org/10.1007/978-981-99-1373-2_21

[4] and the load distribution balance [5], packets are spread over many hops. Routing protocols assess the trade-off between network characteristics such as delay, energy cost, and load balancing [6] in order to ensure optimal routing. The sensor nodes' insufficient power sources are seen as a major problem in WSN. Consequently, node failure causes network failure [7, 8]. The optimal use of energy in WSNs is also necessary to increase the WSN's lifespan and improve its performance [9]. This results in a reduction in network energy consumption and an increase in network scalability [10, 11]. All of the clusters in the network are connected to each other via a cluster head (CH) [12]. As a result, clustered WSNs employ a routing protocol to find the most energy-efficient method of transmitting data from a CH to a BS. [13]. In addition to fault tolerance and dependability, the routing protocols also provide information accumulation and scalability[14].

During data transmission, the goal of the study is to lessen the nodes' energy usage. The sensor nodes' energy consumption is reduced as a result of this, allowing more packets to be sent to BS. The swarm intelligence is employed extensively in this study because of its ability to search, its robustness, and its ability to adapt to new situations.

The following is a list of the research's main findings:

With its excellent stability and low computational complexity, AOA is utilized to choose the CH in the WSN. Remaining energy and distance to the SN are used by AOA to pick the CH in this investigation.

The HBA is used to classify the path from CH to BS since it has the capacity to give speedy discovery of solutions in WSN. In this case, the proposed approach is optimized using the outstanding energy and the coldness travelled.

The network's lifespan is extended due to the service of energy-efficient CHs and the construction of appropriate data transmission routes. In addition, the BS receives more packets because the nodes use less energy when transferring data packets.

2 Related Works

In order to pick the best optimizer (GWO) is presented in [15]. By not selecting cluster chiefs in each round, the protocol aims to extend the network's lifespan even further. Thus, recurrent clustering and network energy consumption are reduced because the setup phase is only executed when the cluster heads' remaining energy in the previous round is insufficient. To make matters worse, the exorbitant power ingesting of cluster heads will be exacerbated. In a large-scale application, the distance among cluster heads and a base station could be more than the sensor node's maximum radio range. Incomplete monitoring data can result if the cluster heads are unable to locate it.

In [16], proposed an energy-efficient clustering route algorithm-based ACO that uses ant colonies. The energy and distance among network nodes are used to design a new way of pheromone update. When using the pseudo-random route discovery approach, sensor node energy consumption is effectively balanced. Another benefit is that the network's energy ingesting is abridged, and its lifespan is extended by using an opportunistic broadcast strategy instead of flooding. The balanced energy consumed is occurred only by considering the random value for path discovery. Non-uniform clustering isused in [17] to handle the issue of cluster heads being distributed unevenly. As a result, designated as cluster leaders, which also to help distribute network load and prevent hot spots, the cluster heads use a combination of single and multiple hops to communicate data to the base station. While solving the localization problem, the accuracy of this algorithm is not affected, but the number of iterations and fitness evaluations is increased.

Reference [18] suggests a clustering routing protocol built on glow-worm swarm optimization that uses less energy than conventional methods. Nodes' residual energy, density, and compactness are all taken into account in this algorithm's clustering process. Nodes' energy consumption is balanced in the communication phase by the use of single and multiple hops. There are many benefits to using the proposed algorithm. The cost function is considered only residual energy of the adjacent CH with communication energy dissipation. By energetically choosing the WSN nodes with the most residual energy as cluster heads, a new procedure is suggested in [19]. Another benefit of using ACO based on distance is that it allows each node to choose its own global best transmission path, resulting in a reduction in the total sum of energy consumed during transmission. In terms of network life and throughput, this approach performs better. A large amount of signaling overhead is occurred during the path failure by re-establishing the network with the control message.

In [20], Crow Whale-ETR optimization is used to construct a protocol based on the WOA and CSA energy and trust factors, which defeated the challenges. The best pathways determined with CWOA are based on a combination of trust and energy. This route is chosen as the best option for transmitting data. The trust and energy levels of every node are updated at the end of every transmission, so that only the most protected nodes boost the network's security. 50 and 100 nodes are tested with or without assaults in order to get the lowest delay and highest energy throughput. Even though CWOA is integrated with the crow search algorithm, its convergence speed is sluggish, its accuracy is low, and it is easy.

In [21], devised an algorithm for routing dubbed glow-worm swarm optimization (LBR-GSO) is proposed. In order to deal with nodes' high energy consumption, this system combined a pseudo-random way discovery method with an enhanced pheromone trail-based update methodology. Route establishment can also be improved by using a cost-effective energy metric to update heuristics. Finally, the control overhead is decreased via the LBR-GSO cast-off energy-based distribution method. The low precision of the GSO algorithm's computations and its ease of falling into local optimums have been exposed as flaws in global search.

In [22], the author uses the butterfly optimization algorithm (BOA) to select a perfect cluster head from a set of nodes. With this information, the node's residual energy, proximity to neighbors, and the base station, as well as its degree and its centrality, are taken into consideration. ACO calculates the most efficient route from the CH to the base station based on distance, outstanding energy, and node degree. BS data packets, live nodes, dead nodes, and other performance parameters are all taken into account when evaluating the suggested approach. In order to discover the best solution, BOA and ACO use a large number of multi-objective fitness functions, which takes a long time.

3 Proposed System

Network energy consumption models and settings for WSN and sensor nodes are introduced in this section to help with the upcoming work.

3.1 Model for Energy Consumption

There are two major components to the energy spent by nodes when they transfer l-bit data: the loss in distribution l-bit data and the power amplifier circuit loss. Only the power used by the receiving circuit is included in the total amount of energy used by a node to obtain data. According to this assumption, we can calculate how much energy it takes for node I to send data across the channel, and how much energy it takes for node *j* to send data across that same channel. l-bit data sent over distance *d* consumes *d* kilowatts of power.

$$E_{Tx}(l, d) = \begin{cases} lE_{\text{elec}} + l\varepsilon_{fs}d^2, d < d_0 \\ lE_{\text{elec}} + l\varepsilon_{mp}d^4, d \geq d_0 \end{cases} \quad (1)$$

Receiving l-bit data necessitates the node to expend is

$$E_{Rx}(l, d) = lE_{\text{elec}} \quad (2)$$

l-bit data fusion requires a large quantity of energy from the cluster head is

$$E_{Fx}(l, d) = lE_{\text{DA}} \quad (3)$$

Energy consumption for conveying 1-bit data (Eelec) and energy consumption for combining 1-bit data (E DA) are shown in the figure. The energy weakening models of the amplifier are shown in the figure. A free space fading the transmission distance is smaller than *d* 0, and the energy obligatory for power intensification is fs. A multipath attenuation model is adopted when transmission distance is higher than *d* 0, and the energy obligatory for power intensification is mp.'

3.2 *System Model*

On network nodes and WSNs, the following assumptions are made:

- As a result of the implementation of the network, all of the sensor nodes in the monitoring area of MM have an exclusive network ID, the base station is centrally situated in the monitoring area, and the nodes themselves are fixed in place.
- Sensor nodes have the same beginning energy and processing and communiqué capabilities; therefore, additional energy cannot be added to any of them. They are all homogeneous.
- Remaining energy may be detected by all nodes and the signal strength used to compute the distance to the signal sending end.
- It is possible for each node to connect directly with the base station, and each node can merge data and select transmission power based on the communication distance.
- The base station's resources are limitless.

3.3 *Proposed Methods for Cluster-Based Routing Protocol*

Cluster heads are selected using the AOA algorithm, and the suggested HBA-based routing algorithm is used to discover the most efficient pathways between each cluster head and the base station during setup. A proposed HBA-based routing method is used during this phase of stabilization to route the data from the memberships of each cluster to the BS. Following the introduction of AOA, we explain the HBA-based routing method in greater detail in the following section. Figure 1 depicts the proposed cluster-based routing protocol's working flow.

3.3.1 Aquila Optimizer Algorithm (AOA)

To begin the optimization process, the population of candidate keys (X) is produced among the given problem in Eq. (4). In each iteration, the best solution that has been found thus far is deemed to be the optimal option [23].

$$X = \begin{bmatrix} x_{1,1} & \cdots & x_{1,j} & x_{1,Dim-1} & x_{1,Dim} \\ x_{2,1} & \cdots & x_{2,j} & \cdots & x_{2,Dim} \\ \cdots & \cdots & x_{i,j} & \cdots & \cdots \\ \vdots & \vdots & \vdots & \vdots & \vdots \\ x_{N-1,1} & \cdots & x_{N-1,j} & \cdots & x_{N-1,Dim} \\ x_{N,1} & \cdots & x_{N,j} & x_{N,Dim-1} & x_{N,Dim} \end{bmatrix} \quad (4)$$

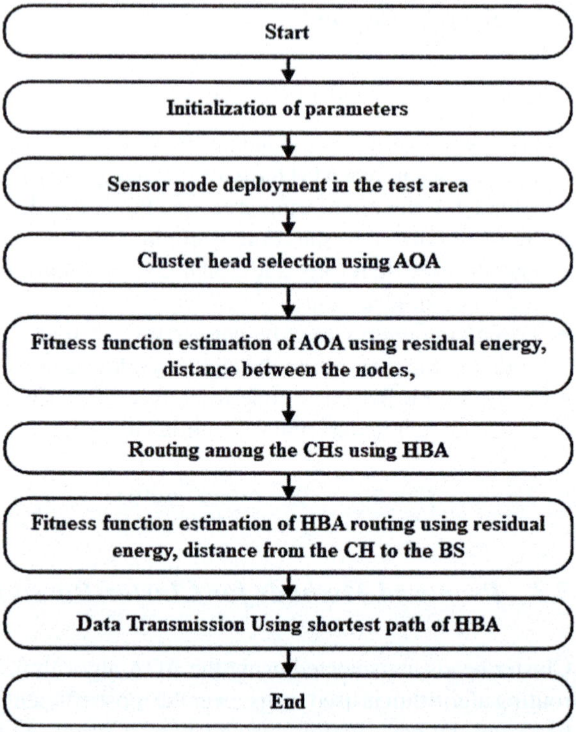

Fig. 1 Flowchart of proposed AOA-HBA-based routing protocol

Expressed in terms of decision values (positions), X i is the position of the ith candidate solution, N is the overall population size of the problem, and Dim denotes how many dimensions there are.

$$X_{ij} = rand \times \left(UB_j - LB_j\right) + LB_j, i = 1, 2, \ldots .N, j = 1, 2, \ldots, Dim \quad (5)$$

In this case, rand is a random number, while LB j and UB j signify the problem's lower and upper limits, respectively.

The four strategies for optimizing the proposed AOA are as follows: fly with sluggish descent walk and seize prey. If $t(23) * T$ is true, the AOA can switch from exploration to exploitation utilizing alternate behaviors; otherwise, the exploitation steps will be carried out.

Step 1: Expanded exploration ($\boldsymbol{X}_1$):

The vertical stoop to choose the optimum hunting location in the first way X_1. Here, the AOA surveys the entire search space from a position of great altitude in order to locate the prey.

$$X_1(t+1) = X_{\text{best}}(t) \times \left(1 - \frac{t}{T}\right) + (X_m(t) - X_{\text{best}}(t) * \text{rand}) \quad (6)$$

where $X_1(t+1)$ represents the outcome of the first search procedure for the next iteration of t. In this case, $X_{\text{best}}(t)$ is the best answer till the tth iteration, which reflects the prey's location. The amount of iterations in the enlarged search (exploration) is controlled by this equation $\left(1 - \frac{t}{T}\right)$. This is the location mean value of the current connected solutions at the tth iteration, which is determined using Eq. (7) and rand generates a random number between 0 and 1. The current repetition is shown by t, while the maximum sum of repetitions is shown by T.

$$X_m(t) = \frac{1}{N}\sum_{i=1}^{N} X_i(t), \forall j = 1, 2, \ldots, Dim \tag{7}$$

Step 2: Narrowed search ($\boldsymbol{X_2}$)

While flying at high altitude, the Aquila circles over its chosen prey and prepare its land before attacking. Aattack is the name given to this technique. AOA narrowly examines the target prey in training for the attack in this instance. Mathematically, this is represented as Eq. (8).

$$X_2(t+1) = X_{\text{best}}(t) \times \text{Levy}(D) + X_R(t) + (y - x) * \text{rand} \tag{8}$$

where $X_2(t+1)$ is the result of the second search technique's next iteration of t. The levy flight distribution function, Levy(D), is derived from Eq. (9) using D as the dimension space. At the ith iteration, $X_R(t)$ is a random solution selected from a range of [1 N].

$$\text{Levy}(D) = s \times \frac{u \times \sigma}{|v|^{\frac{1}{\beta}}} \tag{9}$$

There are three variables: s, the constant value fixed to 0.01 (u), the random sumamong 0–1 (x), and the x-coordinate using Eq. (10) is determined.

$$\sigma = \left(\frac{\Gamma(1+\beta) \times \sin e\left(\frac{\pi\beta}{2}\right)}{\Gamma\left(\frac{1+\beta}{2}\right) \times \beta \times 2\left(\frac{\beta-1}{2}\right)}\right) \tag{10}$$

In this example, where is a constant of 1.5. To show the spiral structure in the search, y and x are calculated as follows in Eq. (8).

$$y = r \times \cos(\theta) \tag{11}$$

$$y = r \times \sin(\theta) \tag{12}$$

where

$$r = r_1 + U \times D_1 \tag{13}$$

$$0 = -\omega \times D_1 + \theta_1 \tag{14}$$

$$\theta_1 = \frac{3 \times \pi}{2} \tag{15}$$

For a set, sum of search cycles r_1 takes a value among 1 and 20, and U is a tiny value fixed at 0.00565. Data 1 (D1) contains integers from 1 to the search space's length (Dim) and a small fixed value of 0.005.

Step 3: Expanded exploitation ($\boldsymbol{X}_3$)

This is the third way, in which an Aquila drops vertically with an initial attack to determine how the target reacts once the prey region has been precisely specified. Low flying with gradual descent assault is the name of this technique. AOA uses this opportunity to close up on its prey and launch an attack. Mathematically, this is represented as Eq. (16).

$$X_3(t+1) = (X_{\text{best}}(t) - X_m(t)) \times \alpha - \text{rand} + ((\text{UB} - \text{LB}) \times \text{rand} + \text{LB}) \times \delta \tag{16}$$

where $X_3(t+1)$ is the result of the third search technique (X 3) for the next iteration of t. Exact location of prey until ith iteration and current solution's mean value at tth iteration are referred to as X_{best} and X_m, , respectively (4). The exploitation adjustment parameters are set to a low value in this paper (0.1). The given problem's lower and upper bounds are denoted by the abbreviations LB and UB, respectively.

Step 4: Narrowed abuse ($\boldsymbol{X}_4$)

When the Aquila reaches the target in the fourth method, they use stochastic motions to attack the prey over land. Walk and grab prey is the name of this technique. Finally, AOA assaults the prey in its last spot. Mathematically, this is represented as Eq. (17).

$$X_4(t+1) = QF \times X_{\text{best}}(t) - (G_1 \times X(t) \times \text{rand}) - G_2 \times \text{Lecy}(D) + \text{rand} \times G_1 \tag{17}$$

In this case, $X_4(t+1)$ is a solution made by the fourth search technique (X 4). Eq. is used to calculate a quality function that is utilized to balance the search approaches (18). When tracking prey during elopement, G_1 specifies a variety of AOA motions that are used to construct the Eq. (19). From 2 to 0, G_2 shows a decreasing flight slope that represents the AOA employed to follow the prey during its escape from the starting site (1) to its final location (t), derived by Eq. (20). This iteration's solution, $X(t)$, is

$$QF(t) = t^{\frac{2\times\text{rand}-1}{(1-T)^2}} \tag{18}$$

$$G_1 = 2 \times \text{rand} - 1 \tag{19}$$

$$G_2 = 2 \times \left(1 - \frac{t}{T}\right) \tag{20}$$

A random sum among 0 and 1 is used to generate the quality function value at iteration t, while $QF(t)$ represents the quality function value at that iteration. Iteration counters t and T show how many times the loop has been iterated so far and how many iterations remain. Using Eq. (9), we can derive the levy flight distribution function Levy(D). Using Eq. (5), the fitness function is determined by taking into account the four best solutions.

AOA's fitness function is used to narrow down the network's pool of potential CHs to a single best candidate. During the clustering process, the residual energy taken into account by the fitness function is employed to CH. In order to find the best CH, the distance between the nodes is taken into consideration.

Residual of the CH

CH performs a variety of responsibilities in a network, including data collection from conventional sensor nodes and transmission of data to BS. Due to the CH's heavy reliance on energy, the node with the most remaining energy is favored to be its source of power. The following Eq. (21) describes the leftover energy (f_1).

$$f_1 = \sum_{i=1}^{m} \frac{1}{E_{CHi}} \tag{21}$$

Distance Between the Sensor Nodes

It establishes the distance among the usual sensor nodes and the CH it has built into itself. The transmission path's length is the primary determinant of the node's energy dissipation. When the designated node has a shorter transmission distance to BS, the node's energy usage is reduced. For the normal sensors' distance from the CH (f_2) sensor, we have the Eq. (22).

$$f_2 = \sum_{j=1}^{m} \left(\sum_{i=1}^{I_j} dis(s_i, CH_j)/I_j \right) \tag{22}$$

where I_j is the amount of nodes in CH, and $\text{dist}(s_i, CH_j)$ is the distance among sensor I and CH j.

Each objective value is assigned a weight value. Multiple objectives are reduced to a single objective function in this situation. δ_1 and δ_2 are the weighted values. The equation depicts the single goal function (23).

$$f = \delta_1 f_1 + \delta_2 f_2 \tag{23}$$

where $\sum_{i=1}^{5} \delta_i = 1, \delta_2 \in (0, 1)$

where the $\delta_1 and \delta_2$ values are 0.35 and 0.25, respectively.

3.3.2 Routing Algorithm Using HBA

The "digging phase" and the "honey phase" are the two phases of HBA. Hypothetically, HBA is a method of global optimization that combines the finding and using phases [24]. The mathematical procedures involved in the proposed HBA are described here. This diagram shows the population of possible HBA solutions:

$$\text{Population of candidate solution} = \begin{bmatrix} x_{11} & x_{12} & x_{13} & \ldots & x_{1D} \\ x_{12} & x_{22} & x_{23} & \ldots & x_{2D} \\ \vdots & \ldots & \ldots & \ldots & \ldots \\ x_{n1} & x_{n2} & x_{n3} & \ldots & x_{2D} \end{bmatrix} \tag{24}$$

$x\,i = [x\,i1, x\,22, \ldots .x\,iD]$ represents the ith position in honey badger or jobs.

The first, introductory stage. Assign each of the initial N honey badgers to one of the slots in Eq. (24):

The formula produces an integer, x I between zero and one. a $x_i = lb_i + r_1 \times (ub_i - lb_i)$.

Step 2: Intensity Definition. The focus of the prey and the distance between it and the honey badger are also factors that affect the ferocity of the pursuit. It is said in Eq. (25) that the prey will move fast if the strength of its smell is high, and slowly if it is low.

$I_i = r_2 \times \frac{S}{4\pi d_i^2}$, r_2 is a random quantity among 0 and 1

$$S = (x_i - x_{i+1})^2 \tag{25}$$

$$d_i = x_{prey} - x_i$$

if S is the concentration strength of the source. As shown in Eq. (25), di represents the separation between the ith prey and the ith badger.

Step 3: Factor of Density updation: The randomization for a seamless transition from inspection to abuse. Make Eq. (26) less unpredictable by adjusting the decreasing factor () as the loop progresses:

$$\alpha = C \times exp\left(\frac{-t}{t_{max}}\right), t_{max} = \text{maximum number of iterations} \quad (26)$$

where C is represent as a constant ≥ 1 (default = 2).

Step 4: Absconding from local optimum. The purpose of this and the next two stages is to allow seepage from local optima regions. So that agents can take their time and thoroughly search the area, and the proposed method incorporates a flag called F to modify the search direction.

Step 5: Updating the positions. As was discussed before, updating an HBA position (x new) consists of two stages: the "digging phase" and the "honey phase." The following text provides an even more thorough description:

Step 5–1: Digging phase. Digging like a honey badger is reminiscent of a cardioid shape. Moving in a cardioid pattern can be mimicked with Eq. (27).

$$x_{\text{new}} = x_{\text{prey}} + F \times \beta \times I \times x_{\text{prey}} + F \times r_3 \times \alpha \times d_i \times \left|\cos(2\pi r_4) \times \left[1 - \cos(2\pi r_5)\right]\right| \quad (27)$$

To rephrase, "x prey" represents the current position of the prey in the global context. The honey badger's food-finding proficiency is indicated by the value 1 (default = 6). There are three random numbers between 0 and 1; we will refer to them as r 3, r 2, and r 1. F is a signal flag that vicissitudes the search direction based on Eq. (28):

$$F = \begin{cases} 1 & \text{if } r_6 \leq 0.5 \\ -1, & \text{else} \end{cases} \quad r_6 \text{ is a random numeral among 0 and 1} \quad (28)$$

The strength of the odor During the digging phase of a honey badger's life, the number of prey items I prey), the distance from the badger (d I and the time-varying search influence factor are all crucial. On top of that, F's disturbance of a badger's digging could lead to the discovery of more prey sites.

Step 5–2: Honey phase. The scenario in which a honey badger follows a honey guide bird to a beehive can be modeled by the following Eq. (29):

$$x_{new} = x_{prey} + F \times r_7 \times \alpha \times d_i, r_7 \text{ is a chance amount among 0 and 1} \quad (29)$$

Explanation: x new represents the new position of the honey badger, x prey refers to the location of the prey, and F are calculated using Eqs. (28 and 26). Equation 29 shows that a honey badger hunts near to the prey position determined so far based on the distance information d_i.

In order to reduce the fitness function values to a single objective, the route cost, the weight value is applied to each fitness value. An adjustment has been made to Eq. (29) in terms of its route cost (C_k). Here is an example of how the route cost is calculated (30).

$$C_k = \varphi_1 E_r + \varphi_2 d_{CH,BS} \quad (30)$$

and where the weighted values are depicted as φ_1, φ_2 and are each equal to half of a full weight of 0. To avoid sensor nodes with insufficient energy, the leftover energy is the top concern. Because the sensor node has a lower energy level, the communication fails. In order to find the quickest route and hence save energy consumption, the distance among BS and CH is only a secondary consideration.

4 Results and Discussion

An core i3 processor and 4 GB of RAM are used in MATLAB R2018a to construct and test the suggested energy-efficient routing protocol. In the 200m200m sensing region, 300 sensor nodes have been randomly planted. Table 1 shows the simulation parameters taken into account for the experiment. Each node in the network's energy consumption is to be reduced as a goal of this research. Consequently, the cluster-based routing is built using AOA for cluster head selection and HBA for cluster head routing. For CH selection and cluster construction, the time required is 10.4841 s and 0.037804 s respectively.

4.1 Performance Evaluation

This section explains the experimental setup and outcomes of the suggested method. The following is a breakdown of the performance metrics:

4.1.1 Energy

This is calculated as the sum of all hops' energy and is expressed as:

$$\text{Energy} = \frac{1}{p}\sum_{n}^{p} E_n \tag{31}$$

Table 1 Simulation strictures

Parameters	Value
Packet size	4000 bits
Area	$200m \times 200m$
E_{elec}	50nJ/bit
E	0.5 J
E_{fs}	$10pJ/bit/m^2$
E_{mp}	$0.001310pJ/bit/m^4$

In multi-hop routing, p represents the number of hops, while E n represents the energy expended on the nth hop.

4.1.2 Packet Delivery Ratio (PDR)

In the network, the proportion of packets that are successfully transported from one node to another.

$$\mathrm{PDR} = \frac{\text{number of packet received succesfully}}{\text{Total number of packets forwarded}} \tag{32}$$

4.1.3 Delay

Total hops (p) energetic for routing to the total network nodes (tn) is given by this ratio.

$$\mathrm{Delay} = \frac{p}{tn} \tag{33}$$

Less delay provides effective network routing.

4.1.4 Network Lifetime

Due to the fact that a node's lifespan must be maximized for effective routing, this is the reason for this.

$$\text{Network Lifetime} = \frac{1}{p} \times \sum_{n=1}^{p-1} \frac{M(N_n, N_{n+1})}{\beta} \tag{34}$$

4.1.5 Scalability

It ensures that no matter how large the network is, the performance will not be compromised.

$$\mathrm{Scalability} = \frac{\text{Performance of the network}}{\text{Network size}} \tag{35}$$

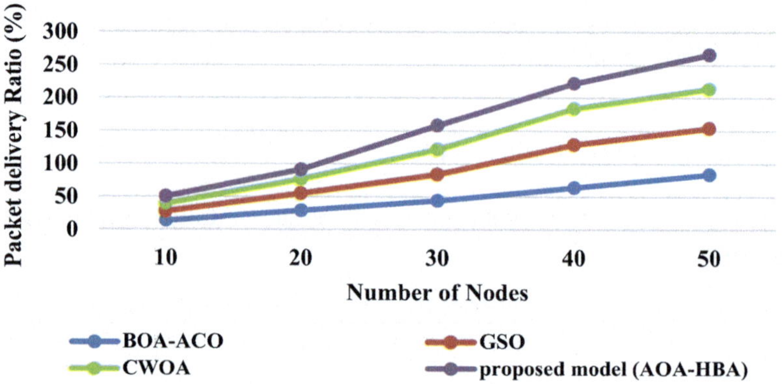

Fig. 2 PDR among existing and projected approaches comparison

4.2 *Performance Analysis of Proposed Model*

In this segment, the AOA-HBA is associated with existing techniques namely BOA-ACO [22], GSO [21], and CWOA [20], where these algorithms are implemented in our system model to test the efficiency of the proposed ideal. The graphical comparison between these algorithms by means of PDR is shown in Fig. 2.

When the number of nodes is less, the PDR for all techniques is less; for instance, BOA-ACO achieved 23%, GSO achieved 35%, CWOA achieved 40%, and proposed model achieved 50%. When the nodes are high, the PDR is increased for all techniques. For instance, proposed model achieved 270% of PDR, CWOA achieved 220% of PDR, GSO achieved 150% of PDR, and BOA-ACO achieved nearly 100%. The reason is that BOA-ACO uses more number of different parameters to select the optimal solution for fitness function. But, the proposed model uses less number of parameters to find the optimal solution for CH selection and shortest path between BS and CH. Therefore, the proposed model achieved better performance than existing techniques. Figure 3 provides the comparative analysis of various techniques in terms of network lifetime.

This section compares the proposed methodology to existing methods in terms of network longevity. The amount of rounds until the total amount of nodes exhausts their energy is used to determine the network's lifetime in this study. When the node is 20, the lifetime of each model includes BOA-ACO, GSO, CWOA, and proposed model achieved 14%, 22%, 26%, and 30%. These same techniques achieved high network lifetime, i.e., 52%, 60%, 70%, and 85%, when the node is 50.We know that the first node to go down has no effect on the network's performance; however, if half of the nodes go down while data is being transmitted, things start to go downhill. Furthermore, when the last node in the network dies, the entire network ceases to function. Due to its energy-efficient CH assortment and optimized path design, the suggested technology has increased network lifetime. The WSN sensor nodes' energy

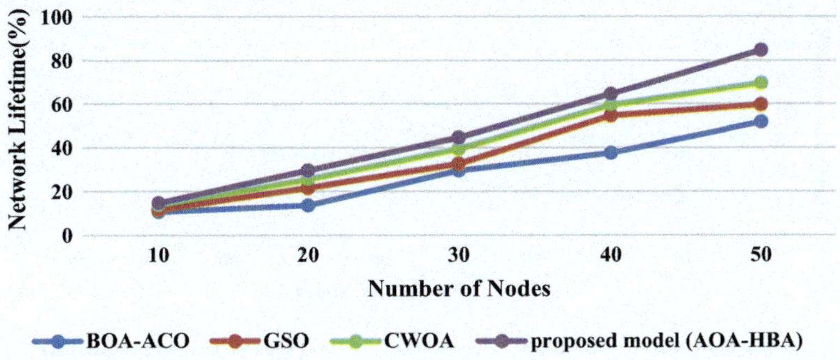

Fig. 3 Plot for network lifetime among existing and proposed approaches comparison

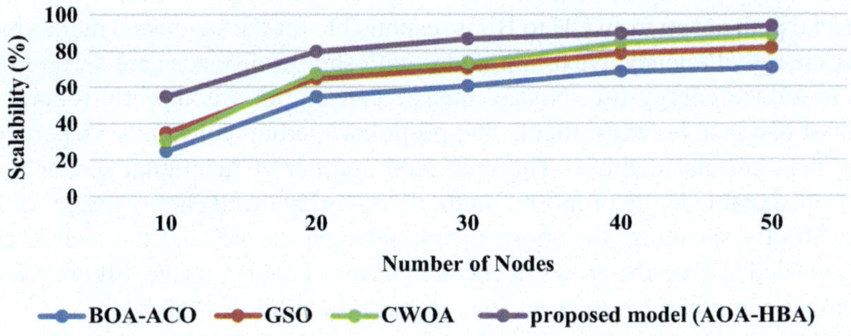

Fig. 4 Comparison of scalability among existing and proposed approaches

is conserved when an optimal CH selection, and path creation is applied during data transmission.. The scalability comparison for all techniques is depicted in Fig. 4.

The number of nodes on the X axis and the percentage of those nodes on the Y axis are shown on the graph. GSO and CWOA, on the other hand, attained about 85 percent accuracy when compared to the existing approach BOA-ACO, while the proposed method achieves 94% for the node 50. The scalability of the BOA-ACO is gradually increased, where the proposed model and other existing techniques increased its stability rapidly, when the node is less. However, the existing CWOA and GSO provide low performance due to local optimum issues and slow convergence speed. The proposed model performs better because the proposed methodology uses a fitness function that minimizes data transmission losses. Energy consumption comparison for proposed model with various existing techniques is provided in Fig. 5.

The proposed methodology with various nodes was used to determine the average energy consumption. Based on these numbers, it can be stated that the proposed attitude is more energy-efficient than BOA-ACO, GSO, and CWOA. Due to its one hop data transfer and random CH selection, GSO requires more energy. Energy consumption is increased because of the lack of consideration of distance while selecting a CH for usage in CWOA. An optimal CH selection from a set of nodes and

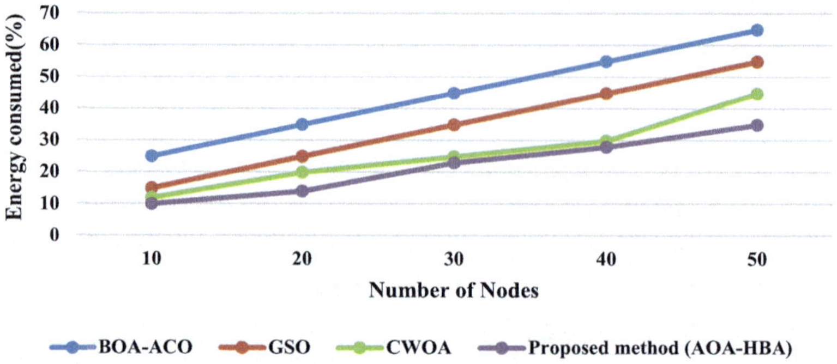

Fig. 5 Comparison of energy between existing and proposed approaches

an optimal path group from CH to BS is responsible for the suggested methodology's greater energy efficiency. Next hop nodes with shorter distances are selected in this study to reduce energy use. Nodes' energy consumption is directly related to the broadcast distance between them. The proposed method consumes 35 percent less energy than current methods. The increased number of functional sensor nodes is mostly attributable to the proposed methodology's balanced energy usage for sensor nodes. Strictly speaking, the shortest link between the BS and the source node is what's needed to keep the nodes in check in terms of energy usage. Figure 6 presents the comparison graph between various techniques in terms of delay.

The delay is directly propositional to the performance of the models. If less delay is achieved, the performance is high. For instance, the BOA-ACO delay is gradually increased, when the node number is increased. BOA-ACO achieved 55%, 65%, 75%, 85%, and 91%, when the node is 10, 20, 30, 40, and 50. CWOA achieved 55%, 68%, and 71% when the node is 20, 40, and 50.However, the delay is high for GSO, while comparing with CWOA. For example, GSO achieved 64%, 78% and 80%, when the node is 20, 40, and 50. While comparing with these techniques, the proposed model achieved less delay that includes 43%, 52%, 59%, 65%, and 69%, when the nodes

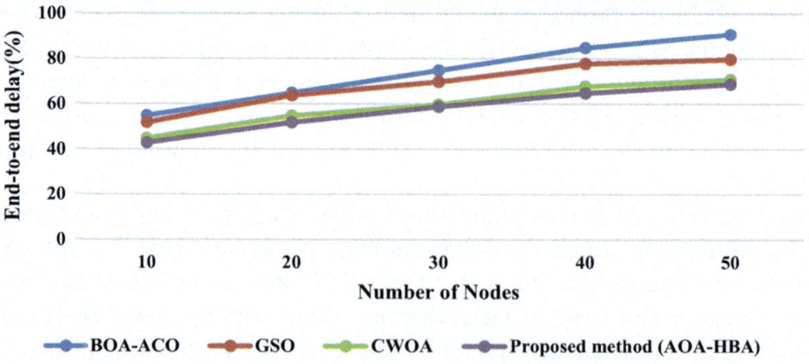

Fig. 6 Comparison of delay among existing and proposed approaches

are 10 to 50. Due to a lack of network lifetime and an incorrect CH selection, the current approaches have a large delay time and routing overhead.

5 Conclusion

In WSN, the collection of the best CH and the development of the most efficient route are considered tough problems. With the help of AOA and HBA, the overall energy consumption is reduced to upsurge the network's lifespan. Only two criteria, such as node residual energy and distance to neighbors, are used in AOA-based CH selection. Using this fitness function, we may narrow down the list of nodes to find the best candidate for a CH. Distance and residual energy can be used to control the most energy-efficient route decision. When compared to existing algorithms like BOA-ACO and GSO and CWOA, the results reveal that the suggested methodology has a longer network life. Using the proposed approach, 25 percent of energy is saved, 78 percent of packet delivery ratio is improved, 34.2% of delay is reduced, 54 percent of network lifetime is increased, and 85 percent of scalability is achieved. It is possible to incorporate more fitness functions into the suggested cluster-based routing algorithm in the future. Hybrid optimization approaches are also useful for increasing search speed.

References

1. Sankaralingam SK, Nagarajan NS, Narmadha AS (2020) Energy aware decision stump linear programming boosting node classification based data aggregation in WSN. Comput Commun
2. Maheshwari P, Sharma AK, Verma K, Energy efficient cluster based-routing protocol for WSN using butterfly optimization algorithm and ant colony optimization. Ad Hoc Netw 110:102317
3. Singh R, Verma AK (2017) Energy efficient cross layer based adaptive threshold routing protocol for WSN. AEU-Int J Electron Commun 72:166–173
4. Elkamel R, Cherif A, Elkamel R, Cherif A, Elkamel R, Cherif A (2017) Energy-efficient routing protocol to improve energy consumption in wireless sensors networks: energy efficient protocol in WSN. Int J Commun Syst 30(6)
5. Sabor N, Abo-Zahhad M, Sasaki S, Ahmed SM (2016) An unequal multihop balanced immune clustering protocol for wireless sensor networks. J Appl Soft Comput 43:372–389
6. Ke W, Yangrui O, Hong J, Heli Z, Xi L (2016) Energy aware hierarchical cluster-based routing protocol for WSNs. J China Univ Posts Telecommun 23(4):46–52
7. Rao PS, Jana PK, Banka H (2017) A particle swarm optimization based energy efficient cluster head selection algorithm for wireless sensor networks. Wireless Netw 23(7):2005–2020
8. Meena G, Dhanwal B, Mahrishi M, Hiran KK (2021) Performance comparison of network intrusion detection system based on different pre-processing methods and deep neural network. In: Proceedings of the international conference on data science, machine learning and artificial intelligence (DSMLAI'21). Association for Computing Machinery, New York, NY, USA, pp 110–115. https://doi.org/10.1145/3484824.3484878
9. Moh'd Alia O (2018) A dynamic harmony search-based fuzzy clustering protocol for energy-efficient wireless sensor networks. Ann Telecommun 73(5–6):353–365

10. Yuste-Delgado AJ, Cuevas-Martinez JC, Triviño-Cabrera A (2019) EUDFC-enhanced unequal distributed Type-2 fuzzy clustering algorithm. IEEE Sens J 19(12):4705–4716
11. Bamasaq O, Alghazzawi D et al (2022) Distance matrix and Markov Chain based sensor localization in WSN. CMC-Comput, Mater Continua 71(2):4051–4068. https://doi.org/10.32604/cmc.2022.023634
12. Mann PS, Singh S (2017) Improved metaheuristic based energy-efficient clustering protocol for wireless sensor networks. Eng Appl Artif Intell 57:142–152
13. Kaur S, Mahajan R (2018) Hybrid meta-heuristic optimization based energy efficient protocol for wireless sensor networks. Egypt Inf J 19(3):145–150
14. Khabiri M, Ghaffari A (2018) Energy-aware clustering-based routing in wireless sensor networks using cuckoo optimization algorithm. Wireless Pers Commun 98(3):2473–2495
15. Daneshvar SMMH, Mohajer PAA, Mazinani SM (2019) Energy-efficient routing in WSN: a centralized cluster-based approach via grey wolf optimizer. IEEE Access 7:170019–170031
16. Li X, Keegan B, Mtenzi F, Weise T, Tan M (2019) Energy-efficient load balancing ant based routing algorithm for wireless sensor networks. IEEE Access 7:113182–113196
17. Saad E, Elhosseini MA, Haikal AY (2019) Culture-based artificial bee colony with heritage mechanism for optimization of wireless sensors network. Appl Soft Comput 79:59–73
18. Xiuwu Y, Qin L, Yong L, Mufang H, Ke Z, Renrong X, Uneven clustering routing algorithm based on glowworm swarm optimization. Ad Hoc Netw 93, Art. no. 101923
19. Xu C, Xiong Z, Zhao G, Yu S (2019) An energy-efficient region source routing protocol for lifetime maximization in WSN. IEEE Access 7:135277–135289
20. Shende DK, Sonavane SS (2020) CrowWhale-ETR: CrowWhale optimization algorithm for energy and trust aware multicast routing in WSN for IoT applications. Springer Wireless Networks, pp 1–9
21. Sampathkumar A, Mulerikkal J, Sivaram M (2020) Glowworm swarm optimization for effectual load balancing and routing strategies in wireless sensor networks. Springer Wireless Networks, vol 21, pp 1–12
22. Maheshwari P, Sharma AK, Verma K (2021) Energy efficient cluster based routing protocol for WSN using butterfly optimization algorithm and ant colony optimization. Ad Hoc Netw 110:10231
23. Abualigah L, Yousri D, Abd Elaziz M, Ewees AA, Al-Qaness MA, Gandomi AH (2021) Aquila optimizer: a novel meta-heuristic optimization algorithm. Comput Ind Eng 157:107250
24. Hashim FA, Houssein EH, Hussain K, Mabrouk MS, Al-Atabany W (2022) Honey badger algorithm: new metaheuristic algorithm for solving optimization problems. Math Comput Simul 192:84–110

Chapter 22
Transfer Learning of Pre-trained CNN Models for Meitei Mayek Handwritten Character Recognition

Deena Hijam and Sarat Saharia

1 Introduction

The introduction of various deep neural networks in recent years has been proven to be a milestone in the field of image classification. One that stands out with regard to HCR especially is the CNN. CNN is a variant of artificial neural networks (ANNs) whose unique architecture and working is proven best for image classification and computer vision tasks in which the input images are two-dimensional. Much like traditional neural networks, CNNs consist of neurons with trainable weights and biases. However unlike the ordinary NNs, the structure of CNN makes it possible for this particular variant of NN to deal with much fewer parameters and to share parameter weights. The architecture of a CNN imitates the pattern in which neurons are connected in the human brain and is analogous to the organization of animal visual cortex [1–3]. Before the advent of CNNs, traditional methods of feature extraction using different techniques and classifiers were used for image classification. Unlike those techniques that rely on handcrafted features, CNNs can learn the features on its own and perform the task of classification in one entity. This provides for a more robust solution for image classification, thereby eliminating the additional problem of feature engineering.

Despite their success, deep neural networks suffer from overfitting problem if sufficient data is not there to train them [4]. They require sufficiently large training data in order to train their multiple layers and large number of parameters. Due to their data-hungry nature, one of the challenges that is facing deep neural networks is the development of large annotated databases which is a tedious and expensive task.

D. Hijam (✉) · S. Saharia
Tezpur University, Napaam, Assam 784028, India
e-mail: deenahi@tezu.ernet.in

S. Saharia
e-mail: sarat@tezu.ernet.in

V. Devedzic et al. (eds.), *Proceedings of the International Conference on Intelligent Computing, Communication and Information Security*, Algorithms for Intelligent Systems, https://doi.org/10.1007/978-981-99-1373-2_22

One accomplishment that the researchers have been able to achieve for mitigating this problem is the introduction of transfer learning [5]. It is the process of exploiting what has been learned for one problem to work for a different problem [6]. The concept behind transfer learning is to transfer the features learned by a base network which is trained on a base dataset, to a new target network. The new target network is then trained on a target dataset using the transferred features of the base network [7]. Popular deep CNN models [8–10] are pre-trained on the very large ImageNet dataset [11]. These pre-trained models serve as the baseline model available to be fine-tuned according to the target dataset. Since its inception, transfer learning has been successfully employed in a number of image classification tasks [12, 13]. The present paper seeks to analyze the performance of five state-of-the-art CNN models using transfer learning on a newly developed handwritten Meitei Mayek character dataset.

The sections that follow are organized as follows: Sect. 2 presents an account of the Meitei Mayek script and Sect. 3 describes CNN and the five pre-trained CNN models. The setup and results of the experiments carried out are shown in Sect. 4. Finally, conclusion of the paper is given in Sect. 5.

2 Meitei Mayek

There are only a handful of scripts for the languages spoken in the north-eastern states of India. Meitei Mayek is one of them and it is used for writing Manipuri language. Manipuri is predominantly spoken in the state of Manipur apart from some parts of Assam and Tripura. Some sections of Manipuri-speaking people can also be found in parts of Myanmar as well. It is believed that the script had been in use from the 11th century to the 17th century. Its existence started to cease in the early 18th century when the Bengali script was introduced. However, the script saw a resurrection in the 1980s. It was considered one of the formerly endangered scripts as per the "Atlas of Endangered Alphabets".[1] According to "The Manipur Official Language (Amendment) Act, 2021",[2] the Government of Manipur allowed the official use of Meitei Mayek concurrently with Bengali script.

Meitei Mayek has a total of 56 letters out of which 55 are used for writing. There are 18 original consonant letters and 9 borrowed ones making it a total of 27 consonant letters. Additionally, there are 8 vowels, 8 final consonants, 3 punctuation marks, and 10 numerals. The Meitei Mayek was added to the Unicode Standard in October, 2009, with the release of version 5.2. The complete character set of Meitei Mayek script is shown in Fig. 1.

---

1 https://www.endangeredalphabets.net/alphabets/meitei-mayek/.

2 http://manipurgovtpress.nic.in/en/details_gazzete/?gazette=658.

	0	1	2	3	4	5	6	7	8	9	A	B	C	D	E	F
U+ABCx	ꯀ	ꯁ	ꯂ	ꯃ	ꯄ	ꯅ	ꯆ	ꯇ	ꯈ	ꯉ	ꯊ	ꯋ	ꯌ	ꯍ	ꯎ	ꯏ
U+ABDx	ꯐ	ꯑ	ꯒ	ꯓ	ꯔ	ꯕ	ꯖ	ꯗ	ꯘ	ꯙ	ꯚ	ꯛ	ꯜ	ꯝ	ꯞ	ꯟ
U+ABEx	ꯠ	ꯡ	ꯢ	ꯣ	ꯤ	ꯥ	ꯦ	ꯧ	ꯨ	ꯩ	ꯪ	꯫	꯬	꯭		
U+ABFx	꯰	꯱	꯲	꯳	꯴	꯵	꯶	꯷	꯸	꯹						

Fig. 1 Complete character set of Meitei Mayek script with the assigned UNICODE values. *Image source* Wikipedia

3 Convolutional Neural Network

A typical CNN consists of three main layers, viz. convolutional, pooling, and fully connected layers which are assembled one after another. The architecture in regard to the number and order of layers in which they are put together is a design choice of the network designer. The three basic layers of a CNN are described below.

Convolutional Layer

As the name suggests, *convolution* operation is carried out in this layer. A defined area of input image (receptive field) is convolved with a two-dimensional array of weights known as filter/kernel. In regard to CNNs, a convolution operation is the element-wise multiplication between the defined area which is the filter-sized region of the input space and the filter. Convolution is carried out systematically on every filter-sized patch of the input image by moving left to right with a defined stride value till it covers the entire width of the input image. It then moves down to the beginning of the image with the same stride value and continues till the entire width of the image is parsed and so on until the whole width and height of the image are covered. Each convolution of the filter with an input patch produces a single value. When this operation is carried out multiple times using the same filter over all the possible patches of the input image, a two-dimensional array is produced. This two-dimensional output array is called an activation map or a feature map. Every filter is applied across the whole image. This allows a filter to detect a specific feature anywhere in the entire input image. This is the most important idea behind convolutional layer. Feature maps for all filters are then stacked to form the output of a particular convolutional layer. These feature maps act as input for the next layer in the network.

Pooling Layer

The pooling layer generally follows a series of one or more convolutional layers. This layer performs down-sampling of its input data. Input to this layer is the output feature maps of the previous layer. Although convolution facilitates in the detection of features anywhere in the input image, the issue with output feature maps is that they are not very robust when there is a change in the position of the features. This issue is addressed by the process of down-sampling the feature maps. Down-sampled feature maps reduce sensitivity toward change in the location of features in the input image. Pooling is carried out in patches of the feature maps. The down-sampled feature maps make the representation invariant to translations and rotations of the input image [6]. The resulting robustness that the pooling layer brings about is termed as *local translation invariance*. There are two common types of pooling: max-pooling [14] and average pooling [15]. For each patch on a feature map, max-pooling takes the maximum value while average pooling takes the average value.

Fully Connected Layer

The layers that come at the end of a CNN architecture are the fully connected layers. They can be one or more in number. This layer is the classification layer which works in the same manner as the traditional feed-forward neural network layer. The extracted features are flattened to a column vector and fed to the classification layer. Learning by the network takes place by applying backpropagation over every iteration of training phase. They have a non-linear activation function or a softmax function to predict the classes of input images.

3.1 State-of-the-Art CNN Models

In the present study, five CNN models are studied which are briefly described below:

- *InceptionV3 (2015)* [10]: InceptionV3 is CNN model developed by Google after its two previous variants namely InceptionV1 (GoogleNet) and InceptionV2. It is popular because of its computational efficiency and fewer parameters compared to earlier models like AlexNet and VGG with a lower error rate. It achieved a top-5 accuracy of 93.7% on ImageNet validation dataset.
- *ResNet50V2 (2016)* [16]: ResNet comes from Microsoft. The main contribution of ResNet architectures is that it reduces vanishing gradient problem in deep neural networks. The idea is to introduce "residual blocks" which have "skip connections" where there are direct connections between layers by skipping some layers in between. The skip connections allow an alternate shorter path for the gradient to flow through thereby solving the problem of vanishing gradient. 93.0% is the top-5 validation accuracy achieved on ImageNet validation dataset.
- *DenseNet121 (2017)* [17]: DenseNets also aim to fight the vanishing gradient problem which is a very common problem of deep neural networks by simplifying the connectivity pattern between the layers. They do so by allowing maximum gradient flow between the layers by directly connecting each layer with all other deeper layers. This way, they require fewer parameters to learn and avoid learning redundant feature maps. Because of its very narrow layers with less number of filters, the layers add only a small number of new feature maps. ImageNet validation dataset shows a top-5 accuracy of 92.3%.

- *MobileNetV2 (2018)* [9]: It is another recent architecture laid out by Google. It is built upon its earlier variant called MobileNetV1. Its main feature is its ability to reduce the complexity cost and network depth for the benefit of devices with low computational power like mobiles devices while giving better accuracy. The authors could achieve this by introducing two new features on top of the depthwise separable convolution introduced in MobileNetV1. The two new features are (1) linear bottlenecks and (2) shortcut connections between the them. It achieves a top-5 accuracy of 90.1% on the ImageNet validation dataset.
- *EfficientNetB3 (2019)* [18]: EfficientNets are the latest CNN architecture from Google. The authors proposed that both accuracy and computational efficiency could be achieved by adopting similar architectures. They claimed that a common architecture with small values of width, depth, and resolution can create a computationally efficient model, and by keeping these parameters bigger, a better accuracy could be achieved. They perform the best for image classification tasks under varying resource availability situations. 95.7% is the top-5 validation accuracy achieved on ImageNet dataset.

4 Experimental Setup and Results

The experimentations are carried out on Google Colab with an allocated RAM size of 13 GB and 2.20 GHz Intel(R) Xeon(R) CPU. The implementations are done in Python using Tensorflow with Keras backend.

4.1 Dataset Details

A publicly available dataset of Meitei Mayek called TUMMHCD is considered for our work [19] which has a total of 85,124 images. The training set comprises 85% of the images selected randomly, and the remaining 15% images forms the test set. The image samples are grayscale in nature. Sample images of some of the character classes are shown in Fig. 2.

Fig. 2 Sample images of TUMMHCD

Table 1 Transfer learning and fine-tuning of the five CNN models

CNN model	Input image size	Total no. of layers in base model	Layer fine-tuning starts
InceptionV3	75 × 75	311	250
ResNet50V2	32 × 32	190	140
DenseNet121	32 × 32	427	370
MobileNetV2	32 × 32	154	100
EfficientNetB3	32 × 32	384	330

4.2 *Transfer Learning of Pre-trained CNN Models*

Layers from the pre-trained model apart from the batch-normalization layers are frozen, and new top layers are added in place of the top fully connected layer. We have added a global average pooling layer, a dropout layer of 0.2, and output layer of 55 nodes since our target dataset has 55 character classes. We then train the base model with newly added top layers on our dataset. We have also carried out fine-tuning as our dataset has significant number of images in each class. For fine-tuning, we unfreeze some of the top layers of the base model to get a new model. The target dataset is then used to re-train the new model. The summary is given in Table 1.

For training and validation purpose, training set of 72,330 images is divided randomly in the ratio 9:1 into training set and validation set. Training set and validation set thus consist of 65,097 and 7233 character images, respectively. Training in deep neural networks faces the problem of overfitting or overtraining the network. This means that when a neural network is overtrained with the training set with more number of epochs than what is required, it tends to suffer from overfitting. Overfitting is undesirable as it causes the network to overfit the training data and resulting to perform badly with the unseen data or the test data. On the other hand, undertraining the network is again undesirable as it might result in an underfit model. In order to mitigate the problem of overfitting or underfitting, we have used the early stopping method during the network training. It is done so in order to train the network long enough on the training set but halt the training when its performance on the validation set begins to degrade. Early stopping achieves this task by allowing us to set an arbitrarily large number of epochs for training the network but letting the training to stop when the validation accuracy starts to decrease or validation loss starts to increase.

For the present work, the validation loss is considered as the criterion for monitoring the model performance. The validation loss is monitored for a set of 10 consecutive epochs each time. When the validation loss is not decreasing for a consecutive 10 epochs, then the model weights for the epoch with minimum validation loss are saved as the best model and the training stops. If, however the validation loss decreases at a certain epoch during this 10-epoch monitoring, then that particular epoch is set as the first epoch for the next 10-epoch set and so on. This process

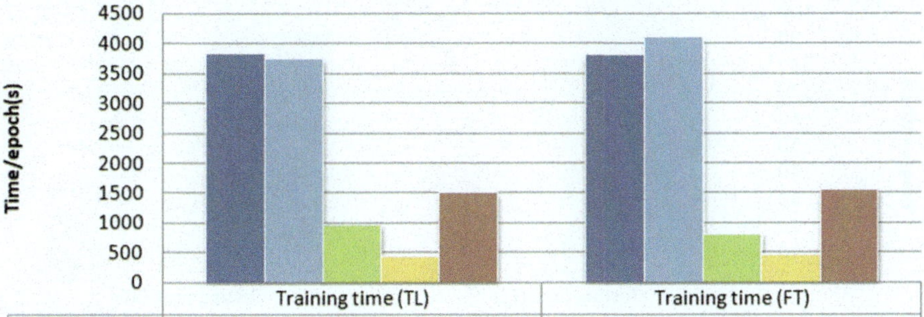

	Training time (TL)	Training time (FT)
InceptionV3	3844	3825
ResNet50V2	3768	4134
DenseNet121	994	826
MobilenetV2	447	465
EfficientNetB3	1523	1574

Fig. 3 Time taken for training the models for transfer learning (TL) and after fine-tuning (FT)

continues until the epoch with minimum validation loss is found else the training continues for a maximum of 30 epochs. Once this best model is found, it is used to test the accuracy on the test set. The models use an initial learning rate of 0.001 during the first 20 epochs using adam optimizer. It is then reduced to 0.00001 for fine-tuning part. For each model, the corresponding pre-processing is applied to the images in dataset before feeding them to the models.

4.3 Experimental Results and Discussion

The amounts of time taken per epoch by each model to train the model on the target dataset using transfer learning and fine-tuning are shown in Fig. 3. It is seen that ResNet50V2 takes maximum time with an average of 3950 s and while MobileNetV2 takes the minimum time with an average of 456 s. The same figures for testing are given in Fig. 4 where it is observed that the average time taken to test on the test set by InceptionV3 is the highest (88 s) and that of the MobileNetV2 is the lowest (11 s). From these two tables, it can be concluded that MobileNetV2 performs relatively better than the other models with respect to the training and testing times.

Figure 5 shows the validation accuracy achieved for transfer learning and fine-tuning by the five models. The highest accuracy of 95.41% is achieved using EfficientNetB3 and the lowest by ResNet50V2 with 88.49%. The same pattern is seen with the test accuracy provided in Fig. 6. EfficientNetB3 and ResNet50V2 give the highest and the lowest test accuracies of 95.21% and 88.49%, respectively. The experimental result is evidence that fine-tuning the model is beneficial for enhancing the test accuracies in all the five models. Infact in case of DenseNet121, there is a huge jump in test accuracy from just 17.18% to 90.81% with fine-tuning.

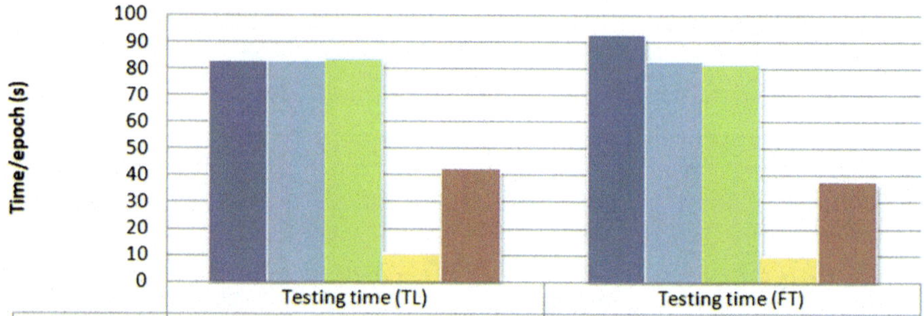

	Testing time (TL)	Testing time (FT)
InceptionV3	83	93
ResNet50V2	83	83
DenseNet121	84	82
MobilenetV2	11	10
EfficientNetB3	43	38

Fig. 4 Time taken by the models on the test set for transfer learning (TL) and after fine-tuning (FT)

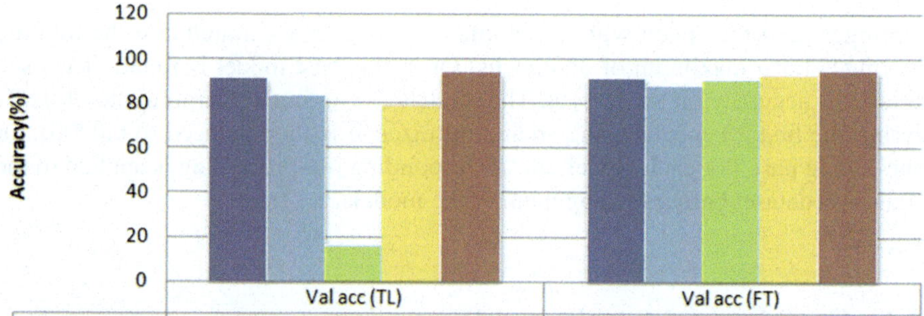

	Val acc (TL)	Val acc (FT)
InceptionV3	91.9	91.9
ResNet50V2	83.9	88.49
DenseNet121	16.6	91.15
MobilenetV2	79.9	93.51
EfficientNetB3	94.64	95.41

Fig. 5 Validation accuracy of the models using transfer learning (TL) and after fine-tuning (FT)

From the experimental results, one can infer that deeper models do not always give better accuracy. MobileNetV2 with a total of only 154 layers outperforms DenseNet121 which has 427 layers with a margin of 5.45% in test accuracy. DenseNet121 also could not perform better than the other shallower models, namely InceptionV3 and EfficientNetB3. One observation is that up-scaling the input-image size increases memory usage during training for large neural networks. For some pre-trained models, there is a minimum input size of the images which should be used in order to use the model. For example, the minimum input size of InceptionV3 pre-trained model is 75×75 which increases the requirement of resources during training.

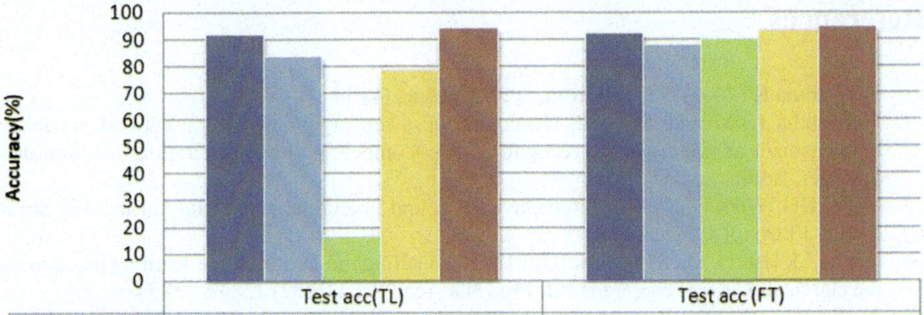

	Test acc(TL)	Test acc (FT)
InceptionV3	92.22	92.56
ResNet50V2	84.13	88.49
DenseNet121	17.18	90.81
MobilenetV2	79.47	93.94
EfficientNetB3	94.36	95.21

Fig. 6 Test accuracy of the models using transfer learning (TL) and after fine-tuning (FT)

It is also observed that MobileNetV2 is the most computational inexpensive while EfficientNetB3 gives the best test accuracy. EfficientNetB3 also is relatively faster compared to the other three models with a testing time of 38 s. Also, MobileNetV2 is more accurate compared to the other three models with a test accuracy of 93.94%. On an average, an inference that can be drawn from the experimental results is that MobileNetV2 and EfficientNetB3 perform the best for our new target dataset with a trade-off between the speed and test accuracy achieved by these two models.

5 Conclusion

The success of deep neural networks has given a boost to many computer vision tasks, image classification being one of its most important sub-tasks. The availability of pre-trained CNN models has only enhanced the performance of such tasks. In this study, the performances of five pre-trained CNN models using tranfer learning and fine-tuning have been tested for a new public image dataset named TUMMHCD. Certain observations are made with respect to the models on the target dataset and it is concluded that MobileNetV2 and EfficientNetB3 provide the best performance with test accuracies of 93.94% and 95.21%, respectively.

References

1. Fukushima K (2007) Neocognitron. Scholarpedia 2(1):1717
2. Fukushima K, Miyake S (1982) Neocognitron: a self-organizing neural network model for a mechanism of visual pattern recognition. In: Competition and cooperation in neural nets. Springer, Berlin, pp 267–285
3. Hubel DH, Wiesel TN (1968) Receptive fields and functional architecture of monkey striate cortex. J Physiol 195(1):215–243
4. Belkin M, Hsu D, Ma S, Mandal S (2019) Reconciling modern machine-learning practice and the classical bias-variance trade-off. Proc Nat Acad Sci 116(32):15849–15854
5. Pan SJ, Yang Q (2009) A survey on transfer learning. IEEE Trans Knowl Data Eng 22(10):1345–1359
6. Goodfellow I, Bengio Y, Courville A (2016) Deep learning. MIT Press, Cambridge
7. Arora A et al (2022) Web-based news straining and summarization using machine learning enabled communication techniques for large-scale 5G networks. Wirel Commun Mobile Comput 2022:3792816
8. Krizhevsky A, Sutskever I, Hinton GE (2012) Imagenet classification with deep convolutional neural networks. Advances in neural information processing systems 25
9. Sandler M, Howard A, Zhu M, Zhmoginov A, Chen LC (2018) Mobilenetv2: inverted residuals and linear bottlenecks. In: Proceedings of the IEEE conference on computer vision and pattern recognition, pp 4510–4520
10. Szegedy C, Vanhoucke V, Ioffe S, Shlens J, Wojna Z (2016) Rethinking the inception architecture for computer vision. In: Proceedings of the IEEE conference on computer vision and pattern recognition, pp 2818–2826
11. Deng J, Dong W, Socher R, Li LJ, Li K, Fei-Fei L (2009) Imagenet: a large-scale hierarchical image database. In: 2009 IEEE conference on computer vision and pattern recognition. IEEE, pp 248–255
12. Mahrishi M et al (2021) Chapter 7 Data model recommendations for real-time machine learning applications: a suggestive approach. De Gruyter, Berlin, Boston, pp 115–128. https://doi.org/10.1515/9783110702514-007
13. Shaha M, Pawar M (2018) Transfer learning for image classification. In: 2018 second international conference on electronics, communication and aerospace technology (ICECA). IEEE, pp 656–660
14. Nagi J, Ducatelle F, Di Caro GA, Cireşan D, Meier U, Giusti A, Nagi F, Schmidhuber J, Gambardella LM (2011) Max-pooling convolutional neural networks for vision-based hand gesture recognition. In: 2011 IEEE international conference on signal and image processing applications (ICSIPA). IEEE, pp 342–347
15. Boureau YL, Ponce J, LeCun Y (2010) A theoretical analysis of feature pooling in visual recognition. In: Proceedings of the 27th international conference on machine learning (ICML-10), pp 111–118
16. He K, Zhang X, Ren S, Sun J (2016) Identity mappings in deep residual networks. In: European conference on computer vision. Springer, Berlin, pp 630–645
17. Huang G, Liu Z, Van Der Maaten L, Weinberger KQ (2017) Densely connected convolutional networks. In: Proceedings of the IEEE conference on computer vision and pattern recognition, pp 4700–4708
18. Tan M, Le Q (2019) Efficientnet: rethinking model scaling for convolutional neural networks. In: International conference on machine learning. PMLR, pp 6105–6114
19. Hijam D, Saharia S (2021) On developing complete character set Meitei Mayek handwritten character database. The visual computer, pp 1–15

Chapter 23
GestureWorks—One Stop Solution

Umesh Gupta, Pransh Gupta, Tanishq Agarwal, and Deepika Pantola

1 Introduction

As of now, Fourth Industrial Revolution is going on, also called Industry 4.0. There has been a strong emphasis on automation, the convergence of physical and digital technologies, and the introduction of web 3.0 technologies such as Artificial Intelligence (AI) [1, 11], Deep Learning, Big Data, Cloud Computing, Robotics, Augmented Reality, IoT, and so on. All of these enhanced digital technologies have begun to integrate totally with our daily activities such as shopping, entertainment, gaming, web search, and advertisements [2–4]. Technologies are more reliant on machines, and as a result, there has been progressed in interactions with technology through a variety of methods such as gesture recognition, speech recognition, and so on. In the 1970s, Krueger introduced a novel kind of HCI [1, 2] called Gesture-based interaction, which is a type of non-verbal/non-vocal communication where the hands or face, can communicate a specific message through movement. Using Human–Computer Interaction (HCI) (2008) [1, 2] to recognize hand gestures might aid to attain the required ease and naturalness. When communicating with others, hand gestures have a significant role to play in transmitting the information. There's something for everyone, from easy hand motions to more typical ones.

U. Gupta (✉) · P. Gupta · T. Agarwal · D. Pantola
SCSET, Bennett University, Greater Noida, UP, India
e-mail: er.umeshgupta@gmail.com; umesh.gupta@ieee.org

P. Gupta
e-mail: pranshgupta18@gmail.com

T. Agarwal
e-mail: tanishq.agrawal15@gmail.com

D. Pantola
e-mail: deepika.pantola@bennett.edu.in

V. Devedzic et al. (eds.), *Proceedings of the International Conference on Intelligent Computing, Communication and Information Security*, Algorithms for Intelligent Systems, https://doi.org/10.1007/978-981-99-1373-2_23

It can utilize our hands of us to point to anything (a target or a person) or hand gestures expressed through physical articulations matched with their syntax & vocabulary, and also sign languages. As a result, individuals may interact more intuitively by employing hand gestures (2017) as a device and then integrating them with computers [10]. Additionally, using voice commands leads to a completely hands-free navigation experience in which a user can communicate easily either through hand gestures or voice navigation with minimal hardware contact. Gesture Controlled Virtual Mouse simplifies human–computer interaction by utilizing hand gestures and Voice Commands. Almost no direct contact is required with the computer [2, 5, 6]. All I/O operations can be controlled remotely via static and dynamic hand gestures, as well as a voice assistant. This project employs cutting-edge Machine Learning and Computer Vision algorithms to recognize hand gestures and vocal instructions, and it works flawlessly without the usage of any additional hardware. It makes use of models such as CNN, which is implemented by Media Pipe and runs on top of pybind11. It is divided into two modules: One that acts directly on the hands by utilizing Media Pipe and is a hand-detecting bot, while the second is a voice recognition bot called "Dragon" [7–9]. Visualization of data of hand movements can aid in obtaining the required comfort and naturalness in HCI. As a result, many academics have been interested in CV-based interpretation and analysis of hand movements, which is a very active study topic. We looked at the research [8–12] on visual analysis of hand movements in the context of its significance in HCI, emphasizing some significant publications by scholars. The goal of this paper is to present the topic of gesture control as a tool for computer interaction.

2 Objectives

Using the results of a decade of research and development in the field of Deep Learning, especially gesture recognition, which is the process of converting motions to representations and subsequently to instructions for some purpose. Our goal is to develop a one-of-a-kind hands-free navigation program that incorporates a variety of input/navigation modalities such as hand gestures, speech recognition, and others (to be incorporated in the future). Hand gesture recognition has been utilized in the GestureWorks module of virtual mouse to identify certain hand movements as input and then process and map these motions as output for devices that operate as a virtual mouse managing the activities of the computer mouse [7].

A voice assistant listens to particular vocal commands and returns relevant information or performs specified actions as directed by the user, using speech recognition, language processing algorithms, and voice synthesis. To construct Voice Assistant in the Dragon, pyttsx3, Speech Recognition, and other modules are used. Our application (Figs. 1 and 2) will initially include three modules: gesture controller, voice assistant, and Bot. Following software execution, two methods of input will be available: direct hand movements and voice commands via the Dragon app. The virtual mouse, as well as other activities like voice control, screen brightness, selection, and

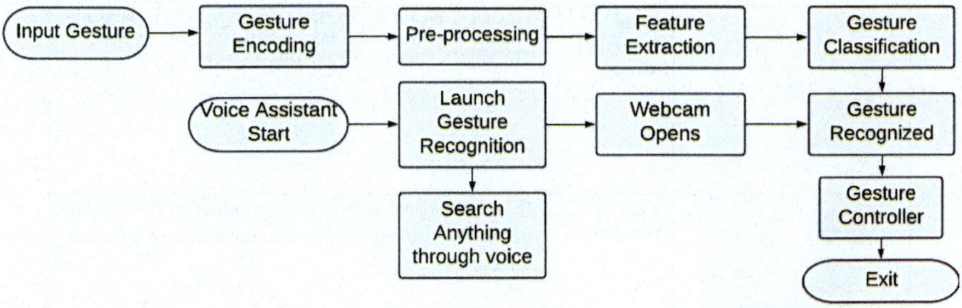

Fig. 1 Proposed methodology

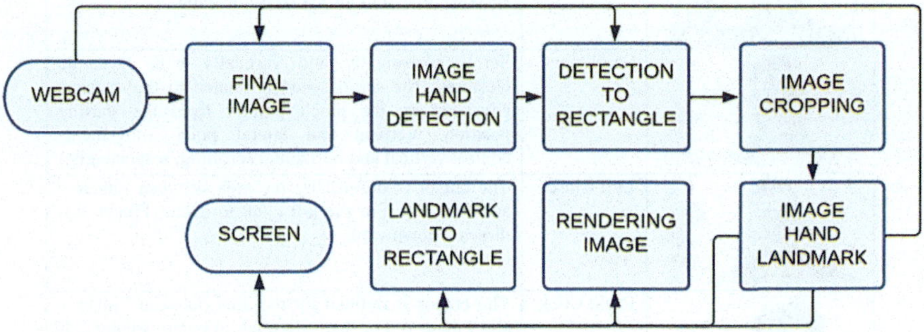

Fig. 2 Functionality of Media Pipe

so on, will be controlled by some hand gestures such as pinching, waving the hand, and synchronized movement of multiple separate fingers as shown in Fig. 3.

3 Proposed Methodology

Following input, either hand images/movements via webcam for virtual mouse or speech-and-text for voice assistant are shown. The needed functions are carried out in line with the requested hand symbols or input voice after pre-processing and feature extraction by the construction of the CNN model.

For the development of our software, various modules/libraries are used such as pyttsx3, pynput, pycaw, speech recognition, comtypes, pyautogui, but primarily OpenCV [6] and Media Pipe [5].

- OpenCV: It is a vision-based software package that covers object recognition and image processing algorithms. OpenCV is associated with an AI-related vision library written that is considered to develop real-time vision-related different applications. This package is extensively considered not only to analyze photos and videos but also to analyze different target detection [6].

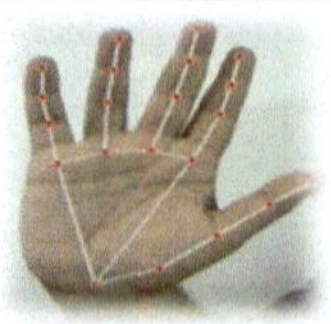	Hand Landmark Model	FINGER_1_MCP, PIP, DIP, TIP. FINGER_2_DIP,TIF, MCP, PIP. FINGER_3_MCP,PI,DIP AND TIP. THUMB_CMC, MCP, IP, TIP. WRIST. PINKY_MCP,PIP,DIF, AND TIP.
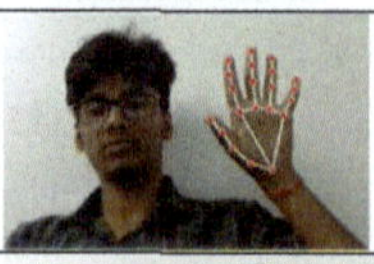	Neutral Gesture	It is used to pause/stop the execution of the current gesture by moving the hand in inwards and outwards direction.
	Move Cursor	The middle of Finger 1 and Finger 2 is designated as the cursor. The pointer is moved to the desired position with this motion. The cursor movement speed is related to the speed of the hand.
	Scrolling	Scroll horizontally and vertically with Dynamic Gestures. The scroll speed is related to the distance advanced by the pinch gesture from the starting position. Vertical and lateral pinch movements control vertical and horizontal scrolling, respectively.
	Left Click	The cursor is defined by the area between Finger 1 and Finger 2. For the left click to occur, Finger 1 is directed downward.
	Right Click	The cursor is defined by the area between Finger 1 and Finger 2. For the right click to occur, Finger 2 is directed downward.
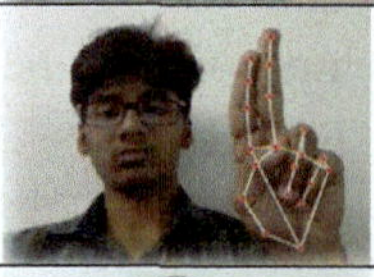	Double Click	For the double click to occur, Finger 1 and Finger 2 are joined together.
	Drag & Drop / Multiple Item Selection	Drag-and-drop functionality is demonstrated with this gesture. It is possible to use this function to move or transfer files from one location to another.
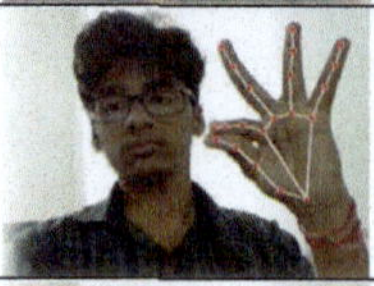	Volume Control	The rate of volume increase/decrease is related to the distance travelled by the pinch motion from the starting position(upwards/downwards).
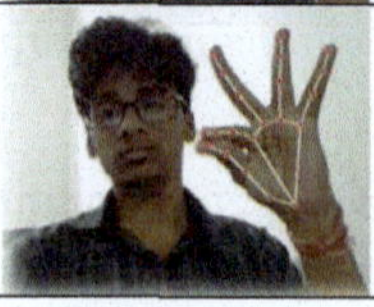	Screen Brightness	The rate of brightness increase/decrease is related to the distance traveled by the pinch motion from the starting point (left/right).

Fig. 3 Gestures representation

- Media Pipe: Developers frequently utilize the Media Pipe [4] framework to build and evaluate systems with graphical plots and also to create processes for application use. The pipeline [2] design is used to carry out the steps in a Media Pipe-based system. The Media Pipe system [7] is consist of 3 primary parts: efficiency measurement, second for gathering sensor data, and a library of reusable calculators.

- Hand Landmark: A hand landmark model is made up of 21 joints or knuckle points in the hand part. The precise placement of 21 important spots was identified using a 3D hand-knuckle [3] points (Fig. 3). It is accomplished inner the selected hand locations by utilizing regression, which will quickly offer the prediction of the location, that is a model of the hand landmark in Media Pipe. Each landmark's hand-knuckle has three coordinates where two are normalized to the range [0–1] by image width and height and the third represents the pointmark's depth.

4 Results

The notion of enhancing interaction with people using computer vision is presented below: Cross-validation of the GestureWorks is problematic due to the smaller number of datasets available. Gestures recognition has been tested in a variety of lighting situations, as well as at varied ranges from the webcam for monitoring and recognition of gestures. To compile the data reported in Table 1, an experimental test has been done. The test was carried out in several lighting conditions and at various distances from the screen, and each participant checked the GestureWorks. The findings of this numerical experiment are summarized in Table 1.

According to Tables 1 and 2 and Figs. 4 and 5, GestureWorks attained an accuracy of around 94% with the proposed work. It may conclude that GestureWorks functioned effectively based on its 94% accuracy. As seen in Table 1, the overall accuracy

Table 1 Result for dıfferent functıons

Fingertip gesture	Mouse functions performed	Success	Failure	Accuracy (%)
TD_1	Mouse movement	95	0	95
TD_2	Left button click	94	1	94
TD_3	Right button click	95	3	95
TD_4	Scroll up function	95	0	95
TD_5	Scroll down function	91	1	91
TD_6	No action performed	94	1	94
Result		564	6	94

TD_1 = UP[TIP = 1/(1&2)]; TD_2 = UP & DIST < 30 [TIP = 0&1]; TD_3 = UP & DIST < 40 [TIP = 1&2]; TD_4 = UP BOTH & DIST > 40 [TIP = 1&2UP]; TD_5 = DOWN BOTH & DIST > 40 [TIP = 1&2UP]; TD_6 = UP[TIP = ALL]

Table 2 Comparison between different models with accuracy

Existing models	Parameter	Accuracy
Virtual mouse system using RGB-D [7]	Tested under the same circumstances	92.13
Hand (Palm, Fingers) [8]	Tested under the same circumstances	78.0
Hand gesture-based virtual mouse [9]	Tested under the same circumstances	78.0
Proposed Gestureworks	Tested under the same circumstances	**94.0**

93.6% uniform background ǁ 75.45% with cluttered background
Bold denotes the highest accuracy of the model among the campared ones.

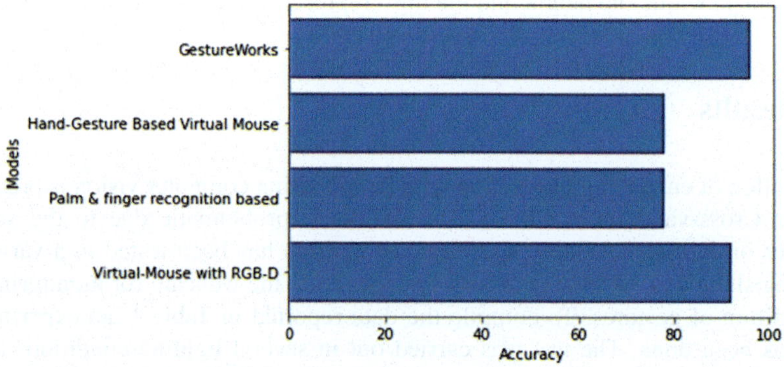

Fig. 4 Comparative performance analysis of different models

for "Right Click" is poor since this is the most difficult motion for a computer to grasp. Because the gesture utilized to achieve the specific mouse function is more difficult, overall accuracy for right-clicking is low. Also, the accuracy for all other motions is good and high. In comparison to earlier techniques for web mouse, our model performed exceptionally well, with 94 percent accuracy (Code is available at https://github.com/GestureWorks2022/GestureWorks2022).

5 Conclusion

The primary goal of GestureWorks is to operate mouse cursor functionalities with hand gestures rather than a hardware mouse. The proposed model may be realized by having a webcam or an in-built camera that identifies hand motions and tips and analyses these frames to run the specific mouse tasks. Based on the model's findings, it concludes that the GestureWorks worked extremely well and also more

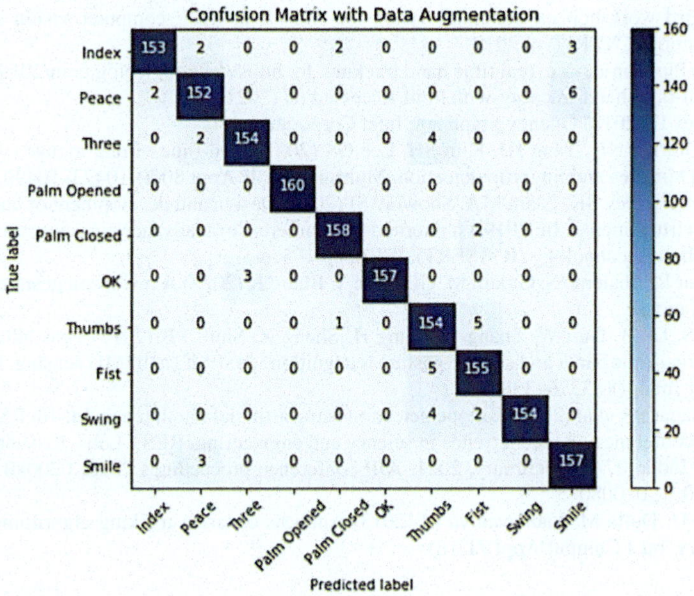

Fig. 5 Confusion matrix with data augmentation

efficient than current models. The majority of the limitations of the present systems are dealt with by the proposed model. GestureWorks may be considered for real-world applications as the proposed model is working more precisely, and it can also be a consideration to minimize the spread of COVID-19 due to it can be operated virtually using hand movements rather than the standard physical mouse. The model has various shortcomings, such as a little loss in precision in the right-clicking mouse operation and some difficulty in simply dragging and dropping to pick text. As a result, it will have to strive next to address these constraints by upgrading the fingertip-detecting algorithm to give more accurate outcomes.

References

1. Pantic M, Nijholt A, Pentland A, Huanag TS (2008) Human-centred intelligent human? Computer interaction (HCI^2): how far are we from attaining it? Int J Auton Adapt Commun Syst 1(2):168–187
2. Zhang F, Bazarevsky V, Vakunov A, Tkachenka A, Sung G, Chang CL, Grundmann M (2020) Mediapipe hands: on-device real-time hand tracking. arXiv preprint arXiv:2006.10214
3. Kumar T et al (2022) A review of speech sentiment analysis using machine learning. In: Kaiser MS, Bandyopadhyay A, Ray K, Singh R, Nagar V (eds) Proceedings of trends in electronics and health ınformatics. Lecture Notes in Networks and Systems, vol 376. Springer, Singapore
4. Lugaresi C, Tang J, Nash H, McClanahan C, Uboweja E, Hays M, Zhang F, Chang CL, Yong M, Lee J, Chang WT (2019) Mediapipe: a framework for perceiving and processing reality.

In: Third workshop on computer vision for AR/VR at IEEE computer vision and pattern recognition (CVPR) (vol 2019)
5. Media Pipe: on-device, real time hand tracking. In: https://ai.googleblog.com/2019/08/on-device-real-time-hand-tracking-with.html Accessed on (2021).
6. Matveev D (2018) Opencv graph api. Intel Corporation 1
7. Tran DS, Ho NH, Yang HJ, Kim SH, Lee GS (2021) Real-time virtual mouse system using RGB-D images and fingertip detection. Multimed Tools Appl 80(7):10473–10490
8. Shibly KH, Dey SK, Islam MA, Showrav SI (2019) Design and development of hand gesture-based virtual mouse. In: 2019 1st ınternational conference on advances in science, engineering and robotics technology (ICASERT). IEEE, pp 1–5
9. Agrekar R, Bhalerao S, Gulam M, Gudadhe V, Bhure K (2022) A review on gesture controlled virtual mouse
10. Jiang S, Lv B, Guo W, Zhang C, Wang H, Sheng X, Shull PB (2017) Feasibility of wrist-worn, real-time hand, and surface gesture recognition via sEMG and IMU sensing. IEEE Trans Industr Inf 14(8):3376–3385
11. Kshirsagar PR et al (2022) Fatigue detection using artificial ıntelligence. In: 4th RSRI ınternational conference on recent trends in science and engineering, REST Labs, Krishnagiri, Tamil Nadu, India, 27–28, February, 2021, AIP conference proceedings 2393, 020080 (2022), pp 020080–1~020080–5
12. Gupta U, Dutta M, Vadhavaniya M (2013) Analysis of target tracking algorithm in thermal imagery. Int J Comput Appl 71(16)

Chapter 24
Statistical and Quantitative Analysis on IoT-Based Smart Farming

G. Dinesh, Ashok Kumar Koshariya, Makhan Kumbhkar, and Barinderjit Singh

1 Introduction

Throughout human history, agricultural advancements achieved to utilize fewer resources and labor. Because of the high population density, it was impossible to maintain a balance between supply and demand. There is a 25% chance that the world's population will reach 10.2 billion by 2060 [1]. Most of the world's population growth will occur in emerging nations [2]. 69% of the world's population will reside in cities by 2060. (now 49% [3]). Increased food consumption will be a side effect of rising wages in emerging nations. As people become more health-conscious and concerned about food quality, their tastes may shift from wheat and grains to legumes and, ultimately, meat. Food output must double to feed a growing, more urbanized, and exploding population by 2060. From 2.1 billion to 3 billion tons, grain and meat output must rise to meet demand [6–10] (Fig. 1).

G. Dinesh (✉)
Department of Computational Intelligence, School of Computing, SRM Institute of Science and Technology, Kattankulathur, India

A. K. Koshariya
Department of Plant Pathology, School of Agriculture, Lovely Professional University Jalandhar Punjab, Jalandhar, India

M. Kumbhkar
Department of Computer Science and Elex, Christian Eminent College and Visiting Lecturer at School of Data Science and Forecasting, School of Biotechnology, DAVV, Indore, India
e-mail: makhan@christianeminent.com

B. Singh
I. K. Gujral Punjab Technical University, Kapurthala, Punjab, India

V. Devedzic et al. (eds.), *Proceedings of the International Conference on Intelligent Computing, Communication and Information Security*, Algorithms for Intelligent Systems, https://doi.org/10.1007/978-981-99-1373-2_24

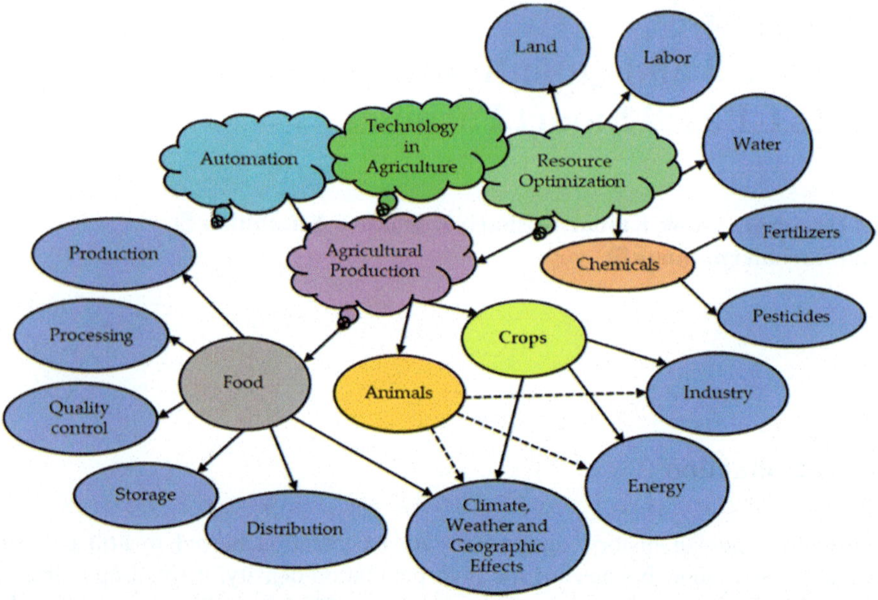

Fig. 1 Major areas of agriculture

Cotton, rubber, and gum are significant economic and food resources for many countries. The use of bioenergy produced from food crops is becoming more popular. By the decade's end, ethanol production alone needed 110 million metric tons of coarse grains [7, 8]. Will improve future food security if more food crops for biofuel, bioenergy, and other industrial uses. The world's already-depleted agricultural supply must stretch even farther to meet these demands. Due to temperature, climate, terrain, and soil quality, only a tiny percentage of the Earth's surface is appropriate for agriculture, and even the most favorable regions are inconsistent. Many small changes across landscapes and plant types. Political and economic factors, such as land and climate patterns and population density, impact agricultural land availability. Over several decades, the land used to grow food has declined. According to the United Nations, about 39.47% of the planet's surface was arable in 1991. According to [10], they covered 18.6 million square miles (377.3%) of land in 2013. A growing chasm separates what people desire from what they have.

Many different types of agricultural fields exist. Each one may depend on its own specific merits. Soil type, nutritional content, irrigation flow, insect resistance, etc., are all factors in determining crop suitability. Even though the same crop over the whole farm, site-specific analyses are necessary for maximum yield production.

Due to temperature, climate, terrain, and soil quality, only a tiny percentage of the Earth's surface is appropriate for agriculture, and even the most favorable regions are inconsistent. Various minor changes may across landscapes and plant types. Political and economic factors, such as land and climate patterns and population

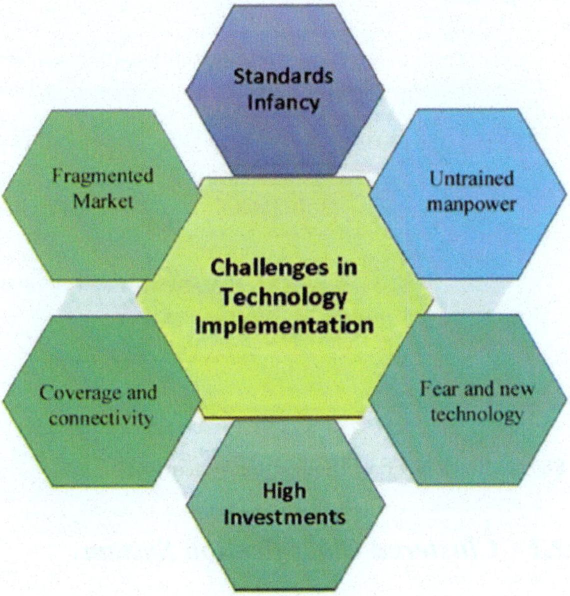

Fig. 2 Major challenges in in technology implementation

density, impact agricultural land availability. Over several decades, the land used to grow food has declined. According to the United Nations, about 39.47% of the planet's surface was arable in 1991. According to [10], 18.6 million square miles (377.3%) of land was covered in 2013. A growing chasm separates what people desire from what they have (Fig. 2).

Agricultural fields come in various shapes, sizes, and characteristics, all of which may be evaluated individually. Crop appropriateness is determined by soil type, nutrient content, irrigation flow, pest resistance, etc. Site-specific evaluations are required for optimal yield production in most circumstances, even if the same crop is planted over the whole farm. The application of intelligent agriculture technology to specific crops is made more difficult by time.

2 Research Methodologies

Various concepts may be shown in Fig. 3. An IoT cloud, an IoT sensor module, an Agri robot, and device security management are all included in this functional block diagram [18]. With centralized servers, the Internet of Things collects sensor data and provides real-time input to green fieldwork equipment. All sensors are integrated into IoT devices. Input/output interfaces for audio and video. Soil or other sensor data is processed optimally by IoT's central processing unit.

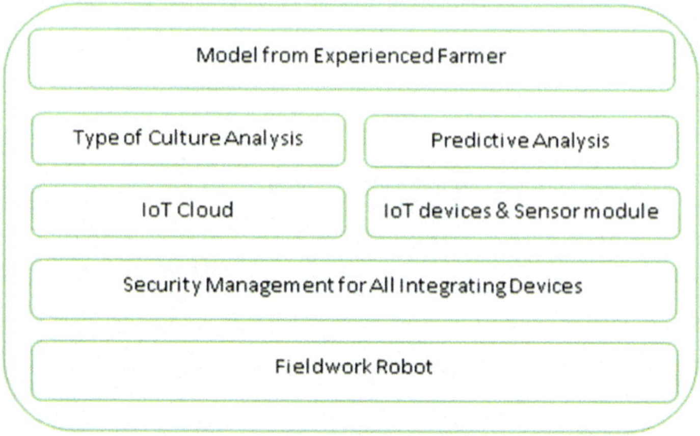

Fig. 3 Block diagram of proposing system

2.1 *Clustered Classification System*

Crop status and pest control for horticulture, floriculture, and citriculture are now activated and ready for use. Citrus fruits and flowers may be listed individually to determine the profit margin. Organic fertilizer may be produced via a process known as vermiculture. The goal of silviculture is to control the land's composition and quality for a variety of plant growths. Cloud-based environmental analysis. In arboriculture, shrubs and woody plants maintain soil nutrient levels. Plants may be predicted by humans consuming them via olericulture.

Cloud computing necessitates a multi-cultural study of vast IoT data for predictive analysis. Monsoon output is boosted via probabilistic methods. It's a predictive system. Traditional agriculture can anticipate soil nutrients, temperature, rainfall, and future climatic factors with the help of a community of experienced farmers. Numerous sector data analyses are used in this framework to anticipate the most probable outcome. Based on historical data, pest and attack treatment trends in traditional agriculture. Optimized a prediction algorithm before doing substantial data analysis. Predicting the vehicle's use to deliver all plugged plant components may increase profit and market sales. Using this strategy, the farmer can better understand his present and future profits and reduce his risk. The new era of framing is formulated and managed by this process.

2.2 *IoT Frame Work*

Data may be stored and transferred across different devices using the Internet of Things (IoT) cloud. Items, plant illnesses, and detailed data analysis are maintained separately. Agro specialists may provide smart agriculture and future projections via

Internet services; the experts can advise on field crop planting, pesticide control, and land management. These services are available to traditional farmers. Because the Internet of Things (IoT) it's easy to utilize.

Sensors, cameras, display units, microcontrollers, routers, switches, etc., are all included in this category [29]. Actuators are used in predictive activities to adjust sensor settings. The central processing unit (CPU) is in charge of transferring data across IoT devices.

Security management is in charge of safeguarding data. Network layer protocol infrastructure was enabled due to the preventive action utilized. Device malfunction, fabrication, destruction, and improper handling may all be prevented by this security management for the Internet of Things (IoT). Wi-Fi, GSM, and CDMA [31] are safe communication methods. Components can communicate data more easily with the aid of Zigbee. It's possible to share through GSM, CDMA, and LTE, among others; The HTTP, WWW, and SMTP protocols are used in Internet authentication and access procedures and systems.

2.3 Robotic-Agriculture

The intelligent agricultural concept includes an Agri robot for fruit picking, autonomous vehicles, and water spraying. After receiving a signal via IoT, the Agri robot will identify the fruit's size and color. The applications mentioned above will employ robots for agricultural tasks. Figure 4 shows Agri robots picking fruit.

Apple, strawberry, and guava fruit picking may be programmed. Compared to handpicking, this harvesting period is short.

3 Result Discussion

Figure 5 displays field 1–3 findings. Moisture, temperature, and humidity irrigation sensor data. The gadget alerts the fieldwork robot when it exceeds the threshold level. Figure 8 exhibits moisture, temperature, and humidity sensor output. This graph displays device performance during feedback processing.

Figure 6 displays smart agriculture's total performance; Robots react to real-time scenarios in sensor fieldwork; our IoT-based intelligent farm monitoring system includes sensors and a CPU. The microcontroller unit to mobile for live field data. Figure 9 provides raw sensor data. IoT fieldwork robots in smart agriculture will offer feedback based on this data.

Wi-Fi fetches the information. Here, IoT soil moisture takes effect. Many fieldwork switching devices are "ON" and "OFF" based on IoT sensor inputs.

Fig. 4 Agri robots picking fruit

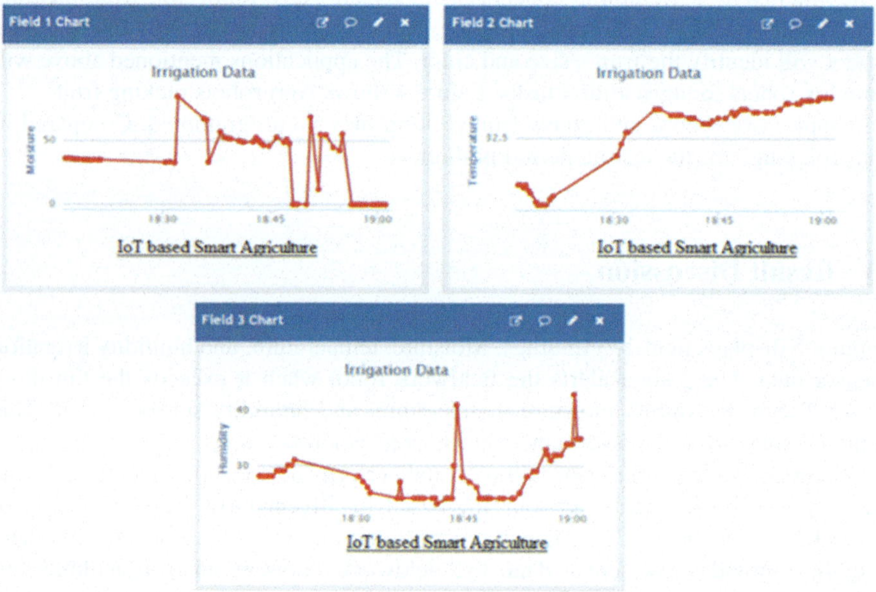

Fig. 5 Moisture, temperature, and humidity irrigation sensor data

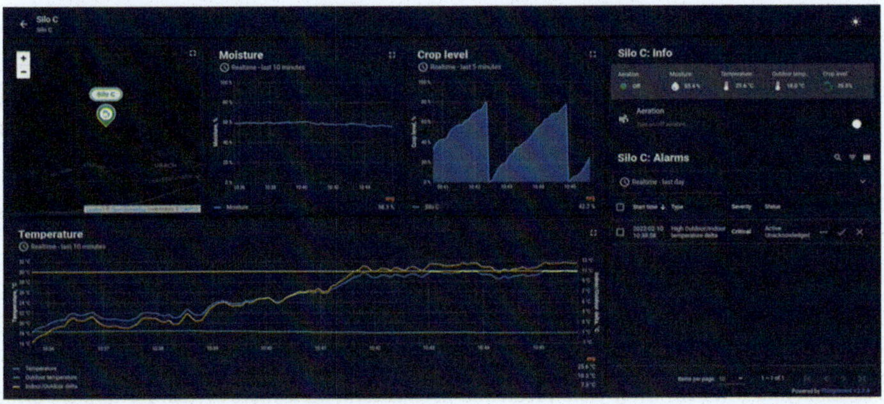

Fig. 6 Smart agriculture's total performance

4 Conclusion

This research proposes a model with several analytical components. Innovative agricultural equipment with IoT modules illustrates various advantages of our integrated systems; This architecture of platform and security has limitations. Obstacles and hurdles confront IoT-enabled smart agriculture. By decreasing hardware and software costs without sacrificing accuracy, IoT devices focus on cost-effectiveness. The lower price of the components used in imported electronics is ignored. Device homogeneity and execution performance may be improved by standardizing the data format for processes. In improving system goods or services, active farmer suppliers are monitored. Web-enabled devices make the proposed integrated system more challenging. The IoT process of heterogeneity enhances precision and system performance. Agricultural IoT production may be boosted by deep learning analysis using a wide range of variables or data.

References

1. Ayaz M, Uddin A, Sharif Z, Mansour A, Aggoune E-H (2019) Internet-of-things (IoT)-based smart agriculture: toward making the fields talk. IEEE Access, pp 1–1. https://doi.org/10.1109/ACCESS.2019.2932609
2. Suma V (2021) Internet-of-Things (IoT) based smart agriculture in India—an overview. J ISMAC 3:1–15. https://doi.org/10.36548/jismac.2021.1.001
3. Navarro E, Costa N, Pereira A (2020) A systematic review of IoT solutions for smart farming. Sensors (Basel). 20(15):4231. https://doi.org/10.3390/s20154231.PMID:32751366;PMCID:PMC7436012
4. Rehman A, Saba T, Kashif M, Fati SM, Bahaj SA, Chaudhry H (2022) A revisit of internet of things technologies for monitoring and control strategies in smart agriculture. Agronomy 12:127. https://doi.org/10.3390/agronomy12010127

5. Köksal Ö, Tekinerdogan B (2019) Architecture design approach for IoT-based farm management information systems. Precision Agric 20:926–958. https://doi.org/10.1007/s11119-018-09624-8
6. Bayih AZ, Morales J, Assabie Y, de By RA (2022) Utilization of internet of things and wireless sensor networks for sustainable smallholder agriculture. Sensors 22, 3273. https://doi.org/10.3390/s22093273
7. Ferehan N, Haqiq A, Ahmad MW (2022) Smart farming system based on intelligent internet of things and predictive analytics. J Food Qual 2022, Article ID 7484088, pp 8. https://doi.org/10.1155/2022/7484088
8. Ayaz M, Ammad-Uddin M, Sharif Z, Mansour A, Aggoune E-H (2019) Internet-of-Things (IoT)-based smart agriculture: toward making the fields talk. IEEE Access, IEEE 7, 129551–129583. ⟨https://doi.org/10.1109/ACCESS.2019.2932609⟩. ⟨hal-02439204⟩
9. Pang H et al (2021) Smart farming: an approach for disease detection implementing IoT and image processing. IJAEIS 12(1):55–67. https://doi.org/10.4018/IJAEIS.20210101.oa4
10. Lambture B (2020) Crop yield prediction in smartfarm agriculture system for farmers using IoT. Int J Adv Sci Technol 29(7):5165–5175. Retrieved from http://sersc.org/journals/index.php/IJAST/article/view/23613
11. Jacob M, Pramod (2018) A comparative analysis on smart farming techniques using internet of things (IoT). HELIX 8, 3294–3302. https://doi.org/10.29042/2018-3294-3302
12. Dadheech et al (2020) Implementation of internet of things-based sentiment analysis for farming system. J Comput Theor Nanosci 17(12):5339–5345 (7). https://doi.org/10.1166/jctn.2020.9426
13. Muangprathub J, Boonnam N, Kajornkasirat S, Lekbangpong N, Wanichsombat A, Nillaor P (2019) IoT and agriculture data analysis for smart farm. Comput Electron Agri 156, 467–474, ISSN 0168-1699. https://doi.org/10.1016/j.compag.2018.12.011
14. Islam N, Ray B, Pasandideh F (2020) IoT based smart farming: are the LPWAN technologies suitable for remote communication? IEEE Int Conf Smart Internet of Things (SmartIoT) 2020:270–276. https://doi.org/10.1109/SmartIoT49966.2020.00048
15. Darshini P, Kumar MS, Prasad K, Jagadeesha SN (2021) A cost and power analysis of farmer using smart farming IoT system. In: Pandian A, Fernando X, Islam SMS (eds) Computer networks, big data and IoT. Lecture notes on data engineering and communications technologies, vol 66. Springer, Singapore. https://doi.org/10.1007/978-981-16-0965-7_21
16. Yang J et al (2021) IoT-based framework for smart agriculture. IJAEIS 12(2):1–14. https://doi.org/10.4018/IJAEIS.20210401.oa1
17. A review on IOT based smart agriculture for sugarcane (2019)
18. Mahrishi M et al (eds) (2020) Machine learning and deep learning in real-time applications. IGI Global. https://doi.org/10.4018/978-1-7998-3095-5
19. Khan Z, Khan MZ, Ali S, Abbasi IA, Rahman HU, Zeb U, Khattak H, Huang J (2021) Internet of things-based smart farming monitoring system for bolting reduction in onion farms. Sci Prog 2021, Article ID 7101983, pp 15. https://doi.org/10.1155/2021/7101983
20. Lytos A, Lagkas T, Sarigiannidis P, Zervakis M, Livanos G (2020) Towards smart farming: systems, frameworks and exploitation of multiple sources. Computer Networks vol 172, 107147, ISSN 1389-1286. https://doi.org/10.1016/j.comnet.2020.107147

Chapter 25
Survey on Secure Encrypted Data with Authorized De-duplication

Punam Rattan, Swati, Manu Gupta, and Shikha

1 Introduction

The quantities of data have significantly increased in the last three decades. In the 1990s we can see that data volumes were expressed during terabytes. Relationship databases and data stores that represent organized information in lines and segments, to store and oversee business information. Data began to work during the next decade with a scope of information sources oversaw by ease of use and distributing advancements, similar to content-oversaw servers and system associated capacity frameworks. Big Data has numerous thoughts. Somebody portrays Big Data as data that is preferably mind boggling over "simpletogo" and it's difficult to procure, store, oversee, and process it since it's "big" One, in some cases stated, is classified "3 V" characterized by three terms (volume, assortment, and speed). Such two definitions and others are referred to and referenced underneath: "Big Data" refers to vast amounts of data with differing degrees of complexity, created at varying rates, and with varying levels of risk. Regular innovation, preparation processes, calculations, or business-ready solutions cannot handle this type of data '[1, 2].

Hug measure of data: Big Data can be trillions of lines and a huge number of segments supposed to thousands or a large number of columns. 'Complexity of data types and frameworks: Big Data speaks to the assortment of new data sources, configurations and structures, as literary follows left on the web and other data files for ensuing investigation [3].

P. Rattan
Lovely Professional University, Phagwara, Punjab, India

Swati (✉) · M. Gupta · Shikha
Sanatan Dharma College, Ambala, Haryana, India
e-mail: bswati002@gmail.com

V. Devedzic et al. (eds.), *Proceedings of the International Conference on Intelligent Computing, Communication and Information Security*, Algorithms for Intelligent Systems, https://doi.org/10.1007/978-981-99-1373-2_25

1.1 De-duplication

Data de-duplication is the process of removing redundant data that. Data de-duplication is a compression strategy that is commonly used to eliminate recurring repetitions of files or data by retaining unique copies in the storage system to conserve space. The "reference pointer" is directly used to locate vital bits of data copied from the data archive or storage systems during de-duplication, also known as storage capacity optimization. Depending on how it is implemented, data de-duplication can occur at the file or block level. By recognizing and eliminating redundant information, de-duplication removes the need for numerous copies of the same file or data.

1.2 Three Ways of De-duplication

De-duplication is carried out mainly in three ways, i.e., it is also referred to as the main de-duplication layers and has been listed below:

1.2.1 Chunking

Chunks are specified in a few systems by the physical layer constraints and whole files are compared in some systems that is called as single instance storage or SIS. One of the best methods to chunk is typically a sliding block, where the size of the window is moved along with the file to split the stream of data and check for more obvious limitations of internal files.

1.2.2 Client-Side Backup De-duplication

In client-side data de-duplication, identical data is found and eliminated by the backup-archive client and server to free up server storage space. The de-duplication hash calculations are performed first on the source, which is the client machine. Identical hashed files are not transferred to the target device. Target devices create correlations inside links to prevent files or data from being copied repeatedly. This form of de-duplication is beneficial since it prevents unnecessary files and data from being transmitted over the network. This reduces traffic congestion.

1.2.3 Primary Storage and Secondary Storage

Although primary storage systems are regarded to be the most efficient, Researchers are also the most expensive. Furthermore, primary storage systems are less forgiving of any applications or activities and are more expensive, both of which can have an impact on performance. So, when developing secondary storage, keep two things in mind: maximum performance and the lowest potential cost. Most of the time, secondary storage systems include replicated and reference copies of data. Because these copies of the data aren't always employed in the actual building process, Researchers're more ready to sacrifice certain results for greater efficiency. The biggest risk of sending data through a network is that few parts of the data will be lost. And data de-duplication solutions alter how data is processed in relation to how it was written. As a result, vendors are concerned about whether or not their ambiguous results are genuine. In the end, the quality of the data will be determined by how the de-duplication method is configured and how well the algorithms are designed.

1.3 De-duplication Techniques

Data de-duplication is typically classified into three groups based on the location of data de-duplication, data unification, and disc placement (Fig. 1). The following Fig. 2 shows three groups of de-duplications.

1.3.1 Data Unit Based

It is divided into three categories based on data unit.

Byte-Level De-duplication

One form of Block-Level De-duplication is Byte-level De-duplication that understands the content or data "Semantics". These systems are also referred as "CAS–Content Aware Systems". Block-Level De-duplication is performed by de-duplication devices. The information is delivered across the network as bytes. Every

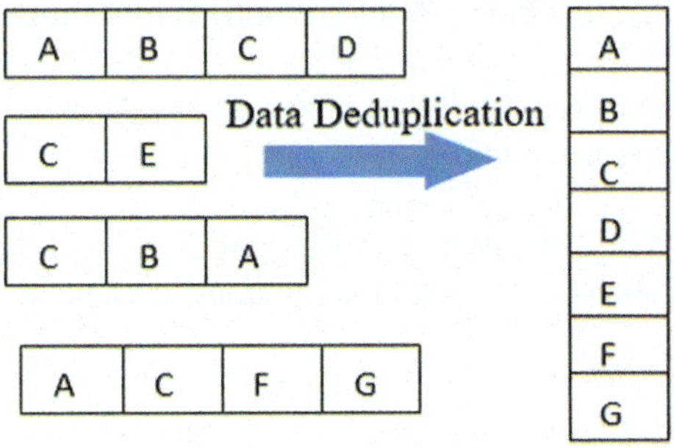

Fig. 1 De-duplication process

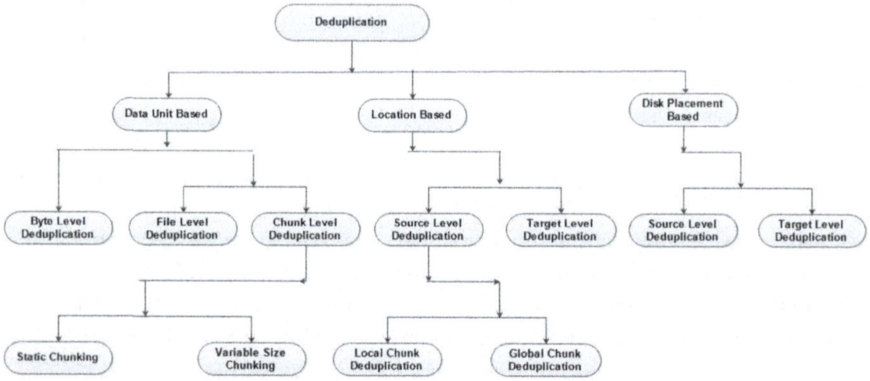

Fig. 2 Data de-duplication techniques

byte of data appears to be checked to see if it is the same as of other data. If a byte of data is identical to the one preceding it, it should not be broadcast over the network. If a data byte differs from the one preceding it, it should be transmitted over the network.

File Level De-duplication

Data file de-duplication is concerned with data files. If the file's hash value is the same, just one copy should be retained. If two files have the same hash value but distinct values, it is not considered identical or unique.

Chunk Level De-duplication

Each file is divided into block numbers in data de-duplication at the Block Level or data at Chunk Level. On the storage device, each item should only have one copy. If a block with the identical ID number exists on the storage device previously, it will not be stored again. If the storage device does not already have a block with the same ID, this block should be added. It is similarly divided into two sections [7].

Static Chunking

Static chunking requires that each block be partitioned into the same size of blocks. The size of each block that needs to be disassembled is uniform.

Variable Chunking Size

Each block must be fragmented into various sizes in the variable chunking size.

1.3.2 Location Based

There are several methods for removing duplicate data. This type of data duplication is referred to as location-based data de-duplication. It is used to remove duplicate data from another location.

Source Level De-duplication

When deleting duplicate data at the client site or where the data was created rather than where it was stored, data de-duplication at the source level is employed. Eliminating duplicate data at the source level must be divided into two steps.

i. *Local Chunk De-duplication*–Data de-duplication redundant data must be eliminated in the local chunk level before it is sent to the destination where the data is to be processed.
ii. *Global Chunk De-duplication*–For each device, redundant data is removed at global level in the global chunk data de-duplication technique.

Target Level De-duplication

The process of removing duplicate or similar data at the site where the data is stored is known as target level data de-duplication. In this type of data de-duplication, the customer is unsure how to get rid of unnecessary data. This type of technology increases processing time while decreasing bandwidth.

1.3.3 Disk Placement Based

The use of either forward reference or backward reference technique is focused on how data is to be put on the disk data de-duplication technique.

- *Source Level*–Recent data chunks are preserved in the forward index, and all the old data chunk is connected with pointers pointing to the current chunks.
- *Target Level*–This introduces maximum fragmentation for the previous data blocks.

1.4 De-duplication Storage System

There are several de-duplication storage systems that are used for various storage purposes. Some of them are listed below:

1.4.1 Venti

The network frequently discusses systems such as Venti. The network de-duplication method organizes data chunks using hash values that are unique to each piece. Overall, this uses less storage space. Venti employs the "write once" method to avoid data collisions, and this storage technology is frequently associated with chunk generation for large storage applications.

1.4.2 Hydrastor

A traditional file interface for secondary storage system, decentralized hash indexes for a grid of storage nodes are used. In the Hydrastor backend, a directional acyclic structure is suited for managing large, diversified, and stable address data blocks. The hash table backs up the conclusion. These systems are primarily intended to serve as backup systems.

1.4.3 Extreme Binning

These storage systems are built to manage unusual workloads. This storage system can be enlarged to use the same de-duplication method as before. This must be accomplished by assembling low-local files. File similarity is more essential than file location, and looking for file blocks only requires one disc access. Similar data is placed in bins for this procedure, and chunks of duplicates are extracted from each bin. In this way, de-duplication in terms of the various bins has been accomplished. Besides this intense binning only retains primary index in memory which reduces the consumption of RAMs.

1.4.4 MAD2

An important feature of MAD2 is its key, its precision. This is an exact duplicate network backup service, not a pure storage system. It often works both at the file level and at the diagnostic level. Various techniques are used to achieve the desired performance, including distributed hash table-based hash compartments matrix and body balance.

1.4.5 Duplicate Data Elimination (DDE)

It is possible to utilize content hashing, write copying, and lazy updates in conjunction with Duplicate Data Elimination to discover and merge identical SAN data blocks (storage area network system). The DDE storage system differs from other storage systems in that it de-duplicates and analyses hash values at the block level.

The rest of the paper is organized in the given order as Sect. 2 includes the literature review related to this work. Challenges faced are presented in Sect. 3. Conclusions appears in Sect. 4.

2 Related Work

In this section, investigation is conducted on various strategies for de-duplication of information. The following are edited collections of those inquiries carried out by different colleagues in the research. Organizations are evolving every day, storing big data in unique ways. However, organizations need to combine large data sets in different formats and evaluate them to improve query processing. There are different methods and techniques used to analyze big data. Certain methods are used to integrate the various data formats. The information is processed and analyzed, and requests for this filtered information are processed. In one of the paper, a de-duplication process was implemented in large scale database of photographs. In the proposed method, an attempt is made to eliminate the duplicate electricity cards from the database using the Block truncation code technique. To speed up the de-duplication process, the entire data is compressed into different block size levels. After BTC is applied to our images we can see the single instances of the images in our database, which would not create further chaos and confusion in our database. In another paper, a de-duplication process was implemented in large scale database of photographs.

Jinbo Xiong et.al. [4] proposed to enable dynamic privilege and revocation via role-based encryption, encryption techniques which were utilized to prevent convergence encryption and privacy leakage, allowing for authoritative deduction and revocation (SRRS). Authors have proposed a mechanism that enables for proprietary authentication while still providing the best proof of ownership for authorized users. Authors established a Role Permission Tree (RAT) to manage duties and essential relationships, as well as a control center to handle permission requests. With the convergent encryption algorithm and position re-encryption technology, only authenticated users with the matching function re-encryption key can be expected to have clear access without data leakage. Researchers employed the convergence encryption algorithm and the role re-encryption algorithm to protect their confidential information. Sumedh Deshpande et.al. [5] provided services through the rearrangement of various Internet tools. Data storage is one facility. Cloud data storage is mostly in encrypted form. So there is some vulnerability in the current system over security issue. So, authors are using hash code creation and double encryption techniques to solve these issues. Researchers devised a strategy for handling encrypted massive data in the cloud, including PRE-based de-duplication and ownership difficulties. This resolves the issue. This approach will make it easier to communicate and update data, even when the data proprietors are not connected to the Internet. Anand Bhalerao and Ambika Pawar [6] researchers discussed how the current need for huge data backups is causing cloud storage space and speed issues. Numerous techniques have been developed to improve the effectiveness of cloud storage and storage space. Data de-duplication is a viable solution to these storage issues. Researchers used a variety of chunking techniques and algorithms to remove unneeded data. One of the most popular solutions for cloud storage issues is to remove duplicate data. Chunking algorithms are classified depending on their position, timing, and granularity. There are various types of chunk algorithms that require upgrades and benefits. AE is one of the technologies that has a lot of potential. AE can be improved by determining how to distribute chunk sizes. Junbeom Hur et.al. [7] poses confidentiality and privacy concerns. The overall outcome of this project is that the suggested system includes a strategy of re-encryption that enables continuous alerts on any adjustments in the cloud storage possession. When the group of people who own the outsourced data changes ownership, the records are re-encoded with a new ownership group file that is sent only to the proper owners. As a result, the new approach improves data privacy and cloud storage security against clients who lack the right to handle their data, as well as against a cloud server that appears trustworthy but may be hiding something.

Wen Xia and Min Fu [8] mentioned that duplicate files are the source of cross-user redundant data to encrypt, the convergent encryption method is utilized. The major goal is to keep cloud storage secure while also conserving network space and traffic. The solution is to obtain less storage space than is now available. Therefore, under this technique, the user was exposed to convergent key encryption and management of multi-level keys. The experimental result is a better performance. Jin Li et. al. [9] introduced a base line solution where the user holds only the master key. The suggested scheme is used to handle key efficiently and with convergence. Using proposed Dekey, researchers use various constraints to attain their goal. Users don't have to handle the key on their own thanks to Dekey. The final outcome of this experiment is the distribution of the convergent keys over multiple servers. Key management has therefore been somewhat successful. Mihir Bellare et.al. [10] indicated that message is locked because it needs to be encrypted to prevent file duplication This encryption ensures that outsourced files remain private and secure. Because researchers knew how it was set up and how large it was, Researchers used plain text data on it. It is utilized in Cloud to provide the best possible answer for the intended research. Pasqualeet.al. [11] Additional encryption operations and access control mechanisms were announced. The objective was to address security and privacy issues. Researchers provided CloudDup to address various restrictions. Researchers have lowered storage space and saved storage space in the past. This experiment's outcome was only half successful. Junbeom Hur et.al. [12] Reduction of replicas with various quality criteria has been presented in order to utilize de-duplication strategy to achieve reasonable performance and efficiency. Control of complex ownerships allowed for the achievement of the performance. Researchers made advantage of the lowest possible frequency and bandwidth. The ownership management is the foundation for the experimental findings.

MihirBellare et.al. [13] From the message, encryption and decryption were implemented. Researchers created it to effectively remove duplicates. Researchers provide models for how to distribute schemes among various contexts and message source kinds. Meanwhile, the result falls short of expectations in terms of de-duplication. It was damaged by brute force attacks. Hua et.al [14] Attribute-Based Encryption (ABE) is a mechanism for efficiently sharing data while using less storage space. Person has been given rights to decipher & compute the enciphered data in system if and only if attributes of particular user is matched.

Vishalakshi et al.[15] Data was encrypted using convergent key encryption prior to outsourcing. In which Researchers discuss the topic of permitted duplicated data and take a different approach than other typical de-duplication methods. Researchers present a previously accepted model test scheme. Minal Bharat et. al. [16] All that everyone wants is to access, store their data without thinking about protection. The IT department will guarantee the confidentiality of the records. The content of the documents is modified by a foreign attacker, so the integrity must be verified. This data is duplicated, and it will consume much of the unused space. Mamta et. al. [17] Researchers have put forth an ABE-based searchable encryption technique in which a user's search ability is determined by access control. This system guarantees that the user's secret key and the keyword's ciphertext are always the same size. Additionally, because there is a fixed amount of pairing processes, their system provides quick searches.

Samant et. al [18] shown major component at the back of the victorious derivation of defect free valid software products from SPL is the standard of FM. Flaw due to redundancy in FMs is one of such imperfections which hamper the derivation of high-quality valid software products in SPL. An ontological FOL rule-based method is proposed to deal with redundancies. The method has been authenticated using 35 models with different sizes up to 30,000 features which conclude that it is coherent, accurate and scalable. The final outcomes of this are refinement in corrections includes eliminating redundant features in model.

Zargar et.al. [19] in her paper stated that, de-duplication process was implemented in large scale photographs database. Block truncation code technique is used to eliminate the duplicate electricity cards from the database. After images have undergone BTC the single instances of the images can be seen. Ghrera et.al. [20] shown various possible approaches for enhancing security in data de-duplication. An algorithm was also proposed. Also a secret key had been derived from user password to offer data origin authentication. In this paper, analysis reveals the proposal is secure in terms of definition quantified in the proposed secure data de-duplication. Ninnisingh et al. [21] had shown the secure authentication in the multi-operator domain. Authentication scheme used is based on ticket.

The following comparisontable1 shows that, the overall authentication cost, system delay throughput & encryption cost is improved as compared to one of the previous proposed techniques. The details must be de-duplicated and transferred onto the cloud server to prevent this. It is used for the stable hashing algorithm. For a block of data, measure the hash value and save it in the cloud. The solution provider should only address the security.

The above mentioned Table 1 comparison of prior research shows that de-duplication has been carried out using a variety of techniques in the same Cloud context. Researchers had made an effort to succeed in the parameters of storage space, privacy, reliability, and efficiency. However, it struggled with authority of correctness and continued to have issues with security and wasted storage space. Therefore, the aforementioned variables must be the main emphasis of the proposed effort.

3 Challenges

It has been observed from the review of literature that this area is facing a number of challenges. When a file is uploaded, evidence of ownership is set for the file duplication check. This proof is saved in the file and will be used to determine who has access to it. It will specify who may determine whether a file has already been opened. The user must give his file and evidence of ownership to transmit duplicate check request. The duplication check request is accepted only.

3.1 System Architecture

The system architecture, which consists of the public cloud, the private cloud, and the user. The first cloud is public, whereas the second is private. The user's files and other data are stored in the public cloud, while the user's credentials are stored in the private cloud. The user must obtain a token for the private cloud for each transaction with the public cloud. If the user credentials are the same in both the public and private clouds, the user can check for duplication. The following procedures must be followed during authenticate duplicate check.

3.2 Encryption of File

The private cloud houses the secret key for encrypting user data. This key is used to convert plain text to cipher text and cipher text to plain text. To encrypt and decrypt, three basic functions were employed:

Table 1 Comparison Various De-duplication Techniques

Authors and Years	Techniques used	Parameter	Result
"Wenxia and Lin Fu in 2015"	"Convergent Encryption"	Security, reliability	The data is hashed to ensure key usage reliability
"Patrick P.C.Lee and Wenjing Lou in 2014"	"Dekey"	Realistic environment, uses of Key	In order to handle all available keys, a ramp secret sharing method is implemented
"Mihir Belare and Sriram keelveedhi in 2013"	"Message locked encryption"	Security and storage space	Dupless encrypts messages using PRF protocol and message-based keys obtained from key servers
Pasquale Puzio and Retikmolva in 2016"	"Convergent key"	Efficiency	Used to determine whether a particular plaintext has previously been saved
"Jun beonHur and Denfoung in 2016"	Data re-encryption	Data Privacy, confidentiality	De-duplication works well when users store their data in the cloud even when the owner changes
"MihirBellare, Sriramkeelveedhi and Thomas Ristenpart in the year 2013"	MLE	Privacy	Decryption and encryption were carried out automatically from the message
"Pyla. Naresh, K. Ravindra and Dr. A. Chandra Sekhar in 2016"	"D-Cloud'	Reliability and security	Before sending the data to Cloud, it encrypts the data
"Arthur Rahumed, Henry C.H chen, Yang Tang, Patrick P.C.Lee and John C.Lui in 2011"	"Fade version"	Cost and security	Layered encryption approach
"Vishalakshi N S and S.Sridevi (2016)"	"Convergent key encryption"	Bandwidth and storage space	Used an authorized duplicate check approach to find redundant information

(continued)

Table 1 (continued)

Authors and Years	Techniques used	Parameter	Result
"Vishalakshi N S and S.Sridevi in the year 2017"	"Cloudedup"		The objective of Cloudedup is to offer secure and effective data storage
"Shweta D. Pochhi, Prof. Pradnya and V. Kasture 2015"	"Data compression technique"	Data compression technique	Authorized duplicate checking is supported
"K.Kanimozhi and N.Revathi 2016"	"Secure Proof of ownership and hash function"	Confidentiality of data	Encryption is done based on content of the data
"Mamta and Brij Gupta 2020"	Constant size secret keys and ciphertexts	Reliability and security	Enhance The functionality by adding features of verifiability of the Search operation

3.2.1 KeyGenSE (K)

In this case, k represents the method for generating a key that may be utilized to generate a secret file utilizing security settings. This will create public and private key using specification of security for each cloud user.

3.2.2 EncSE (k, M)

M is the plaintext message in this approach, and k is the secret key. Researchers combine to form a cipher text.

3.2.3 DecSE (k, C)

C is the cipher text, and k is the encryption key. Using the cipher text and the secret key, one could construct plain text, or the original message.

3.3 Proof of Ownership

De-duplication is accomplished by performing a cryptographic hash function on the data and comparing the hash value to comparable data. If a duplicate copy is discovered, no new data is uploaded. Instead, the file's owner's pointer is changed, which saves space and bandwidth. Client-side de-duplication involves calculating hash values of data at the client and sending them to the server to check for duplicates.

An attacker can claim file de-duplication if he or she obtains access to the hash value of data for which researchers are not authorized. The attacker now has access to the file. Proof of Ownership (PoW) was proposed as a defense against this type of attack in [18, 20]. This strategy was utilized in [18, 19], and other places. PoW is a method that uses two persons to determine who owns a file: a "prover" and a "verifier". The verifier determines the data's short value, but to prove it must determine the data's short value and send it to the verifier in order to claim possession of the data. [17, 18].

3.4 *Encryption of Data*

This ensures that the duplicated data remains confidential. Each piece of original data is used to generate a convergent key by the user. The convergent key is then used to encrypt the data copy. The user additionally tags the data, which aids the tag in locating duplicate data. A key is generated using the convergent key generation technique[21–23]. This key is used to encrypt user information. This ensures that the data is secure, owned, and controlled by the appropriate individuals.

4 Conclusion

The primary objective of this paper is to compare different de-duplication techniques for encrypted data with authorized key. Security and redundant data are main challenges in the environment of big data. As the data are available in heterogeneous formats such as text, video, audio, and therefore, it is important to combine the data from the different sources and analyze the data which can be used for effective query execution. Upon deployment, the data expands and you need to use a variety of techniques to analyze the data. The de-duplication technique was therefore implemented by combining different data in the same file. Furthermore, data consolidation at record level is applied to accumulated data to evaluate large amounts of data and is used to delete similar data forms. Ultimately, the results of the quest agree productivity about runtime, storage, and output. In the proposed method, an attempt can be made to eliminate the duplicate e-cards from the database using the Block truncation code (BTC) technique. To speed up the de-duplication process, the entire data is compressed into different block size levels. After BTC is applied to our images we can see the single instances of the images in our database, which would not create further chaos and confusion in our database.

References

1. Vishalakshi N S and S.Sridevi, "Survey onSecure De-duplication with Encrypted Data for Cloud Storage," international journal of advanced science and research, Vol. 4, Issue 1, January2017.
2. Minal Bharat Pokale and Dr. Sandeep M.Chaware "De-duplication Approach with Enhance Security for Integrity" 2018 Fourth International Conference on Computing Communication Control andAutomation.
3. Halevi, D. Harnik, B. Pinkas, and A. Shulman- Peleg,"Proofsofownershipinremotestoragesystems," in Proc. ACM Conf. Comput. Commun. Security,2011, pp.491–500.
4. JinboXiong and Yuanyuan Zhang Shaohua Tang, XimengLiu , and Zhiqiang Yao, "Secure Encrypted Data With Authorized Deduplication in Cloud" IEEE Transactions June,2019.
5. Jain P et al (2014) Impact Analysis and Detection Method of Malicious Node Misbehavior over Mobile Ad Hoc Networks. International Journal of Computer Science and Information Technologies (IJCSIT) 5(6):7467–7470
6. Anand Bhalerao and Ambika Pawar "A Survey: On Data Deduplication for Efficiently Utilizing Cloud Storage for Big Data Backups" Computer Science and Engineering, Symbiosis Institute of Technology Pune, India anand ICEI2017.
7. JunbeomHur,Dongyoungkoo,YoungjooshinandKyungtae Kang, "Secure Data Deduplication with Dynamic Ownership Management in Cloud Storage," IEEE Transaction on knowledge and data engineering/ TKDE-2016.
8. Wen Xia, Min Fu, Fungting Huang and Chunguang Li, "A UserAware Efficient FineGrained secure de- duplication scheme with Multi-Level key management," Huazhong University of science and technology in the year2015.
9. Li,Jin,YanKitLi,XiaofengChen,PatrickPCLee, and Wenjing Lou. "A hybrid cloud approach for secure authorized deduplication." Parallel and Distributed Systems, IEEE Transactions on 26, no. 5 (2015): 1206- 1216.
10. Mihir Belare, sriramKeelveedhi and Thomas Ristenpart, "Dupless: server-Aided Encryption for de- duplicated Storage," appeared at 2013 USENIX SecuritySymposium.
11. Pasquale Puzio, Refik Molva, MelekOnen, "Secure De-duplication with Encrypted Data for Cloud Storage," secured project supported by French Government in the year2013.
12. JunbeomHur, Dongyoung Koo, YoungjooShinandKyungtae Kang Secure "Data Deduplication with Dynamic Ownership Management in Cloud Storage" 2017 IEEE 33rd International Conference on Data Engineering.
13. Bellare M (2013) Sriram keelveedhi and Thomas Ristenpart"Message Locked Encryption and Secure De- duplication." In EUROCRYPT, LNCS 7881:296–312
14. Hua Ma, Ying Xie, JianfengWang ,Guohua Tian , And Zhenhua Liu, "Revocable attribute-based encryption scheme with efficient deduplication for e-health systems," Volume 7, 2019.
15. Dimitriosvasilopoulos ,MelekOnen , kaoutarElkhiyaoui and Refikmolva , " Message- locked proofs of retrievability with secure deduplication," on October 28,2016.
16. DiptiBansodeandAmarBuchade"studyonsecure data deduplication system with application awareness over Cloud storage system," international journal of advanced computer engineering and networking volume-3, Issue-1 on Jan2015.
17. P. Vijayakumar, et.al. (2022), "Network Security Using Multi-layer Neural Network", 4th RSRI International Conference on Recent Trends in Science and Engineering, REST Labs, Krishnagiri, Tamil Nadu, India, 27–28, February, 2021, AIP Conference Proceedings 2393, 020089 (2022), pp. 020089–1~020089–5, https://doi.org/10.1063/5.0074089
18. Mamta, Gupta, B.B. : An attribute-based searchable encryption scheme for non monotonic access structure. In: Handbook of Research on Intrusion Detection Systems, IGI Global, pp. 263– 283 (2020).
19. P. Samant, M. Bhushan, A. Kumar, R. Arya, S. Tiwari and S. Bansal, "Condition Monitoring of Machinery: A Case Study," in 6th International Conference on Signal Processing, Computing and Control (ISPCC), IEEE, 7 Oct, 2021, pp. 501–505, doi: https://doi.org/10.1109/ISPCC53510.2021.9609512.

20. Zargar, A. J., Singh, N., Rathee, G., & Singh, A. K. (2015, February). Image data-deduplication using the block truncation coding technique. In 2015 international conference on futuristic trends on computational analysis and knowledge management (ABLAZE) (pp. 154–158). IEEE.
21. Ghrera, P. A Novel Encryption Technique for Data De-Duplication. In 2015 international conference on futuristic trends on computational analysis and knowledge management (ABLAZE) IEEE.
22. Ninni Singh, Gunjan Chhabra, Singh, K.P., Hemraj Saini (2017). A Secure Authentication Scheme in Multi-operator Domain (SAMD) for Wireless Mesh Network. In: Satapathy, S., Bhateja, V., Joshi, A. (eds) Proceedings of the International Conference on Data Engineering and Communication Technology. Advances in Intelligent Systems and Computing, vol 468. Springer, Singapore.
23. Pyla. Naresh, K. Ravindra and Dr. A. Chandra Sekhar, "The Secure Integrity Verification in Cloud Storage Auditing with Deduplication," on IJCST vol.7, Issue 4,2016.

Chapter 26
Performance Evolution of OFDM Modulation for G-Distribution Noise Channel Using Pade's Approximation

Rashmi Choudhary, Ankit Agarwal, and Praveen Kumar Jain

1 Introduction

Frequency Division Multiplexing (FDM) technology splits the spectrum into subbands so that many carriers can broadcast simultaneously [1]. For inter-carrier interference, guard bands must be put between any neighboring carriers, resulting in a reduced data rate. A multi-carrier digital communication system, such as Orthogonal Frequency Division Multiplexing (OFDM), provides a solution to both problems. OFDM is based on the principle of orthogonality splitting the available spectrum into several narrow band sub-channels, each of which suffers almost flat fading. A high data rate communication system from multiple low data rate carriers. Orthogonality provides a rationale for the tight spacing of the carriers, even if they overlap. Inter-symbol interference is considerably reduced when the data rate of each carrier is low [2]. However, even though OFDM was first proposed in 1966 [3], it has only recently become the "modem of choice in wireless applications." It's now intrigued enough by OFDM's inner workings to play with some of its features. Recently, OFDM systems have received a lot of attention. Most high data rate wide-band communications now include applications. The use of high data rate transmissions in a multipath environment is common in audio and video broadcasting (DAB and DVB). Using OFDM, we may also employ a single-frequency network (SFN) where one broadcast multiplex can be sent at the same frequency by several transmitters. Figure 1 shows the basic block diagram of OFDM system. In this paper, we are working on channel by applying the G-distribution as a noise model as compared to AWGN. Generally, for the theoretical approach, AWGN noise channel is easy to simulate and implement but for the practical approach, the AWGN noise channel

R. Choudhary (✉) · A. Agarwal · P. K. Jain
Department of Electronics and Communication Engineering, Management and Gramothan, Swami Keshvanand Institute of Technology, Jaipur 302017, India
e-mail: rashmiec87@gmail.com

V. Devedzic et al. (eds.), *Proceedings of the International Conference on Intelligent Computing, Communication and Information Security*, Algorithms for Intelligent Systems, https://doi.org/10.1007/978-981-99-1373-2_26

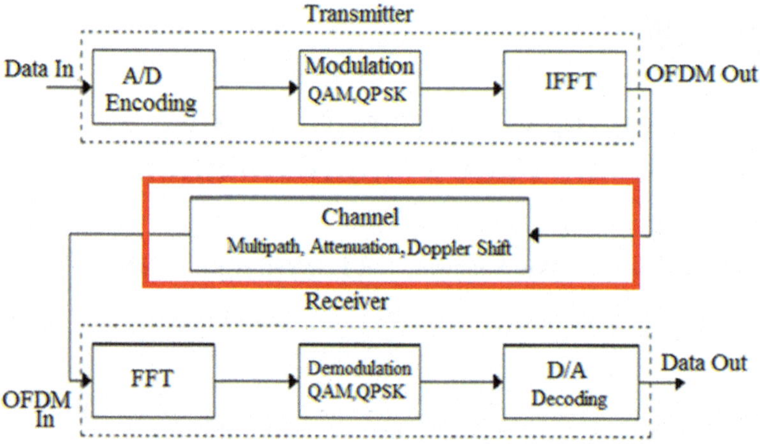

Fig. 1 Basic block diagram of OFDM system

is not considerable for the mobile and wireless communication. In different wireless communication, for different frequency band different types of noise channel is considerable in practical approach.

Section 1 provides an introduction to wireless communication and OFDM systems. Section 2 provides the literature survey about the OFDM system and noise distribution in the channel. This section also shows the classification of fading. Section 3 is purely discussed about mathematical modeling of Pade's approximation with the example to show how Pade's approximation is applied to solve the infinite series function. Section 4 discussed the mathematical modeling of G-distribution. Section 5 shows the results of bit error probability vs signal-to-noise power ratio for OFDM and 16-QAM modulation in the case of a G-distribution noise channel. This section also shows the bit error probability for different shadowing conditions.

2 Literature Review

Wireless is the fastest-growing method of communication that provides high data speeds and worldwide coverage at a low cost. When it comes to wireless communication, there are many distinct scenarios where multipath fading and shadowing might be experienced. A fading signal deviates from the carrier-modulated signal's attenuation transmission of data over a particular medium of propagation. When there is a signal, several versions of the sent signal arrive at the receiver within a short period of time different periods of time [4–6]. Multipath propagation is the term for this phenomenon. There are two types of fading: multipath-induced fading and shadow fading, both of which alter the propagation of waves in the environment. A random process is used to model the fading in terms of time, location, and/or radio

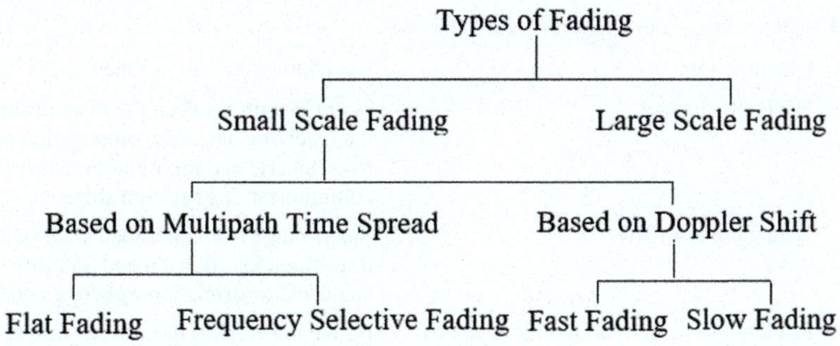

Fig. 2 Different types of fading in a communication system

frequency [7]. A fading channel is one that gradually loses signal strength over time. The lack of a direct line of sight between the mobile antennas and the base station is one of the primary causes of fading. The reflections from the ground and nearby structures cause multipath even when there is a direct line of sight. The arriving radio waves have varied propagation delays because they come from different directions. A receiver traveling at high speed can experience multiple fades in a short amount of time due to the constructive and destructive effects of multipath waves accumulating at distinct points in space. Fading is a spatial phenomenon if the objects in the radio channel are stationary, and only the motion of the mobile is considered. As the receiver passes through the multipath field, the spatial differences of the generated signal are interpreted as temporal fluctuations. Figure 2 shows the classification of fading appears in the communication system.

This different type of fading is classified as small-scale and large-scale fading. Small-scale fading appears when the signal does very rapidly whereas large-scale fading occurs when the signal does not vary rapidly [8, 9]. The reason for small-scale fading is due to the Doppler shifting of the transmitter or receiver and multipath signal transmission. When the channel's coherence time is greater than the delay restriction, slow fading occurs. It is possible to assume that the channel's amplitude and phase change will remain constant during the duration of its use, whereas coherence time is tiny relative to the delay restriction, and fast fading occurs as a result. Over the course of use, the channel's amplitude and phase change might vary greatly. If the mobile radio channel's strength and phase response are stable over a larger bandwidth than the transmitted signals, then flat fading will occur otherwise the signal is suffered from frequency-selective fading [4, 5, 7, 10]. Table 1 shows the different types of fading channels. These channels appear in wireless/Mobile/Satellite communication on the basis of application and condition of the channel. The Rayleigh distribution is made up of two Gaussian random variables with the same mean value and variance (σ^2). In the absence of a significant received component, Rayleigh fading is a widely accepted model for small-scale fast amplitude variations. There is a Rician distribution where a strong path meets a weak path. This strong component may be a line of sight route or one that arrives with reduced attenuation. Nakagami distribution, which is known as the m distribution is preferable for Land-first mobile's

Table 1 Channel type and environmental behavior

S. No	Channel type	Description and environment
1	Rayleigh channel	No LOS path between the transmitting and the receiving antennas, propagation over tropospheric and ionosphere, radio communications between ships
2	Nakagami-q (Hoyt)	Spans range from one sided Gaussian ((q = 0) to Rayleigh (q = 1)) and Satellite links subjected to strong ionosphere scintillation
3	Rayleigh (n = 0) to no fading ($n = \infty$)	Nakagami-n (rice) has a wide variety of spans and Rician K factor (n2 = k) -related One direct LOS component and numerous random weaker components make up the propagation route. Pico and micro-cellular environments can be found in urban and suburban areas, including inside and in factories
4	Nakagami-m (spans range from one Gaussian (m = 1/2), Rayleigh (m = 1) to fading ($m = \infty$))	Often best suited for land mobile, indoor mobile multipath propagation, and radio linkages to the ionosphere [11, 12]
5	Lognormal shadowing	Caused by terrain, building, trees-urban land mobile systems, land mobile satellite systems
6	Composite Gamma/Lognormal	Nakagami-m multipath fading overlaid with Lognormal shadows. Slow-moving individuals and cars in congested downtown districts. In addition, in the land mobile systems are shadowed by vegetation or urban areas

resolvable long-distance (LOS) paths are typically affected by fading. In multipath fading environments, the lognormal distribution is frequently employed to explain large-scale fluctuations in signal amplitude. Long-term fading or shadowing can be modeled using this technique.

3 Mathematical Modeling of Pade's Approximation

Pade's approximation is used to approximate infinite power series in terms of a rational function. This section basically discusses the method of calculating the Pade's approximation of any infinite power [13–15]. For better understanding, solve the e^{-z}function using Pade's approximation:

$$[L/M] = \left[\frac{a_0 + a_1 z + \ldots + a_L z^L}{b_0 + b_1 + \ldots + b_M z^M}\right]$$

$$f(z) = \sum_{i=0}^{\infty} c_i z^i = \left[\frac{a_0 + a_1 z + \ldots + a_L z^L}{b_0 + b_1 + \ldots + b_M z^M}\right] + O\left(z^{(L+M+1)}\right)$$

$$p^{[4/5]}(z) = \frac{a_0 + a_1 z + a_2 z^2 + a_3 z^3 + a_4 z^4}{b_0 + b_1 z + b_2 z^2 + b_3 z^3 + b_4 z^4 + b_5 z^5}$$

Equate the Pade's approximation function with the expansion of infinite function

$$\frac{a_0 + a_1 z + a_2 z^2 + a_3 z^3 + a_4 z^4}{b_0 + b_1 z + b_2 z^2 + b_3 z^3 + b_4 z^4 + b_5 z^5} = 1 - \frac{z}{1!} + \frac{z^2}{2!} - \frac{z^3}{3!} + \frac{z^4}{4!} - \frac{z^5}{5!} + \frac{z^6}{6!} - \frac{z^7}{7!} \ldots$$

We can also write this equation

$$a_0 + a_1 z + a_2 z^2 + a_3 z^3 + a_4 z^4 = \left(b_0 + b_1 z + b_2 z^2 + b_3 z^3 + b_4 z^4 + b_5 z^5\right) \left(1 - \frac{z}{1!} + \frac{z^2}{2!} - \frac{z^3}{3!} + \frac{z^4}{4!} - \frac{z^5}{5!} + \frac{z^6}{6!} - \frac{z^7}{7!}\right)$$

By equating the term of z in this equation

$$a_0 = b_0 = 1$$
$$a_1 = b_1 - b_0$$
$$a_2 = b_2 - b_1 + \frac{b_0}{2!}$$
$$a_3 = b_3 - b_2 + \frac{b_1}{2!} - \frac{b_0}{3!}$$
$$a_4 = b_4 - b_3 + \frac{b_2}{2!} - \frac{b_1}{3!} + \frac{b_0}{4!}$$

Using these equations, we can calculate the value of a_0, a_1, a_2, a_3, a_4 also the values of $b_0, b_1, b_2, b_3, b_4, b_5$. So the Pade's approximated function of e^{-z} in the form of is $P^{[4/5]}$.

$$p^{[4/5]}(z) = \frac{1 - 0.175z + 0.0337z^2 - 0.001z^3 + 2.13 \times 10^{-5} z^4}{1 + 1.062z + 0.483z^2 + 1.14z^3 + 0.148z^4 + 0.009z^5}$$

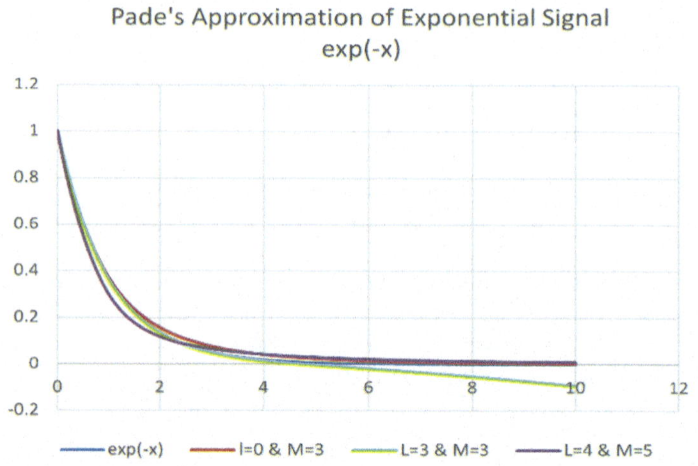

Fig. 3 Exponential signal and Pade's approximation of exponential signal

Figure 3 shows the result of Taylor series expansion of exponential signal and Pade's approximated signal for different value of L and M. From this figure, we can check that Taylor series expansion and Pade's approximated value of exponential signal is almost equal.

4 Mathematical Modeling of G-Distribution

This distribution is more appropriate distribution as compared to other distribution [19, 20]. The pdf of G-distribution is

$$f_x(x) = \left(\frac{\lambda}{\theta^2}\right)^{m+\frac{1}{2}} \sqrt{\frac{\lambda}{2\pi}} \frac{4m^m x^{2m-1} \exp\left(\frac{\lambda}{\theta}\right)}{\Gamma(m)\left(\sqrt{g(x)}\right)^{m+\frac{1}{2}}} \mathrm{K}_{m+1/2} g(x)$$

where

$$g(x) = \frac{2\lambda}{\theta^2}\left(mx^2 + \frac{\lambda}{2}\right)$$

And $k_v(.)$ is the modified Bessel procedure for a higher-order function. At m = 1, this distribution reduces to the Rayleigh-Inverse Gaussian A simple probability density function for the instantaneous composite signal-to-noise power ratio $f_\lambda(\lambda)$ is all that is needed to get the answer.

$$f_\gamma(\gamma) = A\frac{\gamma^{m-1}}{\sqrt{\alpha+\beta\gamma}^{m+\frac{1}{2}}}K_{m+\frac{1}{2}}\left(b\sqrt{\alpha+\beta\gamma}\right)$$

where the following constant have been used:

$$A = \frac{(\lambda\gamma)^{\frac{1+2m}{4}}}{\Gamma(m)}\sqrt{\frac{2\lambda}{\pi\theta}\exp\left(\frac{\lambda}{\theta}\right)\left(\frac{\gamma}{m}\right)^m}$$

$$b = \frac{1}{\theta}\sqrt{\frac{\lambda}{\gamma}}$$

$$\alpha = \lambda\gamma$$

$$\beta = 2m\theta$$

The moment generating function of G-distribution is given as

$$E\left[\gamma^n\right] = \sqrt{\frac{2\lambda}{\alpha\theta}e^{\frac{\lambda}{\theta}}}\left(\frac{\gamma}{m}\right)^m\frac{\Gamma(m+n)}{\Gamma(m)}K_{n-\frac{1}{2}}\left(\frac{\lambda}{\theta}\right)$$

$$M_x(s) = \frac{E\left[\gamma^n\right]}{n!}s^n$$

$$M_x(s) = \frac{\sqrt{\frac{2\lambda}{\alpha\theta}e^{\frac{\lambda}{\theta}}}\left(\frac{\gamma}{m}\right)^m\frac{\Gamma(m+n)}{\Gamma(m)}K_{n-\frac{1}{2}}\left(\frac{\lambda}{\theta}\right)}{n!}s^n$$

λ and θ are the shadowing parameter which is dependent on different shadowing condition. The shadowing environment that are mainly considered is infrequent light shadowing ($\mu = 0.115$ and $\sigma = 0.115$) corresponds to sparse tree cover, average shadowing ($\mu = -0.115$ and $\sigma = 0.161$) corresponds to an average tree cover and frequent heavy shadowing ($\mu = -3.914 \text{and} \sigma = 0.806$) corresponds to dense tree cover, m is the fading parameter.

5 Results and Discussion

The moment generating function for G-distribution is solved for the shadowing parameter (m = 4). The solution of moment generating function is as

$$M_x(s) = 1 - \gamma s + 1.199\gamma^2 s^2 - 1.962\gamma^3 s^3 + 4.189\gamma 4\gamma^4 s^4$$
$$-11.045\gamma^5 s^5 + 34.732\gamma^6 s^6 - 123.33\gamma^7 s^7 + 520.283\gamma^8 s^8 - 2399.4068\gamma^9 s^9$$

Pade's approximation of moment generating function of a [2/4] is calculated by method of calculating of Pade's approximation.

$$M_x(s) = P^{\left[\frac{2}{4}\right]}(s, m, \gamma) = \frac{1 + 5.873\gamma s + 7.061\gamma^2 s^2}{1 + 6.875\gamma s + 12.723\gamma^2 s^2 + 6.463\gamma^3 s^3 + 0.482\gamma^4 s^4}$$

The average bit error probability is as follows:

$$P_b(E) = C_1 M_x(s)$$

The bit error probability is given as

$$P_b(E) = C_1 \frac{1 + 5.873\gamma s + 7.061\gamma^2 s^2}{1 + 6.875\gamma s + 12.723\gamma^2 s^2 + 6.463\gamma^3 s^3 + .482\gamma^4 s^4}$$

For the OFDM, put the value of $C_1 = 1/2$ and $s = 1$

$$P_b(E) = \frac{1}{2}\left[\frac{1 + 5.873\gamma + 7.061\gamma^2}{1 + 6.875\gamma + 12.723\gamma^2 + 6.463\gamma^3 + .482\gamma^4}\right]$$

For the 16-QAM, put the value of $C_1 = 1/2$ and $s = 1/2$

$$P_b(E) = \frac{1}{2}[\frac{\left\{1 + 5.873\gamma\left(\frac{1}{2}\right) + 7.061\gamma^2\left(\frac{1}{2}\right)^2\right\}}{1 + 6.875\gamma\left(\frac{1}{2}\right) + 12.723\gamma^2\left(\frac{1}{2}\right)^2 + 6.463\gamma^3\left(\frac{1}{2}\right)^3 + .482\gamma^4\left(\frac{1}{2}\right)^4]}$$

In this Fig. 4, it is clearly shows that as move toward high value of average SNR bit error probability is less in OFDM modulation scheme $\left(5 * 10^{-4}\right)$ as compared to 16-QAM ($8 * 10^{-3}$). So OFDM is better modulation scheme in G-distribution system.

BER for m = 7 in case of frequent heavy shadowing, moment generating function is

$$M_x(s) = 1 - \gamma s + 1.109\gamma^2 s^2 - 1.56\gamma^3 s^3 + 2.7609\gamma 4\gamma^4 s^4\backslash - 5.51\gamma^5 s^5$$
$$+13.31\gamma^6 s^6 - 35.89\gamma^7 s^7 + 107.58\gamma^8 s^8\backslash - 353.79\gamma^9 s^9$$

Pade's approximation of moment generating function of a [2/4] is calculated by method of calculating of Pade's approximation.

$$M_x(s) = P^{\left[\frac{2}{4}\right]}(s, m, \gamma) = \frac{1 + 4.43\gamma s + 3.89\gamma^2 s^2}{1 + 5.40\gamma s + 8.31\gamma^2 s^2 + 3.89\gamma^3 s^3 + .399\gamma^4 s^4}$$

This Fig. 5 represents the BER versus SNR curve for the OFDM and 16-QAM modulation technique for shadowing parameter (m = 7). From this figure, it is represented that also for the case of heavy shadowing OFDM modulation is better than

Fig. 4 Comparison of BER for OFDM and 16-QAM modulation for m = 4

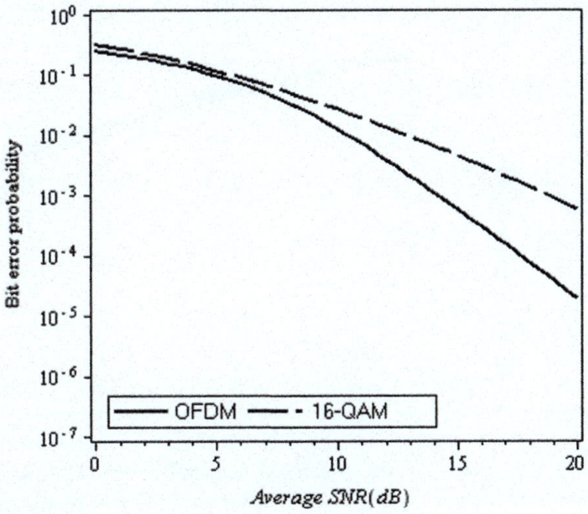

16-QAM modulation scheme. For the SNR = 20 dB, the bit error probability of OFDM is $6 * 10^{-4}$, and for 16-QAM, the BEP is $3 * 10^{-3}$.

Figure 6 shows the comparison graph for OFDM and 16-QAM modulation in the case of different shadowing parameter (m = 2, 4, 5). As we increase the shadowing effect in channel, BER decreases for particular value of SNR. It shows that if the effect of fading/shadowing increases, the error in bits also increases; on the basis of this comparison graph, it is clearly says that the OFDM modulation is better modulation scheme for the G-distribution noise channel.

Fig. 5 BER versus average SNR for OFDM and 16-QAM (m = 7)

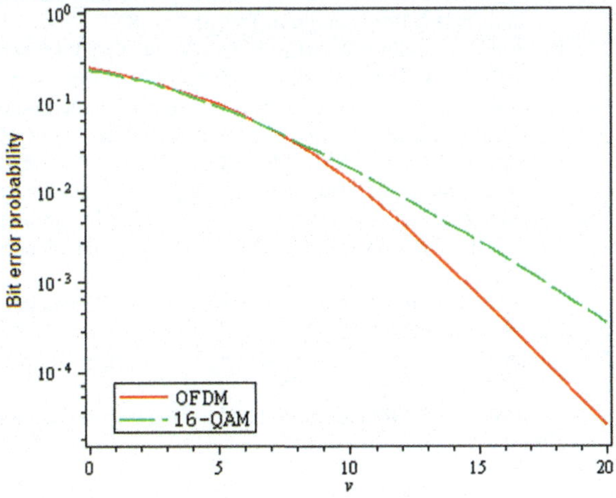

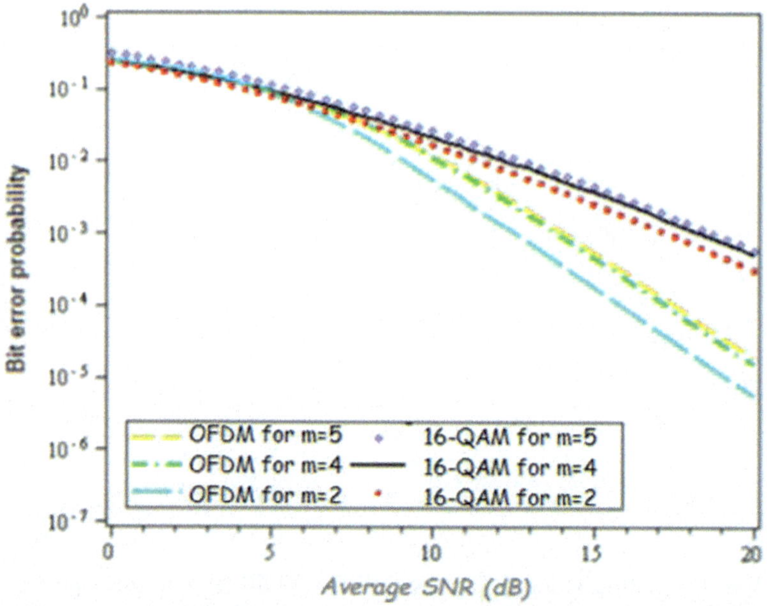

Fig. 6 Comparisons of OFDM and 16-QAM for different value of m = 2, 4, 5

References

1. Schulze H, Luders C (2005) Theory and applications of OFDM and CDMA John Wiley & Sons, Ltd.
2. Lui H, Li G (2005) OFDM-based broadband wireless networks design and optimization Wiley-interscience
3. Mosier RR, Clabaugh RG (1958) Kineplex, a bandwidth-efficient binary transmission system. IEEE Trans 76:723–728
4. Kostic IM (2005) Analytical approach to performance analysis for channel subject to shadowing and fading. IEE Proc Commun 152(6):821–827
5. Laourine A, Alouini MS, Affes S, Stephenne A (2008) On the capacity of generalized-K fading channels. IEEE Trans Wireless Commun 7(7):2441–2445
6. Jain P et al (2014) Impact analysis and detection method of malicious node misbehavior over mobile ad hoc networks. Int J Comput Sci Inf Technol (IJCSIT) 5(6):7467–7470
7. Laourine A, Alouini MS, Affes S, Stephenne A (2009) On the performance analysis of composite multipath/shadowing channels using the g-distribution. IEEE Trans Commun 57(4)
8. Mason S, Anstett R, Anicette N, Zhou S (2007) A broadband underwater acoustic modem implementation using coherent OFDM. In: Proceedings of national conference for undergraduate research
9. T. S. Rappaport, Wireless Communications - Principles and Practice, 2nd ed., Pearson Education India, 2009.
10. Simon MK, Alouini M-S (2004) Digital communication over fading channels, 2nd edn. John Wiley & Sons Inc., New York
11. Lathi BP (2003) Modern digital and analog communication system, 3rd ed. Oxford university press 54

12. Arora A, Gupta A et al (2022) Web-based news straining and summarization using machine learning enabled communication techniques for large-scale 5G networks. Wirel Commun Mobile Comput 2022, Article ID 3792816, 1–15. https://doi.org/10.1155/2022/3792816
13. Baker GA, Gammel JJL (1970) The Pade approximant in theoretical physics, vol 71, Academic Press
14. Amindavar H, Ritcey JA (1994) Pade approximations of probability density function. IEEE Trans Appers Aerosp Electron Syst, pp 416–424
15. Ismail MH, Matalgah MM (2006) On the use of pade approximation for performance evaluation of maximal ratio combining diversity over Weibull fading channels. J Wirel Commun Netw, pp 1–7
16. Sharma G, Agarwal A, Dwivedi VK (2013) Performance evaluation of wireless multi-path/shadowed G-distributed channel. J Eng Technol Res 5(5):139–148

Chapter 27
Smart Living: Safe and Secure Smart Home with Enhanced Authentication Scheme

C. M. Naga Sudha, J. Jesu Vedha Nayahi, S. Saravanan, and B. Jayagokulkrishna

1 Introduction

With the fast improvement of the economy, expectations for everyday comforts and the weakening of environment, individuals' interest for healthy, comfortable and good living condition are more earnest. Future life will have higher insight, lower energy utilization and more effective utilization of sustainable power source. With the development of technology, the security factor of smart home is improved. Smart home coordinates programmed control innovation, portable correspondence innovation and so forth. Smart home hardware has progressively entered the general home. Even though a few advancements have been generally experienced, there is as an expansive space for development in this field. The elderly and disabled people can get help in evaluating their body parameters with the help of smart gadgets of this environment. The support for speed growth in smart home surroundings has been provided by the advancement in the information and communication technology and Internet.

Smart home alludes to any structure incorporated with an imparting domestic gadgets and appliances empowering home proprietors to manage the ability of houses via complex monitor norms. These permit house proprietors to robotize exercises of natural conditions without their contribution utilizing different sensors detector, motion detector, temperature sensors and soon. Smart homes can provide various functionalities extending from automated security systems to alert the users.

C. M. N. Sudha (✉) · B. Jayagokulkrishna
Anna University-MIT Campus, Chennai, India
e-mail: cmsudha30@gmail.com

J. J. V. Nayahi
Anna University, Regional Campus, Tirunelveli, India

S. Saravanan
K. Ramakrishna College of Technology, Trichy, India

V. Devedzic et al. (eds.), *Proceedings of the International Conference on Intelligent Computing, Communication and Information Security*, Algorithms for Intelligent Systems, https://doi.org/10.1007/978-981-99-1373-2_27

The one-to-one checking of the home conditions is limited by the smart home just as giving remote access to the smart machines using sensors on vital locations in houses. Effectiveness and security to the environment is assisted in house automation. Utilizing the technology of IoT, various frameworks are controlled in mechanizing the home.

To get certain smart devices connected (e.g., temperature humidity sensor), initially, registration by the user at the trusted authority is needed. Similarly, all smart devices and the gateway node GWN (which acts as the bridge between the smart device and users, connects smart devices to the external world using the Internet) must be registered at the registration server (RS). The GWN takes the work of controlling network interoperability along with secured managing. The trusted server for registering all the smart device (SD), users and GWN is the registration authority (registration server). After the successful registration of users, smart device and GWN securely, this information is stored in memory of smart phones of user and also in memory of smart device GWN by the RS, which are further used at the time of authentication and key establishment process. To have access to a smart device, user needs to send an authenticated request to GWN as both have already done the registration phase at the registration server. Three categories of mutual authentications occur (1) between the user and the GWN, (2) between the GWN and the smart devices and (3) between the users and the smart device. Moreover, a secret session key is created between users and smart device to protect interchanged messages (Fig. 1).

2 Literature Review

A wireless vision sensor network (WVSN) for such smart home is implemented by Tsai et al. [1], which include the characteristics of VSN and server. The main feature is that complete system-based algorithm has been established to carry out a vision-based analysis. Kumar et al. [2] have implemented utilizing RF-based correspondence in a family unit to make an IoT-enabled smart home security framework and introduced low-cost engineering for a smart entry way sensor and enhance the facet of defense from unauthorized accidents.

Xiao et al. [3] mainly focused on implementing home-shield which is a credential-less authentication framework for smart home systems and advantage is to effectively defend against various attacks namely open-port problem and over-privilege problem. Isai et al. [4] implemented a wireless home automation communication and security with Internet of things and the key threats along with their associated solutions in IoT by identifying various domains susceptible to security along with privacy attacks.

Santoso and Vun [5] implemented a smart home using the Rapid Application Development (RAD) methodology, consists of three categories namely cloud system, the remote system, simulated system. Chottaraj and Subhankar [6] implemented the system which are uniquely identified and securing IoT for smart home system represented virtually configured.

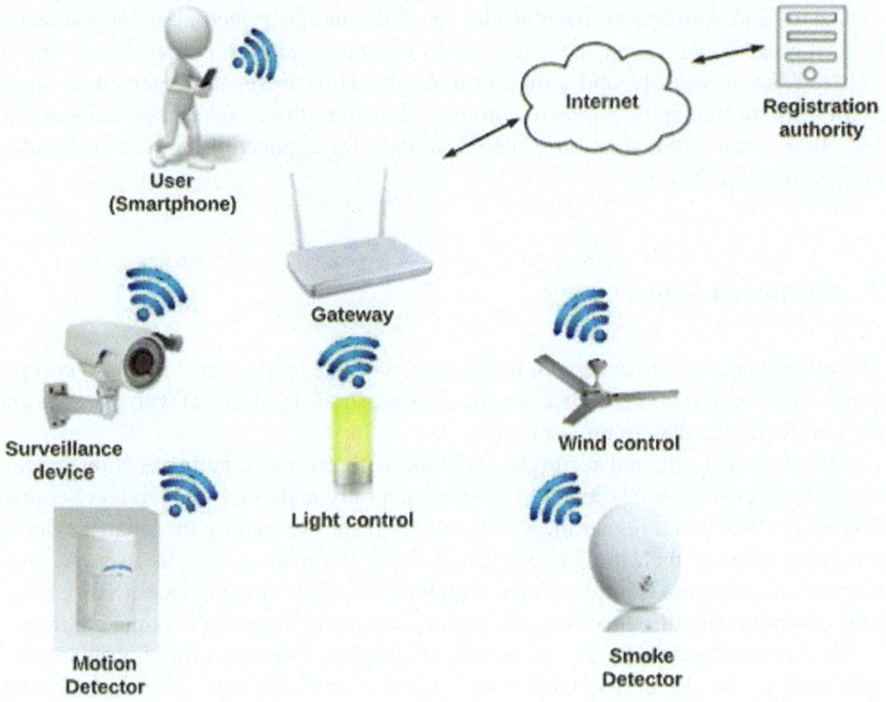

Fig. 1 Smart home network

Smart home automation is demonstrated Soliman and Moataz [7] used different sensors along with Arduino which allows house owners to activate the activities of their surrounding conditions using various sensors without their engagement. It sends digital signals depends on the data retrieved from the sensors, and various functions range from automated security systems like display lights forward email to legal users [8]. House automation describes about the process of implementing micro-controllers or CT to have accessor to manage and control home appliances.

Pirbhulal et al. [9] implemented automation that overcome wild challenges and opportunities, and emphasize every appliance that are supposed to be registered to DCL where authorized users access 100 home gateway through smartphones. It also explains how unauthorized access to the department can be detected and trigger the alarm automatically.

In addition, the lights are switched on at nights, when a motion is detected with the help of hidden web cameras. Also, smart home with enhanced security using a sensor network is implemented to facilitate efficient encryption of data. A triangle-based security algorithm provides safe and secure smart home system—(TBSA) which consumes less energy.

Davies and Anireh [10] have used a set of trusted encrypted keys for protection against attacks and trigger the action when malicious activity is performed. In [11, 12], IoT-based security and congestion mechanisms were implemented in smart home, and mobile applications to automate the smart home were implemented [13, 14]. Two-factor authentication is also described for connecting to the cloud and to update the data [15].

3 Proposed Work

The ultimate point is to design an IoT system with the help of most required components for a secured home such as smart window, smart light, smart web cameras and then allowing the data to the users (Fig. 2).

The proposed safe and secure smart home with enhanced authentication scheme is good for resource-constrained SDs because it only makes use of Wireless Security Protocols (WSP) in a registration authority. A response sent by the smart device to the gateway node then the gateway node sends the response to the user. A smart network-based home is implemented with the ease of the smart devices (SDs). Every SDs communicate through wireless channels by using node of the home gateway.

This system includes many smart objects that are used executing house automation such as garage doors, fans, doors, lawn sprinklers, web cameras and many sensors. The registration server, home gateway are used in constraining the objects and sensors. It gives a better environment for programming which controls the objects

Fig. 2 Smart home functionality

that are in connection and deliver control mechanisms through the registration of home gateway smart devices.

(1) **Home Gateway**: A smart device is directly registered with IoE on a house gateway or network base. The house gateway offers 4 Ethernet ports as well as wireless contact point on channel 6 equipped with the SSID "home gateway." For better safe connections, it is probably possible to configure WEP/WPA-PSK/WPA2companies to wireless links.

It is very easy to manage IoE system. The IP address of home gateway (LAN) is 192.168.25.1, but it can be reached via its IP address in front of Internet (Figs. 3 and 4).

(2) **Registration Server**: In the registration server, the smart devices are registered. The IP address of the registration server is configured.

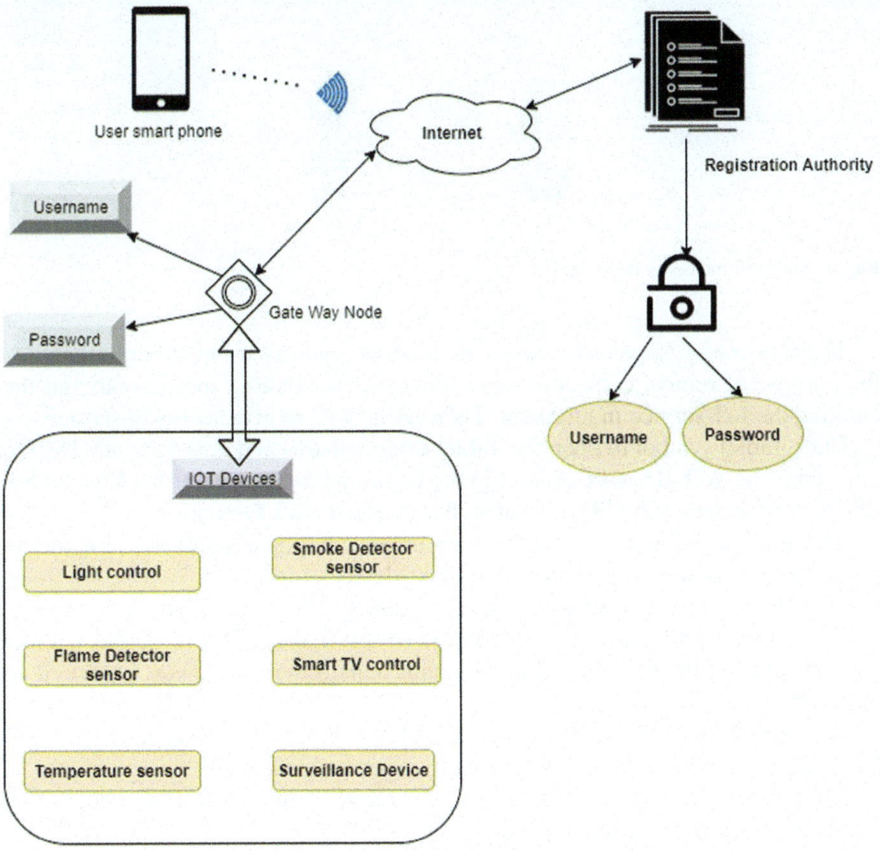

Fig. 3 Block diagram of the proposed safe and secure smart home with enhanced authentication scheme

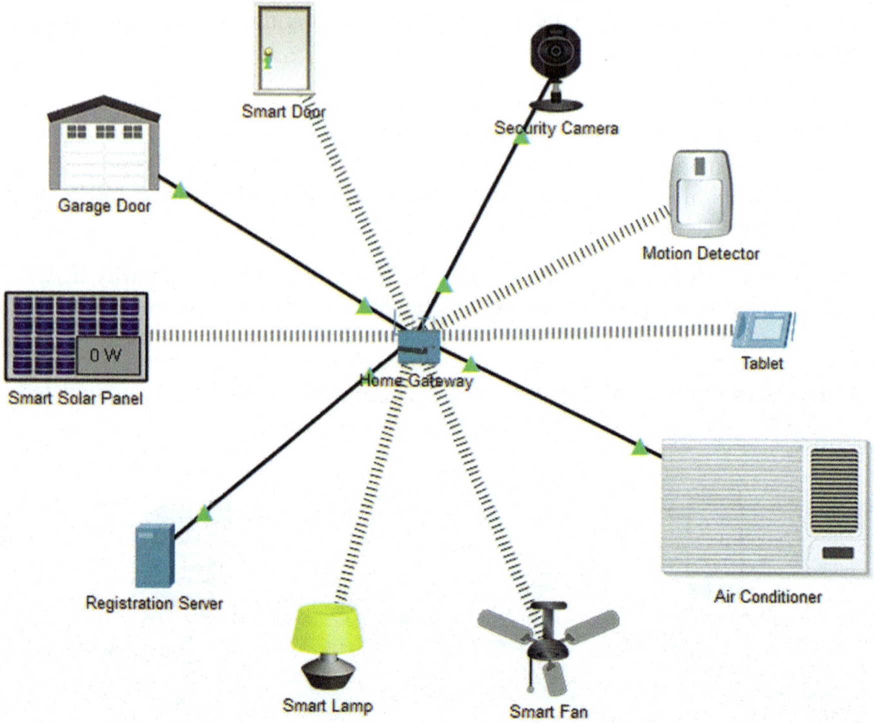

Fig. 4 Smart home network in Packet Tracer

The registration server is connected to home gateway using Ethernet cable for the purpose of remote control of smart devices. By clicking services tab and then clicking the IoT service in left pane. Then click "On" to give access to service.

Click Tablet symbol to open the Tablet setup window and next go to the Desktop tab. Select the Web Browser symbol in the Desktop tab. Then enter the IPv4 address of registration server in URL box in web server and click Go option.

Create another login account by Sign up and then give another username and password. Initially, there will not be any IoT devices enrolled.

Now the smart devices are registered by configuring them. The configuration is done by changing the gateway from static to DHCP. In the IoT server settings, remote server is clicked which will enable three tabs namely server address, password and username.

In server address tab, the IP address of RS has to be given, username and password have to be entered as given in the registration server login.

After registering the smart device, to the registration server, all the devices are retrieved by legitimate users via web.

Thus, all other smart devices are to be registered. After registering all smart devices to registration server, click Tablet option and then go to Desktop tab and then click Web Browser option. In the URL box, enter IP address of the registration

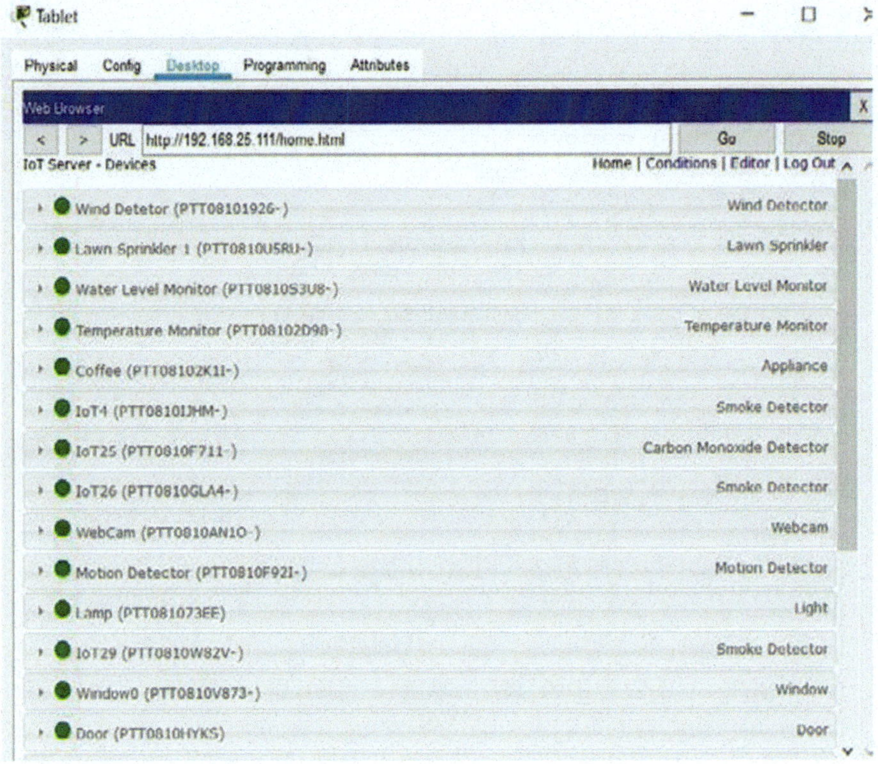

Fig. 5 Smart devices registered to registration server

server. Type the username as well as password and click Sign In. So, in IoT server–devices window, all smart devices that are registered with the registration server will be displayed as shown in Fig. 5.

(3) **Microcontroller (MCU-PT) Board**: The MC board is used for connecting different objects (Fig. 6).

3.1 Flowchart

The flowchart of the proposed safe and secure smart home with enhanced authentication scheme is explained as follows (Fig. 7).

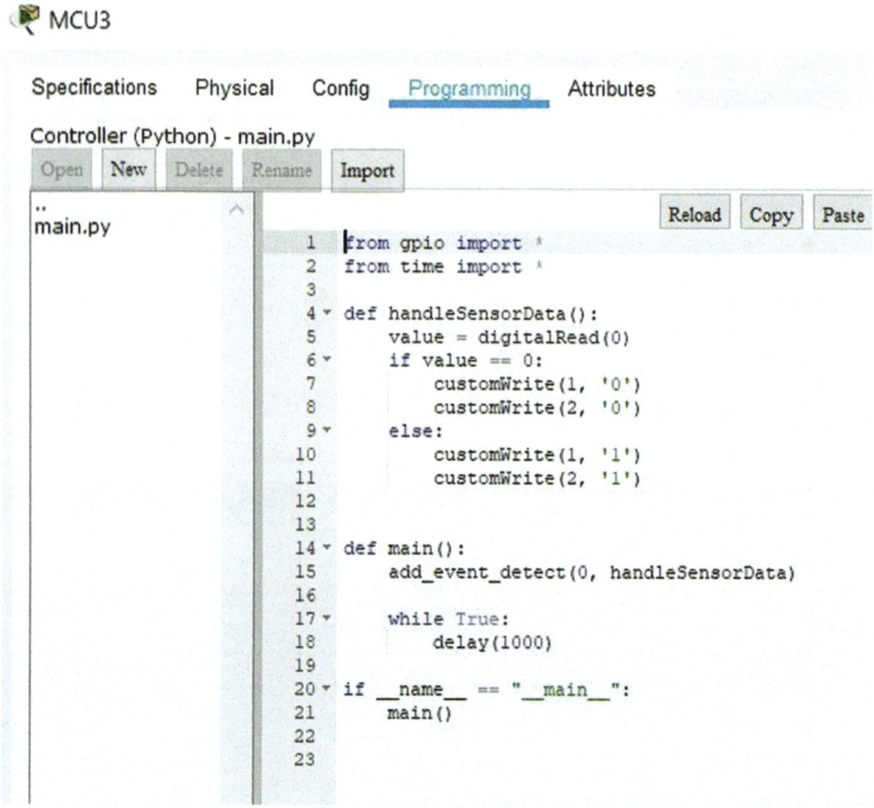

Fig. 6 MCU programing environment

3.2 *Device Used for Design*

The devices used in the proposed safe and secure smart home system are given as follows (Table 1).

3.3 *Architecture Diagram*

The proposed safe and secure smart home with enhanced authentication scheme consists of connected components as shown in Fig. 8. All the data which are fetched from the connected components are transmitted to the cloud services.

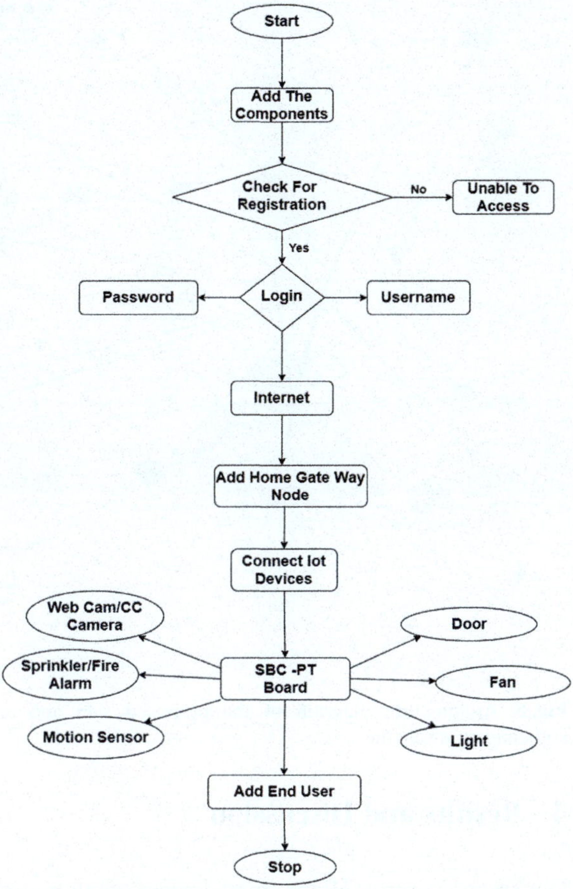

Fig. 7 Flowchart of the proposed safe and secure smart home with enhanced authentication scheme

Table 1 Devices connected in the proposed safe and secure smart home with enhanced authentication scheme

S. No.	Devices	Functions
1	Cable modem	Connect the Internet at home
2	Home gateway	Register smart objects with IP address
3	Registration server	Monitor the intelligent things that are recorded
4	MCU	Connect different intelligent things
5	PC	Link to home destination
6	Ceiling fan	Used for ventilating home environment
7	Webcam	Control the home
8	Light	Provide light
9	Motion detector	Links registration server and detect motion
10	Smart door	Link to registration and provide event based on functions

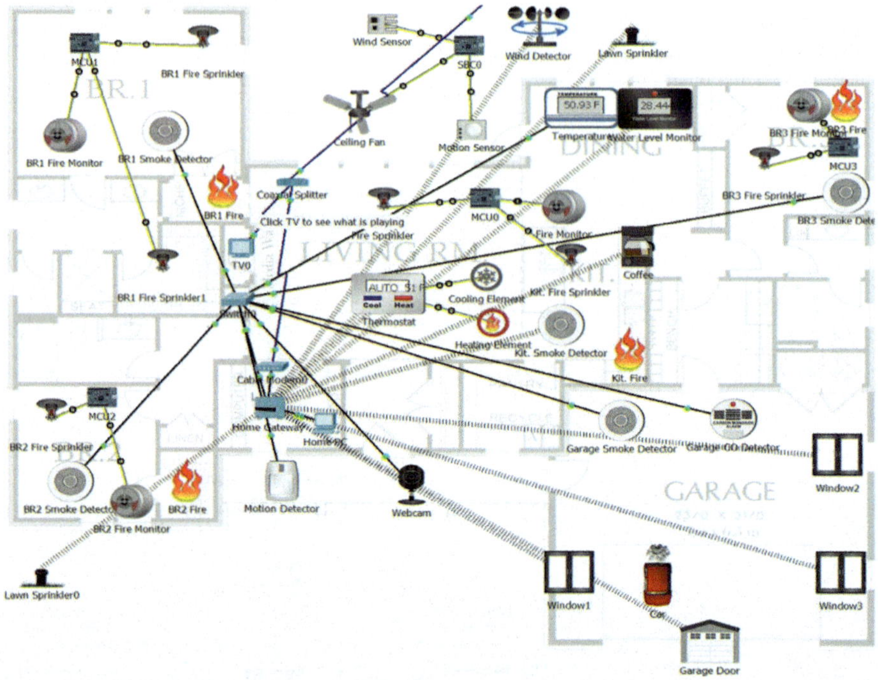

Fig. 8 Architecture diagram of the proposed safe and secure smart home with enhanced authentication scheme

4 Results and Discussion

Our system controls all the smart devices of our house through home PC. The system has improved the security system of proposed safe and secure smart home with enhanced authentication scheme.

Figure 9 demonstrates the graph of the smart home environment which is propositional to the time and values. The vertical and horizontal axes show the temperature values and time taken of the each environment in smart home, respectively. The colored solid line demonstrates every environment which fluctuates in the smart home. The blue solid line speaks to the ambient temperature of the proposed safe and secure smart home with enhanced authentication scheme which ranges 56 °F at 12 noon. The red solid line indicates the sunlight of the smart home environment. The yellow solid line indicates the water temperature; the light green solid line indicates the fire, and the light yellow indicates the wind direction of the environment.

Figure 10 shows the connectivity of the registration server to get the username and password through network provider.

Figure 11 shows watering the plants utilizing the lawn sprinkler. Watering the lawn could be an activity controlled through conditional statements configured in the local IoT server on home gateway. In figure, lawn sprinkler is in on mode dependent in

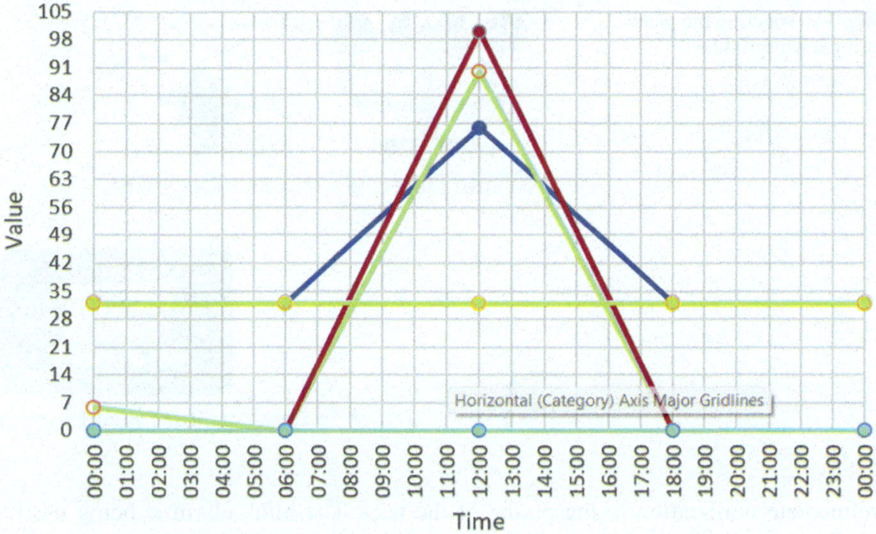

Fig. 9 Graph of the smart home environment

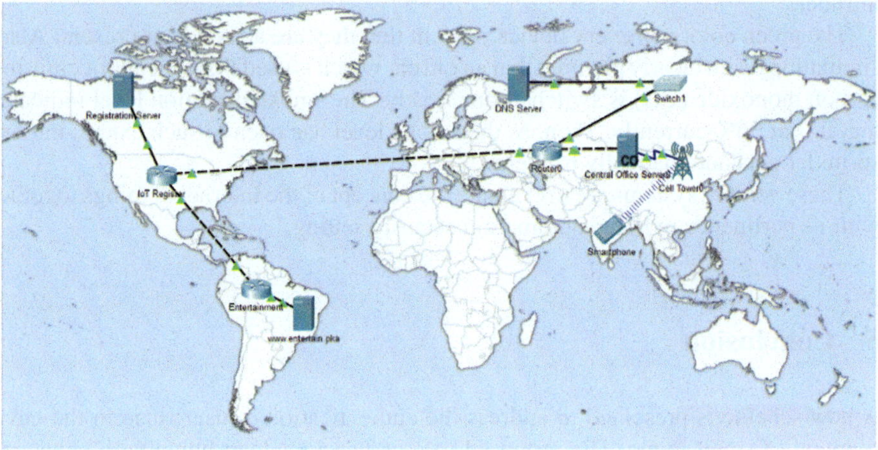

Fig. 10 Register service output

a home gateway condition as well as if a lawn sprinkler is in off, if the water level screen arrives at in excess of 10 cm water level.

The proposed safe and secure smart home with enhanced authentication scheme depicts the IoT devices that were registered on such a gateway of home with status. Red ones indicate that the device is currently down. When the smoke detector level is zero, it actually means that the environment is free from smoke; and it is activated. It is proved that if a stranger tries to have access to the building, it will send an

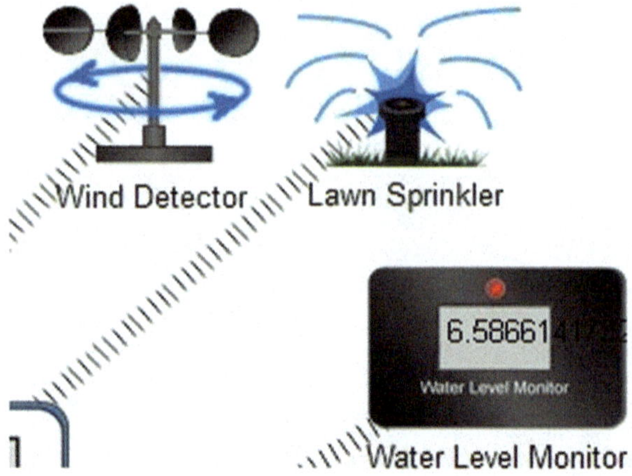

Fig. 11 Watering the plant using lawn sprinkler

immediate notification to the phone of the user. The blink alarm is being used to picturize the notification that will be sent. Also, if user is able to gain access into the building, the motion detector will detect his movement and the motion detected. It will also switch on smart light and triggers the webcam to record the activities of the intruder.

The green colors on every device portrait that they are activated at present. Also, simulation result clearly pictured that an old car which is used like a means of emitting carbon monoxide (CO) is switched on already. The smoke detection level indicates that it is at 36% currently. Then, as the smoke level has risen to such extent, the fire sprinkler gets automatically triggered.

These venture systems are to evaluate the concept of the Internet of things together with its pertinence in home automation security setting.

5 Conclusion

A new scheme is presented to address the authentication of user issue in the environment of smart home. The proposed safe and secure smart home with enhanced authentication scheme provides additional functionality features. IoT is another innovation that is being utilized for the interconnection of gadgets with the help of the web-based associations. It mainly empowers the gadgets in order to detect gadgets remotely. Overall, the proposed safe and secure smart home with enhanced authentication scheme delivers a good trade in security as well as the functionality things and over heads when compared to other pre-existing related schema.

6 Future Work

Smart home automation system services the segregation of objects that are connected through the motion sensor, server and switch connection among things. This system uses a computer PC to monitor the home components. This system is useful from various systems, because they are instinctively changed their state. After a lag time-sensing, components go back to their previous state. This process happens back and forth. Mobile phone or computer is not only a material for communication but also tries to give us better control for automated home. Real-time implementation with all the hardware components is our future work.

References

1. Tsai T-H, Huang C-C, Chang C-H, Hussain MA (2020) Design of wireless vision sensor network for smart home. Department of Electrical Engineeing, National Central University, Taoyuan, Taiwan, p 32001
2. Kumar PP, Krishna M, Ramprakash MR (Sep 2019) Design and implementation of smart home using cisco packet tracer simulator. Int J Innov Technol Explor Eng (IJITEE) 8(11S). ISSN: 2278-3075
3. Xiao Y, YizhenJia CL, ArwaAlrawais MolkaRekik, Shan Z (2020) HomeShield: a credential-less authentication framework for smart home systems. Guang University of Finance andEconomics, Guangzhou, China, School of Information Science
4. Isai U, Karthikeyan G, Harideesh R (2020) Wireless home automation communication and security with internet of things. In: International conference on emerging trends in information technology and engineering (ic-ETITE)
5. Santoso FK, Vun NCH (2017) In securing IoT for smart home system. In: International symposium on consumer electronics (ISCE), Madrid, Spain, pp 1–2
6. Chattoraj S, Subhankar C (2015) Smart home automation based on different sensors and arduino as the master controller. Int J Sci Res Pub 1–4
7. Soliman M, Moataz S (2013), Smart home: integrating internet of things with web services and cloud computing. Cloud Comput Technol Sci (CloudCom) 2
8. Brush A, Lee B, Mahajan R, Agarwal S, Saroiu S, Dixon C (2011), Home automation in the wild: challenges and opportunities. In: Proceedings of the SIGCHI conference on human factors in computing systems, pp 2115–2124
9. Pirbhulal S, Zhang H, Alahi MEE, Ghayvat H, Mukhopadhyay SC, Zhang Y-T, Wu W (2017) A novel secure IoT-based smart home automation system using a wireless sensor network. Sensors 17(1):69
10. Daviess EI, Anireh VIE (2019) Design and implementation of smart home system using internet of things. J Digit Innov Contemp Res Sci Eng Technol 7(1)
11. Volner R, Bore P, Smrz V (2018) A product based security model for smart home appliances. In: 11th International biennial Baltic electronics conference. Tallinn, Estonia, pp 221–222
12. Abdalla M, Fouque P, Pointcheval D (2016) Password-based authenticated key exchange in the three-party setting. In: 8th International workshop on theory and practice in public key cryptography (PKC'05). (Lecture notes in computer science (LNCS)), vol 3386. Les Diablerets, Switzerland, pp 65–84
13. Alghazzawi D, Bamasaq O et al. (2021) Congestion control in cognitive IoT-based WSN network for smart agriculture. IEEE Access 9:151401–151420. https://doi.org/10.1109/ACCESS.2021.3124791

14. Mandula K., Parupalli R., Murty CHAS, Magesh E, Lunagariya R (Dec, 2015) Mobile based home automation using internet of things (IoT). In: International conference on control, instrumentation, communication and computational technologies (ICCICCT), pp 18–19
15. Wazid M, Das AK, Odelu V, Kumar N, Susilo W Secure remote user authenticated key establishment protocol for smart home environment. IEEE Trans Depend Sec Comput. https://doi.org/10.1109/TDSC.2017.2764083

Chapter 28
Role of Internet of Things (IoT) in Preventing and Controlling Disease Outbreak: A Snapshot of Existing Scenario

Manpreet Kaur Dhaliwal, Rohini Sharma, and Naveen Bindra

1 Introduction

The Internet of Things (IoT) is a network of physical objects or "things" that are embedded with sensors, software, and other technologies that enable them to communicate and exchange data with other devices and systems over the Internet. Numerous IoT applications are proposed by researchers in diverse fields like smart manufacturing [1], agriculture [2], education [3], automobile [4], etc., that can be more productive in terms of the potential use of resources, good growth conditions and high-quality yield, safety, efficiency, maintaining discipline, and cleanliness. According to experts, there will be more than 22 billion connected IoT devices by the year 2025, up from more than 7 billion currently [5]. Health care is one of the IoT's application domains that has received a great deal of interest from the health sector, academia, and public sector.

According to the US Centers for Disease Control and Prevention, "An epidemic is an increase in the number of disease cases beyond what is usually expected in a geographic region". In most cases, the surge in cases occurs rapidly. In history, various epidemics are reported like plague, small pox, cholera, measles, Ebola, chikungunya, and COVID-19 which wiped out countries' population. The maximum fatality rate is noticed in Ebola disease, i.e., 50% [6] followed by MERS-CoV34.3% [7] and the minimum fatality rate is of chikungunya 0.1% [8]. In most outbreaks, the mode of transmission is respiratory and fever, cough, shortness of breath, fatigue, etc., are the

M. K. Dhaliwal (✉) · R. Sharma
Department of Computer Science and Applications, Panjab University, Chandigarh, India
e-mail: preet2016@pu.ac.in

R. Sharma
e-mail: rohini@pu.ac.in

N. Bindra
Postgraduate Institute of Medical Education and Research, Chandigarh, India

V. Devedzic et al. (eds.), *Proceedings of the International Conference on Intelligent Computing, Communication and Information Security*, Algorithms for Intelligent Systems, https://doi.org/10.1007/978-981-99-1373-2_28

major symptoms. These symptoms can be detected using wearable devices, and thus remote monitoring of patients is possible. Therefore, the use of wearable devices based on IoT can be quite helpful in checking the spread.

This study is an attempt to understand the significance and usefulness of IoT to control outbreaks and formulate future directions. Research papers published from 2010 to 2022 have been considered. As the use of IoT to control the outbreak is quite naïve, this review study is the first step to the best of our knowledge that can help to understand the current scenario and define the way for future studies in the field. It critically examines every aspect of IoT in disease outbreaks, providing more clarity to researchers and proposed a hybrid architecture for dealing with the limitations of the existing frameworks. The paper has been divided into five main sections. Section 2 discusses the significance of IoT in disease outbreak. Section 3 discusses framework/architectures, for handling epidemic diseases using IoT technology. Techniques used in frameworks are discussed in Sect. 4. Datasets used by researchers are discussed in Sect. 5. Key findings and challenges are covered in Sect. 6. Section 7 proposes a hybrid framework that can be useful to handle disease outbreaks. Finally, the conclusion of the paper is given in Sect. 8.

2 Significance of Internet of Things

IoT-based solutions are proposed to handle various medical conditions including increased temperature, heartbeat [9], stroke detection [10], old age and disability [11], and disease outbreaks like COVID-19. Chamola et al. [12] discussed about numerous IoT technologies to support the battle against COVID-19. These IoT technologies include drones, wearable devices, mobile apps, autonomous vehicles to deliver meals and medical supplies, IoT buttons are used to maintain the cleaning standard, smart thermometers connected to mobile apps trace high fever persons in the region and robots. Pratap et al. [13] discussed IoT-based applications in COVID-19 that help in the superior treatment, timely diagnosis, error-free reports, inexpensive treatment, timely control, monitoring of disease, interconnected hospitals, rapid screening, telehealth consultation, smart tracing of infected patients, wireless healthcare, informing medical staff during an emergency, etc., which are being used in few countries during pandemic. Aman et al. [14] discussed the use of an Internet of Medical Things (IoMT)-based architecture and technology and how IoMT was used by Germany, Taiwan, and South Korea to regulate the spread of infectious diseases. Swayamsiddha and Mohanty [15] discussed that Cognitive Internet of Medical Things (CIoMT) enables real-time tracking, contact tracing, faster diagnosis, and remote health monitoring of patients suffering from COVID-19.

Apart from this, IoT-enabled smart helmets [16], smart glasses [17], IoT-Q-Band [18], and easy band [19] are used for tracking quarantine cases and monitoring social distancing. IoT-based surveillance systems using face images and IoT-enabled healthcare platform for patients in ICU bed [20] are solutions proposed by researchers to handle COVID-19 outbreak. Few solutions have also been implemented by countries.

According to [21], CMED Health offered 1.5 million people in Bangladesh an IoT-linked health monitoring solution through a mobile app. The Ministry of Health and Family Welfare, Government of India [22] and UNDP designed and implemented the electronic vaccine intelligence network (eVIN), an IoT-linked mobile device focused technology that allows control of vaccine cold chain logistics in real time. This IoT capable technology monitors vaccine position, temperature, stock levels and reduces vaccine stock-outs by 80%. In China [23], Internet of things technology is used to monitor temperature and humidity, oxygen density, personnel flow, fire conditions, and other abnormal conditions in the ward. The patients use wearable devices for real-time feedback of body data with the help of 5G or Wi-Fi.

3 Frameworks/Architectures for Handling Disease Outbreak Using IoT

The studies are broadly divided into two categories:

- **Empirical Studies**: These studies are based on simulation.
- **Non-empirical Studies**: These studies consist of frameworks/architecture which are proposed by researchers but are not simulated or experimented.

3.1 Empirical Studies

For the identification and tracking of Ebola-infected patients, Sareen et al. [24] proposed an architecture based on WBAN and RFID to remotely monitor the patient in real time using cloud computing. The J48 decision tree is used to determine whether or not a user is infected. RFID detects close proximity interactions (CPIs) between users, and temporal network analysis (TNA) is used for assessing which regions or users are infected. Synthetic data of 2 million people are used to assess the model. A comparative study of J48, random tree, Naive Bayes, and REP tree classification techniques is performed. J48 achieves higher accuracy. The study's main flaw is that it was conducted on a fictitious dataset.

Sood et al. [25] proposed a fog-based framework that uses fuzzy-C means (FCM) clustering to diagnose users as infected or uninfected from chikungunya and sends diagnostic notifications to the user's mobile device. Body sensors, water quality detector sensors, mosquito sensors, RFID tags, GPS sensors, climate detecting sensors are used for collecting data to form a real dataset. The status of the outbreak of mosquito-borne disease is presented using social network analysis (SNA). It sends out alarm notices to users who are not infected but are traveling or staying in infected areas and GPS locations for rerouting to a safer path. The key benefit of the proposed framework is that it is energy efficient due to its fog design, which generates real-time notifications of risky locations with minimal delay. Priti and Sandeep [26] suggested a cloud network system for the forecasting and inhibition of the chikungunya infection. K-means clustering is used to predict whether the patient is chikungunya infected or not. Data are collected through a mobile application manually, where the user enter all the symptoms and sensors are used for collecting chikungunya spreading locations. The information is stored on the cloud that is accessed by doctors, hospitals, caretakers, and other users. The researchers stated that data is obtained through body worn sensors as well as manually reported symptoms, however, the sensors used to collect data and the process of data collection has not been discussed.

A fog-based architecture is proposed for controlling and mitigating Zika virus by Suseendran et al. [27]. The Naive Bayesian network (NBN) classifier uses symptoms to determine whether a user is infected or not. GPS is used to map the location of the sites specified by the sensors to find risk-prone areas. The study's primary shortcoming is that it is based on a synthetic dataset. Sareen et al. [28] proposed a cloud network system for controlling Zika virus outbreak. Data are collected using an app in which the user fills all the symptoms. Sensors are used for collecting information about the locations where Zika spreads and GPS locations are shared to prevent disease outbreaks. NBN is used to identify if a user is infected or not. The proposed framework achieved 89% accuracy in identifying the risk-prone areas. To detect and resist Zika virus, a fog-based architecture is presented by Sareen et al. [29] where fuzzy k-nearest neighbor classifier is applied over symptoms data to determine if a user is infected or not. Environmental sensors are placed to collect data, and the user's symptoms are provided by the user himself using an application. GPS is used to map the locations of the sites specified by the sensors and finding the risk-prone areas and also reroute the users to safer paths. The proposed system is based on synthetic dataset but achieved 95.5% accuracy in identifying risk-prone areas. It provides security to user's data using information granulation concept, handles large number of users and also provides pathway to handle the epidemic situation.

To recognize the probability of MERS-CoV infection and recovery, Alturaiki et al. [30] used Naive Bayes classifier and J48 decision tree algorithm. The real data is collected from the control and command center of the Saudi Ministry of Health which is quite limited in size, and features are not clearly stated. To predict and avoid the MERS-CoV outbreak, Sandhu et al. [31] also suggested the same kind of approach where Bayesian belief network (BBN) is used to identify if the user is infected or not. The study is based on synthetic dataset and does not provide a detailed description of the body worn sensors used to collect data.

To track and manage H1N1 virus, a cloud-based framework was proposed by Sandhu et al. [32]. The user's data are gathered and sent to the cloud via cell phones. Infected and uninfected people are classified based on their symptoms-response. Users infected with H1N1 virus are closely monitored until they are recovered. The information and recommendation box on the device component is used to gather all data about users, doctors, and hospitals and provides the user with current details and recommendations concerning the disease using text messages or e-mails. An Outbreak Role Index (ORI) is utilized to assess the magnitude of a user's ability to transmit or acquire the virus. The graph for an infected person is created using social network analysis that helps uninfected users in case of local exposure. Security issues are not discussed in the study, and this system is slower than fog-based system in terms of minimum response delay.

Ketu and Mishra [33] proposed multitask Gaussian process (MTGP) regression model for forecasting of COVID-19 virus outbreak. In this paper, 1, 3, 5, 10 and up to 15 days advance forecasting of confirmed cases have been performed worldwide. It also discussed the importance of IoT and how MTGP helps in lessening the effect of COVID-19. The study is not based on real-time dataset. Otoom et al. [34] projected an IoT and cloud-based architecture to collect real-time symptoms from wearable devices and use a machine learning algorithm for predicting and monitoring the COVID-19 cases. These predicted data are used by healthcare physicians to respond quickly to a patient and further investigate the COVID-19 cases. The dataset used in the study has not been discussed much. Bowen Wang et al. [35] proposed a risk aware adaptive identification (RAI) algorithm that helps in disease control and prevention by early identifying COVID-19 cases. This algorithm is based on susceptible-exposed-infected-removed (SEIR) model. Social Internet of Things (SIoT) is used to detect the geographic location and proximity among two users. Collected data from SIoT are converted into minimum weight vertex cover graph which helps to identify high-risk users that lead to minimizing the rate of epidemic propagation. Ahanger et al. [36] proposed a four-layer framework to predict the COVID-19 and also alert users if an infected person is near them. Health sensors, ecological sensors, meteorological sensors are used for collecting data which is then divided into two classes, i.e., normal or infected using fuzzy-C means. In-depth implementation of the framework is missing and does not ensure the security aspect of user's data. Aljumah [37] projected five-phase IoT-fog architecture for early prediction of COVID-19 using various machine learning techniques and simultaneously sharing the information with government agencies, doctors, and healthcare workers. However, the security and privacy concerns of user's data are not ensured.

Vijayakumar et al. [38] proposed a fog-based health monitoring and risk assessment system (F-HMRAS) for predicting whether user is suffering from any mosquito-borne disease like dengue, chikungunya, malaria, Zika virus, and yellow fever. In F-HMRAS, each user is registered through a mobile application by providing basic details like personal information, geographic location, etc. Sensor data and health information are fed into the fog layer to distinguish among diseases spread by mosquitoes based on symptoms. The FKNN classifier is used to determine whether the users are infected or not. The present state of the communicable disease is described using the social network analysis (SNA) and warning notifications are sent to registered users when it detects a risk-prone area. Privacy concerns of user's data are not discussed.

3.2 Non-empirical Studies

Hassan et al. [39] projected a conceptual cloud structure for forecasting the status of patients suffering from dengue. The authors discuss how and how many sensors can be placed for collecting information regarding patients suffering from dengue. Experimentation has not been performed on the suggested framework, the study did not explore implementation challenges or performance. Navin et al. [40] used a simple mobile application installed in the client phone where the epidemiologist generates the survey questions which are sent to the expected mobile clients who have registered. Predictive and preventive steps are taken to monitor the outbreak based on the responses. The survey is conducted on 800 students of the college hostel. The authors have not discussed about the privacy and security issues.

Maghdid et al. [41] projected a framework that uses mobile phone sensors for detecting COVID-19. To determine whether or not a patient has COVID-19, algorithms such as decision trees and *K*-nearest neighbor are applied to sensor data. Pravin et al. [42] proposed a framework where data collected from sensors and apps are processed by the fog system. If a patient is suffering from dengue, her living place is traced to know the environmental conditions by GPS locations. If there is any warning or alert, it is sent to healthcare associates. After identifying dengue-borne places, preventive measures are taken.

Sinharay et al. [43] proposed an architecture using the Arduino board connected to *E*-health shield and to a computer system where the anomaly is detected. The message is then passed to the doctor through the cloud. Phone sensors are used for collecting data but in an epidemic like situation, robots are used to collect data. This architecture is tested on ten participants. The security issues of data are not elaborated. Rani et al. [44] proposed architecture for *S*-health to control the chikungunya virus. The data is sensed and collected using sensor nodes and mobile apps. This data is transferred to edge server. After processing, data is transferred to cloud for analysis. Data is also shared with healthcare departments; however, security issues are not discussed.

Bai et al. [45] discussed the smartphone application, COVID-19 Intelligent Diagnosis and Treatment Assistant Program (nCapp) based on the Internet of Medical Things on a cloud network. This program includes an automatic diagnostic device with eight terminal functions that can be used in real time. Patients submit the questionnaires report, and nCapp automatically generates a test report, and treatment recommendations are provided according to the disease intensity. It also provides information about COVID-19 cases in the user's region and performs real-time updating of intelligent test model to improve overall diagnostic accuracy. The main drawback is automatic devices are required at every step, and undeveloped countries cannot afford this type of program.

Andreas et al. [46] suggested a cloud-based system that collects data from sensors and allows users to input symptom-based information. Hospitals, doctors, and government authorities are notified if a person gets infected. To determine if a user is infected or not, the J48 classifier is employed. However, the privacy of user's data is compromised. Paganelli et al. [47] proposed a three-layer design to monitor the condition of patients in remote locations who are mild to moderately affect by COVID-19. W-kit, which is made up of many sensors, is used to collect data. To ensure the privacy of patient, data blockchain technology is used. Saha et al.

[48] projected the fog-based two-tier architecture to predict severity of the COVID-19 patient. The oxygen level of patients is collected through sensors or entered manually by users, and this data is passed to COVID engine (CovEn). CovEn using probabilistic classification predicts severity of patient.

Mukhtar et al. [49] proposed a rule-based framework to predict if patient is infected or uninfected with COVID-19. Data are collected through the E-health kit which consists of sensors in the presence of a nurse. Ubidots application analyzed and visualized the results in real time. Deep et al. [50] proposed a three-tier architecture that takes data from sensors like health sensors and location sensors. Transferred data is to be processed by fog layer to classify if user is suffering from swine flu or not and passing appropriate messages to users and caregivers. The security, scalability, and network issues are not discussed in the study.

4 Evaluation Techniques

Table 1 shows the evaluation techniques used by various researchers to evaluate the frameworks. It is evident that machine learning techniques have been used widely by researchers to test the proposed methodologies. J48, random tree, Naive Bayes, REP tree, fuzzy-*C* means (FCM), neural network (NN), fuzzy *K*-nearest neighbor (FKNN), linear regression (LR), multilayer perceptron (MLP), etc., are data mining techniques used by the researchers for data analytics. J48, FKNN, NBN, and boosted random forest achieved higher accuracy as compared to other techniques. True positives (TP), false positives (FP), precision, *F*-measure, recall, ROC are performance metrices used to check the accuracy of various classification techniques. TP represents how many cases are appropriately classified by the classifier, and FP represents cases that are wrongly classified by the classifier. Precision and recall represent the relevancy of results. *F*-measure and ROC represent classification accuracy. For data mining, Weka and MATLAB tools are used by researchers in studies.

5 Datasets

The research studies use data collected through wearable sensors, humidity sensors, temperature sensors, and mobile apps along with personal information data, environmental data, and census data available online. However, the symptom-based dataset is not available [24–29]. Few researchers have used synthetic data for the evaluation of models. Synthetic data are produced considering all possible combinations of symptoms. Features are significant part of any study. Use of real data along with characteristic features needs to be explored further to get concrete results. Summary of datasets and architectures used is given in Table 2.

Table 1 Evaluation techniques

Epidemic handled	Technique	Performance	Result
Ebola	J48, random tree, NB, REP tree [24]	TP, FP, *F*-measure, ROC precision, recall	J48 performs better
Chikungunya	FCM, NN, FKNN, NB [25], NNW, NB, FCM, FKNN [26]	Recall, *F*-measure, ROC TP, FP, precision	FCM performs better [25], FKNN performs better [26]
Zika	LR, MLP, NN, NBN [27], NBN [28], MLP, NN, NBN, LR, FKNN [29]	Precision, TP, FP, recall, *F*-measure, ROC, sensitivity	NBN performs better [27], NBN performs better [28], FKNN performs better [29]
MERS-CoV	J48, NB [30] BBN, *K*-NN, LR, NN [31]	Accuracy, precision, recall, TP, FP, MCC, *F*-measure, ROC	NB performs better [30], BBN performs better [31]
H1N1	Random decision tree [32]	TP, ROC, MCC, FP, precision, recall, *F*-measure	Random decision tree
COVID-19	Linear regression, SVM, random forest regression, MTGP and long short-term memory [33], SVM, neural network, NB, *K*-NN, decision table, decision stump, OneR, and ZeroR [34], FCM, NB, FKNN, random decision tree, temporal RNN [36], Dense neural network, decision tables, LSTM and OneR, ANN, *K*-NN, SVM, NB [37] NBN, J48 [46]	Root mean square error and mean absolute percentage error Accuracy, *F*-measure, ROC area, root mean square error, precision, recall	Multitask Gaussian process regression [33], SVM, decision table [34], FCM [36], SVM [37], J48 [46]

6 Key Findings and Challenges

- Only few real datasets are available [51, 52] for asymptomatic detection of COVID-19, otherwise synthetic datasets are generated by researchers using testbeds. Validating the accuracy of these datasets is a challenge.
- Real symptom-based data are required to be collected for accurate analysis. In case of real data privacy of personal records, health conditions, diagnosis notes, and other medical details included in the patient's medical data are required to be maintained. Hackers, intruders can access this information and change it which leads to inaccurate assessment of disease information and treatment.

Table 2 Summary of datasets and architecture

Epidemic handled	Dataset	Type of architecture	GPS	SNA/TNA
Ebola	Synthetic data for symptoms. Real data for social patterns (CPI). https://www.sociopatterns.org/ [24]	Cloud	Yes	Yes
Zika	Synthetic data [28]	Cloud	Yes	No
H1N1	Synthetic data [32]	Cloud	Yes	Yes
Chikungunya	Synthetic data for symptoms. Real data for environmental attributes like climate, temperature, humidity, rainfall [25]. Synthetic data [26]	Fog	Yes	Yes
		Cloud	Yes	No
MERS-CoV	Real data. https://www.moh.gov.sa/en/ccc/2013-15 [30] Synthetic data for symptoms. Real data for the census. https://archive.ics.uci.edu/ml/datasets/Adult. [31]	–	No	No
		Cloud	Yes	No
COVID-19	https://www.who.int/emergencies/diseases/novel-coronavirus-2019/situation-reports/ [33] https://doi.org/10.5281/zenodo.3715506 [34] Synthetic dataset [36]	–	No	No
		Cloud	No	No
		Fog-cloud	Yes	No
Mosquito-borne diseases	Real data of personal and demographic information [38] https://archive.ics.uci.edu/ml/datasets/Adult Real data for social patterns. https://www.sociopatterns.org/datasets/high-school-contact-and-friendship-networks. Synthetic data for symptoms	Fog	Yes	Yes

- The classifiers like Naive Bayesian, *K* clustering, J48, etc., and data mining techniques are used in literature. Deep learning techniques have not been explored yet to identify characteristic features and prediction.
- Techniques require a huge amount of data for accurate analysis. There is no study so far where real symptoms data is available in large amounts. Use of wearables

for collecting the data automatically can help in collecting a huge amount of real data. Classifying the collected data according to its importance and priority can help in efficient analysis.
- Tracing an infected patient using GPS location or call tracing is again a very big challenge because of security and privacy reasons, especially in the countries where the mobile phone number belongs to a different person and is used by a different person.
- Environmental factors also play an important role in the rise of cases.
- Security and privacy issues need to be explored.

7 Proposed Hybrid Framework

We present a hybrid architecture as shown in Fig. 1 and flowchart in Fig. 2 for controlling the limitations of existing proposed frameworks as discussed in Sect. 6. It consists of five layers—sensor layer, database layer, preprocessing layer, application layer, and communication layer. These layers have been proposed in a way to produce conducive outcomes.

The first one is the sensor layer which is responsible for the collection of data from the environmental and wearable sensors like heart rate, temperature, accelerometer, glucose monitor, etc., and transferred to a database in real time. In previous frameworks, synthetic data is used, whereas our framework utilizes real data. Database layer stores the data in cloud platform retrieved from sensor layer and pass this to next layer for further processing. Preprocessing layer is proposed for feature extraction and manipulation. Thus, irrelevant features and inconsistencies in the data will be removed and stationarity check will be performed. The dataset can be preprocessed using resampling and scaling techniques. Further, the application layer will perform real-time analysis, performance analysis, and predict the condition of the patient. It chooses the machine learning algorithm depending on the situation and available dataset size. As the dataset size increases, the training and testing of the same will provide better results. The trained and tested machine learning algorithm will predict

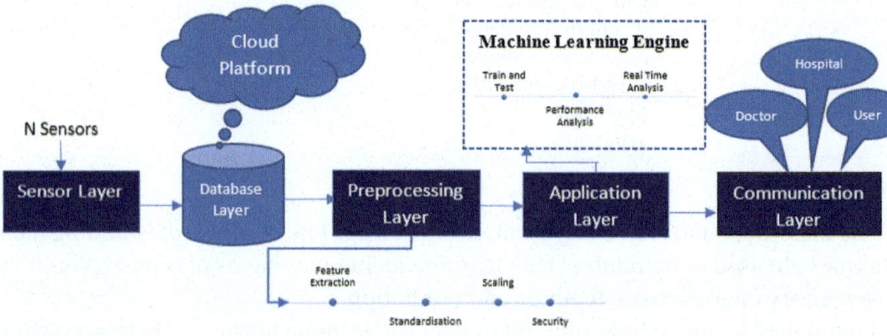

Fig. 1 Hybrid IoT-based system architecture

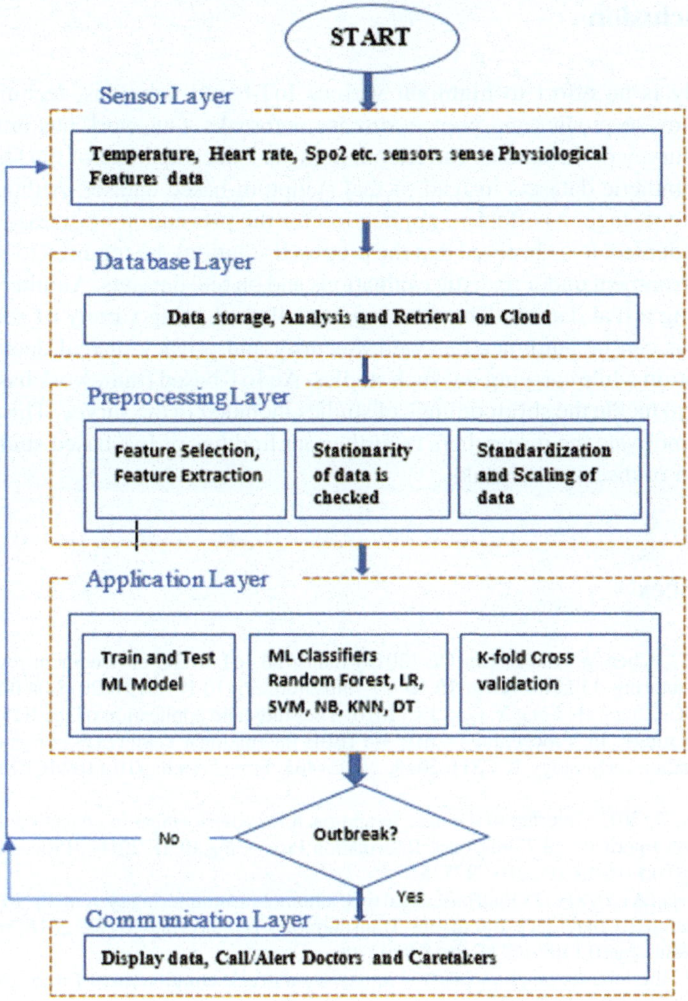

Fig. 2 Flowchart of hybrid IoT-based system

similar diseases or new epidemics. Finally, communication layer will communicate data, reports, and send alerts to doctors and attendants of patients. This is the hybrid model where in security and privacy of user's data can be taken care of by applying hashing technique to user data. Most of the study lacks how to apply technologies of IoT for health care, or in-depth information and implementation. Our framework will take care of these limitations and will be improved gradually.

8 Conclusion

This study is an effort to highlight various IoT-based solutions, techniques, and architectures to predict and prevent disease outbreaks. Our work has presented an in-depth survey of the studies carried out in the last decade. Most of the studies have utilized synthetic datasets instead of real symptom-based dataset. Although IoT is in its nascent stage, its effective application for the prevention of disease outbreaks cannot be denied. It is observed that the proposed solutions are required to be tested in a real environment under realistic evaluations and on real datasets. Another challenge is acquiring a real dataset. The data can be collected using variety of sensors, and the process can be automated as well. Security and privacy of real data also need consideration while carrying out such studies. An IoT-based framework has also been proposed to tackle the shortcomings of studies included in the survey. This work will certainly motivate the researchers to further our findings of IoT-based studies for the prevention of disease outbreaks.

References

1. Cheng J, Chen W, Tao F, Lin CL (2018) Industrial IoT in 5G environment towards smart manufacturing. J Ind Inf Integr 10:10–19. https://doi.org/10.1016/J.JII.2018.04.001
2. Zhao JC, Zhang JF, Feng Y, Guo JX (2010) The study and application of the IOT technology in agriculture. In: Proceeding—2010 3rd IEEE international conference computing science information technology ICCSIT 2010. 2:462–465. https://doi.org/10.1109/ICCSIT.2010.5565120
3. Sharma R (2017) Internet of Things: an approach for advancement in educational institution. In: India International Conference Information Processing IICIP 2016—Proceeding. https://doi.org/10.1109/IICIP.2016.7975351
4. Srinivasan A (2018) IoT cloud based real time automobile monitoring system. In: 2018 3rd IEEE International Conference Intelligent Transportation Engineering ICITE 2018. pp 231–235. https://doi.org/10.1109/ICITE.2018.8492706
5. What Is the Internet of Things (IoT)? https://www.oracle.com/internet-of-things/what-is-iot/. Last accessed Dec 08 2022
6. Ebola virus disease. https://www.who.int/health-topics/ebola#tab=tab_1. Last accessed Nov 26 2021
7. How do SARS and MERS compare with COVID-19? https://www.medicalnewstoday.com/articles/how-do-sars-and-mers-compare-with-covid-19#MERS. Last accessed 01 Dec 2022
8. Chikungunya virus and prospects for a vaccine. https://www.medscape.com/viewarticle/774865_9. Last accessed 03 Mar 2022
9. Valsalan P, Baomar TAB, Baabood AHO (2020) IoT based health monitoring system. J Crit Rev 7:739–743. https://doi.org/10.31838/jcr.07.04.137
10. Ani R, Krishna S, Anju N, Sona AM, Deepa OS (2017) IoT based patient monitoring and diagnostic prediction tool using ensemble classifier. In: 2017 international conference on advances in computing, communications and informatics, ICACCI 2017. pp. 1588–1593. https://doi.org/10.1109/ICACCI.2017.8126068
11. Box IM, Yang G, Xie L, Mäntysalo M, Zhou X, Pang Z, Xu LD, Member S (2014) A health-IoT platform based on the integration of intelligent packaging. Unobtrusive 10:2180–2191. https://doi.org/10.1109/TII.2014.2307795

12. Chamola V, Hassija V, Gupta V, Guizani M (2020) A comprehensive review of the COVID-19 pandemic and the role of IoT, drones, AI, blockchain, and 5G in managing its Impact. IEEE Access 8:90225–90265. https://doi.org/10.1109/ACCESS.2020.2992341
13. Singh RP, Javaid M, Haleem A, Suman R (2020) Internet of things (IoT) applications to fight against COVID-19 pandemic. Diabetes Metab Syndr 14:521–524. https://doi.org/10.1016/J.DSX.2020.04.041
14. Mohd Aman AH, Hassan WH, Sameen S, Attarbashi ZS, Alizadeh M, Latiff LA (2021) IoMT amid COVID-19 pandemic: application, architecture, technology, and security, /pmc/articles/PMC7605812/. https://doi.org/10.1016/j.jnca.2020.102886
15. Swayamsiddha S, Mohanty C (2020) Application of cognitive internet of medical things for COVID-19 pandemic. Diabetes Metab Syndr Clin Res Rev
16. Mohammed MN, Syamsudin H, Al-Zubaidi S, Sairah AK, Ramli R, Yusuf E (2020) Novel covid-19 detection and diagnosis system using iot based smart helmet. Int J Psychosoc Rehabil 24:2296–2303. https://doi.org/10.37200/IJPR/V24I7/PR270221
17. Mohammed MN, Hazairin NA, Syamsudin H (2019) Novel coronavirus disease (Covid-19): detection and diagnosis system using IoT based smart glasses. 29:954–960
18. Singh V, Chandna H, Kumar A, Kumar S, Upadhyay N, Utkarsh K (2020) IoT-Q-Band: a low cost internet of things based wearable band to detect and track absconding COVID-19 quarantine subjects. EAI Endorsed Trans Internet Things 6:163997. https://doi.org/10.4108/eai.13-7-2018.163997
19. Tripathy AK, Mohapatra AG, Mohanty SP, Kougianos E, Joshi AM, Das G (2020) EasyBand: a wearable for safety-aware mobility during pandemic outbreak. IEEE Consum Electron Mag 9:57–61. https://doi.org/10.1109/MCE.2020.2992034
20. Morais IDE, Filho B, Aquino G, Malaquias R, Girão G, Melo S (2021) An IoT-based healthcare platform for patients in ICU beds during the COVID-19 outbreak. 1–171. https://doi.org/10.1109/ACCESS.2021.3058448
21. GSMA | IoT applications in the fight against COVID-19 | Mobile for development, https://www.gsma.com/mobilefordevelopment/blog/iot-applications-in-the-fight-against-covid-19/. Last accessed 01 Dec 2022.
22. Improving vaccination systems—eVIN | UNDP in India, https://www.in.undp.org/content/india/en/home/projects/gavi1.html. Last accessed 03 Mar 2022
23. China's IoT response against coronavirus outbreak Shenzhen Jimi IoT Co., Ltd., https://www.jimilab.com/bolg/iot-against-coronavirus.html. Last accessed 03 Mar 2022
24. Sareen S, Sood SK, Kumar S (2018) IoT-based cloud framework to control Ebola virus outbreak. J Ambient Intell Humaniz Comput 9:459–476. https://doi.org/10.1007/s12652-016-0427-7
25. Sood SK, Mahajan I (2017) Computers in Industry Wearable IoT sensor based healthcare system for identifying and controlling Chikungunya virus. Comput Ind 91:33–44. https://doi.org/10.1016/j.compind.2017.05.006
26. Thakur P, Kaur S (2018) An intelligent system for predicting and preventing Chikungunya virus. In: 2017 International conference on energy, communication, data analytics and soft computing. ICECDS 2017, Institute of electrical and electronics engineers inc., pp 3483–3492. https://doi.org/10.1109/ICECDS.2017.8390109
27. Mahalakshmi B, Suseendran G (2018) Zika virus: A secure system usinf nbn classifier for predicting and preventing zika in cloud. Int J Recent Technol Eng 7:28–32
28. Internet S, Cloud OFT, To F, Zika C (2017) Methods secure internet of things-based cloud framework to control Zika virus. 1. https://doi.org/10.1017/S0266462317000113
29. Sareen S, Gupta SK, Sood SK (2017) An intelligent and secure system for predicting and preventing Zika virus outbreak using Fog computing. Enterp Inf Syst 11:1436–1456. https://doi.org/10.1080/17517575.2016.1277558
30. Al-Turaiki I, Alshahrani M, Almutairi T (2016) Building predictive models for MERS-CoV infections using data mining techniques. J Infect Public Health 9:744–748. https://doi.org/10.1016/j.jiph.2016.09.007
31. Sandhu R, Sood SK, Kaur G (2016) An intelligent system for predicting and preventing MERS-CoV infection outbreak. J Supercomput 72:3033–3056. https://doi.org/10.1007/s11227-015-1474-0

32. Sandhu R, Gill HK, Sood SK (2016) Smart monitoring and controlling of pandemic influenza a (H1N1) using social network analysis and cloud computing. J Comput Sci 12:11–22. https://doi.org/10.1016/j.jocs.2015.11.001
33. Ketu S, Mishra PK (2021) Enhanced Gaussian process regression-based forecasting model for COVID-19 outbreak and significance of IoT for its detection. Appl Intell 51:1492–1512. https://doi.org/10.1007/s10489-020-01889-9
34. Otoom M, Otoum N, Alzubaidi MA, Etoom Y, Banihani R (2020) An IoT-based framework for early identification and monitoring of COVID-19 cases. Biomed. Signal Process Control 62:102149. https://doi.org/10.1016/J.BSPC.2020.102149
35. Wang B, Sun Y, Duong TQ, Member S, Nguyen LD (2020) Risk-aware identification of highly suspected COVID-19 cases in social IoT : A joint graph theory and reinforcement learning approach. 115655–115661. https://doi.org/10.1109/ACCESS.2020.3003750
36. Ahanger TA, Tariq U, Nusir M, Aldaej A, Ullah I, Sulman A (2022) A novel IoT–fog–cloud-based healthcare system for monitoring and predicting COVID-19 outspread. J Supercomput 78:1783–1806. https://doi.org/10.1007/s11227-021-03935-w
37. Aljumah A (2021) Assessment of machine learning techniques in IoT-based architecture for the monitoring and prediction of COVID-19. Electronics 10:1834. https://doi.org/10.3390/ELECTRONICS10151834
38. Vijayakumar V, Malathi D, Subramaniyaswamy V, Saravanan P, Logesh R (2019) Fog computing-based intelligent healthcare system for the detection and prevention of mosquito-borne diseases. Comput Human Behav 100:275–285. https://doi.org/10.1016/j.chb.2018.12.009
39. Hassan NH, Salwana E, Drus SM, Maarop N, Samy GN, Ahmad NA (2018) Proposed conceptual Iot-based patient monitoring sensor for predicting and controlling dengue. Int J Grid Distrib Comput 11:127–134. https://doi.org/10.14257/ijgdc.2018.11.4.11
40. Navin K, Krishnan MBM, Lavanya S, Shanthini A (2017) A mobile health based smart hybrid epidemic surveillance system to support epidemic control programme in public health informatics. In: IEEE International conference on IoT and its Applications, ICIOT 2017. https://doi.org/10.1109/ICIOTA.2017.8073606
41. Maghded HS, Ghafoor KZ, Sadiq AS, Curran K, Rawat DB, Rabie K (2020) A novel AI-enabled framework to diagnose coronavirus COVID-19 using smartphone embedded sensors: design study. In: Proceedings—2020 IEEE 21st international conference on information reuse and integration for data science, IRI 2020, pp 180–187. https://doi.org/10.1109/IRI49571.2020.00033
42. Pravin A, Jacob TP, Nagarajan G (2020) An intelligent and secure healthcare framework for the prediction and prevention of Dengue virus outbreak using fog computing. Health Technol (Berl) 10:303–311. https://doi.org/10.1007/s12553-019-00308-5
43. Pal A, Banerjee S, Banerjee R, Bandyopadhyay S, Deshpande P, Dasgupta R (2016) A novel approach to unify robotics, sensors, and cloud computing through IoT for a smarter healthcare. 536–542. https://doi.org/10.1007/978-3-319-47063-4.
44. Rani S, Ahmed SH, Shah SC (2019) Smart health: a novel paradigm to control the Chickungunya virus. IEEE Internet Things J 6:1306–1311. https://doi.org/10.1109/JIOT.2018.2802898
45. Bai L, Yang D, Wang X, Tong L, Zhu X, Zhong N, Bai C, Powell CA, Chen R, Zhou J, Song Y, Zhou X, Zhu H, Han B, Li Q, Shi G, Li S, Wang C, Qiu Z, Zhang Y, Xu Y, Liu J, Zhang D, Wu C, Li J, Yu J, Wang J, Dong C, Wang Y, Wang Q, Zhang L, Zhang M, Ma X, Zhao L, Yu W, Xu T, Jin Y, Wang X, Wang Y, Jiang Y, Chen H, Xiao K, Zhang X, Song Z, Zhang Z, Wu X, Sun J, Shen Y, Ye M, Tu C, Jiang J, Yu H, Tan F (2020) Chinese experts' consensus on the Internet of Things-aided diagnosis and treatment of coronavirus disease 2019 (COVID-19). Clin eHealth 3:7–15. https://doi.org/10.1016/j.ceh.2020.03.001
46. Andreas A, Mavromoustakis CX, Mastorakis G, Mongay Batalla J, Sahalos JN, Pallis E, Markakis E (2021) IoT cloud-based framework using of smart integration to control the spread of COVID-19. In: IEEE International conference on communications. IEEE, pp. 1–5. https://doi.org/10.1109/ICC42927.2021.9500528

47. Iyda A, Elkind P, Miranda P, Branco A, Alencar P, Cowan D, Endler M, Pelegrini P (2021) A conceptual IoT-based early-warning architecture for remote monitoring of COVID-19 patients in wards and at home. Internet of Things. 100399. https://doi.org/10.1016/j.iot.2021.100399
48. Saha R, Kumar G, Kumar N, Kim TH, Devgun T, Thomas R, Barnawi A (2021) Internet-of-things framework for oxygen saturation monitoring in COVID-19 environment. IEEE Internet Things J 1–11. https://doi.org/10.1109/JIOT.2021.3098158.
49. Mukhtar H, Rubaiee S, Krichen M, Alroobaea R (2021) An iot framework for screening of covid-19 using real-time data from wearable sensors. Int J Environ Res Public Health 18. https://doi.org/10.3390/ijerph18084022.
50. Deep P, Kaur R, Deep K, Dhiman G, Soni M (2021) Informatics in medicine unlocked fog-centric IoT based smart healthcare support service for monitoring and controlling an epidemic of Swine Flu virus. Inf Med Unlocked 26:100636. https://doi.org/10.1016/j.imu.2021.100636
51. Mishra T, Wang M, Metwally AA, Bogu GK, Brooks AW, Bahmani A, Alavi A, Celli A, Higgs E, Dagan-rosenfeld O, Fay B, Kirkpatrick S, Kellogg R, Gibson M, Wang T, Hunting EM, Mamic P, Ganz AB, Rolnik B, Li X, Snyder MP (2020) Smartwatch data. Nat Biomed Eng 4. https://doi.org/10.1038/s41551-020-00640-6
52. Quer G, Radin JM, Gadaleta M, Baca-motes K, Ariniello L, Ramos E, Kheterpal V, Topol EJ, Steinhubl SR (2021) Wearable sensor data and self-reported symptoms for COVID-19 detection. Nat Med 27. https://doi.org/10.1038/s41591-020-1123-x

Chapter 29
Analysis of Recent Query Expansion Techniques for Information Retrieval Systems

Deepak Vishwakarma and Suresh Kumar

1 Introduction

The web is a collection of huge amount of unstructured data. A web user usually inputs a very short and a very general query to satisfy its information needs [1]. If a user searches for a term which has many different meanings, then IR system may produce many unrelated results. For example, if the query is *operating* then this very short and ambiguous query may relate to many different contexts. Most of the times, user has the habit to put this ambiguous kind of query. Other times, user does not know how to formulate the best suitable query. In both the cases, information retrieval (IR) system results into mostly unrelated documents [2].

Extensive research work is being conducted by prominent researchers to tackle with the problem of non-relevant document retrieval by employing many different strategies [3, 4]. Technological advancement is continuous but sudden increment in the complexity and ambiguity in natural language content is also continuous [5]. This study is going to cover important IR system development aspects and help young researchers to work on this area following the right directions.

Following is the organization of this paper. Section 2 presents motivation to conduct this review. Section 3 explains the challenges in the field of IR research activities. Section 4 discusses the formal approach of QE along with categories of QE models

D. Vishwakarma (✉)
CSE Department, Ambedkar Institute of Advanced Communication Technologies & Research (now NSUT East Campus, Geeta Colony, New Delhi 110031, India
e-mail: deepakvishw@gmail.com

Guru Gobind Singh Indraprastha University, Dwarka, New Delhi 110078, India

S. Kumar
CSE Department, Netaji Subhas University of Technology, East Campus (Formerly AIACTR), Geeta Colony, New Delhi 110031, India
e-mail: suresh.kumar@nsut.ac.in

V. Devedzic et al. (eds.), *Proceedings of the International Conference on Intelligent Computing, Communication and Information Security*, Algorithms for Intelligent Systems, https://doi.org/10.1007/978-981-99-1373-2_29

and performance metrics used to evaluate the IR systems. Section 5 presents the literature review with comparison of recent developments in QE techniques. Finally, conclusion and future scope of our study are given in Sect. 6.

2 Motivation

A very short and ambiguous query cannot produce the expected results. To retrieve relevant set of documents, a user must put the best form of query into the search bar. Unfortunately, this is not possible all the times because the user either becomes habitually careless or does not know much about the topic under interest. So, reformulation of original user query is needed to improve its quality and to guarantee the retrieval of relevant information only [6].

User needs an IR systems at different scales like desktop IR [7], web IR [8], domain-specific IR [9, 10], and so on. Query reformulation is required in almost all these cases to get improved results for different applications. Few applications are Question Answering system [11], Information Filtering [12], Cross-language IR [13], Text Summarizing [14], Multimedia IR [2], and so on. Extensive research is being performed to improve the performance of current IR systems.

The data on which IR systems work mostly have an unstructured format. A natural language is very complex and ambiguous from processing point of view. Moreover, in the modern era, the level of ambiguity in NLP tasks is also changing continuously. This makes the processing and analysis part very difficult [15]. The improvement in IR systems is always on demand as the size of data and its complexity are continuously growing. Researchers, their work in this field, technologies, and resources are present and continuously growing. Researchers spend a lot of their time in searching for literature to conduct their research work. Also, there is a need to know every aspect in less time. All these points are the motivating factors to conduct the review of recently developed QE schemes. This study shall help the young researchers compare different QE techniques on the basis of different factors.

3 Challenges in Performing Information Retrieval Through Query Expansion

A user generally puts a textual query to the IR system and expects a ranked list of relevant documents. Practically, this process is not as simple as it seems to be. There are so many challenges which are identified in the field of IR systems [16, 17]. Common challenges are *developing document indices, pre-processing, dealing with language ambiguities, expansion term selection* and *matching, and evaluation of developed system*. Processing of QE activities consumes resources which affect the efficiency of IR. The computational costs to deal with these challenges are usually very high which can be minimized if ideally right direction is followed for developing QE technique [18, 19].

4 Query Expansion: The Formal Definition

To find the relevant information from the web, a user puts a textual query to fulfill its information needs. Since the input query is something in natural language, it generally has some ambiguities. In fact, natural language statements are generally very ambiguous for IR systems. This ambiguity in query results in irrelevant document retrieval. Many researchers proposed different strategies to improve the retrieval process [20]. Ideally, a user must input best suitable query with enough terms and that too without any ambiguity. However, this is not practically possible all the time. This incompleteness in the query can be minimized by augmenting typically suitable terms to the query which make the initial query more clear and meaningful. After this augmentation, the reformulated query might produce better and improved results. This process is called *Query Expansion (QE)*. Figure 1 shows the standard method of QE using pseudo-relevance feedback (PRF) scheme [21].

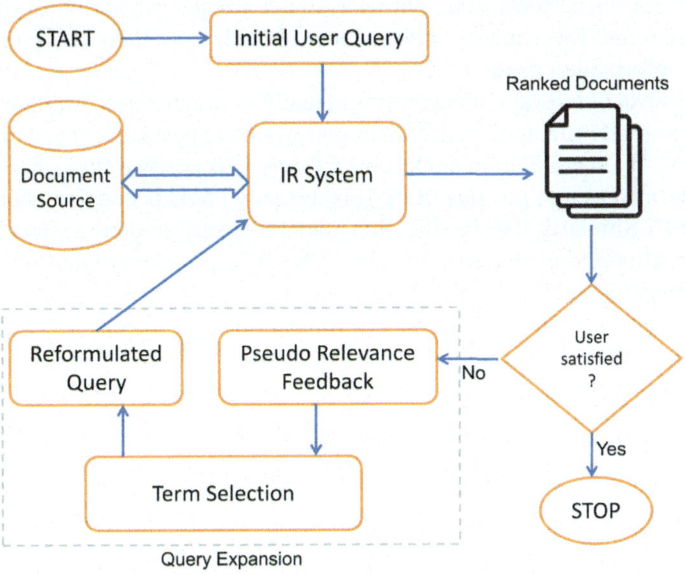

Fig. 1 Standard method of query expansion using PRF

4.1 Categories of QE Models

Based on term selection, QE can be classified into one of the following categories:

- *Manual Query Expansion*: This is completely a manual process in which user selects the candidate expansion terms.
- *Automatic Query Expansion (AQE)* [22]: In this case, the term selection process is completely controlled by the IR system. There is no user involvement in term selection process.
- *Interactive Query Expansion* [23]: The term selection is achieved through the combined efforts of Human and system intelligence. Initially, the retrieval results are produced in front of the user. User marks the relevance results according to its information needs, and those relevant documents are considered for further term selection. Based on the user's choice, an updated set of retrieved documents gets produced and the IR system runs the same term selection process with the help of user.This process loops until the user's information need is satisfied.

4.2 Performance Metrics

There are various performance metrics available for the evaluation of information retrieval systems. These metrics give the idea about how well the IR system satisfies the user's information needs.

The retrieval of relevant documents is actually a classification problem where relevant and non-relevant document classes correspond to positive and negative classes, respectively. If a classification model correctly predicts positive class, then this observation is termed as true positive (TP). If incorrectly predicts positive, then it is false positive (FP). Similarly, if a classification model correctly predicts negative class then this observation is termed as true negative (TN). If incorrectly predicts negative, then it is false negative (FN).

The two most commonly used performance metrics, precision and recall, are derived from these classes. Equations 1 and 2 evaluate precision and recall, respectively.

$$\text{Precision}\ (P) = \frac{\text{TP}}{\text{TP} + \text{FP}} \tag{1}$$

$$\text{Recall}\ (R) = \frac{\text{TP}}{\text{TP} + \text{FN}} \tag{2}$$

F1-score, Mean Average Precision (MAP), P@rank, Mean Reciprocal Rank (MRR), and Normalized Discounted Cumulative Gain (NDCG) are other performance metrics. It is important to understand the advantages and limitations of various metrics. With this knowledge, a combination of metrics should be used for testing a specific IR model. For example, in classification problems the classes are usually imbalanced, and hence, we cannot rely on accuracy. So, precision and recall are useful in case of imbalanced classes. F1-score considers both precision and recall and provides their harmonic mean. F1-score is beneficial when we have fewer samples and precision/recall values are very low. Similarly, the Mean Average Precision (mAP) considers both precision and recall by considering both false positives (FP) and false negatives (FN). This makes mAP a suitable metric for most detection applications. NDGC is very similar to mAP with an additional advantage that it can even differentiate between relevant and more relevant documents. MRR is a good metrics for targeted search where user already knows about the correct result. Clearly, for different types of requirements, different performance metrics should be used for evaluation of IR systems.

5 Literature Review: Recent Developments of QE Techniques

The main objective of our study is to review recent research works intended to improve IR systems by QE methods. Recently, a lot of research is going on to improve the performance of IR systems by QE. In this section, we did comparative analysis of those recent techniques. Following this discussion, Table 1 shows the comparison in tabular form.

Recently, Zheng et al. [24] used BERT (Bidirectional Encoder Representations from Transformers) [25] for QE. The authors developed a QE model using unsupervised chunk selection strategy.

Claveau in [26] performed the query expansion by text generation using neural generative model, OpenAI's GPT-2 which comes with pre-trained model for English language. Author also discusses the possibilities to reduce the text generation time and computational complexities of the work.

Authors in [27] proposed a QE technique that deals with the problem of term mismatch. For term selection and term weighting purposes, they used global statistical properties of term co-occurrence. The authors prepared a graph to show the co-occurrence of terms over entire corpus and applied BM25 to determine the semantic similarity.

According to reference [28], query can be expanded on the basis of related events too. To identify the candidate terms, authors presented both query and events on the vector space. Firstly, they identified the query-related events and then determined the expansion terms based on the identified events.

A lot of researchers worked on ontology-based systems [9, 29] which is a good choice for QE model development. Authors of reference [30] developed a ontology framework based on fuzzy logic. For the development of the fuzzy ontology, they used domain-specific knowledge. In their work, by applying the fuzzy membership the related concepts in a particular domain can be identified easily. Reference [31] presents the survey on the semantic web-based IR systems.

Bouziri et al. [13] proposed QE model based on Learning to Rank (LTR). They took advantage of the association rule mining to develop their model. Model ranks the association rules according to query and then selects the expansion terms. Authors claimed that their model performed well in case of hard and long queries.

Silva et al. [32] developed a supervised query expansion technique based on the Multinomial Naive Bayes. In their work, the top retrieved documents after supplying the initial query are exploited for expansion terms.

Question answering is a widely used application of QE methods. User puts a question before the IR system and expects a precise answer in response. Reference [11] uses a hybrid approach to develop a question answering system based on lexical resources and word embeddings. They tested their work for different combinations like local or global corpus with setting context information set to yes or no.

In reference [33], the authors developed a PRF (pseudo-relevance feedback) framework that performs matching relevance and semantics in two rounds. In first round, they used BERT to determine term relevance along with the similarity of the query and the documents by semantics. Then, in second round they used PRF methods based on probability and language model to process the results obtained in the first round.

Sharma et al. [22] presented a combined QE approach considering various IR-related factors like lexical variation, synonyms, and n-gram pseudo-relevance feedback for query enhancement. Their work performs good with small datasets.

Yusuf et al. [34] used the powerful concept of word embedding to enhance the capability of QE. To capture the semantic similarity, Glove model follows a unigram model and maps various terms with Okapi BM25 results.

Data source plays a very important role in the QE process. There can be many different types of datasets used for QE techniques. Azad and Deepak in [2] used two different data sources Wikipedia and WordNet for extracting candidate expansion terms. In their work, the authors defined different weighing scores at different levels for different data sources.

In Table 1, we enlisted recent QE methods with their brief descriptions. This table also discusses about the various test collection and performance metrics used to test the respective QE methods. Last two columns of the table tell about the advantages, limitations, and future scope of the QE method under discussion. On careful analysis, we observed that there is no general or universal QE method for all types of circumstances. Similarly, for evaluation purposes, different test collections and performance metrics should be selected for different scenarios. If any QE method in combination with some specific test collection and performance metrics produces specific results with one data source, then the same combination may produce different results with other data sources.

Table 1 A review of recent developments in the field of query expansion

Refs.	Year	Brief description	Test collections	Performance metrics	Pros and cons	Future scope
[24]	2021	A QE model performing unsupervised chunk selection based on BERT (Bidirectional Encoder Representations from Transformers)	TREC Robust04, GOV2	MAP, P@rank, NDCG@rank	The model is claimed to outperform BERT-Large. Computational cost is relatively slightly more	Training of model in an end-to-end manner
[26]	2021	Text generation for QE on a well-known neural generative model, OpenAI's GPT-2	Tipster, Robust, GOV2, ohsumed	MAP, P@rank, NDCG@rank, R-precision, Recall@rank	Simple and easy to implement but processing time is slightly high	Text generation process can be researched for better results
[28]	2021	First, query-related events are identified and then expansion terms are identified based on those events	TREC collections (Robust, TREC-12, WSJ, AP)	MAP, P@10, NDCG	Expansion terms are suggested as per events	Approach can be implemented for other IR tasks
[32]	2021	A supervised query expansion technique based on the Multinomial Naive Bayes	TREC2017 Precision Medicine Track	MAP, R-precision, P@rank	Helpful for specialized document collections like medical corpora	The idea can be utilized in topic modeling techniques for better results
[30]	2021	Creation of Fuzzy ontology based on specific domain knowledge	Search Engines: Google, Yahoo, Bing and Exalead	Precision, Mean Average Precision (MAP), Mean Reciprocal Rank (MRR), R-precision and Number of documents retrieved	By applying the fuzzy membership the related concepts in a particular domain can be identified easily	The approach can be researched on future problems
[13]	2020	Learning to rank-based query expansion model using Association Rule Mining	CLEF2003, TREC-Robust and TREC-Microblog	MAP, P@rank, NDCG@rank	Candidate ARs are ranked in order to select more suitable terms	The approach can be applied with word embedding
[11]	2020	A hybrid approach to develop a question answering system based on lexical resources and word embeddings	Ad hoc test collection	Accuracy@1, MRR	A good hybrid approach for QA systems	The proposed approach can be researched for more parts of speech
[33]	2020	PRF framework using BERT combining term matching and semantic matching between query and documents	TREC collections (AP90, AP88-89, DISK4&5, WT10G)	MAP, P@rank, NDCG@rank	The proposed framework combines relevance matching and semantic matching	The more options for combining relevance and semantics can be researched
[22]	2020	A combined QE approach considering various IR-related factors like lexical variation, synonyms, n-gram pseudo-relevance feedback for query enhancement	CACM, CISI, TREC-3	MAP, F-measure, PR Curve	Good approach for small datasets but not suitable for large datasets	Consideration of large datasets
[34]	2019	Enhancement of query expansion by using BM25 (for term matching) and Glove (for semantic match)	Arberry	MAP	Simple and easy approach	Future scope for distributed semantic representations
[2]	2019	Proposed model uses two data sources for QE, In-link score for Wikipedia articles and tf-idf like score for WordNet terms	FIRE	MAP, GM_MAP (Geometric Mean Average Precision), P@rank, bpref (binary preference) and overall recall	Two-level strategy good for individual terms but not for phrases	The approach can be enhanced for phrasal terms consideration

6 Conclusion and Future Scope

We have reviewed recent research works on various QE techniques. Specifically in this paper, we reviewed recent papers on QE techniques. Table 1 shows the different grounds on which review and analysis are performed. This table gives an idea about test collections, performance metrics, advantages, limitations, and possible future scope of different research works. This review suggests different aspects and possibilities for the future works in the field of IR systems using QE.

We observed that the vagueness in natural language is increasing exponentially. In such changing scenarios, one single method for the QE model development and evaluation may not be a good method. In future, different techniques can be combined, if possible, to achieve better results.

An IR method may use a number of test collections to evaluate its capability. We analyzed that the selection of right collection is a very important aspect to achieve expected trust level in developed QE technique. As per observation, the collection should reflect the real-life examples and its size should be large enough to capture all possibilities in evaluation process. We have also mentioned pros and cons of different works which can wisely be used by different researchers to start their research on QE.

References

1. Spink A, Wolfram D, Jansen MB, Saracevic T (2001) Searching the web: the public and their queries. J Am Soc Inf Sci Technol 52(3):226–234
2. Azad HK, Deepak A (2019) A new approach for query expansion using Wikipedia and WordNet. Inf Sci 492:147–163
3. Anand SK, Kumar S (2022) Experimental comparisons of clustering approaches for data representation. ACM Comput Surv (CSUR) 55(3):1–33
4. Mahrishi M, Morwal S, Muzaffar AW, Bhatia S, Dadheech P, Rahmani MKI (2021) Video index point detection and extraction framework using custom yolov4 darknet object detection model. IEEE Access 9:143378–143391
5. Dwivedi A, Kumar S, Dwivedi A, Singh M (2011) Cancellable biometrics for security and privacy enforcement on semantic web. Int J Comput Appl 21(8):1–8
6. Sharma DK, Pamula R, Chauhan DS (2021) Semantic approaches for query expansion. Evol Intell 14(2):1101–1116
7. Pasi G (2010) Issues in personalizing information retrieval. IEEE Intell Inf Bull 11(1):3–7
8. Azad HK, Deepak A, Abhishek K (2020) Query expansion for improving web search. J Comput Theor Nanosci 17(1):101–108
9. Anand SK, Kumar S (2022) Uncertainty analysis in ontology-based knowledge representation. New Gener Comput 40(1):339–376
10. Sharma A, Kumar S (2022) Shallow neural network and ontology-based novel semantic document indexing for information retrieval. Intell Autom Soft Comput 34(3):1989–2005
11. Esposito M, Damiano E, Minutolo A, De Pietro G, Fujita H (2020) Hybrid query expansion using lexical resources and word embeddings for sentence retrieval in question answering. Inf Sci 514:88–105
12. Belkin NJ, Croft WB (1992) Information filtering and information retrieval: two sides of the same coin? Commun ACM 35(12):29–38

13. Bouziri A, Latiri C, Gaussier E (2020) LTR-expand: query expansion model based on learning to rank association rules. J Intell Inf Syst 55(2):261–286
14. Tas O, Kiyani F (2007) A survey automatic text summarization. PressAcademia Procedia 5(1):205–213
15. Dwivedi A (2011) Current security considerations for issues and challenges of trustworthy semantic web. Security 983:978–983
16. Momtazi S, Klakow D (2015) Bridging the vocabulary gap between questions and answer sentences. Inf Process Manag 51(5):595–615
17. Song M, Song I-Y, Hu X, Allen RB (2007) Integration of association rules and ontologies for semantic query expansion. Data Knowl Eng 63(1):63–75
18. Kumar S, Singh M, De A (2012) Owl-based ontology indexing and retrieving algorithms for semantic search engine. In: 2012 7th international conference on computing and convergence technology (ICCCT). IEEE, pp 1135–1140
19. Vishwakarma D, Kumar S (2020) A survey on effective index design schemes in local and distributed IR systems. Test Eng Manag 83(12572):12572–12582
20. Kumar S, Kumar N, Singh M, De A (2013) A rule-based approach for extraction of link-context from anchor-text structure. Intelligent informatics. Springer, Berlin, pp 261–271
21. Rocchio J (1971) Relevance feedback in information retrieval. The smart retrieval system-experiments in automatic document processing, pp 313–323
22. Sharma DK, Pamula R, Chauhan D (2020) A contemporary combined approach for query expansion. Multimedia Tools Appl 1–27
23. Fonseca BM, Golgher P, Pôssas B, Ribeiro-Neto B, Ziviani N (2005) Concept-based interactive query expansion. In: International conference on information and knowledge management, proceedings, pp 696–703
24. Zheng Z, Hui K, He B, Han X, Sun L, Yates A (2021) Contextualized query expansion via unsupervised chunk selection for text retrieval. Inf Process Manag 58(5):102672
25. Malpani P et al (2016) A novel framework for extracting geospatial information using SPARQL query and multiple header extraction sources. In: Proceedings of the international conference on recent cognizance in wireless communication and image processing, New Delhi. Springer India, pp 489–499
26. Claveau V (2021) Neural text generation for query expansion in information retrieval. In: ACM international conference proceeding series, pp 202–209
27. Aklouche B, Bounhas I, Slimani Y (2021) A discriminative method for global query expansion and term reweighting using co-occurrence graphs. J Inf Sci 1–24
28. Rosin GD, Guy I, Radinsky K (2021) Event-driven query expansion, vol 1. Association for Computing Machinery
29. Sharma A, Kumar S (2020) Bayesian rough set based information retrieval. J Stat Manag Syst 23(7):1147–1158
30. Jain S, Seeja KR, Jindal R (2021) A fuzzy ontology framework in information retrieval using semantic query expansion. Int J Inf Manag Data Insights 1(1):100009
31. Sharma A, Kumar S (2020) Semantic web-based information retrieval models: a systematic survey. Communications in computer and information science, 1230 CCIS, pp 204–222
32. Silva S, Vieira AS, Celard P, Iglesias EL, Borrajo L (2021) A query expansion method using multinomial Naive Bayes. Appl Sci (Switzerland) 11(21):1–14
33. Wang J, Pan M, He T, Huang X, Wang X, Tu X (2020) A pseudo-relevance feedback framework combining relevance matching and semantic matching for information retrieval. Inf Process Manag 57(6):102342
34. Yusuf N, Yunus MAM, Wahid N, Wahid N, Nawi NM, Samsudin NA (2019) Enhancing query expansion method using word embedding. In: 2019 IEEE 9th international conference on system engineering and technology (ICSET). IEEE, pp 232–235

Chapter 30
Image Captioning on Apparel Using Neural Network

Vaishali and Sarika Hegde

1 Introduction

Human brain can process the images considerably quicker than text; they can rapidly communicate a product or brand. Additionally, images provide deep insight and perspective to a description or story, making it a much more captivating experience. On the other hand, an image description gives sighted readers another way to interpret visual data. That is how the concept of image captioning was conceived.

The process of dynamically constructing a written description for an image is known as image captioning. Automatically, producing captions for an image demonstrates that computers can understand the visual, which is a fundamental task of intelligence. To caption an image, the model should not only be efficient to identify the items in the image, but also convey the irrelationships in a natural language such as English. As a result, the computer must be trained to recognize the visual information of an image and construct a descriptive text in response. It was formerly thought impossible for a computer to explain an image, but with the advancement of deep learning techniques and the availability of enormous amount of data, it is now possible to create models that generate captions for images.

Automatic generation of caption for an image could have a variety of applications, including recommendations in editing apps, virtual assistants, assisting visually impaired people in understanding an image content, and image indexing [1].

Vaishali (✉) · S. Hegde
Department of CSE, Nitte (Deemed to be University), NMAM Institute of Technology (NMAMIT), Nitte, Udupi, India
e-mail: vaishalibangera98@gmail.com

S. Hegde
e-mail: sarika.hegde@nitte.edu.in

V. Devedzic et al. (eds.), *Proceedings of the International Conference on Intelligent Computing, Communication and Information Security*, Algorithms for Intelligent Systems, https://doi.org/10.1007/978-981-99-1373-2_30

2 Literature Review

In the past few years, many machine learning-based algorithms for autonomously providing visual description have been presented. Captions for images are often created using supervised learning, reinforcement learning, and GAN-based approaches [1].

The supervised learning method is further categorized as encoder–decoder architecture, compositional architecture, attention-based, semantic concept–based, and so on. To encode the input image, encoder–decoder architectural methods use recurrent neural networks (RNNs) [2–4], as well as a deep convolutional neural network (CNN) and a long short-term memory (LSTM) [4–6]. CNN-RNN framework-based image captioning technique has two limitations. The first disadvantage is that each caption is given equal weight regardless of its particular value, and the second disadvantage is that objects may not be correctly detected during caption production [7].

The usage of spatial attention techniques on CNN layers was added into the creation process to incorporate visual context that implicitly conditions the text written thus far. It has been demonstrated and qualitatively observed in [4] those captioning systems that use attention mechanisms result in improved generalization. Since these models can produce creative text descriptions based on the perception of the global and local elements that make up images.

Existing methods are either top-down, in which agis to fan image is transformed into phrases, or bottom-up, in which words describing distinct parts of an image are generated and earlier or later mixed. In [8], a new technique is used that combines both approaches using a semantic attention model.

Long short-term memory with attributes (LSTM-A) is evolutionary architecture with five versions that integrates attributes into the popular CNN and RNN image captioning system by training them from inception to delivery [2]. To facilitate greater picture interpretation in image captioning, top-down techniques have been widely deployed [9]. LSTM networks have the compelling ability to memorize lengthy interactions through a memory module; therefore, they have been considered as the standard for vision-language tasks such as image captioning [10–14], visual question answering [9, 15–17], question generation [18, 19], and visual dialogue [20, 21]. The sophisticated referencing and rewriting method, combined with naturally sequential processing and the large amount of storage required, poses challenge during training. LSTM units are more complicated and, by definition, consecutive with time. Recent research has demonstrated the advantages of convolutional networks for language processing and probabilistic image captioning in addressing this issue. CNN increases entropy in the output probability distribution, improves word prediction accuracy, and is less susceptible to vanishing gradients [3].

The aforementioned research works have conducted experiments on most popular datasets such as MSCOCO, Flickr30 K, and Flickr8 K, etc., that have focused on wide areas with various types of images, portraying people in their daily lives or products they utilize on a daily basis. The field of fashion, on the other hand, is one of the areas where image captioning has not been widely used. The proposed model was tested on the InFashAIv1 dataset, which enclosed nearly 16,000 images of African apparel products.

3 Methodology

Humans are visual creatures; images not only capture their attention but also elicit emotions and draw them in. However, if the quality of image is not good, then it could have a great impact on the people who does online shopping, using social media, websites, etc. In case of online shopping problems such as missing or unclear product information, confusing product features like color, etc., may been countered.

In order to solve this problem, a website is developed using CNN and Django framework to assist people in better understanding the content of images. Figure 1 depicts the model flow that convey the various steps included in training the model using CNN.

- **Load the dataset**: The dataset consists of 16,000 images of apparel of four different parameters such as color, gender, style, and material.
- **Preprocess the image**: To prepare dataset for the model input, preprocessing is required. For example, fully connected layers in CNN needed that all images be the same size arrays. The images in dataset are resized to 48 × 48. Preprocessing reduces model training time and speeds up model inference. If the input is large, reducing them will decrease model training time in half without sacrificing model performance.
- **Define the model**: A basic CNN model with three convolutional layers followed by max-pooling layers is defined. To prevent over fitting, a dropout layer is inserted after the third max pool operation. The model is built with Adam as the optimizer and binary_Cross entropy as the loss function.

Fig. 1 Flow diagram

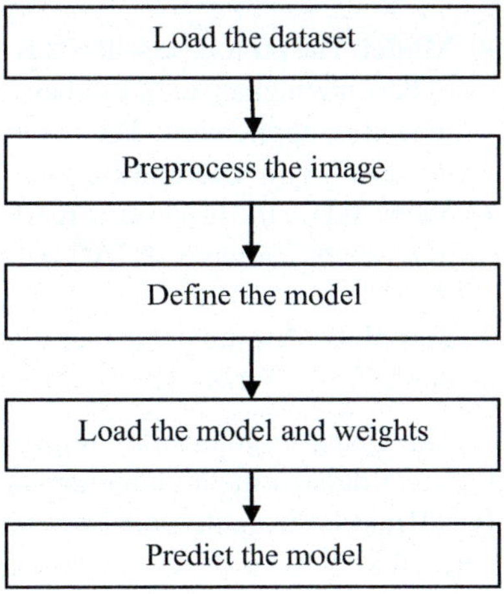

Fig. 2 Block diagram

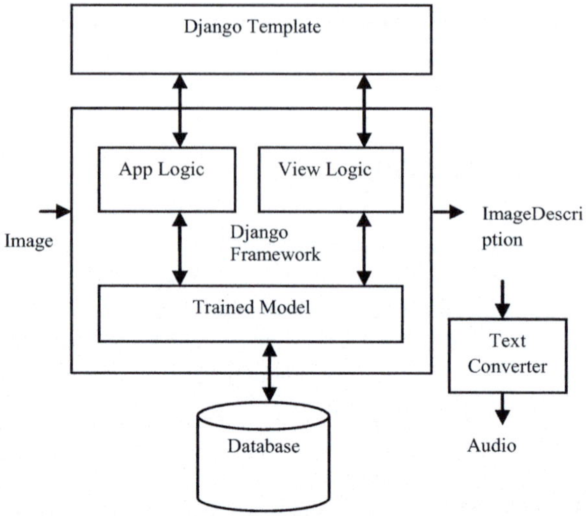

- **Load the model and weights**: The simplest way to save the model in Tensor Flow is to utilize the built-in Tensor Flow function. The save_weights method is part of the "Model saving and serialization APIs." Model weights are all of the model's parameters (trainable and non-trainable), which are all of the parameters utilized in the model's layers. The weights of the models are recorded in h5 format.
- **Predict the model**: The model should be able to predict the color, gender, style, and material of the image accurately. The trained model is then utilized to predict the output using Django framework.

The image is provided as input to the Django framework which is based on MVT architecture, i.e., a design paradigm for creating a web application. Figure 2 explains the block diagram for the proposed model.

MVT structure is made up of three parts:

- **Model**: The prototype will act as a user interface for the data. It is the responsibility of data maintenance. A database is a conceptual data structure that stores supports the entire application. The database used is MySQL, i.e., is mainly used to store the user log in information.
- **View**: When the browser is rendered, the view is the user interface where the user may see the browser. HTML/CSS/Javascript and Jinja files are used to represent it.
- **Template**: A template is made up of static elements of the intended HTML result and unique syntax that specifies how dynamic data are collected.

The Django framework comprises App logic, view logic, and the trained model. Figure 3 depicts the architecture of Django framework. A web application a waits HTTP queries from the web browser in a normal data-driven website. An application determines what is necessary when it receives a request depending on the URL and

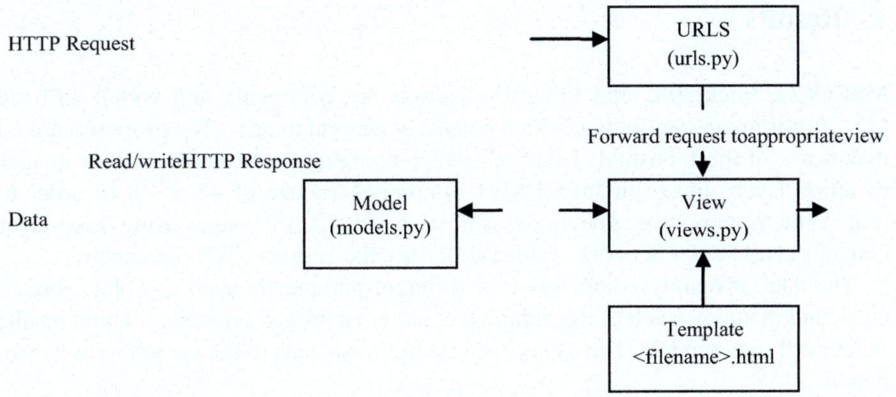

Fig. 3 Django framework [22]

optionally POST data or GET data. It then reads or writes data from a database or conducts other actions necessary to fulfill the demand, depending on what is required. The program will then send a response to the web browser, usually by putting the acquired data into place holder sin an HTML format and instantly constructing an HTML page for the browser to view

- **URLs**: While it is possible to process requests from all URLs with a single function, developing a separate view function for each resource is significantly more maintainable. Based on the URL of the request, a URL mapper is used to direct HTTP requests to the appropriate view. In Django, a request is sent to urls.py, which subsequently sends it to views.py matching function. Views.py methods take the http requests from urls.py and provide it to templates.
- **View**: A view is a request handler function that receives requests from urls.py and responds to it. Views use prototype to access the data required to satisfy requests and delegate response formatting to templates.
- **Models:** Models specify the layout of an application data and provide ways for querying database entries and perform operations such add, delete, modify, and so on.
- **Templates:** A template is a file that contains placeholders for relevant information and defines the framework or layout for HTML page. By using the template and feeding it with data from a prototype, a view can automatically utilizes an HTML page. It is not necessary for a template to be HTML to gain an overview of a file.

The output from the Django framework is a natural language description of the apparel. Further, the text is transformed to audio, which allows the visually impaired individual to grasp the image information.

4 Results

MSCOCO, Flickr30K, and Flickr8K datasets are commonly and widely utilized [23–26] for image captioning, which consist of general images. The proposed model makes use of the InFashAIv1 dataset, which contains over 16,000 African images of attire. Every image in InFashAIv1 is cropped to size of 48 × 48 in order to reduce the training time. The model is trained with 10,000 images using three-layer convolutional neural network. Table1 describes the various CNN parameters.

The dataset mainly comprises four different parameters such as color, gender, style, and material. Each of the parameters has been trained separately which results in four different models. The categorical accuracy for each of the parameter is shown below.

The first parameter is color, and Fig. 4 depicts accuracy for color via graph. The model is trained with 977 images for 19 colors, namely multicolor, Green, Brown, Blue, White, Beige, Purple, Indigo, Gold, Yellow, Mauve, Turquoise, Pink, Gray, Burgundy, Red, Black, Kaki, Orange. The graph shows that by the end of the 25th iteration, the color model had achieved a 75% accuracy.

Material is the second parameter, and Fig. 5 shows the accuracy of material. The model was trained with 5891 different material images, including Silk, Wool, Ankara, Crepe, Veil, Organza, Spandex/stretch, Velvet, Madras, Plastic, Coconut, Cloth, Mesh, Cashmere, polyester, Satin, Polyester, Batik, Lace, Line, Fabric, Cotton,

Table 1 Parameters of CNN

Number of layers	3
Hidden size	512
Batch Size	128
Dropout	0.25
Number of epoche	25

Fig. 4 Color_accuracy

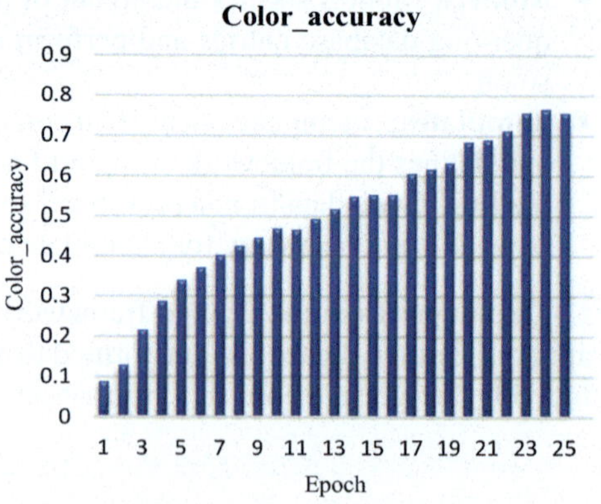

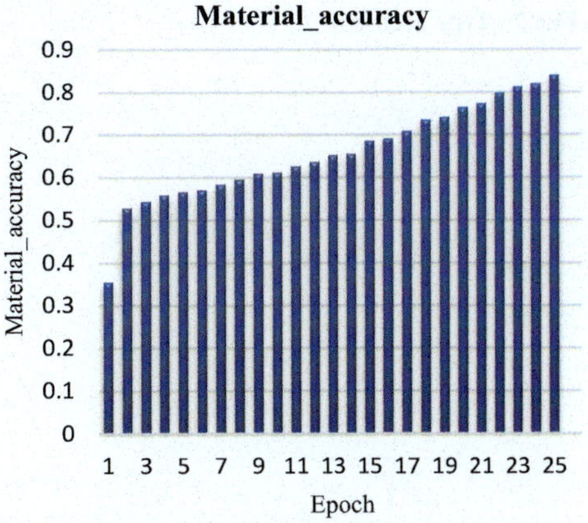

Fig. 5 Material_accuracy

Resin, Prints, Horn, Bazin, Crystal, Mousseline, Leatherette, Bogolan, Glass, Fiber, Kente, Hemp, The model has an accuracy of 84%, as seen in the graph.

Gender is the fourth factor to consider and the accuracy is shown in Fig. 7. A total of 9854 photos were used to train the model for the two categories: He and She. The graph shows that the model gender has a 98% accuracy rate. Type is the third parameter. The model has been given 2772 images to train with, including Casual, Streetwear, A Line, Casual None, None Streetwear, Off shoulder, Sexy Clothes, Streetwear Casual, All Ankara, Elegant fashion, Bomber jacket. Figure 6 shows that the model has achieved a 93% accuracy rate.

Thus, in general, for the proposed model the categorical accuracy for each of the parameter increases with the increase in iteration.

The above Fig. 8 depicts the categorical loss of each of the model. The categorical loss of each model reduces with increase in the number of iterations thus improves the performance and accuracy of the model.

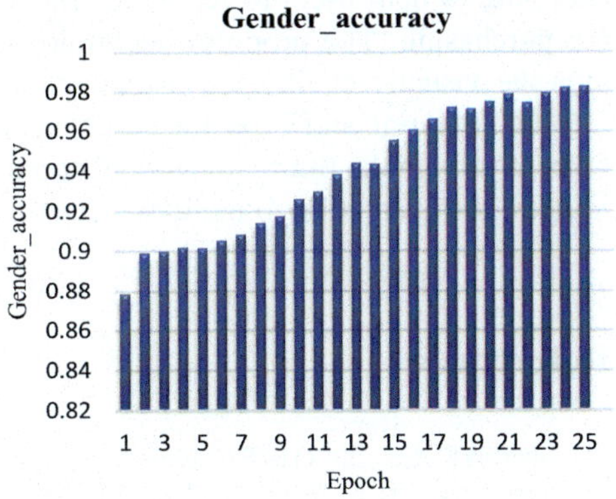

Fig. 6 Gender_accuracy

Fig. 7 Type accuracy

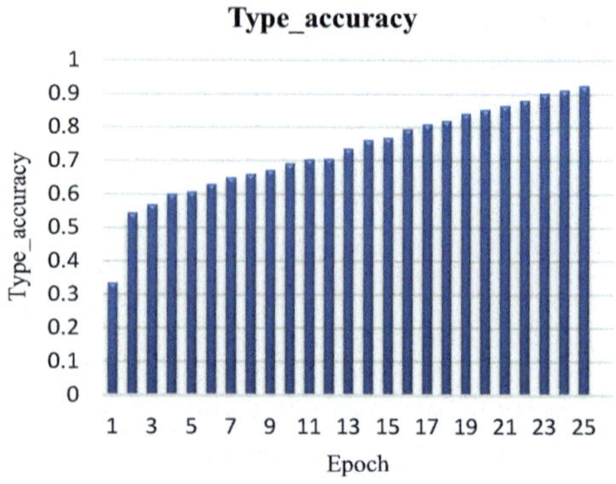

Fig. 8 Categorical_loss

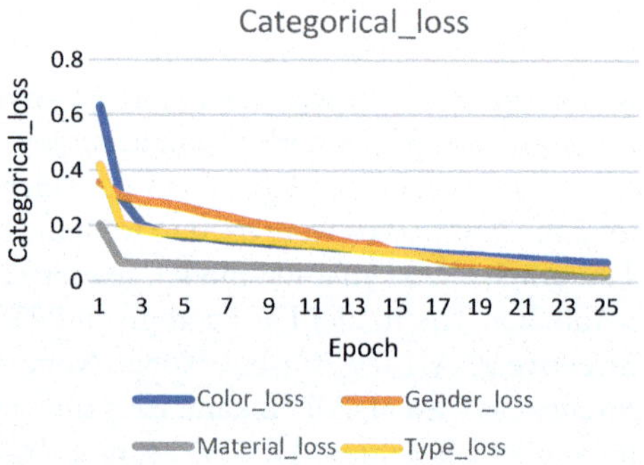

5 Conclusion and Future Work

To offer natural language descriptions of the apparels, a web application was created utilizing CNN and the Django framework. The web application can be used in online shopping to help users to recognize the color, material, and type of clothing they are purchasing. The proposed technology also translates the text to audio, which aids the visually challenged in understanding the image information. For Color, Gender, Material, and Type, the model achieves 75%, 98%, 84%, and 93% accuracy, respectively. In the future, the model could be further trained with a huge number of images of Indian attire to increase accuracy.

References

1. Sharma R et al. (2021) IOP Conference series: material science engineering, vol 1131, p 012001
2. Yao T, Pan Y, Li Y, Qiu Z, Mei T (2017) Boosting image captioning with attributes. In: Proceedings of the IEEE international conference on computer vision, pp 4894–4902
3. Aneja J, Deshpande A, Schwing AG (2018) Convolutional image captioning. In: Proceedings of the IEEE conference on computer vision and pattern recognition, pp 5561–5570

4. Rennie SJ, Marcheret E, Mroueh Y, Ross J, Goel V (2017) Self-critical sequence training for image captioning. In: Proceedings of the IEEE conference on computer vision and pattern recognition, pp 7008–7024
5. Sharma H, Jalal AS (2020) Incorporating external knowledge for image captioning using CNN and LSTM. Mod Phys Lett B 34(28):2050315
6. Kalra S, Leekha A (2020) Survey of convolutional neural networks for image captioning. J Inf Optim Sci 41(1):239–260
7. Sharma H, Agrahari M, Singh SK, Firoj M, Mishra RK (2020, February) Image captioning: a comprehensive survey. In: 2020 International conference on power electronics & IoT applications in renewable energy and its control (PARC). IEEE, pp 325–328
8. You Q, Jin H, Wang Z, Fang C, Luo J (2016) Image captioning with semantic attention. In: Proceedings of the IEEE conference on computer vision and pattern recognition, pp 4651–4659
9. Anderson P, He X, Buehler C, Teney D, Johnson M, Gould S, Zhang L (2018) Bottom-up and top-down attention for image captioning and visual question answering. In: Proceedings of the IEEE conference on computer vision and pattern recognition, pp 6077–6086
10. Chen X, Zitnick CL (June 2015) Mind's eye: a recurrent visual representation for image caption generation. In: 2015 IEEE Conference on computer vision and pattern recognition (CVPR), pp 2422–2431
11. Karpathy A, Fei-Fei L (June 2015) Deep visual-semantic alignments for generating image descriptions. In: 2015 IEEE conference on computer vision and pattern recognition (CVPR), pp 3128–3137
12. Vinyals O, Toshev A, Bengio S, Erhan D (2017) Show and tell: Lessons learned from the 2015 mscoco image captioning challenge. IEEE Trans Pattern Anal Mach Intell 39(4):652–663
13. Mahrishi M, Morwal S et al. (2021) Video index point detection and extraction framework using custom YoloV4 darknet object detection model. IEEE Access 9:143378–143391. https://doi.org/10.1109/ACCESS.2021.3118048
14. Xu K, Ba JL, Kiros R, Cho K, Courville A, Salakhut Dinov R, Zemel RS, Bengio Y (2015) Show, attend and tell: Neural image caption generation with visual attention. In: Proceedings of the 32Nd international conference on machine learning, vol 37. ICML'15, pp 2048–2057. JMLR.org
15. Antol S, Agrawal A, Lu J, Mitchell M, Batra D, Zitnick CL, Parikh D (2015) VQA: visual question answering. In: International conference on computer vision (ICCV)
16. Schwartz I, Schwing AG, Hazan T (2017) High-order attention models for visual question answering. In: Proceeding NIPS
17. Shih KJ, Singh S, Hoiem D (2016) Where to look: Focus regions for visual question answering. In Computer Vision and Pattern Recognition
18. Jain U, Zhang Z, Schwing AG (2017) Creativity: generating diverse questions using variational auto encoders. In: Computer vision and pattern recognition
19. Mostafazadeh N, Misra I, Devlin J, Mitchell M, He X, Vanderwende L (2016) Generating natural questions about an image. In: ACL the association for computer linguistics
20. Das A, Kottur S, Gupta K, Singh A, Yadav D, Moura JMF, Parikh D, Batra D (2017) Visual dialog
21. Jain U, Lazebnik S, Schwing AG (2018) Two can play this game: visual dialog with discriminative question generation and answering. In: Proceeding CVPR
22. https://developer.mozilla.org/en-US/docs/Learn/Server-side/Django/Introduction
23. Herdade S, Kappeler A, Boakye K, Soares J (2019) Image captioning: transforming objects into words. arXiv preprint arXiv:1906.05963
24. Jiang W, Ma L, Jiang YG, Liu W, Zhang T (2018) Recurrent fusion network for image captioning. In: Proceedings of the European conference on computer vision (ECCV), pp 499–515

25. Devlin J, Gupta S, Girshick R, Mitchell M, Zitnick CL (2015) Exploring nearest neighbor approaches for image captioning. arXiv preprint arXiv:1505.04467
26. Pedersoli M, Lucas T, Schmid C, Verbeek J (2017) Areas of attention for image captioning. In: Proceedings of the IEEE international conference on computer vision, pp 1242–1250
27. https://www.researchgate.net/publication/352898415_Neural_Fashion_Image_Captioning_Accounting_for_Data_Diversity

Chapter 31
Fake News Detection: A Study

Sainyali Trivedi, Mayank Kumar Jain, Dinesh Gopalani, Yogesh Kumar Meena, and Yogendra Gupta

1 Introduction

The Internet has emerged as the most effective communication instrument of the twenty-first century. It enables the speedy and reliable transport of media across locations. Social media platforms like Instagram, Facebook, Twitter, Whatsapp, and Telegram are important means of news dissemination in the modern period, and people rely on them without questioning the veracity or source of the information. By weakening freedom of expression, democracy, justice, the truth, and public confidence, social media is utilised by dishonest individuals and puts a great deal of strain on society [1]. Online news or tales that are not real are referred to as fake news. Fake news comes in two varieties:

1. False information that is knowingly distributed or published in an effort to mislead readers or draw a large audience to a Website.
2. Storeys that could be partially true but are not entirely truthful.

There are several Websites that can be utilised to search for pre-verified data, such as TrurthorFiction.com, Media Bias/FactCheck (MBFC News), PolitiFact.com, and

S. Trivedi (✉) · Y. Gupta
Department of Computer Science and Engineering, Swami Keshvanand Institute of Technology Management and Gramothan, Jaipur 302017, Rajasthan, India
e-mail: trivedisainyali13@gmail.com

M. K. Jain · D. Gopalani · Y. K. Meena
Department of Computer Science and Engineering, Malaviya National Institute of Technology, Jaipur 302017, Rajasthan, India
e-mail: 2018RCP9116@mnit.ac.in

D. Gopalani
e-mail: dgopalani.cse@mnit.ac.in

Y. K. Meena
e-mail: ymeena.cse@mnit.ac.in

V. Devedzic et al. (eds.), *Proceedings of the International Conference on Intelligent Computing, Communication and Information Security*, Algorithms for Intelligent Systems, https://doi.org/10.1007/978-981-99-1373-2_31

FullFact.org, and various research types have been done for its proper identification to minimise the horrible consequences of fake news and avoid its terrible outcome. In response to a Cambridge University experiment that was successful, Google plans to display advertisements that advise consumers about misinformation tactics [2]. The most prominent example of this dissemination of fake news in 2020 is "Cancellation of Tokyo Olympics 2020"; later, the Organisers confirmed that the reported cancellation of the Summer Games is untrue [3]. The exploitation of existing photographs or videos is the most frequent method of incorrect information about the events in Ukraine being circulated in the year 2022. News organisations are gathering a list of suspicious social media comments on the fight and revealing fabricated films [4]. Recently, fake news about the government listening in on WhatsApp conversations has circulated throughout India, causing enormous havoc. Later, it became evident that the government had not released any rules [5]. To deal with such actions, preventive measures must advance at the same rate as technology. Consequently, the process of automated data classification constitutes a fascinating and fruitful area of research.

Case study on fake news dissemetion during COVID-19

The conservative Website, the Florida Standard, published an article on August 16, 2021 with the misleading headline "Massacre: Nearly Half of Pregnant Women in Pfizer Trial Miscarried". More than 12,000 people liked a screenshot of the storey that was posted on Instagram in only two days. The "44% statistic is inaccurate", the Daily Clout piece indicated at the bottom of the page by August 19; by August 23, the post had been taken down [6].

2 Literature Review

With substantial specified characteristics for the prediction values, the classic ML techniques learn from the data. Additionally, it is simple to build and does not need for a lot of computer power. In addition, conventional ML typically produces positive outcomes with little datasets as opposed to DL models. The difficult work of feature engineering calls for knowledge gathering for feature extraction [7].

Word embedding models assign a high-dimensional vector for each word in a given language in which the geometry of the vectors represents meaning links between the words. In a recent study, authors stated a deep-learned embedding technique for categorical data encoding on a categorical dataset. Word embedding, which is a component of a DL model, serves as the foundation of their method. So that the use of the distributed word representations could be made possible, each categorical variable was treated as a single word or as a token [8].

In this paper, a superior Arabic text categorisation deep model was introduced by the researchers, a multi-kernel model developed by CNN for categorising Arabic news documents which is enhanced and endowed with n-gramme word embedding (satcdm). Using 15 publicly accessible datasets, the suggested approach is able to achieve extremely high accuracy when compared to recent studies on Arabic text classification. The model's accuracy on the Arabic document categorisation job ranges from 97.58 to 99.90%, which is higher than that of comparable research [9].

The performance of three DL models and five ML models on two genuine and fake news datasets of varying sizes using hold-out cross-validation is analysed in the study. Authors employed term frequency, embedding techniques, and term frequency-inverse document frequency to get text representation for DL and ML models. Then, they demonstrated their ground-breaking stacking model, which on the KDnugget dataset achieved a testing accuracy of 96.05% [10].

In this paper, the authors proposed a hybrid fake news detection system that makes use of both linguistic-based (i.e., number of words, lexical diversity, title, reading ease, and sentiment) and knowledge-based approaches. These features are referred to as fact-verification features, and they include three different types of data: (i) the credibility of the news outlet's Website; (ii) coverage, or the amount of information used to report the news; and (iii) facts verification. Additionally, researchers used four distinct ML algorithms: XGBoost, additional trees discriminant (ATD), logistic regression (LR), and random forest (RF). The experimental findings suggest that their model exceeds the most recent investigations on rumor event identification and achieves a high accuracy of 94.4% [11].

In this exploratory work, researchers suggested a source-based approach that makes use of knowledge about the information's origin and disseminators. The outcomes of the trials demonstrate that two techniques are quite successful in identifying fake news. The three best-performing models are RNN, CATBoost, and Light GBM with accuracy ratings of 98.4%, 98.14%, and 98.6%, respectively [1].

A system comprised of voting classifiers, and feature engineering was proposed in this study. The Fake-or-Real-News dataset had an accuracy of 94.5%, the Media-Eval dataset had an accuracy of 91.2%, and the ISOT dataset had an accuracy of 100%, all of which were greater than before [12].

To address the epidemic of fake news, OPCNN-FAKE was introduced. Four benchmark datasets for fake news were used to evaluate the performance of the OPCNN-FAKE employing the recurrent neural network (RNN), long short-term memory (LSTM), and the six standard ML approaches. The DL and ML characteristics were improved using the hyperopt optimization and grid search. The accuracy, precision, recall, and F1-score were used to validate the results [13].

Researchers introduced a MVAN model to concentrate on the early identification of rumors using deep neural networks. The transmission structure of news is encoded and represented using graph attention networks (GATs). Both the original tweet's content and its distribution pattern were acquired by MVAN. The findings show that MVAN can, on average, offer an explanation and considerably surpass rest approaches by 2.5% in accuracy [14].

Researchers discuss research on context-aware misinformation detection using deep learning architectures in this publication. Multi-class classification was carried out using two text preprocessing pipelines (lemma and aggressive text preprocessing), 10 neural networks, and three context-aware word embeddings (GloVe, FastText, and Word2Vec). Overall, the accuracy, precision, and evaluation of the lemma text preprocessing-trained RCNN model with Word2Vec embedding are 87.56%, 87.55%, and 88.11%, respectively [15].

A thorough analysis and an explored taxonomy of the AI methods used for classifying fake news, and highlighting their developments in the areas covered in this paper. Passive-aggressive LSTM, NB, and random forest algorithms were employed for the classification of Fake news. LSTM achieved the best accuracy 92.34% [16].

EGCN was proposed for Twitter rumor detection on the basis of SR-graphs. Microblog rumor data is transformed into graph data via EGCN. After that, the tagged data is trained using a convolution neural network. The knowledge is handed on to the unlabeled data by altering the node weight in the graph, significantly reducing the burden associated with annotation of hearsay data [17].

This study's objective is to identify fake news on social media by utilising word vector features, stylometric/linguistic data, [18], and the textual content of news items. Here, bagging and boosting techniques are used to apply distinct ML models to stylometric data and word vector characteristics. This study uses the text of news storeys rather than metadata to extract characteristics, which is helpful [19]. Tables 1 and 2 demonstrate the relative comparability of the previous polls for the categorisation of fake news.

3 Methodology

See Fig. 1.

3.1 Data Collection

The first and most important stage is to acquire information on social networking sites like Facebook, Twitter, WhatsApp, and Instagram. In order to identify our fake news, we must examine the collected data.

3.2 Data Preprocessing

The next step after collecting the data is preprocessing it to extract important attributes. Data preparation decreases the dimensions of data by removing stop words, removing white spaces and punctuation, and lemmatizing words, which are all regarded to be standard practises for the identification for fake news.

Table 1 Recent works related to fake news detection

Author	Year	Methodology	Dataset	Accuracy	Drawback
Elhadad et al. [20]	2020	Used combinations BERTSCORE (DA)+BiLSTM BERTSCORE (DA)+SBERT (DA) models for spotting false information in COVID-19	Covid19-Lies	99.68% using NN	Classifiers are limited to the English language and generate poor macro F1 scores
Umer et al. [21]	2020	Used CNN+LSTM layers including PCA and Chisquare	FNC-1	97.80% CNN-LSTM with PCA	Requires additional text capabilities and fusion features
Giuseppe et al. [22]	2020	Used CNN, LSTM, KNN, SVM, SVM-SGD	Twitter (userprofiles and shared news)	91% using SVM-SGD	May utilise current news datasets and user attributes for dependability prediction
Tao et al. [10]	2021	Used stacking approach on DL and ML techniques	ISOT, KDnugget	99.87% RF with TF-IDF on ISOT	Can be used only for english language
Noureddine et al. [23]	2022	Used linguistic-based and knowledge-based approaches on LR, RF, ATD, and XGBoost	BuzzFeed	94.40% with RF	Can used neural network for more accuracy
Vlad-Iulian et al. [15]	2021	Done multiclass-classification using word2vec, GloVe, FasText	GitHub repository	87.56% with RCNN	Absence of hyperparameter adjustment can improve RLANS outcomes
Khubaib et al. [1]	2021	Used source-based fusion(Network+ User profiles) model	Self-aggregated	98.4% with CATBoost	Using multimodal data to identify fake news is still a difficult and unexplored topic that requires more research
Eman et al. [12]	2021	Used ensembled methods and voting classifiers for comparitive analysis	ISOT, the Fake-or-Real-News and the Media-Eval	100% with voting classifier	It is unclear which contextualised embedding will offer the classifier the most beneficial properties
Hager et al. [13]	2021	Applied optimization method for OPCNN-FAKE	dataset1 (Kaggle), FakeNewsNet, ISOT, FA-KES5	98.65% with OPCNN-FAKE	Can use knowledge-based and fact-based approaches and additionally some pre-trained word embedding to increase precision

Table 2 Recent works related to fake news detection continued …

Author	Year	Methodology	Dataset	Accuracy	Drawback
Shiwen et al. [14]	2021	Graph attention networks (GATs) to encode and represent the propagation structure of news	Twitter-15 and Twitter-16	93.65% on Twitter-16 with MVAN	The original tweet and its responses constitute the conversation structure graph, which may be utilised in conjunction with GNN and the attention mechanism to extract the crucial information concealed inside
Abdulqader et al. [11]	2022	Used CNN+BiLSTM and attention networks	ArCov	91.50% with HANN-GloVe	Accuracy on bigger time series datasets might be difficult to achieve
Na et al. [17]	2021	Used SR graph, nodes representing individual tweets	PHEME	84.10% with Text-CNN + GCN	The output dimensions of the two hidden layers are multiplied by the number of parameters in these models, making the calculation costly
Mabrook et al. [24]	2020	Used C4.5 (meta-model), SVM+RF (ensemble-model)	Self-aggregated	97.80% with SVM+RF	DL may be used to increase accuracy and the dataset utilised can be improved
Yue et al. [25]	2019	Used feature-based and text-based classification approach	Self-aggregated (Wiebo, WeChat)	Approx 85% with GBDT	Restricted to chinese language and DL can be used to avoid manual feature extraction
Younghwan et al. [26]	2020	Used ensemble solutions (ESes) bringing multi-ML models together	PHEME and RE2019	Improved F1-score	Still has to be trained using existing rumor data, and additional rumor-related relationships needs to be looked into
Dhiren et al. [16]	2022	Used Ml models for text classification	McIntire	92.34% with LSTM	Early detection of fake news not possible, and it lacks a qualitative and quantitative dataset that would be very useful for analysing the temporal trends in the spread of fake news.

Fig. 1 Standard approach for fake news detection

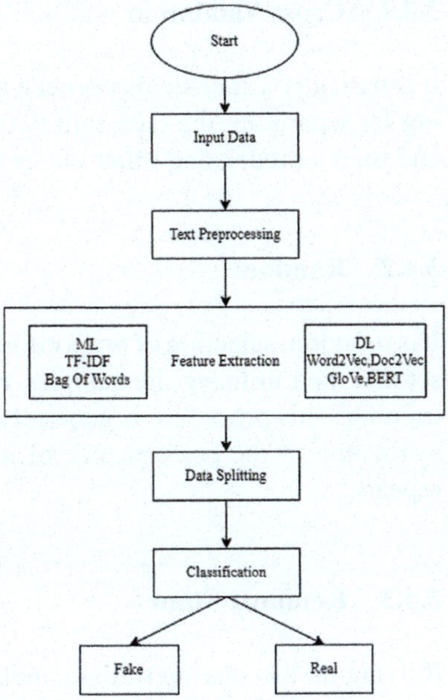

3.3 Feature Extraction

Following data gathering and data preparation, we do feature extraction, sometimes referred to as vectorization. Feature extraction is a process that converts raw data into manageable numerical features whilst preserving the original dataset's content. Bag of words, Word2Vec, Doc2Vec, GloVe, BERT, and TF-IDF are popular ML and DL approaches for feature extraction methods for spotting fake news.

3.4 Data Splitting

Train, test, and validation sets might be created from the dataset. The training set is the subset of data used to modify the parameters. A model's parameters are adjusted using a set of examples referred as validation set. The test set is understood to be a set of instances used solely to evaluate the effectiveness of a fully described model. Such methods could be categorised into three various types which are mentioned as follows:

3.4.1 Cross Validation

It is basically a statistical approach of comparing and evaluating algorithms of learning by separating the data into two major segments. One segment is used to learn and train a model, and other one is used to authenticate the model.

3.4.2 Random

It is a random selection of proportion of samples as well as to retain this as a validation set and then utilising the samples which are remaining samples for the purpose of training. This procedure is generally repeated a number of times, and the concluding estimation of the performance of model is average on the sets of validation of all repeats.

3.4.3 Kennard-Stone

Relying on the sharing of data, methodically choosing a particular number of most delegate samples from the sets of data and utilising the samples which are remaining for the validation is the third method. K-S or Kennard stone algorithm is a good instance of such approach.

3.5 Fake Detection Using Machine Learning Techniques

Machine learning (ML) is a class of algorithms that assist software systems to incorporate more accurate results without directly reprogramming them. To develop predictions, data scientists characterise the changes or characteristics that the model needs to analyse and utilise. The algorithm splits the learned levels into new data when the training is completed.

3.5.1 Decision Trees

The decision node and the leaf node are the two nodes of a decision tree. The leaf nodes, on the other hand, are the results of these choices and do not have any more branches. Decision nodes have several branches and are used to make decisions. It uses a flowchart that resembles a tree, as the name would imply, to display the estimates that result from a series of feature-based splits. This approach for supervised learning categorises the data for both continuous and categorical dependent variables. The unpredictability or impurity of a dataset is measured using E(S). The value of entropy is always in the range between 0 and 1. Its value improves when it equals 0 and degrades when it equals 0. Entropy of the classification of set S in terms of r states if the aim is G with different attribute values as Eq. (1) demonstrates.

$$E(S) = \sum_{i=1}^{r} -m_i \log_2 m_i \tag{1}$$

where "m" stands for probability and "E(S)" for entropy.

$$\text{Gini}(S) = 1 - \sum_{j=1}^{r} (m)_j^2 \tag{2}$$

where pi is the likelihood that an object will be placed in a specific class.

3.5.2 Logistic Regression

Predicting the likelihood of a categorical dependent variable is done using a ML classification technique. When predicting binary values like "fake" and "real", LR, a classification technique, employs the logistic function, often known as the sigmoid function. An estimate of probability is created from the logistic sigmoid function's output. To forecast the values of the predictive variable y, where y is in the range of 0 and 1, in a binary classification task. Positive class is represented by 1, whilst the negative class is shown by 0. A hypothesis $h(\Theta) = \Theta^T X$ will be created to categorise the two classes 0 and 1, and the output threshold for the classifier is when $h(\Theta) = 0.5$. Hypothesis $h(\Theta)$ will forecast $y = 1$ if its value is more than 0.5, indicating that the news is true, and y = 0 if its value is less than 0.5, indicating that the news is fake [10]. So, on the assumption that $0 \leq h\Theta(x) \leq 1$, logistic regression prediction is carried out. Equation (3) may be written as follows to represent the sigmoid function of logistic regression:

$$h\Theta(x) = g(\theta^T X) \tag{3}$$

3.5.3 Support Vector Machine

An SVM, a top-notch ML technique, is built on the idea of structural risk reduction. SVM learning stands out for its independence from the feature space's dimensionality. In SVM, the degree of complexity of a hypothesis is measured by the margin at which the hypotheses are separated. The SVM aims to extend the hyperplane separation of false news data to nonlinear boundaries. In SVM, bogus news is identified using the equations below:

$$\text{If } C_i = +1, zr_i + x \geq 1 \tag{4}$$

$$\text{If } C_i = -1, zr_i - x \leq 1 \quad (5)$$

$$\text{For all } i;\ C_i(z_i + x) \geq 1 \quad (6)$$

In the equations above, C is the news's class label, which may either be 1 or -1, and z is the weight vector [27]. "r" is the vector of fake news data. Each vector of the test data is placed inside the radius r of the training data vector if the training data is appropriate. Now, if the chosen hyperplane is as far away from the data as is physically possible, it maximises the margin between points of classes.

3.5.4 Naive Bayes

The Bayes theorem is the foundation of the supervised ML algorithm. According to the Naive Bayes classifier [28], there is no connection between the inclusion or exclusion of one unique characteristic from a class.

3.5.5 AdaBoost

Since a single weak classifier cannot reliably anticipate an object's categorisation, AdaBoost sklearn combines numerous weak classifiers in progressive learning to create the final strong predictive classifier. AdaBoost is not a model; rather, it is a technique that may be used to improve any classifier's capacity to learn from its mistakes and recommend a more precise and effective model for use in the future.

3.6 Fake Detection Using Deep Learning Techniques

An apparatus or system called a neural network is created to function similarly to the human brain. The basic goal is to create a system that can complete various computing jobs quicker than current systems. Data clustering, approximation, optimization, and pattern detection and classification are some of these activities. Because certain text datasets are exceedingly vast and are not linearly separable, traditional ML cannot properly analyse them. Data that cannot be easily drawn on a hyperplane is simply nonlinear data that cannot be linearly separated. The DL method was suggested as a solution to the issue of forecasting significant trends in linearly non-separable data.

3.6.1 The Convolutional Neural Network (CNN)

It is a DL architecture that maps the characteristics of the input data using convolutional layers. By using various filter sizes, the layers are organised into various feature mappings. Based on the findings of the feature mapping, CNN may learn details about the input data. The convolutional layer is typically followed by a pooling layer to provide accurate output dimensions even with various filters. Additionally, the pooling layer reduces the output dimensions without sacrificing critical information, which lessens the computational strain [14].

3.6.2 Reccurent Neural Network (RNN)

In order to comprehend temporal or sequential data, RNNs are utilised. RNNs may produce more accurate predictions if more data points are included in a sequence. In order to achieve this, they take input, change the output, and then reuse the signals of earlier or later nodes in the sequence. RNNs' internal memory allows them to retain important information, such as the input they received, which enables them to predict what will happen next with astounding accuracy. They are therefore by far the most popular [29].

3.6.3 Long Short-Term Memory (LSTM)

When it comes to NLP issues, LSTM models are leaders. A DL framework for artificial recurrent neural networks is called LSTM. A developed version of RNN is LSTM. Since each word is connected to the one before it and the one after it, the LSTM network can learn long-term dependencies and performs well with textual input. The LSTM layer enables the model to disregard superfluous text and concentrate on specific segments of a sequence [22].

3.7 Fake Detection Using Ensembled Learning Techniques

The methods of ensemble in the machine learning field basically collaborate the imminent which is acquired from the models of multiple learning in order to ease as well as enhance decisions. The three major classes of the respective learning method are stacking, bagging, and boosting. It is vital to both have a thorough understanding of every approach as well as to consider them on the predictive model of the project. A placid introduction to all these approaches as well as the main ideas behind every method is mentioned as follows:

3.7.1 Bagging

It includes fitting of several decision trees on various samples of the similar set of data and also includes averaging of predictions. It acquires it name because it collaborates aggregation as well as bootstrapping in order to develop single model of ensemble.

3.7.2 Stacking

It includes fitting of several sort of models on the similar data and utilising another model in order to learn the ways to combine the predictions in their best form.

3.7.3 Boosting

It involves adding of ensemble members successively that correct the predictions which is made by the previous models as well as outputs a prejudiced average of the forecast.

4 Conclusion

More individuals now regularly consume news on social media than in the past through conventional media. It takes extensive knowledge and skill to spot abnormalities in the text whilst classifying news storeys manually. Prior research on identifying fake news in the field of artificial intelligence focussed mostly on traditional machine learning techniques. Machine learning for text categorisation enables the classification of data based on the previous observations. We came to the conclusion that the majority of research articles employed ensemble learning and neural networks to classify fake news. The issue of categorising fake news items using ML models, DL models, ensembeled learning, and other techniques has been considered because it takes a lot of effort to manually validate one article. An in-depth examination of the most important fake news detection attempts to date was offered by this study.

5 Current Challenges and Future Direction

Numerous studies have been conducted on the topic of identifying fake news, yet there is still room for improvement to advance and research in the future.

- First and foremost, the use of fresh news and user corpora, possibly gathered by an online community. The detection of users based on shared interests and behaviours can also help in identifying fake news.
- On social networking platforms, a real-time fake news detection system will be utilised to benefit a range of new activities, such as preventing the dissemination of fake information during elections or even pandemics.
- According to several research studies, character n-grammes combined with TF-IDF, bagging, and boosting techniques can lead to higher classification accuracy than basic content-related n-grammes and POS tagging.

- Use of an ensemble technique yields superior results in comparison with a straightforward classifier by using DL and ML methods to build an ensemble model. Using CNN, LSTM, and ensemble models, several researchers have achieved great accuracy.
- Graph-based approaches have been extensively studied over the past 20 years in order to show the intricate and complex structures underlying the data model. In cases when it does not get explicit labelling information, the computer must be able to infer patterns between data. This is where semi-supervised learning, which seeks to offer strategies for creating these connections, can be beneficial.
- In order to eliminate the consequences brought on by fake news globally, further studies using multiple datasets and languages should be conducted.

In the future, we would like to include more perspectives and develop a more effective model that runs faster whilst still offering promising results.

References

1. Qureshi KA, Malick RAS, Sabih M, Cherifi H (2021) Complex network and source inspired Covid-19 fake news classification on Twitter. IEEE Access 9:139636–139656. https://doi.org/10.1109/ACCESS.2021.3119404
2. Gerken T (2022) Google to run ads educating users about fake news. BBC. https://www.bbc.com/news/technology-62644550
3. Tokyo Olympics (2021) Organisers say summer games cancellation report fake news-sports news, Firstpost. https://www.firstpost.com/sports/tokyo-olympics-2020-organisers-say-summer-games-cancellation-report-fake-news-9193521.html
4. Fake news is clouding the real stories around the Ukrainian crisis—here's how to spot it (2022). https://www.weforum.org/agenda/2022/03/fake-viral-footage-is-spreading-alongside-the-real-horror-in-ukraine-here-are-5-ways-to-spot-it/
5. Fake alert: will govt monitor your WHATSAPP CHATS? here's the truth (2022). https://economictimes.indiatimes.com/news/new-updates/fake-alert-will-govt-monitor-your-whatsapp-chats-heres-the-truth/articleshow/93712093.cms
6. Mahrishi M, Morwal S, Muzaffar AW, Bhatia S, Dadheech P, Rahmani MKI (2021) Video index point detection and extraction framework using custom yolov4 darknet object detection model. IEEE Access 9:143378–143391. https://doi.org/10.1109/ACCESS.2021.3118048
7. Kumar Mahrishi M, Meena G (2022) A comprehensive review of recent automatic speech summarization and keyword identification techniques. Springer, Cham, pp 111–126
8. Dahouda MK, Joe I (2021) A deep-learned embedding technique for categorical features encoding. IEEE Access 9:114381–114391. https://doi.org/10.1109/ACCESS.2021.3104357
9. Alhawarat M, Aseeri AO (2020) A superior Arabic text categorization deep model (SATCDM). IEEE Access 8:24653–24661
10. Jiang T, Li JP, Haq AU, Saboor A, Ali A (2021) A novel stacking approach for accurate detection of fake news. IEEE Access 9:22626–22639
11. Almars AM, Almaliki M, Noor TH, Alwateer MM, Atlam E (2022) Hann: Hybrid attention neural network for detecting Covid-19 related rumors. IEEE Access 10:12334–12344. https://doi.org/10.1109/ACCESS.2022.3146712
12. Elsaeed E, Ouda O, Elmogy MM, Atwan A, El-Daydamony E (2021) Detecting fake news in social media using voting classifier. IEEE Access 9:161909–161925

13. Saleh H, Alharbi A, Alsamhi SH (2021) Opcnn-fake: optimized convolutional neural network for fake news detection. IEEE Access 9:129471–129489. https://doi.org/10.1109/ACCESS.2021.3112806
14. Ni S, Li J, Kao H-Y (2021) Mvan: multi-view attention networks for fake news detection on social media. IEEE Access 9:106907–106917
15. Ilie V-I, Truică C-O, Apostol E-S, Paschke A (2021) Context-aware misinformation detection: a benchmark of deep learning architectures using word embeddings. IEEE Access 9:162122–162146
16. Rohera D, Shethna H, Patel K, Thakker U, Tanwar S, Gupta R, Hong W-C, Sharma R (2022) A taxonomy of fake news classification techniques: Survey and implementation aspects. IEEE Access 10:30367–30394. https://doi.org/10.1109/ACCESS.2022.3159651
17. Bai N, Meng F, Rui X, Wang Z (2021) Rumour detection based on graph convolutional neural net. IEEE Access 9:21686–21693
18. Jain MK, Garg R, Gopalani D, Meena YK (2022) Review on analysis of classifiers for fake news detection. In: Balas VE, Sinha GR, Agarwal B, Sharma TK, Dadheech P, Mahrishi M (eds) Emerging technologies in computer engineering: cognitive computing and intelligent IoT. Springer, Cham, pp 395–407
19. Jain MK, Gopalani D, Meena YK, Kumar R (2020) Machine learning based fake news detection using linguistic features and word vector features. In: 2020 IEEE 7th Uttar Pradesh section international conference on electrical, electronics and computer engineering (UPCON). IEEE, pp 1–6
20. Elhadad MK, Li KF, Gebali F (2020) Detecting misleading information on Covid-19. IEEE Access 8:165201–165215
21. Umer M, Imtiaz Z, Ullah S, Mehmood A, Choi GS, On B-W (2020) Fake news stance detection using deep learning architecture (CNN-LSTM). IEEE Access 8:156695–156706
22. Sansonetti G, Gasparetti F, D'aniello G, Micarelli A (2020) Unreliable users detection in social media: deep learning techniques for automatic detection. IEEE Access 8:213154–213167
23. Seddari N, Derhab A, Belaoued M, Halboob W, Al-Muhtadi J, Bouras A (2022) A hybrid linguistic and knowledge-based analysis approach for fake news detection on social media. IEEE Access
24. Al-Rakhami MS, Al-Amri AM (2020) Lies kill, facts save: detecting Covid-19 misinformation in twitter. IEEE Access 8:155961–155970
25. Liu Y, Yu K, Wu X, Qing L, Peng Y (2019) Analysis and detection of health-related misinformation on Chinese social media. IEEE Access 7:154480–154489
26. Kim Y, Kim HK, Kim H, Hong JB (2020) Do many models make light work? Evaluating ensemble solutions for improved rumor detection. IEEE Access 8:150709–150724. https://doi.org/10.1109/ACCESS.2020.3016664
27. Wan C, Wang Y, Liu Y, Ji J, Feng G (2019) Composite feature extraction and selection for text classification. IEEE Access 7:35208–35219. https://doi.org/10.1109/ACCESS.2019.2904602
28. Alhashmi SM, Khedr AM, Arif I, El Bannany M (2021) Using a hybrid-classification method to analyze Twitter data during critical events. IEEE Access 9:141023–141035. https://doi.org/10.1109/ACCESS.2021.3119063
29. Mridha MF, Keya AJ, Hamid MA, Monowar MM, Rahman MS (2021) A comprehensive review on fake news detection with deep learning. IEEE Access 9:156151–156170. https://doi.org/10.1109/ACCESS.2021.3129329

Chapter 32
Multiclass Sentiment Analysis of Twitter Data Using Machine Learning Approach

Bhagyashree B. Chougule and Ajit S. Patil

1 Introduction

People regularly spend the majority of their daily time on social media. With the advent of social media, microblogging sites are becoming popular information-sharing platforms for folks. The Internet is expanding rapidly with a vast number of social networking sites, blogs, and different portals where users are generating their feedback, ratings, and suggestions [1]. Thousands of people rely on online reviews for decision making, and it is the purpose of using sentiment analysis to assess sentiments in reviews. In April 2013, it is observed that 90% of client decisions were based on online reviews [2]. Among available social networking sites, Twitter is widely used to express ideas, and opinions on any social issue, event, or product of a company. Millions of tweets are posted daily on Twitter which makes it a rich source of data. Twitter attracted researchers because of its features like short-length messages, a variety of opinions by several people on the same topic, and the availability of a huge amount of data. Their goal is to do sentiment analysis using different machine learning techniques for the classification of tweets as positive or negative [3].

The task of analyzing user behavior from unstructured tweets posted on social media is intriguing but difficult. POS tagging, Word2Vec, unigram feature extraction, and other methods have already been suggested for conducting sentiment analysis on Twitter data. The majority of algorithms currently in use categorize user tweets into positive and negative categories. A multiclass approach reveals more realistic feelings than simply mentioning if something is positive or negative. To improve the performance, it is needed to classify tweets into more than two classes, i.e., multiclass sentiment analysis [4].

B. B. Chougule · A. S. Patil (✉)
Kolhapur Institute of Technology's College of Engineering, Kolhapur, Maharashtra, India
e-mail: patil.ajit@kitcoek.in

V. Devedzic et al. (eds.), *Proceedings of the International Conference on Intelligent Computing, Communication and Information Security*, Algorithms for Intelligent Systems, https://doi.org/10.1007/978-981-99-1373-2_32

Due to limitations on the number of characters (short-length tweets), mining performance degrades as compared to the mining of larger texts. Also, making more classes of polarity rather than two leads to a decrease in accuracy [5]. Another challenge in making more polarity classes is class imbalance. It reduces model accuracy and biases the result toward the majority class [6]. So, very less amount of work is done on multiclass sentiment analysis.

This paper aims to study sentiment analysis of Twitter data using machine learning techniques to deal with ordinal regression (making more classes) problems. In this paper, we followed pre-processing of tweets, feature extraction using term frequency-inverse document frequency (TF-IDF), and finally applying machine learning regressors for the classification of tweets in five classes. Unigram, bigram, and trigram features are used for result analysis. The remaining paper is organized as–Sect. 2 contains a literature review of previous work, Sect. 3 describes the methodology and experimentation, Sect. 4 presents results and discussion, and Sect. 5 provides the conclusion.

2 Literature Review

In recent years, a vast amount of work has been done on sentiment analysis. The main aim behind it is to analyze social media data. Increased use of social media attracted researchers and investigators to retrieve important information from social data. Most of the available work did binary or ternary class sentiment analysis. A very little amount of work is available on multiclass sentiment analysis. This review section is categorized into two subsections. The first section explores multiclass sentiment analysis work, while the second section elaborates on sentiment analysis work with *n*-gram feature extraction.

2.1 Review on Multiclass Sentiment Analysis

In [7], authors constructed a framework with two parts—(1) important feature selection from the text using gain ratio, information gain, CHI statistics, and document frequency algorithms and (2) use of *k*-nearest algorithm (KNN), radial basis function neural network, decision tree, Naïve Bayes (NB), and support vector machine (SVM) algorithm to train multiclass sentiment classifier. Experiments are conducted with three public datasets, and accuracy is checked with 10-fold cross-validation. In multiclass sentiment classification, the Wilcoxon test is used to determine whether there are significant differences between various techniques. Results showed the best performance of gain ratio and SVM among the remaining algorithms.

Bhatia et al. [5] proposed an approach of deeper classification of Twitter text and classified tweets into seven classes. For this, author introduced their tool with an easy GUI called the SENTA tool. To enhance the accuracy of classification, the author introduced writing pattern-based features along with regular features in a text (like unigram). The experiment achieved 60.2% accuracy on multiclass classification. Bouazizi and Ohtsuki [8] extended their work to assign exact sentiment to text rather than labeling overall polarity to user posts. To do this, author added some components to the SENTA tool for performing the "quantification" task. Tweets are categorized into 11 different classes and experiments achieved 45.9% $F1$-score accuracy.

Elbagir and Yang [9] aim to do ordinal regression (multiclass classification) for Twitter sentiment analysis. Tweets are classified into five classes high positive, moderate positive, neutral, moderate negative, and high negative. To study multiclass sentiment analysis classification, decision tree, support vector regressor, SoftMax, and random forest techniques of machine learning are used. Experiments were carried out on the NLTK "twitter_samples" dataset and obtained higher accuracy with the decision tree machine learning classifier. The author introduced a new method named "scoring and balancing" for classifying data into five classes.

2.2 *Review on Sentiment Analysis with N-gram Features*

Iqbal et al. [10] proposed a system for sentiment analysis of data from two different domains namely Stanford Twitter Sentiment-140 and IMDb Movie Review dataset. The study focuses on different feature combinations from unigram and bigram, unigram and stop word filtered word feature, bigram, and stop word filtered word feature, most informative unigram, most informative bigram, and so on. Machine learning classifiers Naive Bayes, support vector machine, max entropy evaluated based on recall, precision, accuracy, and $F1$-score. Results proved that choosing feature combinations increased system accuracy than single random feature selection.

Kaur et al. [11] proposed a method for Twitter sentiment analysis with a combination of N-gram feature extraction and KNN classification technique. Experiments are carried out to check the performance of a system using precision, recall, and accuracy parameters.

The results are compared with the existing support vector machine classifier system which shows a 7% improvement due to the use of n-gram feature extraction as per the author's conclusion.

Alessa and Faezipour [12] built a model for flu disease surveillance using machine learning techniques. To implement a system, author followed pre-processing, feature extraction, and applying machine learning classifier steps. The classifier's performance is compared using accuracy, precision, recall, and F-measure metrics. The outcome proves that the use of the term frequency-inverse document frequency (TF-IDF) weighting scheme for feature extraction improves the system performance.

Brandao and Calixto [13] studied the machine learning classifier—support vector machine for sentiment analysis of tweets in five different datasets—Stanford Sentiment, Senti Strength Twitter Dataset, Sentiment-140, Sanders, Health Care Reform. To obtain better results, author used N-gram and TF-IDF for feature extraction. A combination of unigram, bigram, and trigram with K-fold cross-validation (10, 15, 20-folds) produced accuracy from 63.93 to 81.06%.

Ahuja et al. [14] aim to study the impact of TF-IDF word level and N-gram on Twitter sentiment analysis. An SS-Tweet dataset is used for experimentation with a classifier's decision tree, support vector machine, k-nearest neighbor, random forest, Naïve Bayes, and logistic regression. Performance evaluator's precision, recall, accuracy, and F-score show that TF-IDF gives 3–4% better performance than n-gram features for this dataset. Logistic regression predicts sentiment in a text by giving max output among all performance parameters.

3 Methodology

Social media is a big knowledge hub with a great platform for expressing opinions of all age group people. Sentiment analysis involves taking a piece of text and returning a "score" indicating how positive or negative that text is. This section describes the steps used for the actual implementation of the system. The first module is retrieving tweets as a raw dataset, the second module is pre-processing of tweets, the third module is feature extraction using unigram, bigram, trigram approach, and assignment of five polarity classes, and the fourth module is applying machine learning classifier to classify tweets in five classes as—neutral, high negative, moderate negative, high positive, and moderate positive. Figure 1 depicts the system architecture of multiclass sentiment analysis of the "twitter_samples" dataset.

Data Collection

A publicly available Twitter dataset in NLTK corpora resource is used for experimentation. It contains a total of 10000 tweets which consist of 5000 positive and 5000 negative tweets. While applying a machine learning classifier 7000 tweets (3500 positive and 3500 negative) are used for training and the rest of 3000 tweets (1500 positive and 1500 negative) are used for testing the model.

3.1 Data Pre-processing

Tweets are informal and unstructured English text. Such noisy tweets contain a lot of repeated letters, emoticons, stopwords, etc. It implies unnecessary overhead on sentiment analysis performance. So, some pre-processing is necessary while doing sentiment analysis. The steps followed during pre-processing are as follows:

- Removal of hyperlinks, hashtag (#) symbol, user handles (@user), and retweets,
- Removing stopword,
- Removing repeated letters from words,
- Conversion of words to lowercase,

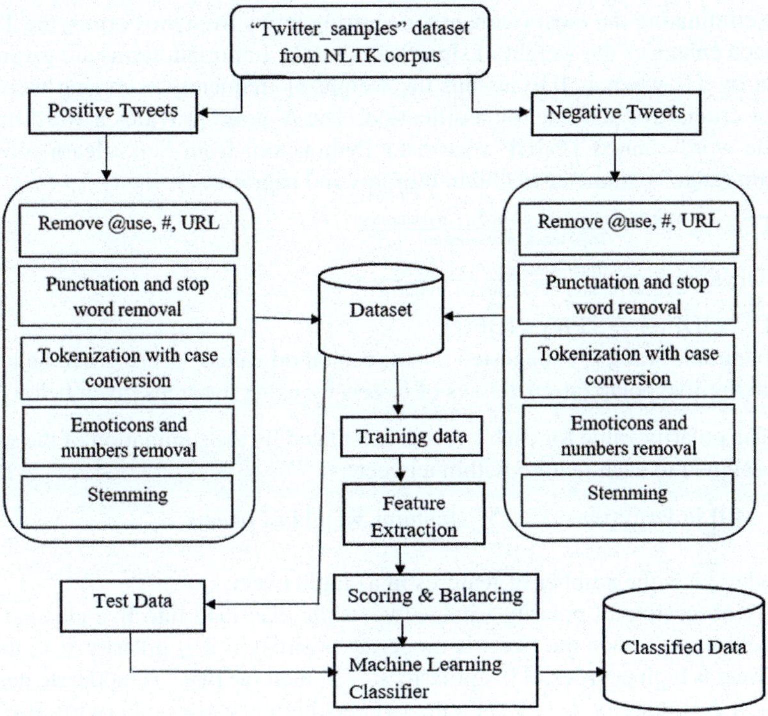

Fig. 1 System architecture

- Tokenization, i.e., converting tweets in to tokens,
- Stemming to get the base form of the word by removal of affix and suffix,
- Emoticon removal,
- Removing non-alphabetic characters from tweets.

3.2 Feature Extraction and Polarity Assignment

Machine learning is a statistical approach that works with numerical data. Before building a sentiment analysis model, it requires converting text into numeric form. There exist lots of feature extraction techniques like term frequency-inverse document frequency (TF-IDF), bag-of-words, *N*-grams, document to vector (DOC2Vec), etc. Extracting important features from a large amount of data leads to scalability, dimensionality reduction, and reduction in computational problems faced while dealing with large volume data [14]. This work uses the term frequency-inverse document frequency (TF-IDF) scheme of the bag-of-words approach. Scikit-learn Python library provides built-in TF-IDF vectorizer functionality.

By eliminating the most frequently occurring terms from the corpus, the TF-IDF approach enhances the weight of significant words. Important terms are given more weight by TF, whereas IDF lessens the weight of frequently occurring words that are not crucial for sentiment classification. The N-gram provides a word order to provide word context.TF-IDF vectorizer Python tool from Scikit-learn offers the "n-gram range" parameter to obtain bigrams and trigrams.

[15] $\mathrm{TF}(t) = \frac{\text{Number of time termt occurs in document}}{\text{Total termt in document}}$

[15] $\mathrm{IDF}(t) = \log \frac{\text{Total Number of Documents}}{\text{Number of Documents with termt}}$

[15] $\mathrm{TF} - \mathrm{IDF}(t) = \mathrm{TF}(t) * \mathrm{IDF}(t)$

Elbagir and Yang [9] suggested one new method called "scoring and balancing" for making five polarization classes of tweets by using the steps given below:

(i) The polarity value for each tweet is calculated by the summation of the weights assigned to each feature within a tweet

$$[9]\ \text{Polarity(tweet)} = \sum_{n=1}^{N} (\text{features weight of tweet})$$

where n is the number of features in a single tweet.

(ii) By observing all polarity values, tweets are classified into five classes: if 0 < polarity < 2, then the tweet is moderate positive, if 2 ≤ polarity ≤ 4, then the tweet is high positive, if 0 < polarity < – 2, then the tweet is moderate negative, if – 2 ≤ polarity ≤ – 4, then the tweet is high negative, otherwise, if polarity = 0, then the tweet is classified as neutral.

3.3 *Applying Machine Learning Classifiers*

This study uses different machine learning algorithms, where a model is trained with training data with known output and tested with new data. Multinomial logistic regression (SoftMax), support vector regressor (SVR), decision tree (DT), and random forest regressor (RF) are used for making multiclass classification with unigram feature extraction. The best performing model among them is further used with bigrams and trigrams to test the effect of n-grams on multiclass sentiment analysis.

(i) Multinomial Logistic Regression (SoftMax)

Logistic regression is extended to multinomial logistic regression for multiclass classification. The loss function in logistic regression is changed to cross entropy loss, and then, it predicts multinomial probability distribution. The algorithm uses a SoftMax function to find out the probability of a target class. SoftMax function of logistic regression is extended for multiclass classification as follows:

[9] $P(y = j|X) = \frac{e^{x^T w_j}}{\sum_{k=1}^{K} e^{x^T w_k}}$

where input to the function is the output of various linear functions of K, the estimated probability of a sample vector x, and a weighting vector w for the j class [9].

(ii) Support Vector Regressor (SVR)

Support vector machine (SVM) is also used for regression along with classification. It maintains all the main features of the original SVM with some minor changes. The main aim of using a regressor is the same as to minimize error, maximizing margin by a distinctive hyperplane. SVR has two variations linear SVR and nonlinear SVR.

(iii) Decision Tree (DT)

A decision tree builds both regression and classification models in the form of a tree. Dataset is divided into smaller subsets by applying decision rules and developing an incremental decision tree. The tree contains decision nodes and leaf nodes. Decision node branches represent the attributes, and the leaf node represents the target class. Decision trees work on both categorical and continuous values.

(iv) Random Forest Regressor (RF)

Random forest is an ensemble method that combines different decision tree algorithms together. It avoids overfitting and provides high accuracy. It uses a bagging technique where all trees run in parallel. Multiple trees are constructed at training time, and the target class is predicted by taking the mean of predictions from all trees. Bias and variance in prediction are handled by bagging and boosting.

4 Results and Discussion

In this study, Twitter sentiment analysis based on multiclass polarization, four machine learning techniques namely SoftMax, support vector regressor, decision tree, and random forest are used. Implementation is done with Python open-source package Scikit-learn. The publicly available "twitter_samples" dataset from NLTK is used as input. Data pre-processing is the very first step of implementation, and its results along with data labeling are shown in Table 1.

Table 1 Pre-processed tweets with labeling

Cleaned tweet	Label
Follow Friday top engaged members community week	High positive
Congrats	Moderate positive
Hopeless tmr	Neutral
Heart sliding waste basket	Moderate negative
Everything kids section ikea cute shamen early months	High negative

Table 2 Accuracy of all models for unigrams

Algorithm	Accuracy (%)
Multinomial logistic regression	66.33
Support vector regression	81.88
Random forest	84.16
Decision tree	91.91

Table 3 Results for mean square error and mean absolute error

Algorithm	Mean square error (MSE)	Mean absolute error (MAE)
Multinomial logistic regression	0.333	0.333
Support vector regression	0.335	0.412
Random forest	0.158	0.157
Decision tree	0.149	0.145

The prediction accuracy score for all classifiers with unigram feature extraction is shown in Table 2. The result shows that random forest and decision tree give better accuracy than support vector regressor and SoftMax. The decision tree achieves a 91.91% accuracy score.

Along with the score, common evaluation metrics for regression is mean square error (MSE) and mean absolute error (MAE). As they are error values, their lower value represents better performance. The results for error values are shown in Table 3.

The score value of each classifier with unigram features is presented in Fig. 2. It shows SoftMax with the lowest score of 66.33%, followed by support vector regressor, and random forest having 81.88% and 84.16% scores, respectively. Decision tree is having the highest score of 91.91% of all other classifiers.

Decision tree is giving best result for multiclass classification with unigram features, so this classifier is further used for multiclass sentiment analysis work with bigram and trigram features. When it is used as a multiclass classifier with bigrams features, gives a 98% score and 99% with trigram feature extraction. It is depicted in Fig. 3. It is observed that the model score increases as we go with increasing context window size.

Along with these error evaluation metrics, MSE and MAE also get decreased for bigrams and trigrams, which is shown in Table 4.

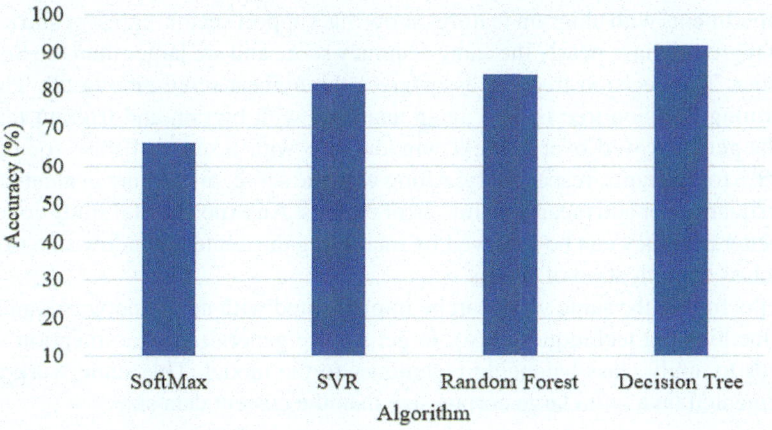

Fig. 2 Accuracy of classifiers with unigram features

Fig. 3 Accuracy of decision tree with unigram, bigram, and trigram

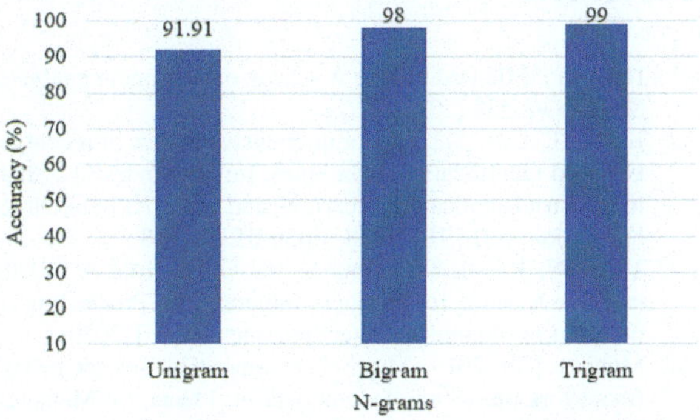

Table 4 MSE and MAE for decision tree with bigrams and trigrams

N-grams	Mean square error (MSE)	Mean absolute error (MAE)
Unigrams	0.149	0.145
Bigrams	0.005	0.005
Trigrams	0.0003	0.0003

5 Conclusion

This study performs multiclass sentiment analysis of Twitter data using different machine learning models and increased context window size. Experiments used NLTK "twitter_samples" dataset with classifiers support vector regressor, multinomial logistic regression, random forest, and decision tree. This study includes the "scoring and balancing" method to get multiple polarity classes followed by applying machine learning classifiers for tweet classification.

Experiments with unigram features show that support vector regressor and random forest regressor have nearly the same accuracy score and are larger than the SoftMax classifier. The decision tree regressor gives the highest score among all. The best performing decision tree is then again analyzed with bigram and trigram features. Results get improved over context window size with a score of 98% for bigrams and 99% for trigrams, respectively. Along with the score, models are evaluated using mean square error and mean absolute error metrics. An experimental study concluded that better accuracy can be achieved by increasing the context window size to do the multiclass classification of tweets.

In the future, the same work can be implemented with neural networks and other machine learning techniques. Also, to get a more generalized classification model, it needs to apply class balancing techniques to the model. This same work can be implemented on a quite large corpus size than the current data size.

References

1. Hussein DME-DM (2016) A survey on sentiment analysis challenges. J King Saud Univ Eng Sci 30:330–338 (2018)
2. Kumar T et al. (2022) A comprehensive review of recent automatic speech summarization and keyword identification techniques. In: Fernandes SL, Sharma TK (eds) Artificial intelligence in industrial applications. learning and analytics in intelligent systems, vol 25. Springer, Cham. https://doi.org/10.1007/978-3-030-85383-9_8
3. Yadav N, Kudale O, Gupta S, Rao A, Shitole A (2020) Twitter sentiment analysis using machine learning for product evaluation. In: Proceedings of the fifth international conference on inventive computation technologies (ICICT-2020)
4. Mohbey KK (2019) Multi-class approach for user behavior prediction using deep learning framework on twitter election dataset. J Data, Inf Manage Springer Nat Switzerland AG
5. Bhatia S et al. (2022) An efficient modular famework for automatic LIONC classification of med IMG using unified medical language. Front Public Health, Sect Digital Public Health Manuscript ID 926229:1–21. https://doi.org/10.3389/fpubh.2022.926229
6. Mukherjee A, Mukhopadhyay S, Panigrahi P, Goswami S (2019) Utilization of over sampling for multiclass sentiment analysis on Amazon review dataset. In: 10th International conference on awareness science and technology (iCAST)
7. Liu Y, Bi J-W, Fan Z-P (2017) Multi-class sentiment classification: the experimental comparisons of feature selection and machine learning algorithms, Elsevier
8. Bouazizi M, Ohtsuki T (2018) Multi-class sentiment analysis in Twitter: What if classification is not the answer. IEEE Access
9. Elbagir S, Yang J (2019) Twitter sentiment analysis based on ordinal Regression. IEEE Access 7
10. Iqbal N, Chowdhury A, Ahsan T (2018) Enhancing the performance of sentiment analysis by using different feature combinations. In: 2018 International conference on computer, communication, chemical, material and electronic Engineering (IC4ME2)
11. Kaur S, Sikka G, Awasthi LK (2018) Sentiment analysis approach based on N-gram and KNN classifier. In: 2018 First international conference on secure cyber computing and communication (ICSCCC)
12. Alessa A, Faezipour M (2018) Tweet classification using sentiment analysis features and TF-IDF weighting for improved flu trend detection. In: International conference on machine learning and data mining in pattern recognition

13. Brandao JdG, Calixto WP (2019) N-Gram and TF-IDF for feature extraction on opinion mining of tweets with SVM classifier. In: International artificial intelligence and data processing symposium (IDAP)
14. Ahuja R, Chug A, Kohli S, Gupta S, Ahuja P (2019) The impact of features extraction on the sentiment analysis. In: International conference on pervasive computing advances and applications–Per CAA 2019
15. Tomer M, Kumar M (2020) Improving text summarization using ensembled approach based on fuzzy with LSTM. Springer-Arab J Sci Eng

Chapter 33
Space–Time Continuum Metric

Anurag Dutta and Pijush Kanti Kumar

1 Introduction

In computational mathematics, algorithms are a set of well-defined, finite though, rules that tell us the step-by-step approach to solving a problem. From the very early ages of developments in the domains of mathematical computing, scientists have been eager enough in their search of best algorithms for each purpose. Now, when some best minds work on something distinctively, a wide variety of good choices would bloom. Now, it becomes very cumbersome to select the best among all. For that reason, several metrics came into vision, like time complexity, space complexity, bit complexity, and many more. Now, after these metrics came into action, scientists developed various asymptotic notations to compare algorithms, like Big–Oh Notation [1], Big–Theta Notation [2], Big–Omega Notation [3]. As of now, there exist several mathematical formulations and techniques to analyze these algorithms keeping in account of their asymptotes. For instance, we have Master's theorem [4], Akra–Bazzi's method [5], etc. But all these analytics were limited to unidimensional accountings, like either they considered into account, the time complexity, or the space complexity individually though. In this work, we would be proposing space–time continuum—a metric built out of pre-existing computational complexity metrices, invoked in the polar coordinate system [6], which would help us in keeping account of the rate of change, or in just terms, the slope attained by the canonicals of the continuum. The best thing about this metric is that the rate of change

A. Dutta (✉)
Department of Computer Science and Engineering, Government College of Engineering and Textile Technology, Serampore, India
e-mail: anuragdutta.research@gmail.com

P. K. Kumar
Department of Information Technology, Government College of Engineering and Textile Technology, Serampore, India

V. Devedzic et al. (eds.), *Proceedings of the International Conference on Intelligent Computing, Communication and Information Security*, Algorithms for Intelligent Systems, https://doi.org/10.1007/978-981-99-1373-2_33

can be justified as per requirements, like if we are to focus more on a space optimized model of mathematical computation, we would keep into consideration the rate of change of $\mathcal{S}(n)$, space complexity, in accordance with $\mathcal{T}(n)$, time complexity, $\frac{\delta \mathcal{S}(n)}{\delta \mathcal{T}(n)}$, while if were to focus more on a time optimized model of mathematical computation, we would keep into consideration the rate of change of $\mathcal{T}(n)$, time complexity, in accordance with $\mathcal{S}(n)$, space complexity, $\frac{\delta \mathcal{T}(n)}{\delta \mathcal{S}(n)}$.

2 Time Complexity

In computer science, time complexity is the complexity of a calculation and represents the amount of computer time it takes to execute an algorithm. Time complexity is usually estimated by counting the number of basic operations performed by the algorithm, assuming that each basic operation takes a certain amount of time to perform. Therefore, it is assumed that the time required and the number of basic operations performed by the algorithm are related by a constant coefficient. The execution time of the algorithm can vary between different inputs of the same size, so it is usually considered the worst time complexity, the maximum time required for a particular size input. What is less common and usually explicitly stated is the complexity of the average case. This is the average time it takes to enter a particular size, which makes sense because the number of possible inputs for a particular size is limited. In both cases, the time complexity is usually expressed as a function of the size of the input. This function is generally difficult to calculate accurately, and execution time is usually not important for small inputs, so we usually focus on complexity behaviors, that is, asymptotic behaviors of complexity, as the input size increases. Therefore, in most cases, the time complexity is usually expressed using the uppercase O notation, better known as Big–Oh Notation. Some of the typical examples are (Table 1).

Now, it's not so that Big–Oh Notations are the typical asymptotic notation for complexity analysis, but they are the most widely used ones.

3 Space Complexity

The space complexity of an algorithm or computer program is the amount of memory needed to solve an instance of a computational problem as a function of input properties. This is the memory required for the algorithm until it is fully executed. In most cases, it is also considered in using the uppercase O notation, better known as Big–Oh Notation. Some of the typical examples are (Table 2).

Table 1 Typical examples of running times of different algorithms

Running time $T(n)$	Examples
$O(1)$	Calculating the sum of the convergent series, like $4\sum_{i=1}^{\infty}\frac{(-1)^{i+1}}{2i-1}$
$O(n)$	Calculating the sum of the series, like $\sum_{i=1}^{n}\frac{cosec^2(i)}{i^3}$ whose convergence is yet to be proven
$O(n\log n)$	The most efficient sorting algorithm [7] $\log_2\left(\sqrt{\pi}\left(\frac{n}{e}\right)^n\left(\sqrt[6]{8n^3+4n^2+n+\frac{1}{30}-\frac{11}{240n}}\right)\right)$ or $\log_2\left(\sqrt{\pi}\left(\frac{n}{e}\right)^n\left(\sqrt[6]{8n^3+4n^2+n+\frac{1-\frac{11}{8x}+\frac{79}{112x^2}}{30}}\right)\right)$
$O(n!)$	Brute Force solution for TSP $\prod_{i=0}^{n-1}(n-i)$

Table 2 Typical examples of running space of different algorithms

Running space $S(n)$	Examples
$O(1)$	Bubble Sort $\exists\mathcal{L}\ni\forall(i,j)\&j\geq i,\mathcal{L}_j\geq\mathcal{L}_i$
$O(n)$	Merge Sort $\exists\mathcal{L}\ni\forall(i,j)\&j\geq i,\mathcal{L}_j\geq\mathcal{L}_i$, with notations at a similar literal meaning as that in Bubble Sort
$O(\log(n+\lvert\alpha_n\rvert))$	LOGSPACE problems [8], solved using Turing machine $\exists(\alpha_n)\ni 1^n\rightarrow\widetilde{\alpha_n}$

4 Space–Time Tradeoff

In the last two sections, we have seen two quite popular metrics for comparing algorithms—"time complexity" and "space complexity." Though in, practice, both metrics are given equal importance, in the aspect of dealing with real-life problems, time complexity is considered at an edge. To maintain a lower magnitude of complexity in terms of time, space complexity is sometimes allowed to vary flexibly, which in turn promotes the efficacy of the algorithm in terms of time complexity. One of the best examples of this kind of tradeoff is the Prime Sieve Tradeoff.

A prime number is the one that has two divisors, unity, and the number itself.

The algorithm for the sieve method [9] is as follows:

1. Make a list of all-natural $\mathbb{N}$ in the range $[2, n]$, where n is the upper limit of the range.
2. Initialize $\rho\sim = 2$ and enumerate the multiples of ρ by counting in increments of $\lambda\rho$ to $(\lambda + 1)\rho$, and mark them down.
3. Find the smallest number in the list greater than ρ which is unmarked. If no such instance exists, then terminate the algorithm, Else, instantiate ρ as the neo, and again enumerate its multiples and with valid incrementalism, mark them down.
4. After the termination, the unmarked numbers are the primes in the house.
5. The pseudocode for this optimal technique is: -

```
function SieveOfEratosthenes

n ← input ∀n > 1
ℒ ←List of booleans (0 or 1) indexed in the range [2,n]

    fori = 2 to√n

      ifℒᵢis true

            forj = i², i² + i, i² + 2i, ...ton

                setℒⱼ := false

            end for

        end if

    end for

    return all i∀ℒᵢis true

end
```

5 Space–Time Continuum

In this section, we tend toward introducing a new metric, as a polynomial combination of space and time complexity. The metric will keep in account of both the complexities—space and time complexities, which at a sure aspect, is a way ahead of all the pre-existing computational complexities.

Let's consider an algorithm, $\mathcal{A}(n)$, with $\mathcal{S}(n) = \Upsilon(\psi(n))$ and $\mathcal{T}(n) = \Upsilon(\phi(n))$, where $\mathcal{S}(n)$ and $\mathcal{T}(n)$ are the respective computational complexities in terms of space and time, respectively, and $\Upsilon()$ is the appropriate asymptotic notation.

Accordingly,

$$\mathcal{S}(n) = \Upsilon(\psi(n)) = \alpha(\psi(n)) + \beta \tag{1}$$

and

$$\mathcal{T}(n) = \Upsilon(\phi(n)) = \delta(\phi(n)) + \mu \tag{2}$$

with the quadruple $(\alpha, \beta, \delta, \mu)$ being constant.

Now, $\exists\varphi(n)\ni\varphi(n) = \left(\frac{\max(\psi(n),\phi(n))}{\min(\psi(n),\phi(n))}\right)$

where,

$$\max(\psi(n), \phi(n)) = \frac{(\psi(n) + \phi(n)) + |\psi(n) - \phi(n)|}{2}$$

and

$$\min(\psi(n), \phi(n)) = \frac{(\psi(n) + \phi(n)) - |\psi(n) - \phi(n)|}{2}$$

Now, since these functions, $\max(\psi(n), \phi(n))$ and $\min(\psi(n), \phi(n))$ are complementary to each other, so,

$$((\psi(n), \phi(n))\ni \max(\psi(n), \phi(n))) = \overline{((\psi(n), \phi(n))\ni \min(\psi(n), \phi(n)))}$$

Due to which, only two cases arise

CASE I: $\max(\psi(n), \phi(n)) = \frac{(\psi(n)+\phi(n))+|\psi(n)-\phi(n)|}{2} = \psi(n)$.

Fetching $\psi(n)$ from Eq. (1) in terms of the tuple $\langle\alpha, \beta\rangle$, we get

$$\psi(n) = \left\lfloor \frac{\mathcal{S}(n) - \beta}{\alpha} \right\rfloor$$

and fetching $\phi(n)$ from Eq. (2) in terms of the tuple $\langle\delta, \mu\rangle$, we get

$$\phi(n) = \left\lfloor \frac{\mathcal{T}(n) - \mu}{\delta} \right\rfloor$$

As mentioned earlier,

$$\varphi(n) = \left(\frac{\max(\psi(n), \phi(n))}{\min(\psi(n), \phi(n))}\right) = \left\lfloor \frac{\psi(n)}{\phi(n)} \right\rfloor = \frac{\left\lfloor \frac{\mathcal{S}(n)-\beta}{\alpha} \right\rfloor}{\left\lfloor \frac{\mathcal{T}(n)-\mu}{\delta} \right\rfloor}$$

$$\therefore \mathcal{S}(n) = \alpha \times \left(\left\lfloor \frac{\psi(n)}{\phi(n)} \right\rfloor \times \left\lfloor \frac{\mathcal{T}(n)-\mu}{\delta} \right\rfloor\right) + \beta \tag{3}$$

CASE II: $\max(\psi(n), \phi(n)) = \frac{(\psi(n)+\phi(n))+|\psi(n)-\phi(n)|}{2} = \phi(n)$[1]

$$\varphi(n) = \left(\frac{\max(\psi(n), \phi(n))}{\min(\psi(n), \phi(n))}\right) = \left\lfloor \frac{\phi(n)}{\psi(n)} \right\rfloor = \frac{\left\lfloor \frac{\mathcal{T}(n)-\mu}{\delta} \right\rfloor}{\left\lfloor \frac{\mathcal{S}(n)-\beta}{\alpha} \right\rfloor}$$

$$\therefore \mathcal{S}(n) = \alpha \times \left(\left\lfloor \frac{\psi(n)}{\phi(n)} \right\rfloor \times \left\lfloor \frac{\mathcal{T}(n)-\mu}{\delta} \right\rfloor\right) + \beta \tag{4}$$

On a general note, for either of the two cases,

$$\mathcal{S}(n) = \alpha \times \left(\left\lfloor \frac{\psi(n)}{\phi(n)} \right\rfloor \times \left\lfloor \frac{\mathcal{T}(n)-\mu}{\delta} \right\rfloor\right) + \beta \forall n \geq 1$$

Since n is constant during execution time of the program (or in broader terms, of the algorithm), we can generate a 2D plot of $\mathcal{S}(n)$ versus $\mathcal{T}(n) \forall n \geq 1$ in the polar coordinate system.

6 Space–Time Continuum in Action

6.1 Bubble and Selection Sort Tug

In this section, we would be seeing space–time continuum in action. For that, we have selected two naïve sorting algorithms, Bubble Sort [10] and Selection Sort [11], which are undistinguishable via normal notion of asymptotic bounds.

For Selection Sort,

Space complexity $= \Upsilon(\psi(n)) = 1 = \mathcal{S}(n)$.

Time complexity $= \Upsilon(\phi(n)) = \frac{n\times(n-1)}{2} = \mathcal{T}(n)$.

Comparing these with Eqs. (1) and (2)

$$(\alpha, \beta, \delta, \mu) = \left(1, 0, +\frac{1}{2}, -\frac{1}{2}\right)$$

[1] Every fraction of the form $\frac{p}{q}$ has been considered as $floor\left(\frac{p}{q}\right)$ or $\frac{p}{q}$, to replicate the integral division operated internally in the computational units.

and $\psi(n) = 1$, $\phi(n) = n^2$

$$\max(\psi(n), \phi(n)) = \left(\frac{(\psi(n) + \phi(n)) + |\psi(n) - \phi(n)|}{2}\right) = \phi(n) = n^2 \forall n \geq 1$$

From Eq. (4),

$$\mathcal{S}(n) = \alpha \times \left(\left\lfloor \frac{\psi(n)}{\phi(n)} \right\rfloor \times \left\lfloor \frac{\mathcal{T}(n) - \mu}{\delta} \right\rfloor\right) + \beta$$

$$\mathcal{S}(n) = \left(\left(\frac{\mathcal{T}(n) - (-\frac{1}{2})}{(\frac{1}{2})}\right) \times (1) \times \left(\frac{1}{n^2}\right)\right) + 0$$

$$\mathcal{S}(n) = (2\mathcal{T}(n) + 1) \times \left(\frac{1}{n^2}\right)$$

$$\mathcal{S}(n) = \frac{2\mathcal{T}(n)}{n^2} + \frac{1}{n^2}$$

Transforming, in accordance with the polar transformation [12], $(\mathcal{T}(n), \mathcal{S}(n)) \equiv (\lambda \cos(\theta), \lambda \sin(\theta))$, where $\lambda = \sqrt[2]{\mathcal{S}(n)^2 + \mathcal{T}(n)^2}$, we get

$$\lambda \sin(\theta) = \frac{2\lambda \cos(\theta)}{n^2} + \frac{1}{n^2}$$

$$\Rightarrow \sin(\theta) = \frac{2\cos(\theta)}{n^2} + \frac{1}{\lambda n^2}$$

$$\Rightarrow \frac{1}{\lambda n^2} = \sin(\theta) - \frac{2\cos(\theta)}{n^2}$$

$$\Rightarrow \frac{1}{n^2 \times \sqrt[2]{\mathcal{S}(n)^2 + \mathcal{T}(n)^2}} = \sin(\theta) - \frac{2\cos(\theta)}{n^2}$$

$$\Rightarrow \sin(\theta) - \frac{2\cos(\theta)}{n^2} = \left\lfloor \frac{1}{n^2 \times \sqrt[2]{\mathcal{S}(n)^2 + \mathcal{T}(n)^2}} \right\rfloor$$

which is the polar characteristic equation for the canonical—"Selection Sort."

For Bubble Sort,

Space complexity $= \Upsilon(\psi(n)) = 1 = \mathcal{S}(n)$.

Time complexity $= \Upsilon(\phi(n)) = \frac{n \times (n+1)}{2} = \mathcal{T}(n)$.

Comparing these with Eqs. (1) and (2)

$$(\alpha, \beta, \delta, \mu) = \left(1, 0, +\frac{1}{2}, +\frac{1}{2}\right)$$

and $\psi(n) = 1$, $\phi(n) = n^2$

$$\max(\psi(n), \phi(n)) = \left(\frac{(\psi(n) + \phi(n)) + |\psi(n) - \phi(n)|}{2}\right) = \phi(n) = n^2 \forall n \geq 1$$

From Eq. (4),

$$\mathcal{S}(n) = \alpha \times \left(\left\lfloor \frac{\psi(n)}{\phi(n)} \right\rfloor \times \left\lfloor \frac{\mathcal{T}(n) - \mu}{\delta} \right\rfloor\right) + \beta$$

$$\mathcal{S}(n) = \left(\left(\frac{\mathcal{T}(n) - (\frac{1}{2})}{(\frac{1}{2})}\right) \times (1) \times \left(\frac{1}{n^2}\right)\right) + 0$$

$$\mathcal{S}(n) = (2\mathcal{T}(n) - 1) \times \left(\frac{1}{n^2}\right)$$

$$\mathcal{S}(n) = \frac{2\mathcal{T}(n)}{n^2} - \frac{1}{n^2}$$

Transforming, in accordance with the polar transformation, $(\mathcal{T}(n), \mathcal{S}(n)) \equiv (\lambda \cos(\theta), \lambda \sin(\theta))$, where $\lambda = \sqrt[2]{\mathcal{S}(n)^2 + \mathcal{T}(n)^2}$, we get

$$\lambda \sin(\theta) = \frac{2\lambda \cos(\theta)}{n^2} - \frac{1}{n^2}$$

$$\Rightarrow \sin(\theta) = \frac{2\cos(\theta)}{n^2} - \frac{1}{\lambda n^2}$$

$$\Rightarrow \frac{1}{\lambda n^2} = \frac{2\cos(\theta)}{n^2} - \sin(\theta)$$

$$\Rightarrow \frac{1}{n^2 \times \sqrt[2]{\mathcal{S}(n)^2 + \mathcal{T}(n)^2}} = \frac{2\cos(\theta)}{n^2} - \sin(\theta)$$

$$\Rightarrow \frac{2\cos(\theta)}{n^2} - \sin(\theta) = \left\lfloor \frac{1}{n^2 \times \sqrt[2]{\mathcal{S}(n)^2 + \mathcal{T}(n)^2}} \right\rfloor$$

which is the polar characteristic equation for the canonical—"Bubble Sort."

It is evident from the characteristic equations of both the Bubble and Selection Sorts, that the rate of change of $\mathcal{S}(n)$, space complexity, in accordance with $\mathcal{T}(n)$, time complexity, $\frac{\delta \mathcal{S}(n)}{\delta \mathcal{T}(n)}$ for both the Sorting Techniques are same. Similarly, the rate of change of $\mathcal{T}(n)$, time complexity, in accordance with $\mathcal{S}(n)$, space complexity, $\frac{\delta \mathcal{T}(n)}{\delta \mathcal{S}(n)}$ for both the Sorting Techniques are same.

The legends being, BLUE for Bubble Sort, and RED for Selection Sort. The polar form of the canonicals is $s(\theta) - \frac{2c(\theta)}{n^2} = \left\lfloor \frac{1}{n^2 \times \sqrt[2]{\mathcal{S}(n)^2 + \mathcal{T}(n)^2}} \right\rfloor$ for Selection Sort, and $\frac{2c(\theta)}{n^2} - s(\theta) = \left\lfloor \frac{1}{n^2 \times \sqrt[2]{\mathcal{S}(n)^2 + \mathcal{T}(n)^2}} \right\rfloor$ for Bubble Sort, where $(s(\theta), c(\theta), t(\theta)) \equiv (\sin\theta, \cos\theta, \tan\theta)$. For all $n \geq 1$, it is evident from the canonicals, that $\tan(\theta)|_{\text{Bubble Sort}} = \tan(\theta)|_{\text{Selection Sort}}$.

For both Bubble and Selection Sorting techniques, the value of slope, $\tan(\theta)$ is equal to

$$\left\lfloor \frac{-\frac{4}{n^2} \pm \sqrt[2]{\frac{16}{n^4} + 4 \times \left(\left(\left\lfloor \frac{1}{n^2 \times \sqrt[2]{\mathcal{S}(n)^2 + \mathcal{T}(n)^2}} \right\rfloor\right)^2 - 1\right) \times \left(\frac{4}{n^4} - \left(\left\lfloor \frac{1}{n^2 \times \sqrt[2]{\mathcal{S}(n)^2 + \mathcal{T}(n)^2}} \right\rfloor\right)^2\right)}}{2\left(\left(\left\lfloor \frac{1}{n^2 \times \sqrt[2]{\mathcal{S}(n)^2 + \mathcal{T}(n)^2}} \right\rfloor\right)^2 - 1\right)} \right\rfloor$$

Though, for the instance when $\theta \to \frac{\pi}{2}$, $\tan(\theta)|_{\text{Bubble Sort}} \approx \tan(\theta)|_{\text{Selection Sort}} \to 0$ which have been shown in the Last Fig. 1a–d, the traversal being in Row Major Order. Though $\left.\frac{\delta \mathcal{S}(n)}{\delta \mathcal{T}(n)}\right|_{\text{Bubble Sort}} = \left.\frac{\delta \mathcal{S}(n)}{\delta \mathcal{T}(n)}\right|_{\text{Selection Sort}}$ $\forall n \geq 1$ and $\left.\frac{\delta \mathcal{T}(n)}{\delta \mathcal{S}(n)}\right|_{\text{Bubble Sort}} = \left.\frac{\delta \mathcal{T}(n)}{\delta \mathcal{S}(n)}\right|_{\text{Selection Sort}}$ $\forall n \geq 1$, an algorithmist can easily conclude from the plot that Selection Sort in having an edge over Bubble Sort for smaller values of the cardinality in the aspect of time. In fact, it has also been proven practically [13] that, $\mathcal{T}$(Bubble Sort) $\geq \mathcal{T}$(Selection Sort) $+ \mathcal{T}$(Insertion Sort) $+ \mathcal{T}$(Merge Sort), where $\mathcal{T}$ is the dimensionality of time on computerized domain.

6.2 Merge and Heap Sort Tug

In this section, we would be seeing space–time continuum in action. For that, we have selected two naïve sorting algorithms, Merge [14] and Heap Sort [15], which are undistinguishable via normal notion of asymptotic bounds.

For Merge Sort,

Space complexity $= \Upsilon(\psi(n)) = n = \mathcal{S}(n)$.

Time complexity $= \Upsilon(\phi(n)) = n \log_2 n = \mathcal{T}(n)$.

Comparing these with Eqs. (1) and (2)

$$(\alpha, \beta, \delta, \mu) = (1, 0, 1, 0)$$

and $\psi(n) = n$, $\phi(n) = n \log_2 n$

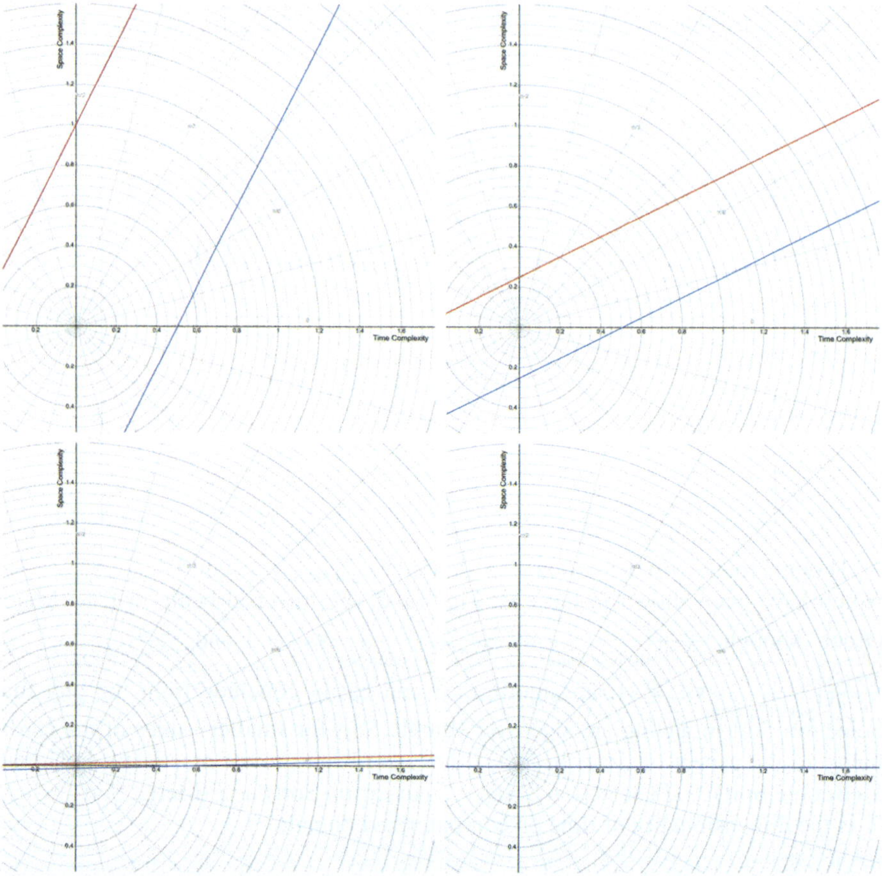

Fig. 1 **a–d** Space–time continuum plot for the tug of Bubble and Selection Sort

$$\max(\psi(n), \phi(n)) = \left(\frac{(\psi(n) + \phi(n)) + |\psi(n) - \phi(n)|}{2}\right) = \phi(n) = n \log n \forall n \geq 1$$

From Eq. (4),

$$\mathcal{S}(n) = \alpha \times \left(\left\lfloor \frac{\psi(n)}{\phi(n)} \right\rfloor \times \left\lfloor \frac{\mathcal{T}(n) - \mu}{\delta} \right\rfloor\right) + \beta$$
$$\mathcal{S}(n) = \left(\left(\frac{\mathcal{T}(n) - (0)}{(1)}\right) \times (1) \times \left(\frac{1}{\log_2 n}\right)\right) + 0$$
$$\mathcal{S}(n) = \mathcal{T}(n) \times \log_n 2$$
$$\mathcal{S}(n) = \log_n 2 \times \mathcal{T}(n)$$

Transforming, in accordance with the polar transformation,

$(\mathcal{T}(n), \mathcal{S}(n)) \equiv (\lambda\cos(\theta), \lambda\sin(\theta))$, where $\lambda = \sqrt[2]{\mathcal{S}(n)^2 + \mathcal{T}(n)^2}$, we get

$$\lambda\sin(\theta) = \log_n 2\lambda\cos(\theta)$$
$$\Rightarrow \sin(\theta) = \left(\log_n 2\right) \times \cos(\theta)$$
$$\Rightarrow \sin(\theta) - \log_n 2\cos(\theta) = 0$$

For Heap Sort,
Space complexity $= \Upsilon(\psi(n)) = \log_2 n = \mathcal{S}(n)$.
Time complexity $= \Upsilon(\phi(n)) = n\log_2 n = \mathcal{T}(n)$.
Comparing these with Eqs. (1) and (2)

$$(\alpha, \beta, \delta, \mu) = (1, 0, 1, 0)$$

and $\psi(n) = \log_2 n$, $\phi(n) = n\log_2 n$

$$\max(\psi(n), \phi(n)) = \left(\frac{(\psi(n) + \phi(n)) + |\psi(n) - \phi(n)|}{2}\right) = \phi(n) = n\log n \forall n \geq 1$$

From Eq. (4),

$$\mathcal{S}(n) = \alpha \times \left(\left\lfloor\frac{\psi(n)}{\phi(n)}\right\rfloor \times \left\lfloor\frac{\mathcal{T}(n) - \mu}{\delta}\right\rfloor\right) + \beta$$
$$\mathcal{S}(n) = \left(\left(\frac{\mathcal{T}(n) - (0)}{(1)}\right) \times (1) \times \left(n^{-1}\right)\right) + 0$$
$$\mathcal{S}(n) = \mathcal{T}(n) \times n^{-1}$$
$$\mathcal{S}(n) = n^{-1} \times \mathcal{T}(n)$$

Transforming, in accordance with the polar transformation,
$(\mathcal{T}(n), \mathcal{S}(n)) \equiv (\lambda\cos(\theta), \lambda\sin(\theta))$, where $\lambda = \sqrt[2]{\mathcal{S}(n)^2 + \mathcal{T}(n)^2}$, we get

$$\lambda\sin(\theta) = n^{-1}\lambda\cos(\theta)$$
$$\Rightarrow \sin(\theta) = n^{-1} \times \cos(\theta)$$
$$\Rightarrow \sin(\theta) - n^{-1}\cos(\theta) = 0$$

The Legends being, RED for Merge Sort, and BLUE for Heap Sort. The polar form of the canonicals is $s(\theta) - \log_n 2c(\theta) = 0$ for Merge Sort, and $s(\theta) - \frac{c(\theta)}{n} = 0$ for Heap Sort, where $(s(\theta), c(\theta), t(\theta)) \equiv (\sin\theta, \cos\theta, \tan\theta)$. For all $n \geq 1$, it is evident from the canonicals, that $\tan(\theta)|_{\text{Merge Sort}} \geq \tan(\theta)|_{\text{Heap Sort}}$.

For both Merge and Heap Sorting techniques, the value of slope, $\tan(\theta)$ is equal to $\log_n 2$ and $\frac{1}{n}$, respectively. Though, for the instance when $\theta \to \frac{\pi}{2}$, $\tan(\theta)|_{\text{Merge Sort}} \approx \tan(\theta)|_{\text{Heap Sort}} \to 0$ which have been shown in the Last Fig. 2a–d, the traversal

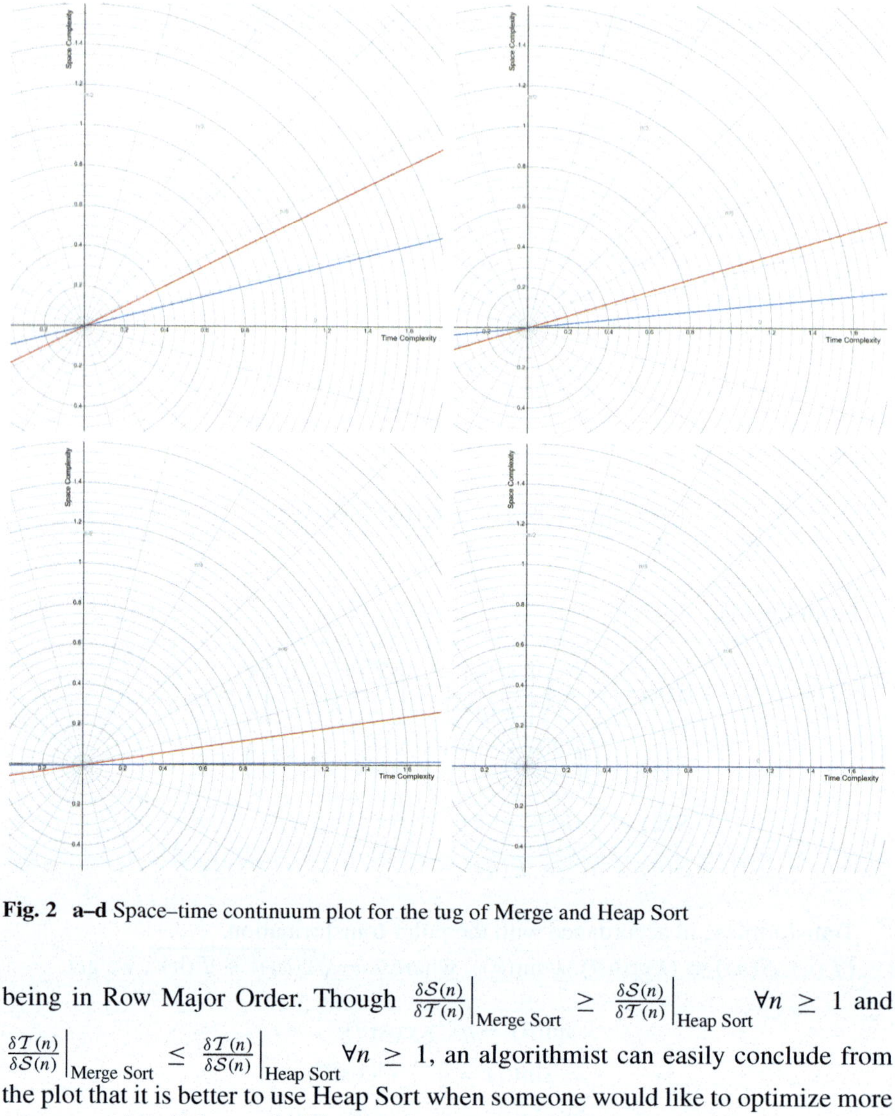

Fig. 2 **a–d** Space–time continuum plot for the tug of Merge and Heap Sort

being in Row Major Order. Though $\left.\frac{\delta\mathcal{S}(n)}{\delta\mathcal{T}(n)}\right|_{\text{Merge Sort}} \geq \left.\frac{\delta\mathcal{S}(n)}{\delta\mathcal{T}(n)}\right|_{\text{Heap Sort}} \forall n \geq 1$ and $\left.\frac{\delta\mathcal{T}(n)}{\delta\mathcal{S}(n)}\right|_{\text{Merge Sort}} \leq \left.\frac{\delta\mathcal{T}(n)}{\delta\mathcal{S}(n)}\right|_{\text{Heap Sort}} \forall n \geq 1$, an algorithmist can easily conclude from the plot that it is better to use Heap Sort when someone would like to optimize more on space, while if the need of the hour is time optimization, one should go for Merge Sort.

7 Conclusion

In this section, we will be concluding our thoughts on the continuum and would invoke some scopes of further research. Space–time continuum, unlike other comparators, is a metric to mark the algorithms, but the merit of this metric is that, unlike other metric, it is not Uno–Dim, that is it's not stuck to a particular metric, like be time or space, rather it keeps in account of both the dimensions parallelly on the same plot. It is going to schlep the algorithmists, to get a better idea of the algorithms in almost all the dimensions. Like, there are instances, when a certain algorithm works well when fused in terms of space, while some algorithms work well when fused in terms of time. In Sect. 6 of the work, we have shown the application of this continuum, applying it on the tugs of Bubble and Selection, Merge, and Heap.

Now, there are some scopes of further research, which could be considered later. Like, one could make use of the third dimension as well to introduce the Bit Complexity as well. In fact, the cardinality, n could also be considered in one dimension. Further, for higher dimensions, one could make use of the Spherical Coordinate System, as that would help in visualizing the concept, quite well.

References

1. Bachmann P (1894) Analytische Zahlentheorie [Analytic number theory], vol 2. Teubner, Leipzig (in German)
2. Landau E (1909) Handbuch der Lehre von der Verteilung der Primzahlen [Handbook on the theory of the distribution of the primes]. B. G. Teubner, Leipzig, p 883 (in German)
3. de Bruijn NG (1958) Asymptotic methods in analysis. North-Holland, Amsterdam, pp 5–7. ISBN: 978-0-486-64221-5
4. Bentley J, Haken D, Saxe J (1980) A general method for solving divide-and-conquer recurrences. ACM SIGACT News 12(3):36–44. https://doi.org/10.1145/1008861.1008865
5. Bamasaq O, Alghazzawi D et.al. (2022) Distance matrix and Markov chain based sensor localization in WSN. CMC-Comput Mater Continua 71(2):4051–4068. https://doi.org/10.32604/cmc.2022.023634
6. Brown RG, Gleason AM (ed) (1997) Advanced mathematics: precalculus with discrete mathematics and data analysis. McDougal Littell, Evanston. ISBN: 0-395-77114-5
7. Dutta A et al. (2022) Validation of minimal worst-case time complexity by Stirling's, Ramanujan's, and Mortici's approximation. In: 2022 3rd International conference for emerging technology (INCET), pp 1–4. https://doi.org/10.1109/INCET54531.2022.9824687
8. Greenlaw R, James Hoover H (1998) Fundamentals of the theory of computation: principles and practice
9. Horsley S (1772) Κόσκινον Ερατοσθένους or, The Sieve of Eratosthenes. Being an account of his method of finding all the Prime Numbers, by the Rev. Samuel Horsley, FRS. Philos Trans (1683–1775) 62:327–347
10. Jain P et al. (2014) Impact analysis and detection method of malicious node misbehavior over mobile ad hoc networks. Int J Comput Sci Inf Technol (IJCSIT) 5(6):7467–7470
11. Ullah S et al. (2016) Optimized selection sort algorithm for two-dimensional array. In: 2015 12th International conference on fuzzy systems and knowledge discovery (FSKD), pp 2549–2553. https://doi.org/10.1109/FSKD.2015.7382357
12. Ford KR, Haug AJ (2022) The probability density function of bearing obtained from a Cartesian-to-Polar transformation. IEEE Access 10:32803–32809. https://doi.org/10.1109/ACCESS.2022.3161974.(2022)

13. Dutta A et al. (2022) A unified vista and juxtaposed study on sorting algorithms. Int J Comput Sci Mob Comput 11(3):116–130. https://doi.org/10.47760/ijcsmc.2022.v11i03.014
14. Zhang J, Jin R (2021) In-situ merge sort using hand-shaking algorithm. In: Atiquzzaman M, Yen N, Xu Z (eds) Big data analytics for cyber-physical system in smart city. BDCPS 2020. Advances in intelligent systems and computing, vol 1303. Springer, Singapore. https://doi.org/10.1007/978-981-33-4572-0_33
15. Chandra OR, Istiono W (2022) A-star optimization with heap-sort algorithm on NPC character. Indian J Sci Technol 15(35):1722–1731. https://doi.org/10.17485/IJST/v15i35.857

Chapter 34
Fuzzy Assessment of Infrastructure Construction Project Performance

Savita Sharma and Pradeep K. Goyal

1 Introduction

The construction industry plays an important role in the infrastructure development and socioeconomic growth of any nation. Although the sector is one of the largest and its contribution to the global economy is more than 10%, this input contribution could only be achieved when projects complete their specified goals successfully and effectively. It is, however, observed that this industry is plagued by the chronic issues of cost overrun, time overrun, low quality, low productivity, etc., due to the dynamic, time-consuming, and complex nature of construction activities involved with infrastructure construction projects. The construction industry has been criticized very much for its poor performance and inadequate performance measurement evaluation system [1]. A large number of research studies have exposed the very poor performance of infrastructure and building construction projects [2–4]. Hence, the performance of the construction projects must be improved, and this is not possible without their evaluation. The "iron triangle" (time, cost, and quality) have traditionally been the most important evaluation criteria for the success of construction projects [5]. Though these three parameters can help determine whether the project is successful or unsuccessful, they do not provide a balanced and true performance, therefore the industry requires to focus on all other alternative methods for measuring and evaluating the project's performance. Approaches based on critical success factors [6] and key performance indicators (KPIs) [7] have been proposed by the researchers for assessing the performance of the project. Therefore, the objective of this research is to present a methodology to assess the performance of construction projects by

S. Sharma (✉)
Government Polytechnic College, Ajmer, Rajasthan, India
e-mail: savitasugotra@gmail.com

P. K. Goyal
Delhi Technological University, Delhi, India

V. Devedzic et al. (eds.), *Proceedings of the International Conference on Intelligent Computing, Communication and Information Security*, Algorithms for Intelligent Systems, https://doi.org/10.1007/978-981-99-1373-2_34

considering the key performance indicators (KPIs). A model is developed using fuzzy inference system for qualitative assessment of the performance of infrastructure construction projects. The fuzzy theory has been found to be useful in managing the imprecise, uncertain, and subjective data inherent in construction projects. It is applied to solve various complex problems related to construction engineering and management.

2 Literature Survey

In the research studies, key performance indicators (KPIs) approach has been shown to be one of the most effective and significant techniques for measuring the performance and progress of construction projects toward the set objectives to make more improvements. Through KPIs, the quality of results associated with the important characteristics of the project could be accurately revealed. Over the past few years, various concepts related to KPIs have been suggested to develop benchmarks in the area of construction management. Some of the studies are presented here in this section.

Cox et al. [8] employed key performance indicators (KPIs) for analyzing and measuring the performance of construction projects. A set of KPIs was proposed by Yeung et al. [9] to measure the performance of construction projects in terms of success in Hong Kong. Shen et al. [10] identified 20 important KPIs related to social, economic, and environmental groups for the infrastructure project's sustainability evaluation. Six important assessment indicators were recognized in hotel buildings by Xu et al. [11] for measuring the sustainability of building energy efficiency retrofit (BEER) projects. Ten important KPIs such as cost; quality; time; safety; owner's satisfaction; communication's effectiveness; effective planning; functionality; environmental performance; and end user's satisfaction were recognized by Yeung et al. [12] to evaluate the success of construction projects. Four main KPIs were proposed by Samra et al. [13] for managing water, wastewater, and road infrastructure system using a multi-objective framework system. Praticò and Giunta [14] suggested a list of KPIs for assessing the performance of railway tracks. Budayan et al. [15] proposed 63 stage-level key project indicators for performance measurement of BOT projects at eight phases. Important KPIs were evaluated as "comprehensiveness of project technical feasibility," "effectiveness of concessionaires' technical knowledge/capability evaluation," "detailed tendering procedure," "effectiveness of facility management," "good relationships between government and concessionaire," "effectiveness of quality control," "technology transfer," and "effectiveness of hand-back management" at each phase. Rathnayake and Ranasinghe [16] proposed a framework for performance measurement, based on the key performance indicators (KPI) by taking into account the local background of contractors in Sri Lanka. It was also revealed in the study that nearly 50% of the research studies had considered the completion time of the project, budget, health and safety issues, quality of the project, client's satisfaction, and productivity as important key project indicators. KhanzadI et al. [17] conducted a research study in Iran, based on BIM system to drive the KPIs during the construction phase of project life cycle (PLC) and it was found that quality enhancement, reduction in cost of construction, and sustainable construction were the important key project indicators during the construction phase Elshaikh et al. [18] revealed that the important KPIs in the construction projects in Khartoum(Sudan) were "experience level of the team," "end user's satisfaction," "training," "safety," "completion time of the project," and "cost deviation in the project," "budget," "risks involved," "resources," "planning period," and "cost of implementation." He et al. [19] discovered five groups of key indicators efficiency of the project, satisfaction of main stakeholders', innovation and development of the construction industry, organizational strategic aims, and comprehensive impact on society. Further, nine KPIs among these groups were recognized for assessing the success of megaprojects.

It was concluded from the literature review that nearly half of the research studies had considered traditional indicators such as budget, quality of the project, completion time, productivity, health and safety, owner's satisfaction, and environmental effects as the preferred indicators with 83%, 73%, 87%, 47%, 83%, 63%, and48%, respectively for analyzing the performance of any project. Therefore, these indicators have been selected in this study to develop the model.

3 Methodology

To develop the model for qualitative assessment of the performance of the construction projects using fuzzy theory, data were first collected to obtain the opinion of construction project participants on the relevance of the KPIs, and furthermore, to determine frequency index, severity index, and the importance index for various KPIs. The model was then developed using the fuzzy inference process by fuzzifying the input members (selected KPIs), forming the rules with the help of experts, and then defuzzifying the rules to get the results from the model. Detailed methodology is presented here in the following sections.

3.1 Data Collection

A questionnaire survey method was employed to collect the data from 50 Indian construction practitioners who had vast experience in the execution and management of construction projects. Cost, quality, time, productivity, health and safety, client satisfaction, and environmental impacts are the important KPIs influencing the performance of the project. Hence, the project performance is based on their respective performance. The participants were asked to assess the frequency and severity index of cost-associated performance, time-associated performance, quality-associated performance, health and safety-associated performance, client satisfaction-associated performance, productivity-associated performance, and environmental-associated performance index on a scale of 1–5 to calculate the importance index of the above performances. Frequency, severity, and importance index have been calculated using relations (1–3), respectively.

$$\text{Frequency Index (f.i.) (\%)} = \sum w(f/N) * 100/5 \tag{1}$$

where *w* expresses the weighting assigned to the responses (ranging from 1 to 5), *f* expresses the frequency of each responses, and *N* denotes the number of response for the particular KPI.

$$\text{Severity Index (s.i.) (\%)} = \sum w(f/N) * 100/5 \tag{2}$$

where *w* expresses the weighting assigned to each response (ranging from 1 to 5), *f* expresses the frequency of each responses, and *N* denotes the number of response for the particular KPI.

$$\text{Importance Index (IMI) (\%)} = (\text{f.i.} * \text{s.i.})/100 \tag{3}$$

Frequency, severity, and importance index of cost-associated performance, time-associated performance, quality-associated performance, health and safety-associated performance, client satisfaction-associated performance, productivity-associated performance, and environmental-associated performance have been presented in Table 1.

Table 1 Importance index for KPIs performance

Performance index	Abbreviated form	Severity index(s.i)	Frequency index(f.i)	Importance index(IMI)	Rank
Cost-associated performance index	CPI	0.821	0.822	0.674	1
Time-associated performance index	TPI	0.786	0.786	0.617	2
Quality-associated performance index	QPI	0.648	0.693	0.449	4
Health and safety-associated performance index	HPI	0.655	0.679	0.444	5
Client satisfaction-associated performance index	CPI	0.724	0.742	0.537	3
Productivity-associated performance index	PPI	0.643	0.634	0.407	6
Environmental-associated performance index	EPI	0.654	0.400	0.261	7

3.2 *Model Development*

Based on the above data, a fuzzy model is proposed for qualitative assessment of the performance of construction projects. The fuzzy theory has been applied to solve many problems related to construction engineering and management that requires qualitative analysis, as it is very difficult to obtain quantitative data in construction industry. Fuzzy theory-based techniques are capable to solve the complex problems intrinsic to construction projects. The theory was first introduced by Zadeh [20]. The main feature of the theory is its membership function $\mu A(x)$ which describes the degree (0 to1) to which any element x is a member of the fuzzy set A as follows:

$$A = \{(x, \mu A(x)) / x \in A, \mu A(x) \in [0, 1]\}$$

The following steps are used to develop the qualitative model for assessing the performance of the construction projects:

Identification of Key Variables

Cost-associated performance index, time-associated performance index, quality-associated performance index, health and safety-associated performance index, client satisfaction-associated performance index, productivity-associated performance index, and environmental-associated performance index are selected as key variables for model development. The input variables are abbreviated as CPI, TPI, QPI, HPI, CSPI, PPI, and EPI, respectively, for cost-associated performance index, time-associated performance index, quality-associated performance index, health and safety-associated performance index, client satisfaction-associated performance index, productivity-associated performance index, and environmental-associated performance index. The overall project performance index abbreviated as OPPI is the output variable.

Membership Function

The crisp input members are fuzzified using the membership function editor. There are several kinds of membership functions, such as triangular, trapezoidal, gbellmf. In this study, trapezoidal membership function are used because of their wide use in the literature. Five membership functions extremely low, low, moderate, high, and extremely high are employed here in this study. Figure 1 shows the membership function editor for cost-associated performance index. The membership function for variables: time-associated performance index, quality-associated performance index, health and safety-associated performance index, client satisfaction-associated performance index, productivity-associated performance index, environmental-associated performance index and output variable project performance index can also be defined similarly.

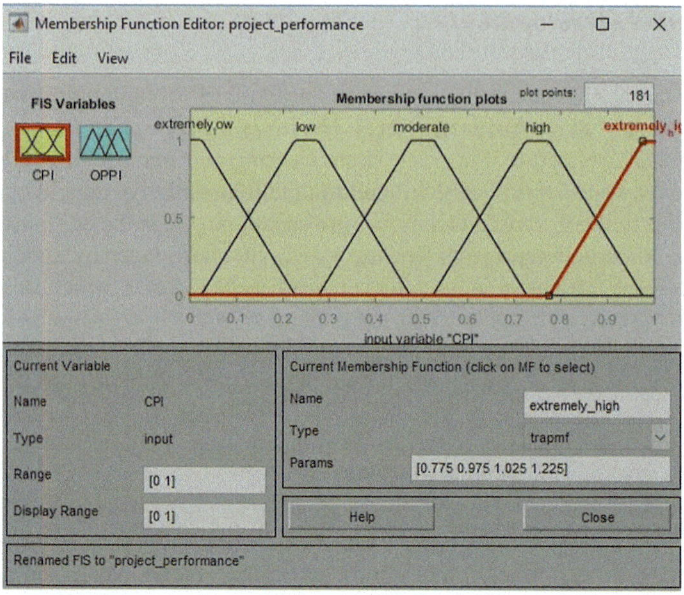

Fig. 1 Membership function for cost-associated performance index

Rules Formation

To develop the knowledge base rules were formed with the help of experienced practitioners from Indian construction industry. Total 35 rules were generated for seven input variables. The rules were weighted as per their importance index. Some of the rules are shown here for illustration purpose:

1. If cost-associated performance index is very low, then overall project performance level is very low.
2. If time-associated performance index is very low, then overall project performance level is very low.
3. If quality-associated performance index is very low, then overall project performance level is very low.
4. If health and safety-associated performance index is very low, then overall project performance level is very low.
5. If client satisfaction-associated performance index is very low, then overall project performance level is very low.
6. If environmental-associated performance index is very low, then overall project performance level is very low.
7. If productivity-associated performance index is very low, then overall project performance level is very low.

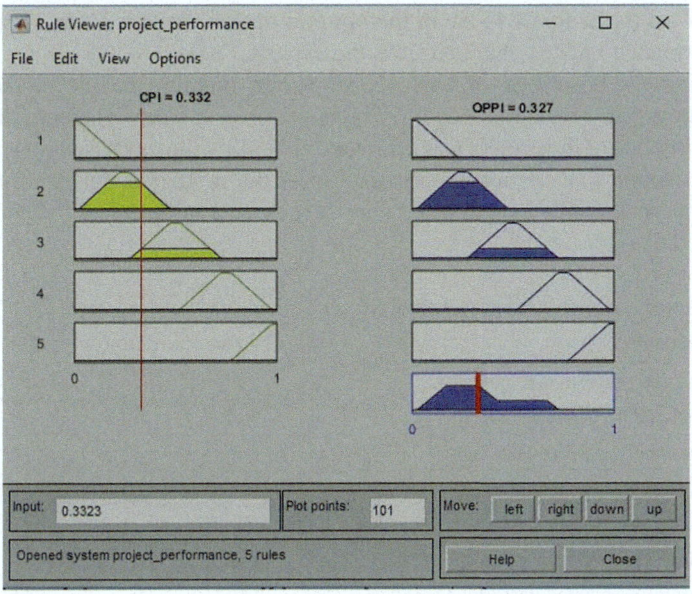

Fig. 2 Defuzzification process

Defuzzification

Rules are defuzzified now to obtain the overall project performance index as shown in Fig. 2.

4 Implementation of Proposed Methodology in Real Case Study

The applicability of the proposed methodology is illustrated here in this section for assessing the performance of a real project. The following steps were performed for the implementation process.

1. A project related to water supply works in Rajasthan (India) was selected for the study.
2. A committee of expert members, including project managers, executive engineers, and others with more than 15 years of experience, was then constituted.
3. On a scale of 1–10, project experts were asked to describe the project's specifics in terms of cost-associated performance, time-associated performance, quality-associated performance, health and safety-associated performance, client satisfaction-associated performance, productivity-associated performance, and environmental-associated performance. They were also asked to evaluate the overall performance of the project. The performance of this project was rated

as a 7 on the scale of 1–10, in the opinion of the experts. Table 2 presents the performance indices obtained from the experts.

4. The indices described in Table 2 were then put into the developed model, and the rules were defuzzified to assess the project's performance. The overall performance of the project was evaluated as 6.41 by the model as shown in Fig. 3.
5. It was found that the results obtained from the model as well as from experts were almost similar, and hence the model can be used to assess the performance of model.

Table 2 Performances indices provided by project participants

Performance index	Abbreviated form	Value
Cost-associated performance index	CPI	7
Time-associated performance index	TPI	7
Quality-associated performance index	QPI	6
Health and safety-associated performance index	HPI	8
Client satisfaction-associated performance index	CSPI	8
Productivity-associated performance index	PPI	8
Environmental-associated performance index	EPI	8

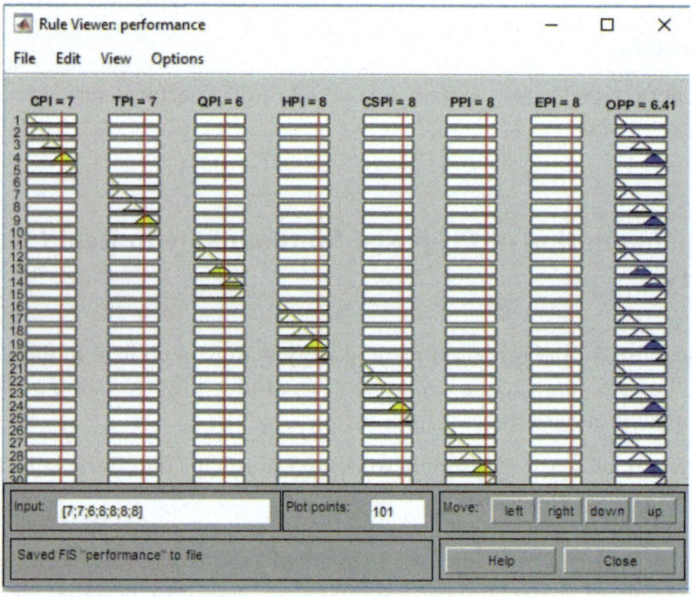

Fig. 3 Overall performance of the project by model

5 Conclusion

This research study was carried out for qualitative assessment of infrastructure construction project performance to make required improvements. For this purpose, a questionnaire survey was conducted and data were obtained from the practitioners of the Indian construction industry. The questionnaires were analyzed to calculate the importance indices of cost-associated performance, time-associated performance, quality-associated performance, health and safety-associated performance, client satisfaction-associated performance, productivity-associated performance, and environmental-associated performance index on the basis of severity and frequency index on the scale of 1–5 provided by the participants of survey. The results of the analysis showed that cost-associated performance was the primary cause influencing the overall performance of the project followed by time-associated performance, client satisfaction-associated performance, quality-associated performance, health and safety-associated performance, productivity-associated performance, and environmental-associated performance. A fuzzy model was developed based on the above-mentioned performance indices using MATLAB software R2015a. The fuzzy inference process was conducted by defining input and output variables, providing membership function, constructing rules, and defuzzifying the rules. A real case study was considered to validate the model developed. It was found that the results obtained from model were nearly same as assessed by projects experts. This methodology can be used for improving and assessing the performance of infrastructure construction projects and may contribute to the body of knowledge in the areas of construction engineering and management.

Conflict of Interest Statement On behalf of all authors, the corresponding author states that there is no conflict of interest.

References

1. Kagioglou M, Cooper R, Aouad G (2001) Performance management in construction: a conceptual framework. Constr Manag Econ 19(1):85–95. https://doi.org/10.1080/01446190010003425
2. Davis K, Ledbetter WB, Buratti JL (1989) Measuring design and construction quality costs. ASCE J Constr Eng Manag 115:389–400
3. Georgy ME, Chang LM, Walsh KD (2000) Engineering performance in industrial construction. In: Construction Congress VI, ASCE (2000), Orlando, Florida. https://doi.org/10.1061/40475(278)96
4. Nitithamyong P, Skibniewski M (2006) Success/failure factors and performance measures of web-based construction project management systems: professionals' viewpoint. J Constr Eng Manag 132(1):80–87. https://doi.org/10.1061/(ASCE)0733-9364(2006)132:1(80)
5. Mohsini RA, Davidson CH (1992) Determinants of performance in the traditional building process. Constr Manag Econ 10(4):343–359

6. Omran A, AbdalRahman S, Pakir AK (2012) Project performance in Sudan construction industry: a case study. Acad Res J (India) 1(1):55–78
7. Mahmoud SY, Scott S (2002) The development and use of key performance indicators by the UK construction industry. In: Greenwood D (ed) 18th Annual ARCOM conference (2002), University of Northumbria, Association of Researchers in Construction Management
8. Cox RF, Issa RRA, Ahrens D (2003) Management's perception of key performance indicators for construction. J Constr Eng Manag 129(2):142–151. https://doi.org/10.1061/(ASCE)0733-9364(2003)129:2(142)
9. Yeung J, Chan A, Chan W (2008) Establishing quantitative indicators for measuring the partnering performance of construction projects in Hong Kong. Constr Manag Econ 26(3):277–301. https://doi.org/10.1080/01446190701793688
10. Shen LY, Wu YZ, Zhang XL (2010) Key assessment indicators for the sustainability of infrastructure projects. J Constr Eng Manag 137(6):441–451. https://doi.org/10.1061/(ASCE)CO.1943-7862.0000315
11. Xu P, Chan EHW, Queena QK (2012) Key performance indicators (KPI) for the sustainability of building energy efficiency retrofit (BEER) in hotel buildings in China. Facilities 30(9):432–448. https://doi.org/10.1108/02632771211235242
12. Yeung FY, Chan PC, Chan WM, Chiang YH, Yang H (2012) Developing a benchmarking model for construction projects in Hong Kong. J Constr Eng Manag 139(6):705–716. https://doi.org/10.1061/(ASCE)CO.1943-7862.0000622
13. Samra SA, Ahmed M, Hammad A, Zayed T (2018) Multi-objective framework for managing municipal integrated infrastructure. J Constr Eng Manag 144(1):04017091. https://doi.org/10.1061/(ASCE)CO.1943-7862.0001402
14. Praticò FG, Giunta M (2018) Proposal of a key performance indicator for railway track based on LCC and RAMS analyses. J Constr Eng Manag 144(2):04017104. https://doi.org/10.1061/(ASCE)CO.1943-7862.0001422
15. Budayan C, Okudan O, Dikmen I (2020) Identification and prioritization of stage-level KPIs for BOT projects—evidence from Turkey. Int J Manag Proj Bus 13(6):1311–1337. https://doi.org/10.1108/IJMPB-11-2019-0286
16. Rathnayake A and Ranasinghe M (2020) A KPI based performance measurement framework for Sri Lankan construction projects. In: Moratuwa engineering research conference (MERCon, 2020), IEEE, Moratuwa, Sri Lanka. https://doi.org/10.1109/MERCon50084.2020.9185304
17. Khanzadi M, Sheikhkhoshkar M, Banihashemi S (2020) BIM applications toward key performance indicators of construction projects in Iran. Int J Constr Manag 20(4):305–320. https://doi.org/10.1080/15623599.2018.1484852
18. Elshaikh EAM, Mahmoud SYM, Omar A, Noureldin A, Alkamil K (2021) Key project indicators in the construction industry in Khartoum, Sudan. Int J Sudan Res 142–155. https://doi.org/10.47556/J.IJSR.11.2.2021.2
19. He Q, Wang T, Chan APC, Xu J (2021) Developing a list of key performance indictors for benchmarking the success of construction megaprojects. J Constr Eng Manag 147(2):04020164. https://doi.org/10.1061/(ASCE)CO.1943-7862.0001957
20. Zadah LA (1965) Fuzzy sets. Inf Control 8(3):338–353

Chapter 35
Smart Phone-Centric Deep Nutrient Deficiency Detection Network for Plants

K. U. Kala

1 Introduction

Adequate quantity of various nutrients is important for plants to their growth, germination, immunity, and reproduction. These provide optimum health and prosperity at each stage of plant growth like flowering or bearing fruit. There exist 17 nutrients for the growth and survival of plant. Each nutrients have their own functions during the lifetime of the plant. Primarily, the plant growth hinge on the nutrients [1].

The key plant nutrients are of two classes: micronutrients and macronutrients. Nutrients of both classes are vital for the growth of the plants. The nutrients which need in large quantities for the plants are called macronutrients. If any of the nutrients are deficient, it has to be supplied into the soil. The primary macronutrients are nitrogen (N), phosphorus (P), and potassium (K), whereas calcium (Ca), magnesium (Mg), and sulfur (S) are the secondary macronutrients. The nutrients which required in very small quantities are called micronutrients. Although their required quantity is less, they are crucial for the health of plant and yield. They consist of boron (B), iron (Fe), zinc (Zn), manganese (Mn), molybdenum (Mo), and copper (Cu) [1, 2].

The optimum range of nutrient requirements is differed according to the type of plants. If the amount is below the minimum level, the plant will show the symptoms of nutrient deficiency. Also, excessive nutrients may lead to toxicity which leads to poor growth. Therefore, the farmers are recommended to apply the required nutrients in adequate amount to their crops whenever the symptoms of infections, malnourishment, or other problems found in the plants [3]. Nutrients are supplied either from soil minerals and soil organic matter or by organic or inorganic fertilizers [4].

K. U. Kala (✉)
Lovely Professional University, Punjab, India
e-mail: kala.29057@lpu.co.in; kalasarin.uiet@gnuindia.org

Guru Nanak University, Punjab, India

V. Devedzic et al. (eds.), *Proceedings of the International Conference on Intelligent Computing, Communication and Information Security*, Algorithms for Intelligent Systems, https://doi.org/10.1007/978-981-99-1373-2_35

For assessing the nutrient contents of the plants and soil, the exists various soil and plant tissue tests. Experts can determine the nutrient need of a plant in a particular soil by analyzing the result of these tests. Timely and accurate identification of nutrient deficiency in crops helps the farmers to save their plants from various nourishment problems.

1.1 Background

In the daily life of Indians, the banana and plantain trees are one of the most common fruit and crop. Not only in India, but also it is popular in all over the world. Also, banana is the most marketable fruit in the world [5]. The growth and development of banana are affected by numerous factors. Nutrients are one of the most important factors among them. The adequate amount of nutrients will lead to get high yields, since they make the plants healthy [6].

The main constraint to the ideal growth plantain tree and banana production is low soil fertility [7]. Fertilization may help to manage soils fertility, but the farmers must be aware of the nutrient problems to make the decision about amount and type of fertilizers. Balanced nutrition supports to improve grading, reduce the maturation time, increase the quantity and quality of the bananas in the bunch. It will increase the overall yield of the farmers and leads to high return [8]. This research work focuses on the most common nutrients which effect banana trees: potassium and boron, since they have vital role in plantain tree cultivation [9, 10]. The role of these nutrients in plantain tree is listed in Table 1.

1.2 Potassium and Boron in Plantain Tree Cultivation

Potassium is one of the essential macronutrients in plantain tree cultivation which has a large effect on banana yield, with peak demand being around flowering. Potassium is known as the quality element of crop production because it has a crucial role in the quality parameters of the crops such as size of the fruit, color, appearance, soluble solids, vitamin content, acidity, taste, and shelf-life [11]. It also helps to regulate the water and to tolerate the stress from the environment such as excess water, drought, high, and low temperature and wind. The juice and vitamin C contents are highly influenced by the potassium. Moreover, the consistency and quickening of banana ripening and bruising resistance are under the control of this element. The common plant reactions such as respiration, photosynthesis, chlorophyll formation, and water regulation are catalyzed by the potassium element. Low level of potassium may lead to float the fruit and creating difficulties like detachment of fruits while cleaning and packing [9].

Table 1 Role of potassium and boron in plantain tree [1, 2]

Nutrient name	Effects of nutrient on plantain tree	Sample nutrient-deficient image
Potassium (K)	• Stimulates early shooting • Shorten the time of fruit maturity • Improves bunch grade, fingers size and fruit quality	
Boron (B)	• Increased fruit number, weight and yield • Activate uptake of other nutrients	

It is essential for appropriate plant growth, consistency of the pulp, and the development of suckers [12]. But it is the most commonly deficient micronutrient found in most of the in the banana plantations of India. The most common symptoms are leaf malformation such as narrow, rolled, and incompletely developed. Also, it is essential for the firmness, skin strength, storage life of the fruits. Low-level boron may lead to fungal diseases and decreases tolerance to stresses from the environments [13, 14]. Plantains of regions with humid climates such as Latin America, the Caribbean, and parts of Asia are suffering with the deficiency of boron. Frequent cultivation of plantain trees reduces the amount of boron from the soil, so farmers are recommended to test and correct their soils [10].

Timely and proper identification of the nutrient deficiency is necessary for balancing them by applying adequate amount of the particular nutrient at the relevant point of plant growth cycle [15]. Traditionally farmers rely on human agricultural specialists for identifying and diagnosing nutritional deficiency. Developing countries are lacking of human experts in agricultural industry. In order to overcome the challenges in farming, the empirical knowledge of the farmers is less effective [16]. Unawareness of the farmers for the identifying the deficient nutrient from the symptoms on the plants necessitates the development of automated systems for providing diagnostic services. The proper identification of deficient nutrients in crops and recommendation of proper fertilizers helps the farmers to increase their yields.

The most symptoms of nutritional deficiencies can be identified by visual appearance and behavioral changes of the plant parts such as leaves, stems, fruits. To learn and to perform in depth analysis of these properties, deep learning (DL) techniques are found to be suitable. The image classification using convolutional neural network (CNN, is a type of DL architecture for processing spatially correlated data (e.g., images)) can accurately identifying deficient nutrients which effects the productivity of plantain trees. Smart phone-centric nutrient deficiency monitoring applications will help to identify the deficient nutrients on real-time basis and are easily accessible for individuals anywhere. This application will assist the common people in identifying the deficient banana nutrients, deciding whether the plant is healthy with balanced nutrients or needs adequate fertilizations.

This research work is focused on build, train, and deploy fast, flexible smart phone applications based on CNNs for boron and potassium deficiency detection in plantain trees. This article is organized as: Sect. 2 describes the works related to nutrient deficiency identification, Sect. 3 presents the materials and methods such as dataset, DL architecture, etc., used for nutrient deficiency classification and detection and Sect. 4 draws the conclusions and points out the future enhancements.

2 Related Work

Generally, the farmers are carrying out the surveillance of the nutrient deficiency by inspecting the visual symptoms of the plants regularly to make sure whether the nutrient needs of the plants are satisfied or not. This method is not practical in large agricultural fields, since it requires intensive attention and effort. Sometimes, laboratory tests such as soil testing, plant analysis, and visual observations are required to check whether the symptoms are of nutrient deficiency or not. Visual observation is the analysis of the appearance of the crop such as leaf discoloration and a stunted growth as a result of nutrient deficiency.

Nowadays, several approaches based on artificial intelligence (AI) have been emerged to assist the control or monitoring tasks in agriculture. Donné et al. [17] proposed a binary segmentation model for 3D reconstruction of maize plant images using CNN. It is useful for monitoring the growth of the maize plant. Grinblat et al. [18] proposed another deep CNN model for plant species identification from leaf vein patterns. The experiment is carried out using the legume leaf dataset which contain the images of white bean, red bean, and soybean. This method is useful for identifying plant species without human intervention. Also, CNN is widely used for disease detection in plants as like in humans. Grinblat et al. [19] proposed CNN model for identifying Northern Leaf Blight (NLB), a disease in maize plant. This model exhibits 96.7% accuracy while classifying 768 healthy and 1796 NLB-infected images of maize leaves.

Mohanty et al. [20] utilized GoogleNet and AlexNet architecture for diagnosing diseases of 26 diseases found in 14 species of crops using 54,306 image dataset and attains classification accuracy of 99.35%. Later, Ferentinos [21] conducted a deep study on the various CNN architectures such as AlexNet [22], AlexNetOWTBn [23], GoogleNet [24], Overfeat [25], and VGG [26], for detecting plant diseases. The experimented dataset consists of 87,848 images of 58 classes of diseased images and healthy plant images data. VGG attains highest accuracy of 99.48% while comparing with others.

Moreover, AI, particularly, DL techniques, plays a vital role in plant disease monitoring and diagnosing nutrient deficiencies. Story et al. [27] proposed a computer vision model for detecting calcium deficiency of lettuces in green based on morphological and color variations of the plants. While compared to human vision, earlier identification of calcium deficiency became possible with this method. Another case study is carried out on optical sensor data for yield and sulfur deficiency prediction [28]. Hengl et al. [29] proposed a model using principal component analysis (PCA) for predicting the contents of micro and macro nutrients from the spatial data of Sub-Saharan Africa (SSA). Mahrishi et al. [30] proposed a CNN model for finding the calcium deficiency in lettuce plants. This model utilizes the concepts such as using Inception-Resnet architecture with transfer learning and fine tuning for increasing the accuracy.

Deficiency of the nutrients is major concern in the Indians agricultural field. Till now, there is no efficient solution existing for diagnosing this problem. Xu et al. [31] compare the state-of-the-art CNNs such as Inception-v3, ResNet, and DenseNet for detecting nutrient deficiency in rice plants. DenseNet121 achieves better performance in terms of accuracy. Recently, many researchers are started to focus on nutrient deficiency detection in various crops [31–35].

Plantain trees are facing nutrient deficiency problems all over India. Previous researchers showed anomer view of how AI can use for the automatic monitoring and detecting the conditions of the plant. However, the area of nutrient deficiency is not that much active, whereas it has vital influence on the quality of the plant. Recently, some researchers focused on the detection of nutrient deficiency identification in plantain trees. This research proposes a technique with DL for detecting the deficiency of selected nutrients based on image input data.

3 Materials and Methods

The nutrition deficiency symptoms reflect on various parts of the plants, and it can be identified visually. In the case of plantain trees, the deficiency of most of the nutrients especially boron and potassium can be identified by analyzing its leaves. This paper introduces a mobile application that uses deep learning techniques to identify potassium (K) and boron (B) deficiency from the images of the symptoms found in banana leaves. The following are the objectives of this project:

1. Creation of the nutrient-deficient image dataset.
2. Feature extraction and train the DL classifier model with the dataset for classifying the user image into healthy, potassium (K) and boron deficient.

Fig. 1 Sample images of the banana nutrient-deficient dataset

3. Develop a real-time and user-friendly smart phone application for detecting the deficient nutrients from the leaves.

3.1 Dataset

The banana nutrient deficiency dataset consists nutrient-deficient leaf images of plantain trees collected from various cultivation fields of Tamil Nadu. It consists of images of potassium-deficient, boron-deficient, and healthy leaves. Each category contains 800 images. The sample images are shown in Fig. 1.

3.2 Feature Extraction and Classification Using CNN

3.2.1 Experimental Settings

The experiments are conducted on a Windows 10 laptop equipped with one intel core i7 CPU with 8 GB RAM, accelerated by GeForce 1050 Ti GPU with 6 GB memory. The implementation of the pre-trained state-of-the-art architectures using the banana nutrient deficiency dataset are powered by MATLAB 2021a software library. For experimentation work, the banana nutrient deficiency dataset is spitted into two: 70% training, 30% testing. The data augmentations such as 'RandXReflection,' 'RandXTranslation,' 'RandYTranslation,' 'RandXScale,' and 'RandYScale' are used in the experiments.

Table 2 Hyperparameter settings

No. of epochs	10
Batch size	8
Momentum	0.9
Optimizers	Adamax, Adam
Learning rate	0.001
Regularizer	L2 with factor as 0.01

3.2.2 Hyperparameter Tuning

Hyperparameters affect the learning of the model during the training process. Several attempts with different combination of these parameters are needed to fix their values for the experiment. The configuration of hyperparameters which are apt for the experiments and architecture of this work is shown in Table 2.

3.2.3 Architecture Comparison

In this work, state-of-the-art pre-trained models: DenseNet-201, ResNet-50, NasNet-mobile are used for image-based nutrient-deficient image classification and the best model based on the evaluation metrics are used for smart phone-centric Mobile App Development. ResNet 50 has a depth of 50 layers, and it has a total of 23,665,574 parameters. This ResNet-50 model has 23,612,454 trainable and 53,120 non-trainable parameters. DenseNet 201 has a depth of 201 with 6,992,806 as trainable parameters and 83,648 as non-trainable parameters. NasNet-mobile version has 4,273,144 trainable parameters and 36,738 non-trainable parameters.

During experimentation, the performance of ResNet and NasNet-mobile has best and almost same performance in terms of accuracy. The performance comparison in terms of accuracy for Adam and Adamax optimizer are shown in Fig. 2. Figures 3 and 4 depict the training graphs and four sample validation images with predicted labels and the predicted probabilities of the images having those labels of ResNet-50 and NasNet-mobile models. Although NasNet-mobile and ResNet-50 performance is better, the number of trainable parameters in ResNet-50 is quite high when compared with NasNet-mobile.

The NasNet-mobile architecture has shown better performance in terms of validation accuracy and sample predicted probability. This performance motivated us to use NasNet-mobile for classification purpose in real-time application development for detecting boron and potassium deficiency in plantain trees.

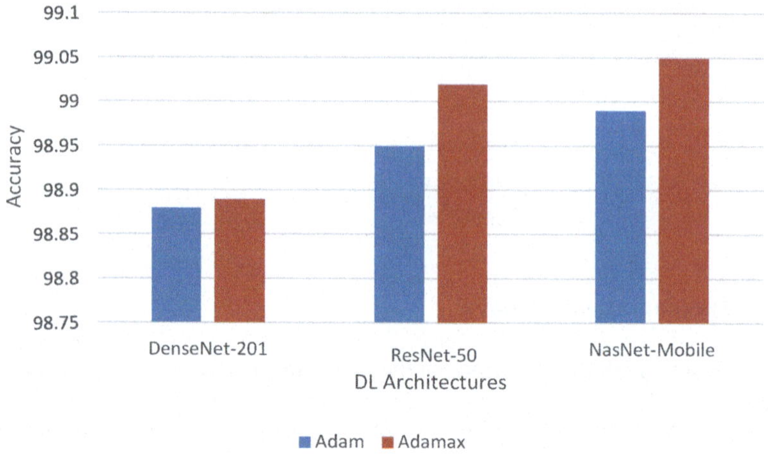

Fig. 2 Accuracy comparison for both Adam and Adamax optimizer

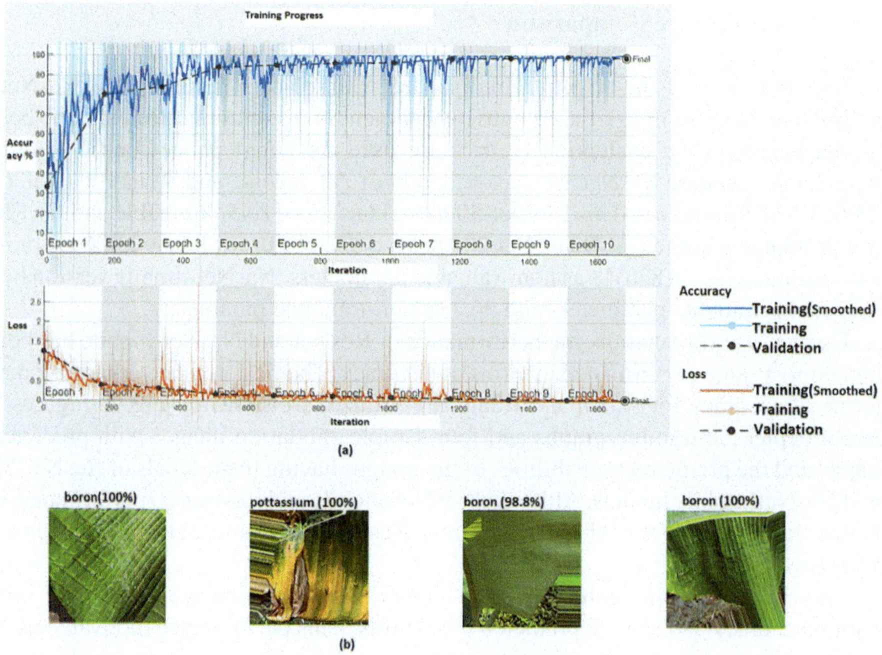

Fig. 3 **a** Training graph of NasNet-mobile, **b** sample validation images and their predicted probabilities

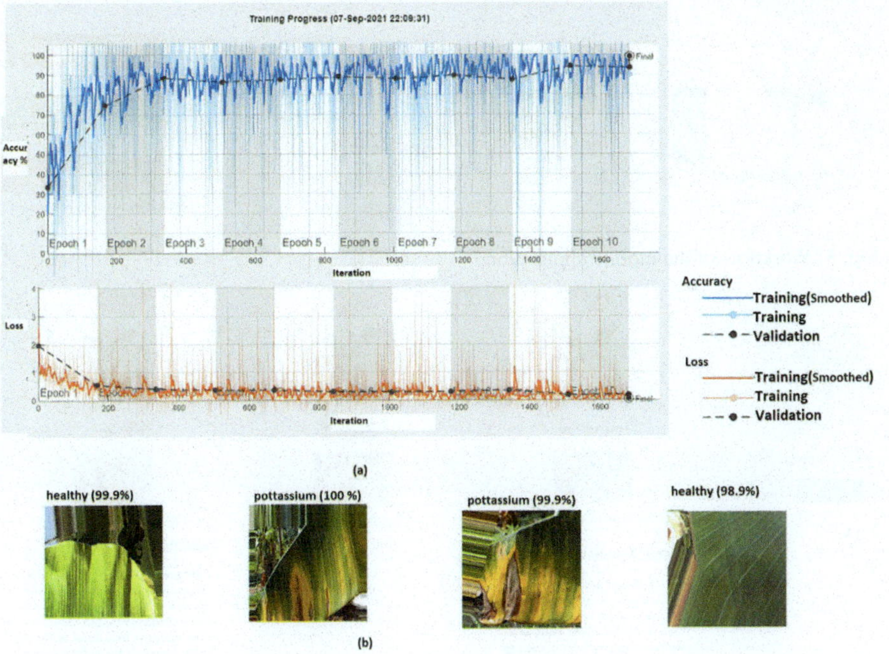

Fig. 4 **a** Training graph of ResNet-50, **b** sample validation images and their predicted probabilities

3.3 *Smart Phone-Centric Application*

AI-based nutrition deficiency diagnosis and detection research works are inaccessible to the common people, especially in developing countries. Most of the research works are always limited in paper works. We are converting this work to smart phone-centric applications for the real-time detection of nutrients. This app is user-friendly and easily accessible for the common people. The workflow of this app is shown in Fig. 5. The people can take the photo of the infected image using their smart phones and upload to the application, and then the app will detect its deficient nutrient. The screen shots of the developed app are shown in Fig. 6. This is an Android app, and it can be utilized by the common people easily.

4 Conclusion and Future Work

Plant malnutrition affects the growth of the plantain trees and quality of the fruit. The detection of nutrient deficiencies is important for increasing the yield of plantain tree cultivation. Along with diseases, nutrient deficiency will also adversely affect the crops. In this work, the pre-trained models: DenseNet-201, ResNet-50,

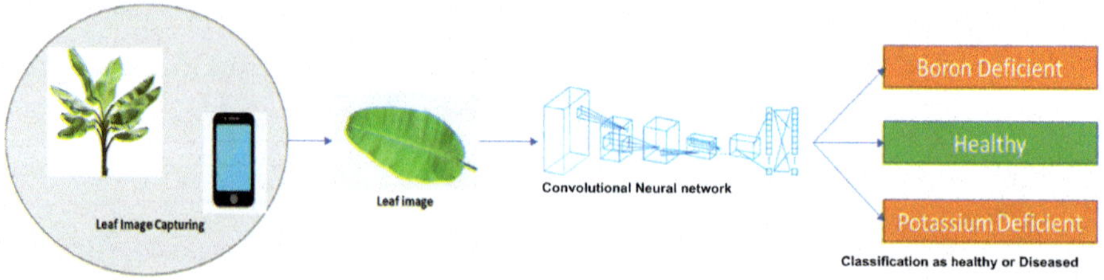

Fig. 5 Workflow of the mobile application

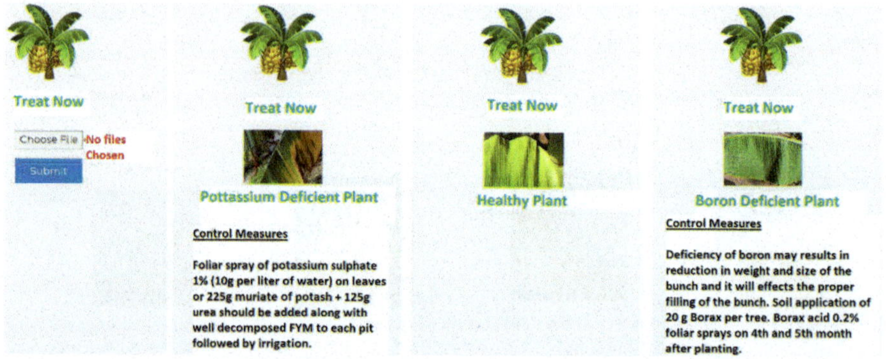

Fig. 6 Screen shots of the mobile application

NasNet-mobile, are used for image-based nutrient-deficient image classification and the NasNet architecture attains better performance in terms of accuracy. Therefore, it is used for the smart phone-centric Mobile App Development. The experiments are carried out only for the boron and potassium nutrients deficiency of the plantain trees. The application presents the nutritional status of the plants, when getting the plantain leaf images through the classification performed by the CNN. In future, the dataset will expand by collecting the images of other nutrient-deficient plants and will improve the classification accuracy with a greater number of images for obtaining more advantageous and expressive results for the real-time use of the farmers.

References

1. Danish M, OndokuzT, Üniversitesi M, Adnan M, Rehman FU, Khan AU (2021) Nutrients and their importance in agriculture crop production: a review. https://doi.org/10.18782/2582-2845.8527
2. Fageria NK (2016) The use of nutrients in crop plants. https://doi.org/10.1201/9781420075113
3. Tripathi DK, Singh VP, Chauhan DK, Prasad SM, Dubey NK (2014) Role of macronutrients in plant growth and acclimation: recent advances and future prospective. In: Improvement of crops in the era of climatic changes, pp 197–216. https://doi.org/10.1007/978-1-4614-8824-8_8
4. Graham RD (2008) Micronutrient deficiencies in crops and their global significance. In: Micronutrient deficiencies in global crop production, pp 41–61. https://doi.org/10.1007/978-1-4020-6860-7_2
5. Lahav E (1995) Banana nutrition. In: Bananas and plantains, pp 258–316. https://doi.org/10.1007/978-94-011-0737-2_11
6. Röös E et al. (2018) Risks and opportunities of increasing yields in organic farming: a review. Agron Sustain Dev 38(2):1–21. https://doi.org/10.1007/S13593-018-0489-3
7. Havlin JL (2020) Soil: fertility and nutrient management. In: Landscape and land capacity, pp 251–265. https://doi.org/10.1201/9780429445552-34
8. Shuen YS, Arbaiy N, Jusoh YY (2017) Fertilizer information system for banana plantation. JOIV: Int J Inf Vis 1(4-2):204–208. https://doi.org/10.30630/JOIV.1.4-2.69
9. Ganeshamurthy AN, Gc S, Patil P (2011) Potassium nutrition on yield and quality of fruit cropswith special emphasis on banana and grapes. Karnataka J Agric Sci. Potassium and Phosphorous view project. Microbial interventions for nutrient mobilization and improving soil health view project. Retrieved from https://www.researchgate.net/publication/277191647. Accessed on 14 Oct 2021
10. George JA (2019) Bunch yield in banana var. Nendran as influenced by application of boron. J Pharmacognosy Phytochem 8(4):998–1000
11. Fratoni MMJ, Moreira A, Moraes LAC, Almeida LHC, Pereira JCR (2017) Effect of nitrogen and potassium fertilization on banana plants cultivated in the humid tropical amazon. Commun Soil Sci Plant Anal 48(13):1511–1519. https://doi.org/10.1080/00103624.2017.1373791
12. Nelson SC, Ploetz RC, Kepler AK (2021) Musa species (banana and plantain) in brief. Retrieved from www.traditionaltree.org. Accessed on 15 Oct 2021
13. Oyeyinka BO, Afolayan AJ (2020) Potentials of musa species fruits against oxidative stress-induced and diet-linked chronic diseases: in vitro and in vivo implications of micronutritional factors and dietary secondary metabolite compounds. Molecules 25(21):5036. https://doi.org/10.3390/MOLECULES25215036
14. Freitas AS et al. (2015) Impact of nutritional deficiency on Yellow Sigatoka of banana. Australas Plant Pathol 44(5):583–590. https://doi.org/10.1007/S13313-015-0371-6
15. Moreira A, Castro C, Fageria NK (2011) Effects of boron application on yield, foliar boron concentration, and efficiency of soil boron extracting solutions in a Xanthic Ferralsol cultivated with banana in Central Amazon. Commun Soil Sci Plant Anal 42(18):2169–2178. https://doi.org/10.1080/00103624.2011.602447
16. Nandhini M, Kala KU, Thangadarshini M, Madhusudhana Verma S (2022) Deep learning model of sequential image classifier for crop disease detection in plantain tree cultivation. Comput Electron Agric 197:106915. https://doi.org/10.1016/J.COMPAG.2022.106915
17. Donné S et al. (2016) Machine learning for maize plant segmentation. In: BENELEARN 2016 proceedings of the 25th Belgian-Dutch conference on machine learning. Retrieved from http://hdl.handle.net/1854/LU-8132925. Accessed on 07 Oct 2021
18. Grinblat GL, Uzal LC, Larese MG, Granitto PM (2016) Deep learning for plant identification using vein morphological patterns. Comput Electron Agric 127:418–424. https://doi.org/10.1016/J.COMPAG.2016.07.003

19. Berlin MA, Upadhayaya N et.al. (2021) Novel hybrid artificial intelligence-based algorithm to determine the effects of air pollution on human electroencephalogram signals. J Environ Prot Ecol 22(5):1825–1835. https://scibulcom.net/en/article/yTwhXn6CjGPSx46i2rmY
20. Mohanty SP, Hughes DP, Salathé M (2016) Using deep learning for image-based plant disease detection. Front Plant Sci. 1419. https://doi.org/10.3389/FPLS.2016.01419
21. Ferentinos KP (2018) Deep learning models for plant disease detection and diagnosis. Comput Electron Agric 145:311–318. https://doi.org/10.1016/J.COMPAG.2018.01.009
22. Krizhevsky A, Sutskever I, Hinton GE (2012) ImageNet classification with deep convolutional neural networks. In: Advances in neural information processing systems, vol 25. Retrieved from http://code.google.com/p/cuda-convnet/. Accessed on 08 Oct 2021
23. Krizhevsky A (2014) One weird trick for parallelizing convolutional neural networks. Retrieved from https://arxiv.org/abs/1404.5997v2. Accessed on 08 Oct 2021
24. Szegedy C et al. (2015) Going deeper with convolutions. In: Proceedings of the IEEE computer society conference on computer vision and pattern recognition, vol 07, pp 1–9. Retrieved from https://arxiv.org/abs/1409.4842v1. Accessed on 08 Oct 2021
25. Sermanet P, Eigen D, Zhang X, Mathieu M, Fergus R, LeCun Y (2013) OverFeat: integrated recognition, localization and detection using convolutional networks. In: 2nd International conference on learning representations, ICLR 2014—conference track proceedings. Retrieved from https://arxiv.org/abs/1312.6229v4. Accessed on 08 Oct 2021
26. Simonyan K, Zisserman A (2014) Very deep convolutional networks for large-scale image recognition. In: 3rd International conference on learning representations, ICLR 2015—conference track proceedings. Retrieved from https://arxiv.org/abs/1409.1556v6. Accessed on 08 Oct 2021
27. Story D, Kacira M, Kubota C, Akoglu A, An L (2010) Lettuce calcium deficiency detection with machine vision computed plant features in controlled environments. Comput Electron Agric 74(2):238–243. https://doi.org/10.1016/J.COMPAG.2010.08.010
28. Sharma LK, Bali SK, Dwyer JD, Plant AB, Bhowmik A (2017) A case study of improving yield prediction and sulfur deficiency detection using optical sensors and relationship of historical potato yield with weather data in Maine. Sensors (Basel, Switzerland) 17(5) https://doi.org/10.3390/S17051095
29. Hengl T et al. (2017) Soil nutrient maps of Sub-Saharan Africa: assessment of soil nutrient content at250 m spatial resolution using machine learning. Nutr Cycl Agroecosyst 109(1):77–102. https://doi.org/10.1007/S10705-017-9870-X
30. Mahrishi M et al. (ed) (2020) Machine learning and deep learning in real-time applications. IGI Global
31. Xu Z et al. (2020) Using deep convolutional neural networks for image-based diagnosis of nutrient deficiencies in rice. Comput Intell Neurosci 2020. https://doi.org/10.1155/2020/7307252
32. Singh Manhas S, Randive R, Sawant S, Chimurkar P, Haldankar G (2021) Nutrient deficiency detection in leaves using deep learning, pp 1–6. https://doi.org/10.1109/ICCICT50803.2021.9510093
33. Rajasekar T (2020) Machine learning based nutrient deficiency detection in crops. Int J Recent Technol Eng (IJRTE) 8(6):2277–3878. https://doi.org/10.35940/ijrte.F9322.038620
34. Jose A, Nandagopalan S, Ubalanka V, Viswanath D (2021) Detection and classification of nutrient deficiencies in plants using machine learning. J Phys: Conf Ser 1850(1):012050. https://doi.org/10.1088/1742-6596/1850/1/012050
35. Barbedo JGA (2019) Detection of nutrition deficiencies in plants using proximal images and machine learning: a review. Comput Electron Agric 162:482–492. https://doi.org/10.1016/J.COMPAG.2019.04.035
36. Murray DB (1960) The effect of deficiences of the major nutrients on growth and leaf analysis of the banana. Trop Agric Trinidad Tobago 37:97–106

Chapter 36
Modified Method of Diagnosis of Arrhythmia Using ECG Signal Classification with Neural Network

Monika Bhatt, Mayank Patel, Ajay Kumar Sharma, Ruchi Vyas, and Vijendra Kumar Maurya

1 Introduction

People worry when they learn someone has heart problems since the heart is the most important organ in the human body [1]. Cardiopathy, also known as vas disease, primarily affects older people or people in higher age groups and indicates that the heart and blood veins in these people are not functioning properly.

2 Cardiovascular Disease

The heart is the vas system's most delicate and crucial organ. The body's blood arteries assist the heart in pumping blood to each and every cell. The oxygen that the cells need to function is carried by the blood and transferred to the heart [2]. Cardiovascular disease or condition is the term used to describe when the heart and blood arteries are not functioning as they should. The first degree of heart illness that requires diagnosis is cardiovascular disease.

M. Bhatt (✉) · M. Patel · A. K. Sharma · R. Vyas · V. K. Maurya
Geetanjali Institute of Technical Studies, Udaipur, Rajasthan, India
e-mail: monikabhatt10@gmail.com

V. K. Maurya
e-mail: Vijendra.maurya@gits.ac.in

V. Devedzic et al. (eds.), *Proceedings of the International Conference on Intelligent Computing, Communication and Information Security*, Algorithms for Intelligent Systems, https://doi.org/10.1007/978-981-99-1373-2_36

3 Arrhythmia

Everybody's heart beats at the same rate. But occasionally, irregular heartbeats are observed, such as those caused by the heart beating too quickly (scientifically referred to as tachycardia), too early (scientifically referred to as premature contraction), too slowly (scientifically referred to as bradycardia), or too irregularly (scientifically referred to as fibrillation), and this type of disease is referred to as arrhythmia [3]. Due to the electrical impulses to the heart's lack of coordination, there is a significant rate of variation in heartbeats, which causes this heart rhythm issue. This type of sickness, which causes our hearts to flutter or race, is typically innocuous.

3.1 Signs and Symptoms of Arrhythmia

Patients may or may not exhibit arrhythmia symptoms. Some people might simply be missing it altogether [4]. A basic physical examination or ECG recording by a physician may reveal any sign or symptom of arrhythmia (that is either the patient feels some symptoms like abnormal heartbeat or doctor detect a sign or the symptom related to heart disease).

3.2 Causes of Arrhythmia

There ought to be an exact and unmistakable pathway for the electrical motivations that train the heart to contract. Any aggravation to these driving forces will be the explanation of arrhythmia [5]. We as a whole realize that the engineering human heart involves four chambers—the chambers on every portion of the heart structure two bordering siphons, with the chamber (upper chamber) and the ventricle (lower chamber). When talking about the heartbeat, during each heartbeat, the relaxed ventricles are filled with blood when smaller and less muscular atria contract [6]. The electrical signals are passes by the small cells present in right atrium (the sinus node) which jointly causes the contraction in right and left atria reason behind reduction. Then, the center of the heart called the atria-ventricular node receives this electrical impulses and after this these impulses move to the ventricles and causing them to pump blood by contractions a serious problem (Table 1).

4 Basic Key Techniques

Electrocardiogram (ECG) is an assessment which is prescribe by the specialists to record and break down the working of heart as an electrical movement. ECG has been

Table 1 Types of arrhythmia

Size	Type	Effect
1	Tachycardias	Rapid arrhythmias
2	Bradycardias	Slow arrhythmias
3	Fibrillations	Irregular arrhythmias
4	Premature contraction	Premature contraction

successfully involved when till today as a significant measure to decide if the heart is working appropriately [5]. ECG recording is not the last step while diagnosing the heart infections in light of the fact that after the assessment, right examination is vital.

4.1 *Electrocardiogram (ECG)*

Heart is a solid organ which supplies blood all through the body and during this movement it agreement and stretch with the each heartbeat. At the point when the muscles of the heart contract, it makes a mood which is estimated by an assessment during any heart issue, known as electrocardiogram (ECG), it estimates each electrical action of the heart during withdrawal that is made by a hub called sino-atrial hub [6]. This hub normally makes this action as a characteristic pacemaker of the heart.

4.2 *Neural Network*

Totally unique in relation to the customary advanced PCs, cerebrum PC is extremely phenomenal gadgets imagined based on mind. Fake brain networks are persuaded by the cerebrum PCs. They work on neurons and their exercises. The possibility of brain networks was presented [7]. He investigated this branch of knowledge as well as demonstrated that neurons are the fundamental unit of human cerebrum and how they control the exercises of human body.

However, on contrasting neurons and fake neurons (silicon rationale entryways), neurons were viewed as a lot slower than the last option. This is for the most part since neurons containing a property that they are having a greatness of request differs from 5 to 6, and occasions occurred in neurons are in milliseconds (3–10 s) range. Then again, in silicon chip occasions occur in the scope of nanoseconds (9–10 s). Be that as it may, there are broad interconnections between wavering quantities of nerve cell (neurons) in human body causes generally sluggish pace of activity as well as response by the mind.

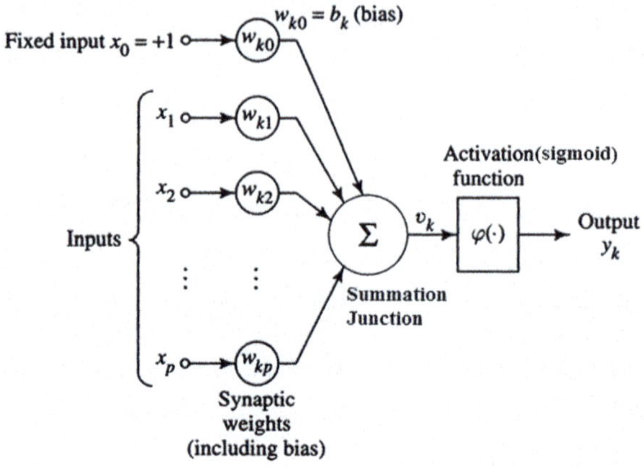

Fig. 1 Nonlinear model of a neuron [11]

4.3 *Models of a Neuron*

The information processing capability on neural network is due to its basic unit called neuron. There are three basic elements of neuron model can be analyzed:

1. **Synapses (Interconnection Between Neurons)**: synapses are the interconnection between neurons and can be differentiated by their weights that are their strength. Specifically, a neuron named k receives a signal x_n at the input and connected with a synapse h with its weight w_{kn} where the subscript of synaptic weight refers to the neuron and input end of synapse, respectively [8]. The weight w_{kn} is negative if the synapse is inhibitory and positive if the associated synapse is excitatory.as shown in Fig. 1.
2. **Adder**: The next part is to calculate the summation of input signal and weight of respective synapses of that particular neuron.
3. **Activation Function**: this is a specific function applied to each neuron to limit its output value is also known as threshold function. Specifically, the normal range of amplitude for activation function for output of neuron varies from 0 to 1 or −1 to 1 [9]. The model of a neuron has the ability to lowering or increasing the net input of the activation function includes an externally applied bias (threshold).

4.4 *Backpropagation Networks*

Error backpropagation algorithms, when trained by supervised learning methods, can be used to solve some difficult various problems and can be successfully applied to multilayer perceptron's [10]. The error correction learning rule based on the error backpropagation algorithm is as follows. Basically, the process of the BP algorithm involves two passes of the entire network (from the input layer to the output layer): forward pass and backward pass [11]. Forward pass, an activation function is applied

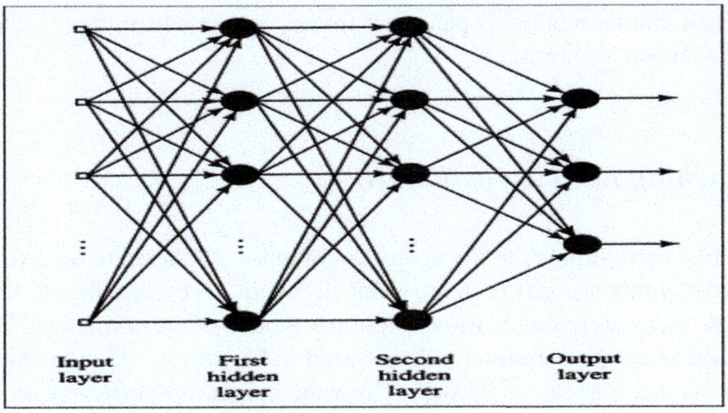

Fig. 2 Feed-forward backpropagation neural network [11]

to form an activity pattern (input vector), applied to the sensory nodes of the network, the effect is propagated throughout the network, as shown in Fig. 2

Without focusing on the analysis part, it is difficult to reach the disease. So, for this purpose, automatic analysis tool is required. This project emphasizes on this automatic analysis and detection of arrhythmia. We have

1. Data pre-processing
2. Feature extraction
3. Training and debugging
4. Backpropagation learning algorithm
5. Arrhythmia classification.

5 Methodology

In this project, ECG data is employed for aggregation heart disease data. A preprocessing during electrocardiogram (ECG) sampling the fundamental task is to notice and method the *R*-peaks throughout the ECG signal process there are some difficulties arises owing to the presence of low frequency elements due to patient breathing, irregularity in peaks formation, and, unsteady and to avoid fluctuating data we tend to solely needed high peak worth's and find only most values by ignoring all alternative minima's initial condition is to organize the information for mistreatment by the MATLAB within the sort of matrix. The absolute value of 201 samples is taken from MLII data base during which eight are lead data. Ehen the cavity beats are derived we tend to get *R*-peaks values from the information base in tabular form. Then, we take a hundred samples from the mitt aspect of thought of *R*-peak known as ventricular beat and 100 samples from manus side from one zero one sample input. Thus, we got the 201×1 matrix of sample data which might use by the NNET tool chest of

MATLAB. A similar method is perennial for any reasonably all inputs whether it is ventricular, fusion, or normal.

5.1 Teaching ECG Arrhythmias

Teaching ECG arrhythmias to the neural community. As there are ten extraordinary sorts of arrhythmia are gift then we want to layout 10 extraordinary ANN shape and educate them accordingly. Most important motive of the normalized statistics is used for education the community that it need to be with inside the variety of zero to at least one because the ANN will deliver the output most effective on this form. The accuracy of the method is checked through splitting statistics into education and finally trying out statistics set. When studying is hereafter finished or stopped, the community is evaluated with the statistics this is retrieved from the check statistics set. Again the dataset derived for the specific ANN is split into units, i.e., education set and trying out set. If we speak approximately education a neural community, then it is far the technique of putting in the high-quality weights at the synapse of inputs to reduce the mistake rate (Fig. 3).

It might be stated as beat-by-beat type is performed at the RR-c program language period sign with the assist of policies described. These policies pave the manner for detecting specific in addition classifications of arrhythmic activities from the RR-periods [11]. An RR-c program language period sliding window is the only in order to classify center sign. The type worries the second one beat of the center RR-c program language period. We may want to classify beats into 4 categories: specifically (Fig. 4)

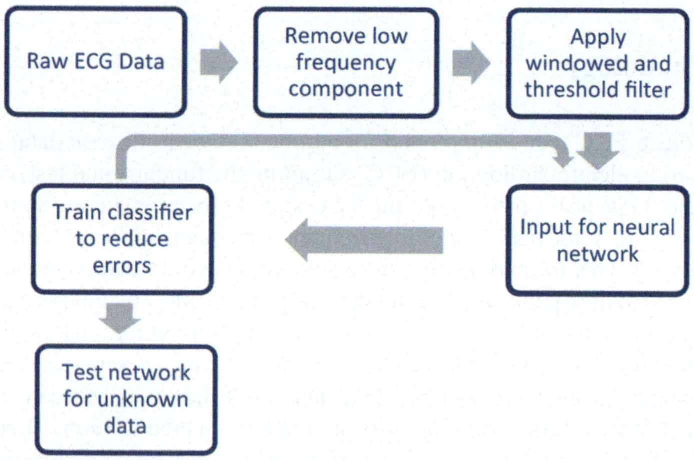

Fig. 3 Initial algorithm

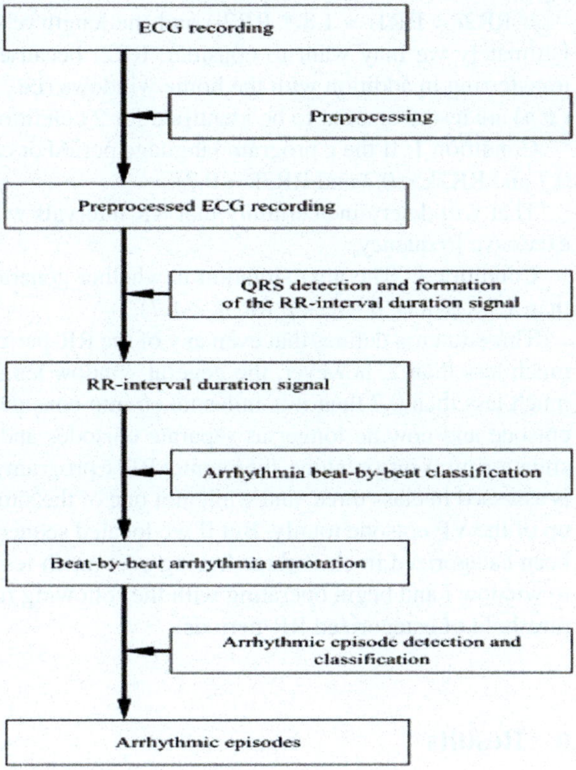

Fig. 4 Proposed algorithm

(1) Regular sinus beats (*N*) and 3 arrhythmic ones:
(2) Untimely ventricular contractions (PVC),
(3) Ventricular flutter/traumatic inflammation (VF) and
(4) 28 coronary heart block (BII).

It is believed that the type might be termed as regular if it does now no longer fits with any of the categories. Talking approximately the operating method, this proposed set of rules begins off evolved with the window this is I additionally includes c program language period specifically RR1i, RR2i, and RR3i. The RR2i c program language period could be described and termed because the prori and regular and could be termed as class 1. Now we are able to be describing the following policies in detail. Rule 1: working at the ventricular flutter/traumatic inflammation (that simply classifies the RR-periods from the ventricular flutter/traumatic inflammation episodes those will define—class three). This specific rule is primarily based totally on defining or can say classifying the full episode of VF now no longer simply on unmarried beat. Hence forth, it need to be stored in thoughts that

If RR2i < RR1i > 1.8 * RR2i) and the length of RR2i is much less than 0.6 s. Ultimately we may want to consider, RR2i because the starting VF episode and transferring in addition with the home windows (i.e., home windows *i* þ 1, *i* þ 2,..., *i* þ *n*) are tested or want to be identified for 2 conditions:

Condition 1: If the c program language period or can say the length lies (RR1k < 0.7 and RR2k < 0.7 and RR3k < 0.7).

That will determine certainly that RR intervals with inside the VF episode have excessive frequency.

Condition 2: The next situation is whether general window length is much less than 1.7 s (RR1k þ RR2k þ RR3k < 1.7).

This situation defines that even any of the RR-periods are not having the intervals much less than 7, however, the general window length is coming in the variety of much less than 0.7 then it is miles taken into consideration as persevering with VF episode and now no longer as separate episodes and additionally if anyone of the situation turns into genuine, then center RR-c program language period of the window is assessed in class three, and while not one of the situation is genuine, actually give up of the VF episode totally. But if we located sequential RR-periods (*n*) which are been categorized in class three having cost much less than 4, it returns lower back to window *i* and begin operating with the following rule. As we had been the use of threshold of categorized RR-periods.

6 Results

This assignment describes the approach of pre-prognosis of various sorts of arrhythmia so that you can lessen probabilities of coronary heart attack, coronary heart failure, stroke, and different excessive coronary heart sicknesses and offers the satisfactory end result at the specific pattern facts set. This gadget has evolved to assist an everyday individual so can he might be capable of diagnose his ailment in in advance level in addition to docs so can they are able to without difficulty diagnose the ailment and plan a higher remedy method for the affected person and might be capable of shop his lifestyles in addition to can lessen the dying price via way of means of coronary heart sicknesses.

7 Conclusion

This undertaking classifies handiest atypical coronary heartbeat to the everyday one that may be elevated among kinds of arrhythmia in addition to severity and levels particularly form of arrhythmia. This segment of the undertaking accomplished a higher accuracy, however, for in addition enhancement may be carried out with the aid of using the use of a few higher algorithms like *k*-nearest neighborhood and assist vector machines. Expand undertaking for identification, class of diverse form of arrhythmia diseases. To layout higher set of rules to boom efficiency.

References

1. Carr JJ, Brown JM (2001) Introduction to biomedical equipment technology. Pearson Education Inc
2. Li C, Zheng C, Tai C (1995) Detection of ECG characteristic points using wavelets transforms. IEEE Trans Biomed Eng 42(1):21–28
3. MIT-BIH arrhythmia database, 3rd edn (1997) Harvard-MIT Division of Health Science Technology, Biomedical Health Centre, Cambridge
4. Sordo M (2002) Introduction to neural networks in healthcare—a review
5. Silipo R, Marchesi C (1998) Artificial neural networks for automatic ECG analysis. IEEE Trans Signal Process 46(5):1417–1425
6. Jadhav SM, Nalbalwar SL, Ghato A (2010) Artificial neural network based cardiac arrhythmia classification using ECG signal data. In: International conference on electronics and information engineering (ICEIE), pp V1-228—V1-231
7. NNet toolbox tutorial for Mat lab
8. Nankani H et al. (2021) A formal study of shot boundary detection approaches—comparative analysis. In: Sharma TK, Ahn CW, Verma OP, Panigrahi BK (eds) Soft computing: theories and applications. Advances in intelligent systems and computing, Springer. https://doi.org/10.1007/978-981-16-1740-9
9. Physionet WFDB toolkit for MATLAB. http://physionet.org/physiotools/matlab/
10. Nazmy TM, Messiry HE, Bokhity BA (2010 Adaptive neuro-fuzzy inference system for classification of ECG signals. J Theor Appl Inf Technol 12(2)
11. Babikier MA, Izzeldin M, Ishag IM, Lee DG, Ryu KH (2005) Classification of cardiac arrhythmias using machine learning techniques based on ECG signal matching. Artif Intell Med 33:237–250

Author Index

V. Devedzic et al. (eds.), *Proceedings of the International Conference on Intelligent Computing, Communication and Information Security*, Algorithms for Intelligent Systems, https://doi.org/10.1007/978-981-99-1373-2

The manufacturer's authorised representative in the EU is Springer Nature Customer Service Centre GmbH, Europaplatz 3, 69115 Heidelberg, Germany. If you have any concerns regarding our products, please contact ProductSafety@springernature.com

Printed and bound by CPI Group (UK) Ltd, Croydon, CR0 4YY
01/10/2026
02225870-0018